Land Development Calculations

Land Development Calculations

Walter M. Hosack

McGraw-Hill

New York Chicago San Francisco Lisbon
London Madrid Mexico City Milan
New Delhi San Juan Seoul
Singapore Sydney Toronto

Cataloging-in-Publication Data is on file with the Library of Congress

McGraw-Hill 

*A Division of The **McGraw-Hill** Companies*

5 6 7 8 9 0 DOC/DOC 0 9 8 7 6

P/N 136256-8

ISBN 0-07-136255-X

The sponsoring editor for this book was Wendy Lochner, the editing liaison was Steven Melvin, and the production supervisor was Sherri Souffrance. It was set in Matt Antique by North Market Street Graphics.

R. R. Donnelley & Sons Company was printer and binder.

This book is printed on recycled, acid-free paper containing a minimum of 50% recycled, deinked fiber.

Contents

Preface

I have long felt that there must be a better way to forecast the capacity of land to shelter the activities of a growing population, and that the potential exists to increase economic yield and conserve much of what is now consumed. We currently forecast the development capacity of land—the amount of building area that can be placed on a given land area—with trial and error experiments at the drawing board, with rules of thumb, with comparisons to existing projects, and with intuitive development regulations. The full range of options is rarely, if ever, evaluated because the options cannot be comprehensively forecast in a brief, cost-effective manner. This means that we have little ability to gauge how wisely we are using the land to generate revenue and maximize profit; nor how wisely we are regulating the use of land to protect our future.

This book describes a *development forecast collection,* included on an attached CD-ROM, that introduces mathematical forecasting to the process of development capacity evaluation. Each forecast model within the collection contains a *design specification panel*. The values assigned to each component of the design specification establish a project "recipe" that is capable of producing anticipated results without dictating final form or appearance. We can learn about the implications of these specification values by studying existing projects using a *context record system* that is also included on the CD-ROM and described in this book.

This system is written to calculate design specification values from a series of project measurements that are entered in the project record. These specification values represent the "pressure point" statistics that produced the project under study, and that can be used or adjusted to produce future projects based on the knowledge gained. The development forecast equations that are based on these specification values simply forecast, they do not pass judgment; but they have the potential to improve our ability to use land wisely when we better understand the implications of the values entered.

This topic is also relevant to city, state, and national economic development efforts. The equations within each forecast model make it possible for an individual to forecast development capacity options for any land area without producing site plan drawings, and in a fraction of the time it would take to produce one sketch. A new option is produced every time one or more values is adjusted in the design specification. These models also make it possible to forecast the construction costs involved, the population accommodated, the traffic generated, the investment potential present, and the revenue implied by the results produced when additional panels are attached, as described in Chaps. 3 and 6. This can assist the planning and policymaking decisions of many disciplines, including the evaluation of economic development opportunities from both public and private perspectives.

Development decisions are made by professionals such as, but not limited to, real estate developers, accountants, attorneys, city planners, architects, civil engineers, landscape architects, bankers, economists, geographers, economic developers, landowners, and government policymakers at the local, state, and national levels. These decisions directly affect our lifestyle and quality of life. The development forecast collection and the context record system are offered as new tools that can assist in making these decisions through objective development capacity comparison and evaluation. This assistance will improve as those entrusted with these decisions learn more about the design specification components involved and the physical, social, and economic implications associated with the values assigned to each. The alternatives chosen from this comparative process can then be used to focus expensive, time-consuming drawing-board evaluation of design and construction details on the most promising possibilities. This integrated process and common language should help to improve the land use allocation, development design, construction investment, land conservation, and regulatory decisions that will affect our sustainable future.

Acknowledgments

This work began with a slide rule, gathered momentum with an electronic calculator, and became feasible with a personal computer. It would not have been possible, however, without the love, support, and patience of Elizabeth Fanning Hosack and our two daughters, Brooke Adams and Kelly Elizabeth Hosack. They stand in my personal pantheon of heroes, along with my grandparents, Svea and Martin Johnson and my mother, Volborg Anna Christina Johnson. Svea Anderson and Martin Johnson were Swedes who had the courage to come and write part of the American story. Fortunately, I was in it, and their great-grandchildren carry on the tradition. They are my strength and inspiration, and I have made every effort to make this book worthy. If it is, then it will represent the contribution I have always wanted to make on their behalf.

An effort like this does not stand alone. It rests on the shoulders of many who have contributed along the way with different skills and from different perspectives. The following list is not complete, but is a brief word of thanks to some of my greatest benefactors.

Frank Vegh: The first on my list. A man and his family whom I am truly blessed to have known.

Barbara and Jim Fanning: They were kind enough to let me share their world.

Carl, George and Olga Johnson: Carl was Martin's brother. George was Carl's son. George and Olga Johnson made the summer shine on Devil's Lake.

Raymond, Helga, Judy, Larry, and Molly Peterson: Helga was Carl's daughter and built a Peterson legacy with Raymond that I have been fortunate to call part of my family.

Rudolph Frankel: An architect and professor at Miami University, Oxford, Ohio, with a vision he called City Design.

Jamie Greene: An architect, planner, and principal with the firm ACP Visioning & Planning, Ltd., who believed in the possibility and builds visions for a living.

John Simpson: A landscape architect and professor at The Ohio State University whose advice was invaluable.

Gary Schmidt: A landscape architect and principal of the firm Schmidt Land Design whose encouragement gave me confidence to carry on.

Tom McCash: An architect, attorney, and politician with the firm Meacham and Aple who helped someone he hardly knew.

Wendy Lochner: An editor who saw the possibilities.

John Joy: An old friend.

Dave Bonk: I'm proud to have known him.

Kent Brandt: An architect and principal of the firm Brubaker/Brandt who saw something years ago and moved me from a struggling city to an office of great potential.

Gordon and Brenda Proctor: Peace Corps volunteers are special people. They will always be remembered.

Brian Higgins: I advertised for a research assistant, and fate smiled upon me when Brian arrived.

The Great Spirit: Called by many names and proven by mathematicians, who call him infinity.

The Development Forecast Collection

This chapter is intended to help those familiar with the planning, design, and construction process to quickly begin using a collection of forecast models. These models are written to predict the development capacity of land based on values entered in each model's design specification. The collection of forecast models is included on the CD-ROM attached to this book. It is possible to turn to the conclusion of this chapter to immediately locate a relevant model within the collection, but a skilled practitioner should avoid this temptation and at least read this chapter before using these models. The ensuing chapters and case studies have been written for those who wish more in-depth information about the organization, content, and use of this collection. For those attempting a quick start, a glossary of terms is included and may prove helpful. Expanded glossaries including descriptions and explanations are included at the ends of Chaps. 2 and 3.

The Concept

Landowners wonder how much can be built on their property. Real estate investors wonder how much land is required for a given building area. The answers are elusive, however. The questions are only partially answered with trial-and-error experiments at the drawing board, with rules of thumb, with comparisons to existing projects, and with intuitive development regulations. The full range of options is rarely, if ever, evaluated because the options cannot be comprehensively forecast in an efficient, cost-effective manner. As a result, the solutions chosen cannot be compared to the complete realm of possibilities, nor can they be compared to a complete catalog of prior results that are graded for effectiveness. This means that we have little ability to gauge how wisely we are using the land to maximize our return on investment; nor how wisely we are regulating the use of this land to protect our future. Using a land area that is too small produces congestion, intensity, and overdevelopment. A land area that is too large produces sprawl. The first option threatens our quality of life. The second threatens our sustainable future. Development professionals have historically determined the development capacity[1] of a given land area, and the amount of land needed to support a given building area, based on their intuition, rules of thumb, and past experience; but the process is like baking bread without a recipe. The results can be unreliable, even with extensive experience. It is possible, however, to translate this intuition and experience into equations and forecast models. These models can then be used to improve the scope of comparisons possible, the knowledge available, and the results achieved by the entire audience that confronts these issues.

In other words, buildings are constructed to shelter activities from an unstable environment. This effort is preceded by at least one of the following two questions:

- How much can be built on the land area that is available?
- How much land is needed to accommodate a given building area objective?

These questions can be quickly and easily evaluated without drawing-board analysis by choosing the right forecast model within a collection attached to this book.[2] These models produce comprehensive forecasts of development options for any given land area, or of the land needed for any given building area objective, without producing site plan drawings, and in a fraction of the time it would take to produce one sketch. This model format removes drawing prerequisites from the preliminary evaluation process. It also reduces time, produces many more options for comparison, and can improve the results produced by development decision makers as a consequence.

The Forecast Collection

The forecast models included in this collection fall into one of six generic design categories intuitively used by architects throughout the world. Models within the collection evaluate development options based on values entered in a design specification.[3] This format permits an unlimited number of development options to be evaluated with a few keystrokes.

Single model analysis is based on one design category and focuses on comparing the results produced by different component values within the same design specification. If more than one model is chosen, the results produced by design specification changes within each design category can be compared to evaluate the options represented. Time-consuming drawing-board analysis can then be reserved for an evaluation of the options that survive.[4]

The Decision Guide

Figures 1.1 and 1.2 have been created as decision guides to locate a forecast model within the collection. Both read from left to right and are designed to lead the user along branches to the appropriate model. Figure 1.1 applies to the residential land use family of building types and Fig. 1.2 applies to the nonresidential land use family. Figure 1.3 refers to mixed-use alternatives that are discussed in later chapters.

Figure 1.1 Decision guide - residential land use family.

DECISION 1 [1] Choose between residential, nonresidential and mixed-use land use groups.

DECISION 2 Choose parking category and follow to right.

DECISION 3 Choose building type and follow to right.

DECISION 4 Choose parking system and follow to right.

DECISION 5 Identify row that corresponds to the information known. Do you know the land area available, the building area desired, or the net density objective to be evaluated? After decision, follow the branch that pertains to the right.

DECISION 6 Identify model.

[2] PARKING CATEGORY	[3] BUILDING TYPE	[4] PARKING SYSTEM	[5] GIVEN	FORECAST	[6] MODEL NEEDED
No parking	Apartment houses	No parking required	Gross building area	Minimum buildable land area needed	RG1B
			Gross land area	Maximum dwelling unit capacity	RG1L
			Land and density	Minimum number of building floors needed	RG1D
Surface parking	Apartment houses	Grade parking around but not under building	Gross building area	Minimum buildable land area needed	RG1B
			Gross land area	Maximum dwelling unit capacity	RG1L
			Land and density	Minimum number of building floors needed	RG1D
	Apartment houses	Grade parking around and under building	Gross building area	Minimum building height and buildable area needed	RG2B
			Gross land area	Maximum dwelling unit capacity	RG2L
			Land and density	Minimum number of building floors needed	RG2D
	Suburb houses	No shared or common parking lots	Land and density	Maximum number of lots that may be created	RSFD
			Land area	Maximum no. of lots and max. dwelling unit sizes possible	RSFL
			No. of lots	Min. buildable acres and max. dwelling unit sizes possible	RSFN
	Urban houses	Grade parking lot and/or private garages	Land and density	Dwelling unit characteristics to meet net density objective	RGTD
			Gross land area	Maximum dwelling unit capacity	RGTL
Structure parking	Apartment houses	Parking structure adjacent to building on same premise	Gross building area	Minimum buildable land area needed	RS1B
			Gross land area	Maximum dwelling unit capacity	RS1L
			Land and density	Minimum number of parking levels needed	RS1D
	Apartment houses	Parking structure underground	Gross building area	Minimum buildable land area needed	RS2B
			Gross land area	Maximum dwelling unit capacity	RS2L
			Land and density	Min.number of building floors and parking levels needed	RS2D
	Apartment houses	Parking structure partially or completely above grade under building	Gross building area	Minimum buildable land area needed	RS3B
			Gross land area	Maximum dwelling unit capacity	RS3L
			Land and density	Min. number of builidng floors and parking levels needed	RS3D

Figure 1.2 Decision guide - nonresidential land use family.

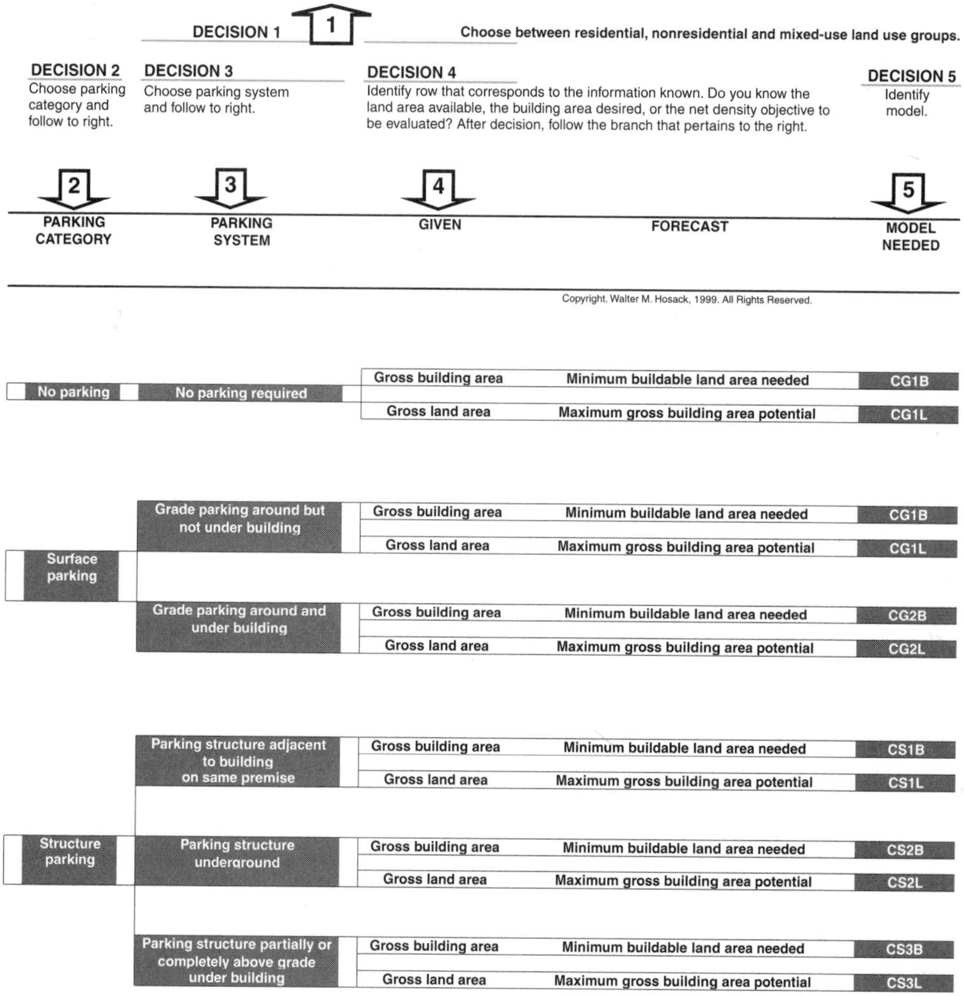

DECISION 1 — Choose between residential, nonresidential and mixed-use land use groups.

DECISION 2 — Choose parking category and follow to right.

DECISION 3 — Choose parking system and follow to right.

DECISION 4 — Identify row that corresponds to the information known. Do you know the land area available, the building area desired, or the net density objective to be evaluated? After decision, follow the branch that pertains to the right.

DECISION 5 — Identify model.

2 PARKING CATEGORY	3 PARKING SYSTEM	4 GIVEN	FORECAST	5 MODEL NEEDED
No parking	No parking required	Gross building area	Minimum buildable land area needed	CG1B
		Gross land area	Maximum gross building area potential	CG1L
Surface parking	Grade parking around but not under building	Gross building area	Minimum buildable land area needed	CG1B
		Gross land area	Maximum gross building area potential	CG1L
	Grade parking around and under building	Gross building area	Minimum buildable land area needed	CG2B
		Gross land area	Maximum gross building area potential	CG2L
Structure parking	Parking structure adjacent to building on same premise	Gross building area	Minimum buildable land area needed	CS1B
		Gross land area	Maximum gross building area potential	CS1L
	Parking structure underground	Gross building area	Minimum buildable land area needed	CS2B
		Gross land area	Maximum gross building area potential	CS2L
	Parking structure partially or completely above grade under building	Gross building area	Minimum buildable land area needed	CS3B
		Gross land area	Maximum gross building area potential	CS3L

If Fig. 1.3 is ignored for the moment, the first decision is to choose either Fig. 1.1 or Fig. 1.2. If Fig. 1.1 is chosen, column 2 asks for a definition of the parking category involved and offers three choices: (1) no parking required, (2) surface parking, or (3) structure parking. If the surface-parking category is chosen, the branch to column 3 asks for a building type definition. Three choices are again offered:

Figure 1.3 Decision guide - mixed use.

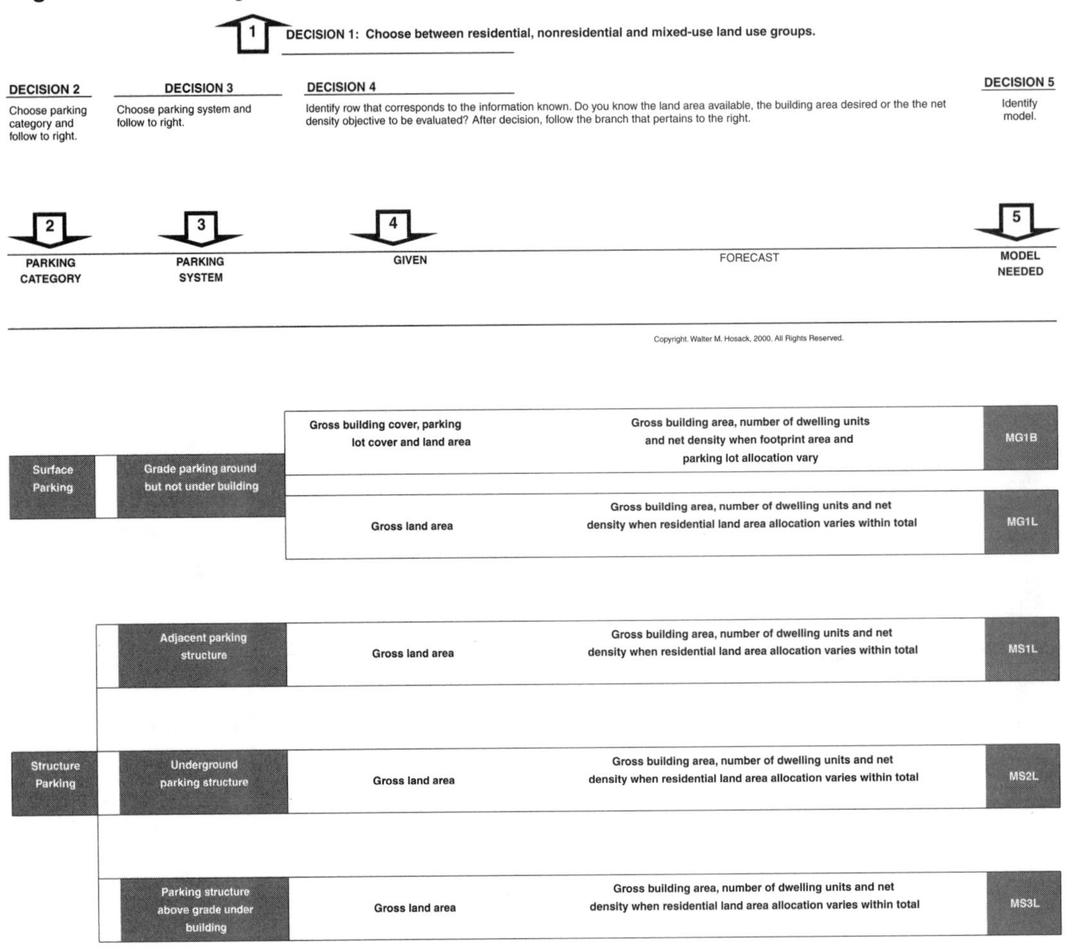

(1) apartment, (2) single-family detached suburb houses, or (3) single-family attached urban houses. If apartment is chosen, the branch to column 4 asks for a parking system definition and offers two choices: (1) grade parking lot around but not under the building, or (2) grade parking lot around and under the building. If a traditional grade parking lot around but not under the building is chosen, the branch to column 5 asks what is known or given and presents three

Table 1.1 Characteristics of Forecast Model RG1L

Objective	Applicability	
	Topic	**Category**
Given: Gross land area	Land use family	Residential
To find: The total number of dwelling units that can be placed on the land area given	Parking category Building type Parking system	Surface Apartments Grade parking around but not under the building

choices: (1) gross building area objective, (2) gross land area, or (3) gross land area and net density objective. If the gross land area is known, the branch to column 6 gives a brief indication of the forecast content and identifies forecast model RG1L[5] as the appropriate tool. The choices just described are summarized in Table 1.1.

If Fig. 1.2 had been chosen, a similar decision-making path would be followed. However, Fig. 1.2 does not offer building type choices. Building areas in the nonresidential group may be used for an office, drugstore, or hospital, but the capacity of land to accommodate building area is not a function of the building appearance, operation, or configuration contemplated. It is a function of the parking requirement associated with the activity, the method used to provide this parking, and the values assigned to design specification components within the forecast model chosen to represent this activity. In Fig. 1.2, column 2 asks for a definition of the parking category involved. The options are no parking, surface parking, or structure parking. If surface parking is chosen, the branch to column 3 asks for a definition of the parking system under consideration and offers two choices: (1) grade parking around but not under the building, and (2) grade parking around and under the building. If grade parking around but not under the building is chosen, the branch to column 4 asks what is known or given and presents two choices: (1) gross building area desired, or (2) gross land area available. If the gross land area available

Table 1.2 Characteristics of Forecast Model CG1L

	Applicability	
Objective	**Topic**	**Category**
Given: Gross land area	Land use family	Non-residential
To find: The maximum gross building area that can be placed on the gross land area given	Parking family Building family Parking system	Surface Non-residential Grade parking around but not under the building

is known, the branch to column 5 gives a brief indication of the forecast content and identifies forecast model CG1L for use in predicting the maximum gross building area potential of a given gross land area. The choices just described are summarized in Table 1.2.

If the structure-parking category had been chosen instead of the surface-parking category, three additional parking choices would have been available. The first would have been adjacent structure parking; the second, underground structure parking; and the third, structure parking partially or completely above grade under the building. These three choices, when combined with the three surface-parking choices mentioned previously, represent the six generic design premise categories that encompass 35 models within the forecast collection.[6] Figure 1.4 illustrates the generic G1 premise. It shows a surface parking lot wrapping around two sides of an office building. It is an arrangement we are all familiar with as one of the most common forms of land development today. It also displays the basic elements of a site plan. Figure 1.5 illustrates a variation of the G1 premise that requires no parking and is often found in central business district areas. It is considered a variation of G1 because it uses the same forecast models and simply requires that zero be entered in the design specification where a parking requirement value is requested. Figure 1.6 illustrates the more unusual G2 premise that places a grade parking lot under as well as around a building. Figure 1.7 illustrates another

(a)

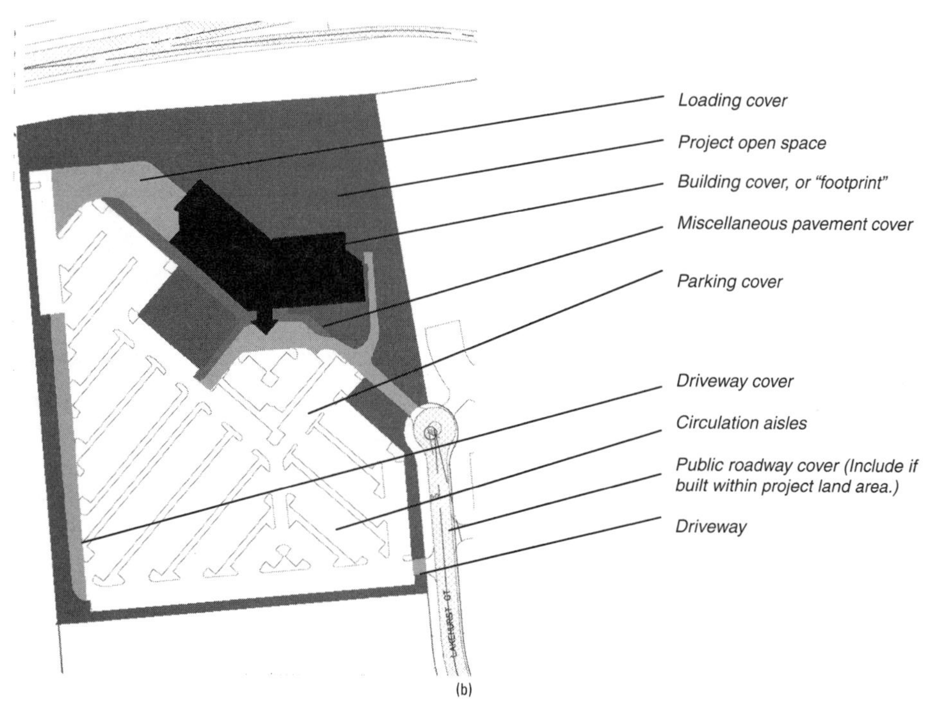

(b)

Figure 1.4 G1—grade parking around building. (*a*) Every parking space shown in the photograph serves a fraction of the gross building area present. This fraction is called a *parking ratio,* and is a function of the activity sheltered and the population density produced. Most local zoning ordinances contain parking ratio requirements that are listed in relation to land use categories. (*b*) As a parking lot grows, more spaces can be provided and a larger building can be justified. This enlargement either can consume project open space or the building footprint can be reduced to compensate, and building height can be increased to produce the additional building area justified. This second alternative preserves the original open space contemplated.

(Photo by W. M. Hosack)

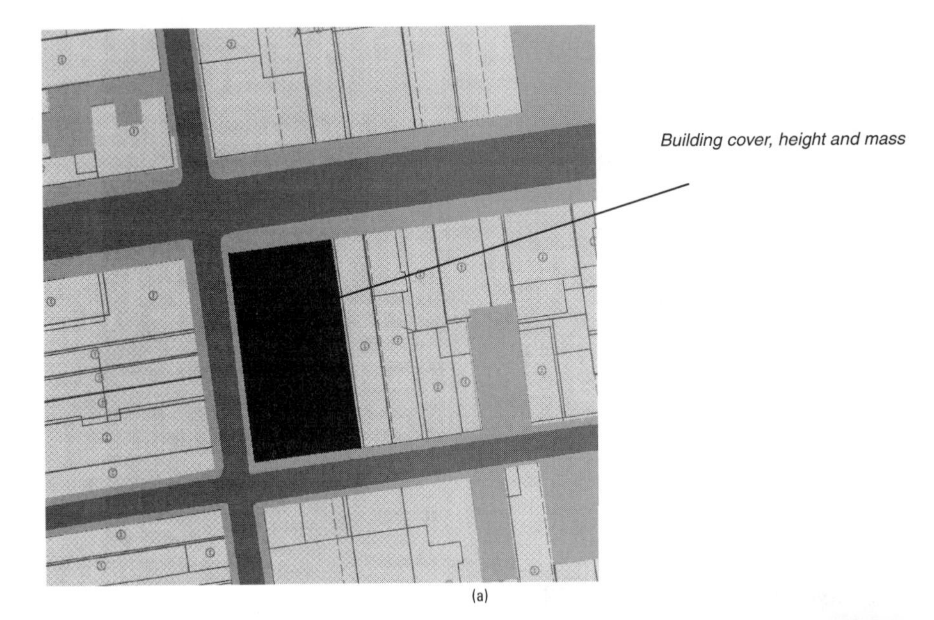

(a)

(b)

Figure 1.5 G1—no parking. Height is the only limitation to development capacity in this example. Responsibility for parking, miscellaneous pavement, roadways, and project open space is assumed by others, or by the public. (*a*) Site plan; (*b*) street view.

(Photo by W. M. Hosack)

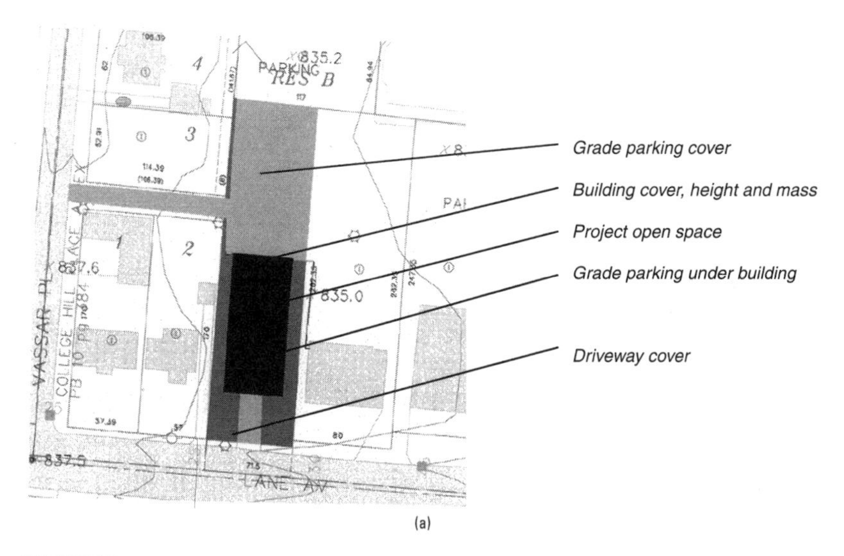

(a)

(b)

Figure 1.6 G2—grade parking around and under building. Project open space in this example is minimal and oriented to curb appearance. (*a*) Site plan; (*b*) street view. (Photo by W. M. Hosack)

familiar sight in most cities and represents the S1 category of parking structures that are placed adjacent to buildings. Design specification values are used to change its capacity, but the generic premise remains the same. Figure 1.8 illustrates the S2 category of underground parking structures and represents the favorite of all who can afford to entirely remove the automobile from pedestrian view. Figure 1.9 illus-

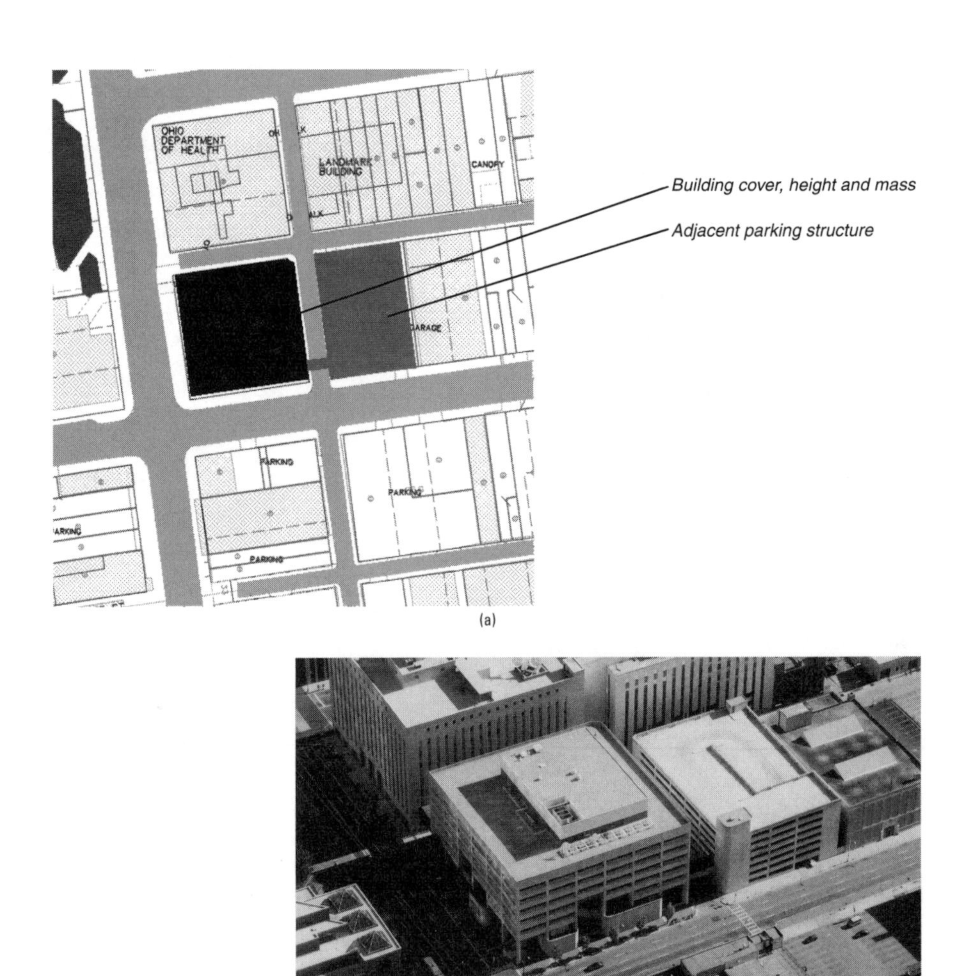

Figure 1.7 S1—adjacent parking structure. Development capacity is limited by the height and area allocated to the adjacent parking structure in this example. Responsibility for miscellaneous pavement, roadways, and project open space is assumed by others, or by the public. (*a*) Site plan; (*b*) overview.

(Photo by W. M. Hosack)

trates the S3 category of parking structures. This premise uses building height to increase development capacity by placing the parking structure above grade under the building. This approach has the ability to maximize the capacity of small sites. In the example shown, this compact approach has also reserved some street-level areas for pedestrians.

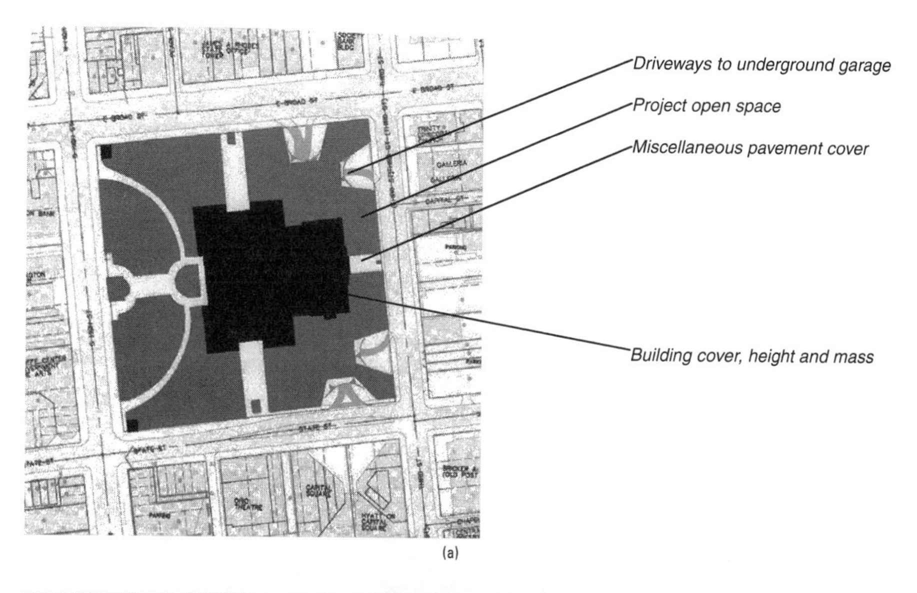

(a)

(b)

Figure 1.8 S2—underground parking garage. Underground parking garages can be used to conserve project open space and provide urban amenity. (*a*) Site plan; (*b*) street view.
(Photo by W. M. Hosack)

Selected Forecast Models

The earlier exercise locates two forecast models in the decision guide that are included in this text as Exhibits 1.1 and 1.2. Both represent G1 parking systems, but one addresses residential land use while the other addresses nonresidential land use.[7] The explanation panel directly below the exhibit title block[8] in each identifies the building type anticipated and the parking system addressed. This panel also itemizes the

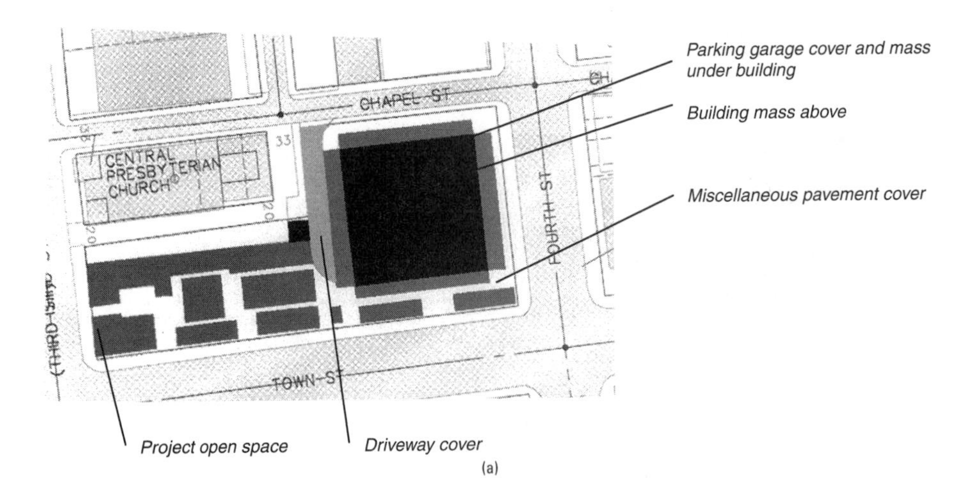

(a)

(b)

Figure 1.9 S3—parking garage above grade under building. Urban gardens are not often provided with this parking system, but in this case they provide welcome relief to the neighborhood context. (*a*) Site plan; (*b*) overview.

(Photo by W. M. Hosack)

Exhibit 1.1 Residential Forecast Model Sample Format

Forecast Model RG1L

Development capacity forecast for ***APARTMENTS*** *based on the use of an adjacent* ***GRADE PARKING LOT*** *located on the same premises. When s and u equal zero in the design specification below, the forecast pertains to conditions when* ***NO PARKING*** *is required.*

Given: *Gross land area.* ***To Find:*** *Maximum dwelling unit capacity of the land area given based on the design specification values entered below.* ***Premise:*** *All building floors considered equal in area.*

DESIGN SPECIFICATION

Enter values in boxed areas where text is bold and blue. Express all fractions as decimals.

Given:	**Gross land area**	**GLA=**	**10.204**	acres	444,486	SF
Land Variables:	Public/ private right-of-way & paved easements	W=	**0.150**	fraction of GLA	6,534	SF
	Net land area	NLA=	8.673	acres	377,813	SF
	Unbuildable and/or future expansion areas	U=	**0.100**	fraction of GLA	4,356	SF
	Gross land area reduction	X=	0.250	fraction of GLA	10,890	SF
	Buildable land area remaining	**BLA=**	**7.653**	acres	333,365	SF
Parking Variables:	Est. gross pkg. lot area per pkg. space in SF	s =	**400**	ENTER ZERO IF NO PARKING REQUIRED		
	Parking lot spaces planned or required per dwelling unit	u=	**1.5**	ENTER ZERO IF NO PARKING REQUIRED		
	Garage parking spaces planned or required per dwelling unit	Gn=	**0.5**	ENTER ZERO IF NO PARKING REQUIRED		
	Gross building area per garage space	Ga=	**240**	ENTER ZERO IF NO PARKING REQUIRED		
	No. of loading spaces	l =	**1**			
	Gross area per loading space	b =	**1,000**	SF	1,000	SF
Site Variables:	**Project open space as fraction of BLA**	**S=**	**0.300**		100,009	SF
	Private driveways as fraction of BLA	R=	**0.050**		16,668	SF
	Misc. pavement as fraction of BLA	M=	**0.100**		33,336	SF
	Loading area as fraction of BLA	L=	0.002		1,000	SF
	Total site support areas as a fraction of BLA	Su=	0.452		150,764	SF
Core:	**Core development area as fraction of BLA**	**C=**	0.548	C=Su must = 1	182,601	SF
Building Variables:	Building efficiency as percentage of GBA	Be=	**0.600**			
	Building support as fraction of GBA	Bu=	0.400	Be + Bu must = 1		

Dwelling Unit Mix Table:

DU dwelling unit type	GDA gross du area	CDA=GDA/Be comprehensive du area	MIX du mix	PDA = (CDA)MIX Prorated du area
EFF	**500**	833	**10%**	83
1 BR	**750**	1,250	**20%**	250
2 BR	**1,200**	2,000	**40%**	800
3 BR	**1,500**	2,500	**20%**	500
4 BR	**1,800**	3,000	**10%**	300
			Aggregate avg. dwelling unit area (AGG) =	**1,933**
			GBA sf per parking space a=	1,289

PLANNING FORECAST

no. of floors FLR	CORE minimum lot area for BCG & PLA	density per net acre dNA	dwelling units **NDU**	pkg. lot spaces NPS	parking lot area **PLA**	garage spaces GPS	garage area **GAR**	gross bldg area GBA no garages	footprint **BCA**	density per dBA bldable acre
1.00	182,601	7.93	**68.8**	103.2	**41,292**	34.4	**8,258**	133,051	**133,051**	8.99
2.00		12.48	**108.3**	162.4	**64,957**	54.1	**12,991**	209,305	**104,653**	14.15
3.00		15.43	**133.8**	200.7	**80,297**	66.9	**16,059**	258,734	**86,245**	17.49
4.00		17.50	**151.7**	227.6	**91,047**	75.9	**18,209**	293,375	**73,344**	19.83
5.00		19.02	**165.0**	247.5	**99,000**	82.5	**19,800**	319,001	**63,800**	21.56
6.00	*NOTE:*	20.20	**175.2**	262.8	**105,122**	87.6	**21,024**	338,726	**56,454**	22.89
7.00	*Be aware when*	21.13	**183.3**	274.9	**109,979**	91.6	**21,996**	354,378	**50,625**	23.95
8.00	*BCA becomes too small to be feasible.*	21.89	**189.9**	284.8	**113,928**	94.9	**22,786**	367,100	**45,887**	24.81
9.00		22.52	**195.3**	293.0	**117,200**	97.7	**23,440**	377,645	**41,961**	25.52
10.00		23.05	**199.9**	299.9	**119,957**	100.0	**23,991**	386,527	**38,653**	26.12
11.00		23.50	**203.9**	305.8	**122,310**	101.9	**24,462**	394,111	**35,828**	26.64
12.00		23.89	**207.2**	310.9	**124,343**	103.6	**24,869**	400,662	**33,389**	27.08
13.00		24.23	**210.2**	315.3	**126,117**	105.1	**25,223**	406,378	**31,260**	27.47
14.00		24.53	**212.8**	319.2	**127,679**	106.4	**25,536**	411,409	**29,386**	27.81

WARNING: These are preliminary forecasts that must not be used to make final decisions.
1) These forecasts are not a substitute for the "due diligence" research that must be conducted to support the final definition of "unbuildable areas" above and the final decision to purchase land. This research includes, but is not limited to, verification of adequate subsurface soil, zoning, environmental clearance, access, title, utilities and water pressure, clearance from deed restriction, easement and right-of-way encumbrances, clearance from existing above and below ground facility conflicts, etc.
2) The most promising forecast(s) made on the basis of data entered in the design specification from "due diligence" research must be verified at the drawing board before funds are committed and land purchase decisions are made. Actual land shape ratios, dimensions and irregularities encountered may require adjustments to the general forecasts above.
3) The software licensee shall take responsibility for the design specification values entered and any advice given that is based on the forecast produced.

Exhibit 1.2 Nonresidential Forecast Model Sample Format

Forecast Model CG1L

*Development capacity forecast for **NONRESIDENTIAL BUILDINGS** based on the use of an adjacent **GRADE PARKING LOT** located on the same premises. When s and a equal zero in the design specification below, the forecast pertains to conditions when **NO PARKING** is required.*

***Given:** Gross land area. **To Find:** Maximum development capacity of the land area (gross building area potential) based on the design specification values entered below. **Premise:** All building floors considered equal in area.*

DESIGN SPECIFICATION

Enter values in boxed areas where text is bold and blue. Express all fractions as decimals.

Given:	**Gross land area**	**GLA=**	**5.882**	acres	256,220	SF
Land Variables:	Public/ private right-of-way & paved easements	W=	**0.150**	fraction of GLA	38,433	SF
	Net land area	NLA=	5.000	acres	217,787	SF
	Unbuildable and/or future expansion areas	U=	**0.000**	fraction of GLA	0	SF
	Gross land area reduction	X=	0.150	fraction of GLA	38,433	SF
	Buildable land area remaining	**BLA=**	5.000	acres	217,787	SF
Parking Variables:	Est. gross pkg. lot area per space in SF	s =	**375**	ENTER ZERO IF NO PARKING REQUIRED		
	Building SF permitted per parking space	a =	**250**	ENTER ZERO IF NO PARKING REQUIRED		
	No. of loading spaces	l=	**1**			
	Gross area per loading space	b =	**1,000**	SF	1,000	SF
Site Variables:	**Project open space as fraction of BLA**	**S=**	**0.300**		65,336	SF
	Private driveways as fraction of BLA	R=	**0.030**		6,534	SF
	Misc. pavement as fraction of BLA	M=	**0.020**		4,356	SF
	Loading area as fraction of BLA	L=	0.005		1,000	SF
	Total site support areas as a fraction of BLA	Su=	0.355		77,225	SF
Core:	**Core development area as fraction of BLA**	**C=**	0.645	C+Su must = 1	140,562	SF

PLANNING FORECAST

no. of floors **FLR**	CORE minimum land area for BCG & PLA	gross building area **GBA**	parking lot area PLA	pkg. spaces NPS	footprint **BCA**	bldg SF / acre SFAC function of BLA	flr area ratio FAR function of BLA
1.00	140,562	**56,225**	84,337	224.9	**56,225**	11,246	0.258
2.00		**70,281**	105,421	281.1	**35,140**	14,057	0.323
3.00		**76,670**	115,005	306.7	**25,557**	15,335	0.352
4.00		**80,321**	120,481	321.3	**20,080**	16,065	0.369
5.00		**82,683**	124,025	330.7	**16,537**	16,538	0.380
6.00	*NOTE: Be aware when BCA becomes too small to be feasible.*	**84,337**	126,505	337.3	**14,056**	16,868	0.387
7.00		**85,559**	128,339	342.2	**12,223**	17,113	0.393
8.00		**86,499**	129,749	346.0	**10,812**	17,301	0.397
9.00		**87,245**	130,868	349.0	**9,694**	17,450	0.401
10.00		**87,851**	131,776	351.4	**8,785**	17,571	0.403
11.00		**88,353**	132,529	353.4	**8,032**	17,672	0.406
12.00		**88,776**	133,164	355.1	**7,398**	17,756	0.408
13.00		**89,137**	133,705	356.5	**6,857**	17,828	0.409
14.00		**89,448**	134,172	357.8	**6,389**	17,891	0.411
15.00		**89,720**	134,580	358.9	**5,981**	17,945	0.412

WARNING: These are preliminary forecasts that must not be used to make final decisions.
1) These forecasts are not a substitute for the "due diligence" research that must be conducted to support the final definition of "unbuildable areas" above and the final decision to purchase land. This research includes, but is not limited to, verification of adequate subsurface soil, zoning, environmental clearance, access, title, utilities and water pressure, clearance from deed restriction, easement and right-of-way encumbrances, clearance from existing above and below ground facility conflicts, etc.
2) The most promising forecast(s) made on the basis of data entered in the design specification from "due diligence" research must be verified at the drawing board before funds are committed and land purchase decisions are made. Actual land shape ratios, dimensions and irregularities encountered may require adjustments to the general forecasts above.
3) The software licensee shall take responsibility for the design specification values entered and any advice given that is based on the forecast produced.

"givens" required by the model and provides a synopsis of the forecast that will be produced in a "to find" notation. The design premise of the forecast is also noted. The only difference between the title blocks illustrated in Figs. 1.1 and 1.2 and the CD-ROM versions is that the space used by the title line in the illustrations is a data-entry block in the electronic versions that can be used to identify company, client, date, and so forth.

Design Specification

The design specification appears directly below the explanation panel in each model. This specification contains a number of data-entry cells identified by bold numbers within box outlines. If the box does not exist or if the numbers are not bold, it is not a data-entry cell. (On the screen, these numbers are also blue.) These data-entry cells have been referred to as the *design components* of a design specification. Any change to a value in one of these cells will produce a different forecast in the lower half of the page. These design components have been grouped under the titles Land Variables, Parking Variables, Site Variables, and Building Variables. All specification groups may not occur in all models, depending on the calculations needed. In the case of all residential models, a dwelling unit mix table is also included to accurately define the dwelling unit areas and mix involved, since this mix directly affects the amount of land needed.[9] A complete glossary of design specification terms, with explanations, is provided at the end of Chap. 2.

Core Value

Each model also includes a line referring to the core value *C* calculated. This value indicates the fractional amount of buildable area[10] exclusively available for building cover[11] and parking cover.[12] It is used to calculate the core development area[13] CORE available on a given site. The core value and resulting core area directly affect the development capacity of land, and both are calculated fields.[14] The core value is found in the design specification panel of each model within a box that has a heavy outline. It can be mistaken for a data-entry cell, but remember that a box alone does not indicate a data-entry location. The number must also be bold (and blue in the electronic version).

Data Entry

Exhibit 1.1 requires that the gross land area be given, and 10.204 acres has been entered in the first line of the design specification panel. The remaining variables in the design specification have also been entered and provide information needed to calculate the development capacity of the land area given. These numbers can be changed at will to test alternate development strategies, since most have a broad range of potential values that can produce radically different design and development results.

Planning Forecast

The planning forecast section of Exhibit 1.1 answers the questions posed in the explanation panel and flags these answers within a column or area of the forecast panel that is bordered with heavy lines. For example, the model is designed to find the maximum dwelling unit capacity of the land area given based on the design specification values entered. The model forecasts the *number of dwelling units* possible in the NDU column. The forecast is expressed as a range of options that are a function of the number of building *floors* under consideration in the left-hand FLR column. In this example, the range extends from 68.8 to 212.8 dwelling units. If the objective were to produce a 10-story apartment building, the forecast predicts that 199.9, or 200, dwelling units could be accommodated by the site based on the design specification values entered. The building floor numbers in the FLR column may be changed at will to produce additional forecasts since zoning ordinance regulations or increased height objectives may demand greater flexibility than the initial list included. Additional development information is also forecast in this panel to assist both development analysis and future drawing-board plan evaluation. This information varies by development model and is explored later with each case study examination.

Single-Model Comparative Analysis

The strength of development capacity forecasting includes not only quantitative data and quick response time—it also makes broader comparisons feasible. As an example, Exhibit 1.2 pertains to nonresidential development and forecasts the gross building area that can be accommodated on a gross land area of 5.882 acres. In the *gross building area* (GBA) forecast column, this potential ranges from 56,225 to 89,720 sq ft, depending on the number of building floors chosen in the left-hand FLR column. For instance, a 3-story building could produce 76,670 sq ft of gross building area based on the values entered in the design specification panel. Project open space[15] is one of the design components that produced this forecast, and 30% was entered as a design objective. This means that 30% of the buildable land area available would remain as lawn or natural setting to counterbalance development cover[16] and building height introduced. If this open space design objective were increased to 50%, Exhibit 1.3 forecasts that the same 3-story building could only produce 52,911 sq ft of gross building area. This 23,759-sq-ft reduction in potential building area[17] occurs because more of the buildable land area available is used for project open space and less is available for construction.[18]

Multiple-Model Comparative Analysis

It is also possible to compare the development capacity results produced by two entirely different parking systems. Most designers intuitively understand that a parking struc-

Exhibit 1.3 Effect of Increased Project Open Space on Development Capacity

Forecast Model CG1L

Development capacity forecast for ***NONRESIDENTIAL BUILDINGS*** *based on the use of an adjacent* ***GRADE PARKING LOT*** *located on the same premises. When s and a equal zero in the design specification below, the forecast pertains to conditions when* ***NO PARKING*** *is required.*

Given: *Gross land area.* ***To Find:*** *Maximum development capacity of the land area (gross building area potential) based on the design specification values entered below.* ***Premise:*** *All building floors considered equal in area.*

DESIGN SPECIFICATION

Enter values in boxed areas where text is bold and blue. Express all fractions as decimals.

Given:	**Gross land area**	**GLA=**	**5.882**	acres	256,220	SF
Land Variables:	Public/ private right-of-way & paved easements	W=	**0.150**	fraction of GLA	38,433	SF
	Net land area	NLA=	5.000	acres	217,787	SF
	Unbuildable and/or future expansion areas	U=	**0.000**	fraction of GLA	0	SF
	Gross land area reduction	X=	0.150	fraction of GLA	38,433	SF
	Buildable land area remaining	**BLA=**	5.000	acres	217,787	SF
Parking Variables:	Est. gross pkg. lot area per space in SF	s =	**375**	ENTER ZERO IF NO PARKING REQUIRED		
	Building SF permitted per parking space	a =	**250**	ENTER ZERO IF NO PARKING REQUIRED		
	No. of loading spaces	l=	**1**			
	Gross area per loading space	b =	**1,000**	SF	1,000	SF
Site Variables:	**Project open space as fraction of BLA**	**S=**	**0.500**	⬅	108,893	SF
	Private driveways as fraction of BLA	R=	**0.030**		6,534	SF
	Misc. pavement as fraction of BLA	M=	**0.020**		4,356	SF
	Loading area as fraction of BLA	L=	0.005		1,000	SF
	Total site support areas as a fraction of BLA	Su=	0.555		120,783	SF
Core:	**Core development area as fraction of BLA**	**C=**	0.445	C+Su must = 1	97,004	SF

PLANNING FORECAST

no. of floors **FLR**	CORE minimum land area for BCG & PLA	gross building area **GBA**	parking lot area PLA	pkg. spaces NPS	footprint **BCA**	bldg SF / acre SFAC function of BLA	flr area ratio FAR function of BLA
1.00	97,004	**38,802**	58,202	155.2	**38,802**	7,761	0.178
2.00		**48,502**	72,753	194.0	**24,251**	9,701	0.223
3.00		**52,911**	79,367	211.6	**17,637**	10,583	0.243
4.00		**55,431**	83,146	221.7	**13,858**	11,087	0.255
5.00		**57,061**	85,592	228.2	**11,412**	11,413	0.262
6.00	*NOTE: Be aware when BCA becomes too small to be feasible.*	**58,202**	87,304	232.8	**9,700**	11,641	0.267
7.00		**59,046**	88,569	236.2	**8,435**	11,810	0.271
8.00		**59,695**	89,542	238.8	**7,462**	11,940	0.274
9.00		**60,209**	90,314	240.8	**6,690**	12,043	0.276
10.00		**60,628**	90,941	242.5	**6,063**	12,126	0.278
11.00		**60,974**	91,461	243.9	**5,543**	12,196	0.280
12.00		**61,266**	91,899	245.1	**5,105**	12,254	0.281
13.00		**61,515**	92,272	246.1	**4,732**	12,304	0.282
14.00		**61,730**	92,595	246.9	**4,409**	12,347	0.283
15.00		**61,918**	92,876	247.7	**4,128**	12,384	0.284

WARNING: These are preliminary forecasts that must not be used to make final decisions.

1) These forecasts are not a substitute for the "due diligence" research that must be conducted to support the final definition of "unbuildable areas" above and the final decision to purchase land. This research includes, but is not limited to, verification of adequate subsurface soil, zoning, environmental clearance, access, title, utilities and water pressure, clearance from deed restriction, easement and right-of-way encumbrances, clearance from existing above and below ground facility conflicts, etc.

2) The most promising forecast(s) made on the basis of data entered in the design specification from "due diligence" research must be verified at the drawing board before funds are committed and land purchase decisions are made. Actual land shape ratios, dimensions and irregularities encountered may require adjustments to the general forecasts above.

3) The software licensee shall take responsibility for the design specification values entered and any advice given that is based on the forecast produced.

ture can provide more parking spaces than a surface parking lot on the same given land area. They may not be as fully aware, however, that the number of parking spaces that can be provided on a given land area is a major factor determining the maximum building area that can be constructed on this land area. The implications of parking design options and space requirements have been difficult and time-consuming to calculate. This has limited the evaluation, but the attached development capacity forecast collection reduces this level of difficulty to several keystrokes.

Model CG1L is outlined in Fig. 1.2 and illustrated in Exhibit 1.2. This model pertains to nonresidential land uses served by grade parking lots around but not under the building. The surface parking capacity forecast in Exhibit 1.2 can be compared with the results that would be produced by an underground parking structure on the same land area by choosing model CS2L.[19] The CG1L design specification values for land, parking, and site can then be introduced in CS2L for comparison.[20] The characteristics of forecast model CS2L are summarized from the decision guide in Table 1.3.

Exhibit 1.4 illustrates the results produced by model CS2L when the CG1L values for land, parking and site variables are introduced. If a 3-story building were located in the FLR column of each exhibit, Exhibit 1.2 forecasts that 76,670 sq ft of gross building area could be constructed

Table 1.3 Characteristics of Forecast Model CS2L

	Applicability	
Objective	**Topic**	**Category**
Given: Gross land area	Land use family	Non-residential
To find: The maximum gross building area that can be placed on the gross land area given	Parking family Building family Parking system	Structure Non-residential Underground parking structure

Exhibit 1.4 Underground Parking Development Capacity Comparison

Forecast Model CS2L

Development capacity forecast for ***NONRESIDENTIAL BUILDINGS*** *based on the use of an* ***UNDERGROUND PARKING STRUCTURE.***
Given: *Gross land area.* ***To Find:*** *Maximum development capacity of the land area given (gross building area potential) based on the design specification values entered below.*
Design Premise: *Underground parking footprint may be larger, smaller or equal to the building footprint above. All similar floors considered equal in area.*

DESIGN SPECIFICATION

Enter values in boxed areas where text is bold and blue. Express all fractions as decimals.

Given:	**Gross land area**	GLA=	**5.882**	acres	256,220	SF
Land Variables:	Public/ private right-of-way & paved easements	W=	**0.150**	fraction of GLA	38,433	SF
	Net land area	NLA=	5.000	acres	217,787	SF
	Unbuildable and/or future expansion areas	U=	**0.000**	fraction of GLA	0	SF
	Gross lot area reduction	X=	0.150	fraction of GLA	38,433	SF
	Buildable lot area remaining	BLA=	5.000	acres	217,787	SF
Parking Variables:	Estimated net pkg. structure area per parking space	s =	**375**	SF		
	Building SF permitted per parking space	a =	**250**	SF		
	No. of loading spaces	l =	**1**			
	Gross area per loading space	b =	**1,000**	SF	1,000	SF
Site Variables at Grade:	**Project open space as fraction of BLA**	S=	**0.300**		65,336	SF
	Private driveways as fraction of BLA	R=	**0.030**		6,534	SF
	Misc. pavement as fraction of BLA	M=	**0.020**		4,356	SF
	Loading area as fraction of BLA	L=	0.005		1,000	SF
	Total site support areas at grade as a fraction of BLA	Su=	0.355		77,225	SF
Core:	**Core development area as fraction of BLA**	C=	0.645	C+Su must = 1	140,562	SF
Below Grade:	Gross underground parking (UNG) as fraction of BLA	G=	**0.900**			
	Pkg. support within parking structure as fraction of UNG	Pu=	**0.200**			
	Net pkg. structure area for parking & circulation as fraction of UNG	Pe=	0.800	Pe+Pu must = 1		

PLANNING FORECAST

NOTE: p=1 is one level below grade and is not a surface parking lot based on design premise

no. of bldg. flrs. FLR	no. of pkg. levels p	development area CORE at grade	gross building area GBA	net pkg area NPA	pkg. spaces NPS	gross pkg area GPA	bldg SF / acre SFAC	flr area ratio FAR function of BLA
1.00	**1.34**	140,562	**140,562**	210,842	562.2	263,553	28,114	0.645
2.00	**2.69**		**281,123**	421,685	1124.5	527,106	56,228	1.291
3.00	**4.03**	footprint at grade	**421,685**	632,527	1686.7	790,658	84,342	1.936
4.00	**5.38**	BCA = CORE	**562,246**	843,369	2249.0	1,054,211	112,456	2.582
5.00	**6.72**	140,562	**702,808**	1,054,211	2811.2	1,317,764	140,570	3.227
6.00	**8.07**		**843,369**	1,265,054	3373.5	1,581,317	168,684	3.872
7.00	**9.41**	net pkg. area / underground level	**983,931**	1,475,896	3935.7	1,844,870	196,798	4.518
8.00	**10.76**	NPL	**1,124,492**	1,686,738	4498.0	2,108,423	224,912	5.163
9.00	**12.10**	156,807	**1,265,054**	1,897,580	5060.2	2,371,975	253,026	5.809
10.00	**13.45**		**1,405,615**	2,108,423	5622.5	2,635,528	281,140	6.454
11.00	**14.79**	*NOTE: p=1 is one level below grade and is not a surface parking lot based on the design premise for this forecast.*	**1,546,177**	2,319,265	6184.7	2,899,081	309,254	7.099
12.00	**16.14**		**1,686,738**	2,530,107	6747.0	3,162,634	337,368	7.745
13.00	**17.48**		**1,827,300**	2,740,949	7309.2	3,426,187	365,482	8.390
14.00	**18.82**		**1,967,861**	2,951,792	7871.4	3,689,740	393,596	9.036
15.00	**20.17**		**2,108,423**	3,162,634	8433.7	3,953,292	421,710	9.681

WARNING: These are preliminary forecasts that must not be used to make final decisions.
1) These forecasts are not a substitute for the "due diligence" research that must be conducted to support the final definition of "unbuildable areas" above and the final decision to purchase land. This research includes, but is not limited to, verification of adequate subsurface soil, zoning, environmental clearance, access, title, utilities and water pressure, clearance from deed restriction, easement and right-of-way encumbrances, clearance from existing above and below ground facility conflicts, etc.
2) The most promising forecast(s) made on the basis of data entered in the design specification from "due diligence" research must be verified at the drawing board before funds are committed and land purchase decisions are made. Actual land shape ratios, dimensions and irregularities encountered may require adjustments to the general forecasts above.
3) The software licensee shall take responsibility for the design specification values entered and any advice given that is based on the forecast produced.

based on a surface parking lot and the design specification values entered. Exhibit 1.4 predicts that a 3-story building with an underground parking garage using the same design specification would produce 421,685 sq ft of gross building area and require 4.03 underground parking levels to support its parking need.[21] This is 5.5 times the building area that can be supported with a surface parking lot. A developer may not be interested in constructing this much building area on a given site because of many other factors, such as construction cost and market demand, but the forecast models permit the option to be quickly explored. If a developer wanted more building area than a surface parking lot could support however, but not the maximum amount possible, the developer could scale back the number of building floors contemplated. For instance, 1 building floor would require 1.34 underground parking levels and produce 140,562 gross sq ft of building area on the land area given.[22] A developer would have to evaluate the financial merits of each option before reaching a decision to proceed with further drawing-board evaluation.

It should also be mentioned that the gross building area of 76,670 sq ft that was supported by a surface parking lot could be supported with an underground parking garage. To investigate this option, simply change the open space percentage *S* in the CS2L design specification until the number 76,670 appears in the gross building area forecast column GBA opposite a desired number of building floors. This will identify the amount of open space that can be produced with this design strategy.

The development options covered here are a few of the design alternatives available for a given land area. The forecast model collection permits these planning policy options to be evaluated and outlined ahead of the drawing process, which can make both more efficient and productive.

Parking Requirements

The value a entered as a parking requirement in the design specification of Exhibit 1.2 is a major factor affecting development capacity. The planning forecast within this exhibit is based on a parking requirement of 1 space for every 250 sq ft of building area ($a = 250$). This is often expressed as 4 parking spaces required per 1000 sq ft of gross building area permitted. If the parking provision were increased to 5 spaces per 1000 sq ft of building area, the value a would equal 200. Exhibit 1.5 illustrates the impact of this parking increase by changing the a value in Exhibit 1.2 from 250 to 200. A comparison of the two exhibits shows that this increased parking provision reduces the 5-story gross building area forecast from 82,683 to 67,740 sq ft.[23] This 14,943-sq-ft reduction is an 18% decrease in potential building area and obviously reduces the potential economic yield from the property for both private and public interests. Since parking requirements are not an exact science, it is no wonder that planning and zoning boards throughout the country see many requests for parking variances that are not always justified.

Project Open Space

All values entered in a forecast model's design specification have development capacity, construction cost, building population, energy consumption, property management, traffic generation, tax revenue, and return on investment implications. The value assigned to project open space is a major limitation, and it also plays a significant environmental role. Table 1.4 lists a series of empirical observations regarding the impact of open space when provided in different quantities within a project area. The amount of open space provided obviously reduces the land available for construction. From a commercial standpoint, this either reduces the eco-

Exhibit 1.5 Effect of Increased Parking Requirement on Development Capacity

Forecast Model CG1L

Development capacity forecast for ***NONRESIDENTIAL BUILDINGS*** *based on the use of an adjacent* ***GRADE PARKING LOT*** *located on the same premises. When s and a equal zero in the design specification below, the forecast pertains to conditions when* ***NO PARKING*** *is required.*

Given: *Gross land area.* ***To Find:*** *Maximum development capacity of the land area (gross building area potential) based on the design specification values entered below.* ***Premise:*** *All building floors considered equal in area.*

DESIGN SPECIFICATION *Enter values in boxed areas where text is bold and blue. Express all fractions as decimals.*

Given:	**Gross land area**	GLA=	**5.882**	acres	256,220	SF
Land Variables:	Public/ private right-of-way & paved easements	W=	**0.150**	fraction of GLA	38,433	SF
	Net land area	NLA=	5.000	acres	217,787	SF
	Unbuildable and/or future expansion areas	U=	**0.000**	fraction of GLA	0	SF
	Gross land area reduction	X=	0.150	fraction of GLA	38,433	SF
	Buildable land area remaining	BLA=	5.000	acres	217,787	SF
Parking Variables:	Est. gross pkg. lot area per space in SF	s =	**375**	ENTER ZERO IF NO PARKING REQUIRED		
	Building SF permitted per parking space	a =	**200**	ENTER ZERO IF NO PARKING REQUIRED		
	No. of loading spaces	l=	**1**			
	Gross area per loading space	b =	**1,000**	SF	1,000	SF
Site Variables:	**Project open space as fraction of BLA**	S=	**0.300**		65,336	SF
	Private driveways as fraction of BLA	R=	**0.030**		6,534	SF
	Misc. pavement as fraction of BLA	M=	**0.020**		4,356	SF
	Loading area as fraction of BLA	L=	0.005		1,000	SF
	Total site support areas as a fraction of BLA	Su=	0.355		77,225	SF
Core:	**Core development area as fraction of BLA**	C=	0.645	C+Su must = 1	140,562	SF

PLANNING FORECAST

no. of floors **FLR**	CORE minimum land area for BCG & PLA	gross building area **GBA**	parking lot area PLA	pkg. spaces NPS	footprint **BCA**	bldg SF / acre SFAC function of BLA	flr area ratio FAR function of BLA
1.00	140,562	**48,891**	91,671	244.5	**48,891**	9,779	0.224
2.00		**59,184**	110,970	295.9	**29,592**	11,837	0.272
3.00		**63,650**	119,345	318.3	**21,217**	12,731	0.292
4.00		**66,147**	124,025	330.7	**16,537**	13,230	0.304
5.00		**67,740**	127,013	338.7	**13,548**	13,549	0.311
6.00	*NOTE: Be aware when BCA becomes too small to be feasible.*	**68,846**	129,087	344.2	**11,474**	13,770	0.316
7.00		**69,659**	130,610	348.3	**9,951**	13,933	0.320
8.00		**70,281**	131,776	351.4	**8,785**	14,057	0.323
9.00		**70,772**	132,698	353.9	**7,864**	14,155	0.325
10.00		**71,170**	133,444	355.9	**7,117**	14,235	0.327
11.00		**71,499**	134,062	357.5	**6,500**	14,301	0.328
12.00		**71,776**	134,580	358.9	**5,981**	14,356	0.330
13.00		**72,012**	135,022	360.1	**5,539**	14,403	0.331
14.00		**72,215**	135,403	361.1	**5,158**	14,444	0.332
15.00		**72,392**	135,735	362.0	**4,826**	14,479	0.332

WARNING: These are preliminary forecasts that must not be used to make final decisions.
1) These forecasts are not a substitute for the "due diligence" research that must be conducted to support the final definition of "unbuildable areas" above and the final decision to purchase land. This research includes, but is not limited to, verification of adequate subsurface soil, zoning, environmental clearance, access, title, utilities and water pressure, clearance from deed restriction, easement and right-of-way encumbrances, clearance from existing above and below ground facility conflicts, etc.
2) The most promising forecast(s) made on the basis of data entered in the design specification from "due diligence" research must be verified at the drawing board before funds are committed and land purchase decisions are made. Actual land shape ratios, dimensions and irregularities encountered may require adjustments to the general forecasts above.
3) The software licensee shall take responsibility for the design specification values entered and any advice given that is based on the forecast produced.

Table 1.4 Empirical Forecast of Environments Created by Various Open Space Allocations

Open Space Allocation, %	Environmental Description
0	Typical downtown office block where no setbacks are provided. Building and parking cover entire lot. Storm detention can only be provided underground or in parking lot. Building covers entire lot when parking not provided. Paved surfaces absorb, reflect, and intensify outdoor air temperatures. Environment presents canyonlike places for people.
10	Setbacks often cannot be provided. Not enough space for multibuilding separation. Storm detention must be provided in the parking lot. Paved surfaces absorb, reflect, and intensify outdoor air temperatures. A building and its support parking dominate the environment created.
20	Setbacks partially accommodated. Multibuilding separation still inadequate. Storm detention must be provided in the parking lot. Paved surfaces absorb, reflect, and intensify outdoor air temperatures. Environment still dominated by building and parking.
30	Setbacks primarily accommodated. Multibuilding separation possible but tight. Storm detention can be provided by a combination of open space and parking lot solutions. Paved surfaces absorb, reflect, and intensify outdoor air temperatures. Environment still dominated by parking and building(s), but some minimal "people places" begin to appear.
40	Setbacks accommodated. Multibuilding separation possible without overwhelming sense of crowding. Storm detention can be constructed within open space provided. Paved surfaces absorb, reflect, and intensify outdoor air temperatures. Environment begins to provide small people places.
50	Setbacks accommodated. Multibuilding separation possible. Storm detention provided within open space. Outside air temperature increases moderated by open space provided. Environment begins to take on appearance of office park with balance provided by people places, lawns, and landscaping.
60	Setbacks accommodated. Multibuilding separation possible. Storm detention provided within open space. Outside air temperature increases moderated by open space provided. Office park environment begins to benefit neighbors as well as on-site users.

nomic yield of the land or increases the revenue that must be generated from each remaining square foot of building area. From a residential standpoint, increased open space generally implies higher value, but also produces greater sprawl. These relationships represent a complex balancing act that receives greater attention in future chapters.[24]

Table 1.4 is included to paint a brief picture of the counterbalancing role that open space plays within a development design composition. It has been created from empirical observation and primarily addresses balance rather than intensity, since it avoids specific references to building height. These relationships between development cover, building height, and open space are important to understand, however, so that we can begin to shelter more with less in order to improve rather than threaten the relationship between our natural world and our built environment.[25]

Design Principles

Development capacity forecasts can also reveal design principles that help to expand our understanding of the complex development relationships involved within our built environment. A simple example is illustrated in Exhibit 1.6. The gross building area forecast presented in Exhibit 1.2 has been plotted in Exhibit 1.6 to produce what is referred to as *Design Principle 1.1*. This principle suggests that the *rate* of increase in gross building area declines at an accelerating rate as the number of building floors increase when surface parking is used and all other design specification values remain constant.

The chart in Exhibit 1.6 illustrates this design principle. It reveals that increasing building height above 3 stories produces a rapidly diminishing rate of increase in gross building area when a surface parking lot is used and all other design specification values remain constant. Reducing the amount

Exhibit 1.6 Rate of Growth in Gross Building Area as Height Increases

Forecast Model CG1L

Development capacity forecast for ***NONRESIDENTIAL BUILDINGS*** *based on the use of an adjacent* ***GRADE PARKING LOT*** *located on the same premises. When s and a equal zero in the design specification below, the forecast pertains to conditions when* ***NO PARKING*** *is required.*

Given: *Gross land area.* ***To Find:*** *Maximum development capacity of the land area (gross building area potential) based on the design specification values entered below.* ***Premise:*** *All building floors considered equal in area.*

DESIGN SPECIFICATION *Enter values in boxed areas where text is bold and blue. Express all fractions as decimals.*

Given:	**Gross land area**	GLA=	**5.882**	acres	256,220	SF
Land Variables:	Public/ private right-of-way & paved easements	W=	**0.150**	fraction of GLA	38,433	SF
	Net land area	NLA=	5.000	acres	217,787	SF
	Unbuildable and/or future expansion areas	U=	**0.000**	fraction of GLA	0	SF
	Gross land area reduction	X=	0.150	fraction of GLA	38,433	SF
	Buildable land area remaining	BLA=	5.000	acres	217,787	SF
Parking Variables:	Est. gross pkg. lot area per space in SF	s =	**375**	ENTER ZERO IF NO PARKING REQUIRED		
	Building SF permitted per parking space	a =	**250**	ENTER ZERO IF NO PARKING REQUIRED		
	No. of loading spaces	l=	**1**			
	Gross area per loading space	b =	**1,000**	SF	1,000	SF
Site Variables:	**Project open space as fraction of BLA**	S=	**0.300**		65,336	SF
	Private driveways as fraction of BLA	R=	**0.030**		6,534	SF
	Misc. pavement as fraction of BLA	M=	**0.020**		4,356	SF
	Loading area as fraction of BLA	L=	0.005		1,000	SF
	Total site support areas as a fraction of BLA	Su=	0.355		77,225	SF
Core:	**Core development area as fraction of BLA**	C=	0.645	C+Su must = 1	140,562	SF

PLANNING FORECAST / **RATE of GBA INCREASE**

no. of floors **FLR**	**CORE** minimum land area for BCG & PLA	gross building area **GBA**	total GBA increase as hgt. increases	rate of GBA increase as hgt. increases
1.00	140,562	**56,225**	0.0%	0.0%
2.00		**70,281**	25.0%	25.00%
3.00		**76,670**	36.4%	9.09%
4.00		**80,321**	42.9%	4.76%
5.00		**82,683**	47.1%	2.94%
6.00	*NOTE:*	**84,337**	50.0%	2.00%
7.00	*Be aware when BCA becomes*	**85,559**	52.2%	1.45%
8.00	*too small to be feasible.*	**86,499**	53.8%	1.10%
9.00		**87,245**	55.2%	0.86%
10.00		**87,851**	56.3%	0.69%
11.00		**88,353**	57.1%	0.57%
12.00		**88,776**	57.9%	0.48%
13.00		**89,137**	58.5%	0.41%
14.00		**89,448**	59.1%	0.35%
15.00		**89,720**	59.6%	0.30%

Design Principle 1.1

The RATE of increase in gross building area declines at an accelerating rate as the number of building floors increase when surface parking is used and all other design specification values remain constant.

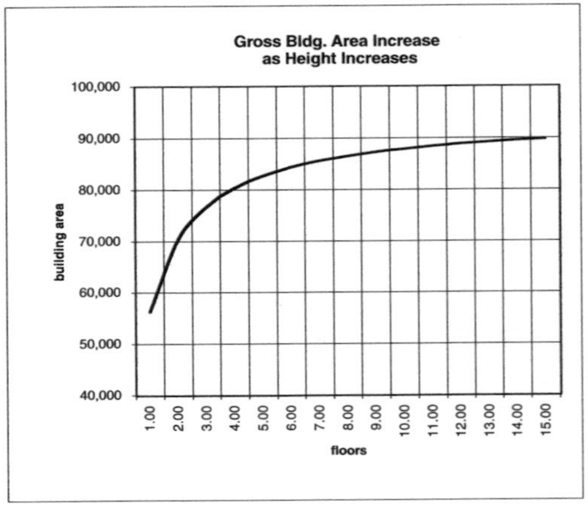

Gross Bldg. Area Increase as Height Increases

building area: 100,000; 90,000; 80,000; 70,000; 60,000; 50,000; 40,000

floors: 1.00; 2.00; 3.00; 4.00; 5.00; 6.00; 7.00; 8.00; 9.00; 10.00; 11.00; 12.00; 13.00; 14.00; 15.00

WARNING: These are preliminary forecasts that must not be used to make final decisions.
1) These forecasts are not a substitute for the "due diligence" research that must be conducted to support the final definition of "unbuildable areas" above and the final decision to purchase land. This research includes, but is not limited to, verification of adequate subsurface soil, zoning, environmental clearance, access, title, utilities and water pressure, clearance from deed restriction, easement and right-of-way encumbrances, clearance from existing above and below ground facility conflicts, etc.
2) The most promising forecast(s) made on the basis of data entered in the design specification from "due diligence" research must be verified at the drawing board before funds are committed and land purchase decisions are made. Actual land shape ratios, dimensions and irregularities encountered may require adjustments to the general forecasts above.
3) The software licensee shall take responsibility for the design specification values entered and any advice given that is based on the forecast produced.

of project open space retained on site can offset this diminishing rate of return by increasing the core area available for building cover and parking cover. This decreases the project open space balance provided and increases the intensity created. This tendency to increase development capacity by increasing height and expanding parking lots into open space creates a condition we often refer to as *overdevelopment*.[26] Unfortunately, we know overdevelopment when we see it, but the level at which relationships between development cover, building height, and project open space become unsatisfactory remains to be defined.

Indexing Development Balance and Intensity

Open space counterbalances development cover. Development cover includes building cover, which combines with building height to create building mass. The combination of building mass and remaining development cover produces the level of development intensity present. These two design components, open space and building height, either limit or enhance the development capacity of any given lot, depending on the quantities of each that are introduced. This means that these quantities can be also used to index the underlying development signature of a project. This signature is simply an index, however, and is not a substitute for a complete design specification.

Balance

The amount of open space provided on a lot, when expressed as a percentage of the buildable lot area, represents a *balance index*. This can be expressed with the following postulate:

> **Postulate 1.1.** Development balance can be indexed using the relative relationship between the total development cover introduced and the project open space that

remains within the buildable lot area. It can be expressed by the percentage of project open space provided.

Intensity

When building cover is combined with building height, three-dimensional building mass is created in addition to the surface cover provided. This mass, when combined with other surface cover, produces *development intensity*. The percentage of development cover provided on a buildable lot area, when combined with the building height introduced, can be used as an index of the intensity planned or constructed. This can be expressed with the following postulate:

Postulate 1.2. Development intensity can be indexed using the building height h introduced and the total development cover percentage D present.

This index INX can be expressed with a combined expression separated by a decimal point that first expresses the building height h present or proposed in feet and then expresses the development cover percentage D present or proposed to the nearest whole number.[27] This can be expressed with the following equation:

$$\text{INX} = h.D$$

In the absence of accurate building height information, a simplified development intensity index can be defined as follows:

Postulate 1.3. Development intensity can be indexed using the number of building floors f introduced and the total development cover percentage D present.

This index INX can be expressed with a combined expression separated by a decimal point that first expresses the number of building floors f present and then expresses the development cover percentage D

present or proposed to the nearest whole number.[28] This can be expressed with the following equation:

$$\text{INX} = f.D$$

Since intensity is also a function of building mass, which is produced by the combination of building cover and height, these indexes could be improved, but also complicated, by including the building cover percentage. The resulting postulate would become:

Postulate 1.4. Development intensity can be indexed using the number of building floors f introduced, the total building cover percentage B present, and the total development cover percentage D present to the nearest whole numbers:

$$\text{INX} = f.B.D$$

An index expressed as 5.15.50 using this format would represent a 5-story building with 15% building cover and 50% total development cover.

The postulates proposed[29] can help index and catalog the design specifications, field observations, and scientific examinations of existing projects. When these index catalogs expand with a scientific awareness of the environmental and ecological impact produced, and the quality of life sustained, these same indexes may be used to guide future development toward a duplication of success without dictating specific project appearance. A sample catalog of research that addresses a limited series of suburban development projects in a very simple manner is presented as Table 1.5 to illustrate this evaluation concept.

Land Use Activities

There are hundreds of land use activities, but it is not necessary to develop separate equations and models for each

Table 1.5 Catalog of Design Specification Values Recorded at Various Development Projects

Design Specification												Resulting Balance and Intensity					
Site Characteristics				Building Characteristics				Pavement Areas				Site Allocation				Intensity	
Total Lot Area	Acres	Parking Spaces		Dwelling Units	Bldg. Cover	Stories	Height	Private Road	Parking Area	Misc. Pave-ments	Total Pave-ment	Bldg. Cover	Devel-opment Cover	Open Space		Index	
TLA	TAC	NPS	POND	NDU	BCA	*f*	ft	RDA	PLA	MSP	PVT	*B*	*D*	*S*	POND	INX	Impressions
322,036	7.39	144			34,595	1	25		67,028	16,988	84,016	10.7%	36.8%	**63.2%**		**1.37**	Quality office park environment.
617,477	14.18		25,315		86,295	3					257,210	14.0	55.6	**44.4**	4.1%	**3.56**	Superior. Parking lot presence more noticeable, but good balance.
1,840,896	42.26		117,227		73,248	4	55				320,448	4.0	21.4	**78.6**	6.4	**4.21**	Superior office park with water.
1,556,007	35.72		117,389		160,901	2	30				145,978	10.3	19.7	**80.3**	7.5	**2.20**	Superior office park with water.
146,097	3.35				12,551	3	40				83,327	8.6	65.6	**34.4**		**3.66**	Pleasant. Height, massing and pavement not great enough to be unpleasant.
54,221	1.24				9,289	1	20				16,750	17.1	48.0	**52.0**		**1.48**	Quality. Massing and development cover offset by open space.
257,846	5.92				33,428	3	45				146,440	13.0	69.8	**30.2**		**3.70**	Pleasant. Massing and parking offset by surrounding office park.
131,227	3.01				18,015	2	35				51,626	13.7	53.1	**46.9**		**2.53**	Pleasant. Contributes to Metro Place office park environment.
632,477	14.52				34,103	9	110				375,603	5.4	64.8	**35.2**		**9.65**	Acceptable. Massing and parking offset by surrounding office park.
1,847,727	42.42		225,204		100,612	1, 3, 5	Varies				745,585	5.4	45.8	**54.2**	12.2		Quality. Open space and water contributes to Metro Place environment.
118,036	2.71				13,105	1	20				43,555	11.1	48.0	**52.0**		**1.48**	Quality. Small building contributes to Metro Place environment.
297,807	6.84				19,511	5	65				142,600	6.6	54.4	**45.6**		**5.54**	Pleasant. Massing and parking offset by open space on site.
177,770	4.08				41,868	1	20				56,817	23.6	55.5	**44.5**		**1.58**	Pleasant. Suburban office environment.
182,805	4.20				39,044	1	20				86,227	21.4	68.5	**31.5**		**1.69**	Pleasant. Suburban office environment.
107,016	2.46				15,297	2	30				41,020	14.3	52.6	**47.4**		**2.53**	Pleasant. Suburban office environment.
124,046	2.85				19,090	1	20				40,896	15.4	48.4	**51.6**		**1.48**	Pleasant. Suburban office environment.
227,949	5.23				21,648	3	45				116,290	9.5	60.5	**39.5**		**3.61**	Pleasant. Suburban office environment.
71,302	1.64				9,583	2	30				16,683	13.4	36.8	**63.2**		**2.37**	Pleasant. Suburban office environment.
49,757	1.14				9,461	2	30				16,904	19.0	53.0	**47.0**		**2.53**	Pleasant. Suburban office environment.
46,106	1.06				7,915	1	25				15,700	17.2	51.2	**48.8**		**1.51**	Pleasant. Suburban office environment.
28,607	0.66				4,248	1	25				9,397	14.8	47.7	**52.3**		**1.48**	Pleasant. Suburban office environment.
38,157	0.88				4,148	1	25				8,240	10.9	32.5	**67.5**		**1.32**	Pleasant. Suburban office environment.
44,714	1.03				6,775	1	25				17,993	15.2	55.4	**44.6**		**1.55**	Pleasant. Suburban office environment.
40,742	0.94				6,773	1	25				10,257	16.6	41.8	**58.2**		**1.42**	Pleasant. Suburban office environment.
89,094	2.05				10,394	2	35				20,066	11.7	34.2	**65.8**		**2.34**	Pleasant. Suburban office environment.

119,021	2.73				12,483	1	30				23,424	10.5	30.2	**69.8**		**1.30**	Pleasant. Suburban office environment.
311,637	7.15				20,807	2	35				173,250	6.7	62.3	**37.7**		**2.62**	Pleasant. Suburban office environment.
144,896	3.33				15,995	2	30				50,948	11.0	46.2	**53.8**		**2.46**	Quality office environment.
66,754	1.53				9,814	2	35				20,906	14.7	46.0	**54.0**		**2.46**	Quality office environment.
698,385	16.03				28,510	2	40				86,402	4.1	16.5	**83.5**		**2.16**	Superior office park.
324,995	7.46	405	18,105		29,177	4		26,261	135,867		162,128	9.0	58.9	**41.1**	5.6	**4.59**	Pleasant. Massing offset by open space and water.
187,503	4.30	187			24,249	2		21,168	54,336		75,504	12.9	53.2	**46.8**		**2.53**	Pleasant suburban office environment.
117,531	2.70	120			14,431	2			47,626		47,626	12.3	52.8	**47.2**		**2.53**	Pleasant suburban office environment.
560,682	12.87				106,377	2	22				54,140	19.0	28.6	**71.4**		**2.29**	Quality multifamily residential environment.
1,460,012	33.52				257,373	1	Varies				258,978	17.6	35.4	**64.6**		**1.35**	Superior retirement community.
73,236	1.68	25		53	11,398	3	39		14,797	2,280	17,077	15.6	38.9	**61.1**		**3.39**	Quality. High speed traffic and small setbacks prevent superior rating.
65,471	1.50	54		29	13,749	2	31				27,498	21.0	63.0	**37.0**		**2.63**	Not built. Considered overdeveloped by planning commission.
65,471	1.50	50		29	13,042	3	31				22,260	19.9	53.9	**46.1**		**3.54**	Not built. Considered overdeveloped by planning commission.
44,404	1.02				4,188	1	25				12,875	9.4	38.4	**61.6**		**1.38**	Pleasant.
66,394	1.52				5,204	1	20				37,895	7.8	64.9	**35.1**		**1.65**	Acceptable suburban setting. Major building and parking presence.
116,842	2.68				10,079	1	20				67,417	8.6	66.3	**33.7**		**1.66**	Acceptable suburban setting. Major building and parking presence.
280,117	6.43				60,211	2	30				69,061	21.5	46.1	**53.9**		**2.46**	Pleasant, but significant building, parking, and traffic presence.
145,895	3.35				22,209	2	30				69,140	15.2	62.6	**37.4**		**2.63**	Acceptable. Significant building, parking, and traffic presence.
474,255	10.89				82,153	4	50				212,976	17.3	62.2	**37.8**		**4.62**	Pleasant. Office park open space reduces parking lot impact on site.

in order to forecast the ability of land to accommodate their needs. All of these activities have building characteristics in common that allow them to be collected into groups; this grouping makes development capacity forecasting possible.

Group forecasting is differentiated by land use family, parking category, building type, and parking system. This hierarchy can be seen in the Fig. 1.1 decision guide.[30] Within each parking system there are several forecast model choices based on what is known. Since many land use activities take place within the same building type and use the same parking system, these models are not differentiated by land use activity, such as office or hospital, but by the more generic building type and parking system characteristics mentioned. For instance, most if not all land use activities employ one of the five parking systems identified in Table 1.6, and most use a grade parking lot. Table 1.6 illustrates that a limited number of land use families, building type categories, and parking systems shelter most land use activities. This makes it possible to forecast the development capacity of land to shelter these activities without writing separate equations for each. If this were not the case, the development forecasting collection would be next to impossible to create and even more difficult to use.

Land with Irregular and Difficult Shapes

When the forecast models are used with land that cannot be fully utilized because of its boundary shape, there are two options. Both options involve the percentage of open space assigned in the design specification panel of a forecast model. The first option is simply to look at the survey of property boundaries, topography, and physical features and assign an open space percentage based on experience, intuition, setback requirements, and the environmental design objectives for the buildable area of the site. If the land shape is difficult, the open space percentage will have to include

Table 1.6 Generic Land Use, Building Type, and Parking System Combinations

Land Use	Building Type	Parking System
Nonresidential	Nonresidential*	Surface parking Structure parking No parking
Residential	Apartment house†	Surface and/or garage parking Structure parking No parking
	Urban house‡	Surface and/or garage parking Structure parking No parking
	Suburb house§	No shared parking
Mixed Use	Apartment and nonresidential combination	Surface and/or garage parking Structure parking No parking

*A building intended to shelter activities and not family units.

†Apartment houses contain single-family dwelling units without direct and immediate access to the land. This topic is discussed in detail in Chap. 14.

‡An *urban house* is a grade-oriented residential building that may contain twin-single, three-family, four-family, rowhouse, or townhouse attached floor plan configurations. This category also includes single-family detached homes with less than 40 ft of frontage, and often includes homes with less than 30 ft of frontage. An urban house is directly entered from, and has a direct relationship with, the land. Each dwelling is attached to at least one other dwelling within the parent building, except for the narrow lots mentioned, and the parent building does not extend beyond one dwelling unit in height. This topic is discussed in detail in Chap. 12.

§A *suburb house* is a grade-oriented, freestanding building that shelters one family unit. It is distinguished more by the lot area, open space, and street system that surrounds it than by its own building area and style. The amount of detachment establishes the context in which it resides. This topic is discussed in detail in Chap. 13.

the waste produced by the inability to use all of the irregular shape. The land remaining after the open space percentage is subtracted from the buildable area is then used by the forecast model equations to predict the development capacity options available.

The second and more accurate approach is to lay out the most intense potential site plan at the drawing board and measure the amount of open space remaining. This will establish the minimum open space percentage required.

This value and other values can then be entered in the design specification panel to produce any number of options for comparison in less time than it took to draw the first. Each design specification, or recipe, has different implications. The comparative process allows these implications to be assessed before further commitments are made.

Conclusion

The entire development forecasting collection is contained on the attached CD-ROM and is listed in Table 1.7. These workbooks are identified by the parking system involved. For instance, the G1 series workbook pertains to both residential and nonresidential building types that use tradi-

Table 1.7 Spreadsheet Workbook Series on the CD-ROM

	Workbooks		
Parking Category	**Series**	**Residential Variations**	**Description**
Surface parking	G1		Surface parking around but not under the building or buildings
	SF	GT	Surface parking and/or garages pertaining to townhouse forecasts
		SF	Private surface parking and/or garages pertaining to single-family residential forecasts
	G2		Surface parking around and under the building or buildings
Structure parking	S1		Adjacent parking structure located on the same premise
	S2		Underground parking structure on the same premise
	S3		Parking structure either partially or completely above grade under the building or buildings.
No parking	G1		Uses the G1 series of spreadsheets for development forecasts

NOTE: Several other series are introduced later in this book.

tional surface parking lots. To locate the forecast model needed, refer to the decision guide in either Fig. 1.1 or 1.2. Figure 1.1 pertains to the residential land use family and Fig. 1.2 pertains to the nonresidential family. If Fig. 1.1 is chosen and model RG1L is selected by following the decision guide to its location, the G1 designation in the center of this four-place identifier indicates that the G1 series workbook should be selected on the CD-ROM. After opening this workbook, the RG1L tab can be selected at the bottom of the screen to bring the required model into use.

Table 1.7 itemizes each workbook series, and each model also identifies this series within its four-place name. The GT and SF groups are both found in the SF series workbook since both pertain to grade-oriented residential housing using surface-parking systems.

This summary should make it possible to find and use a development forecast model within the collection, but the more detailed review of design specification components, forecast topics, single-family forecasting, and case study applications in the following chapters should help to improve the application of these tools.

Loading the CD-ROM on Your Hard Drive

If you wish to load the CD-ROM included with this book on your hard drive, please follow the instructions in the Read Me file listed as a menu selection on the CD. The text of this file is also included in Appendix B. There is *only one way* to load this program successfully, so please follow these instructions carefully.

Notes

1. *Development capacity* means the capacity of land to accommodate building area.

2. A CD-ROM is included that contains the complete collection of forecast models discussed in this book.
3. A *design specification* is a collection of design components and assigned values that control development results without dictating appearance. Each component can be assigned a broad range of values that have direct development capacity implications.
4. This method of analysis, generically referred to as *development capacity analysis,* can improve the selection process and lay the foundation for a common language of evaluation in the continuing debate over the planning, design, and regulation of current and future land use.
5. This notation indicates that a residential land use *R* is involved; that parking system *G1* (surface parking around but not under the building) is involved; and that the gross land area *L* is given.
6. Three single-family detached residential models represent exceptions.
7. Forecast models RG1L and CG1L.
8. Referred to as the identification panel and customer/project identification panel in Chap. 2, Fig. 2.1.
9. Exhibit 1.1 is an illustration of a residential forecast model, and the dwelling unit mix table can be seen in the middle of the page.
10. *Buildable area* means the total land area available for improvement on a given land area and excludes only unbuildable areas of the site such as public or private rights-of-way, ravines, lakes, etc. This term should not be confused with the more common zoning expression that often means the land area located within the building setback lines on a given lot. This topic is discussed in more detail in later chapters.
11. *Building cover area* (BCA) means the amount of land under roof on a given lot. Building cover calculation is often simplified by measuring the area within the building foundation perimeter. This is also referred to as the building *footprint*.
12. *Parking cover area* (PCA) means the amount of land covered by parking structures, or the amount of land covered by *parking lot area* (PLA), including internal parking lot landscaping and circulation aisles. Public/private *roadway area* cover (RDA) means the amount of land covered by public and private rights-of-way and paved easements that are part of a primary circulation system that may lead to individual lots. It does not include driveways and parking lot circulation aisles serving

parking spaces. *Driveway cover area* (DCA) means miscellaneous internal project roadway surfaces that are separate from parking lot circulation aisles but intended to provide access to them. *Miscellaneous pavement* cover (MSP) includes walks, patios, terraces, fountains and other impervious landscape amenities on a lot. *Loading area* cover (LDA) means the amount of land devoted to the delivery and pickup of materials, supplies and equipment associated with building services and operations. It is often referred to as a *loading dock area.*

13. *Core development area* means the land area exclusively available for building cover and parking cover.
14. Core area is discussed in more detail in ensuing chapters.
15. *Project open space area* (OSA), unless specifically referred to as *gross project open space* or *gross open space,* means "net project open space" or "net open space." *Net project open space* is the area among buildings within the buildable lot area that is unpaved and available for landscape improvement. Net project open space includes restricted setback areas, mounds, gardens, and other unpaved landscape features. *Gross project open space* includes net project open space and unbuildable areas such as, but not limited to, ecologically fragile areas, unstable soil areas, ravines, ponds, marshes, and existing improved areas within the gross land area available that are to remain.
16. *Development cover* is the sum of all impervious cover introduced, including, but not limited to, building cover, parking cover, driveway cover, and miscellaneous pavement within the buildable area of a given land area. Within the gross land area available, development cover also includes all public and private roadway cover. *Development balance* is the relative relationship between the total development cover introduced and the project open space that remains. *Development intensity* is a function of the building height introduced and the total development cover present in relation to the project open space that remains.
17. Equal to a 31% reduction
18. This example has held all other design specification values constant for this comparison.
19. This notation indicates that a nonresidential land use *C* is involved, that parking system *S2* (underground parking) is involved, and that the gross land area *L* is given.

20. Exhibit 1.5 also requires values for three below-grade design variables listed at the bottom of the design specification panel.

21. Locate 3 in the FLR column and look at the companion parking level column p to the right to find the number of parking levels required to support the building area forecast. Also note that precise mathematics produces a number like 4.03 parking levels. There should be enough flexibility during the schematic phase of architectural plan preparation to convert this to 4 parking levels.

22. This is the gross building area capacity of the site. More than one building could be constructed or the project could be built in phases over time as long as the total is not exceeded.

23. Ibid.

24. At a global scale, this relationship between open space (our natural world) and development intensity (our built environment) affects our sustainable future. At a smaller scale within our built environment, this relationship between open space and development intensity affects our ability to shelter an expanding population without producing excessive sprawl. The two issues are related and very complex. We need to reduce sprawl to preserve open space, but also need to limit building intensity to ensure that it does not degrade our ability to enjoy the open space we preserve.

25. See App. A for a suggested outline of the divisions of our built environment that are attempting to coexist with the divisions of our natural environment.

26. *Overdevelopment* is an unsatisfactory level of development intensity, or an unsatisfactory relationship between development cover, building height, and project open space. This relationship may involve a given building type within a given land area, a given neighborhood within a community, a given city within a region, etc. The level at which these relationships become unsatisfactory remains to be defined.

27. For instance, the intensity of a 25-ft-tall building with 70% development cover and 30% open space would be expressed with an intensity index INX = 25.70. This is a finely calibrated system that is differentiated by every foot of potential building height.

28. For instance, the intensity of a 2-story building with 70% development cover and 30% open space would be expressed with an intensity index

INX = 2.70. The intensity index of a 6-story building with 70% development cover and 30% open space would be equal to 6.70. Half-stories would be rounded down to the next even story. This is a less finely calibrated index system that does not require as much building research and may be all that is needed given the typical ranges of most floor-to-floor building heights.

29. Either postulate can also be useful in specifying the permitted level of development intensity on any given site, since all other characteristics of development are significantly influenced by these two pressure points.

30. The Fig. 1.2 decision guide for the nonresidential land use family does not differentiate among building types because they do not affect development capacity. This is different from the residential land use family, where several building types are differentiated because of unique characteristics that limit development capacity.

Design Specifications

Values assigned to the components of a development design specification make it possible to forecast the development capacity of land. These values are used by equations within each forecast model to predict an array of development opportunities that are influenced by these values, and that are more fully discussed in Chap. 3. The elements of a design specification have been referred to as *components*. These components represent a group of specification topics that can be used to forecast development capacity without drawing a single line, and without control of architectural details associated with building form, function, style,[1] and program.[2]

Location

The design specification panel is located directly below the explanation panel in every development forecast model. Six to eight panels of information are contained within a model, depending on the topic identified in the explanation panel, and a typical panel arrangement is diagrammed in Fig. 2.1. Exhibit 2.1 illustrates model CG1L, which is arranged according to this format.[3] The design specification for this

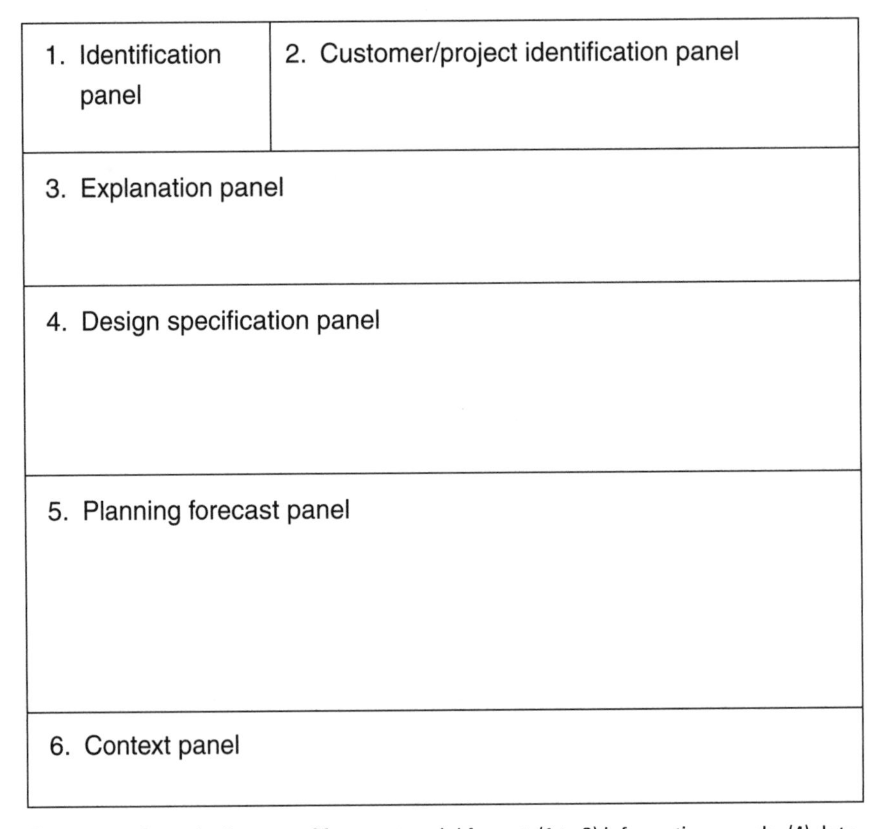

Figure 2.1 Generic diagram of forecast model format: (1 to 3) information panels, (4) data-entry panel, (5) forecast panel, and (6) context panel.

model is clearly labeled and can be found in the fourth panel position. This panel contains 10 values that must be entered by the user, and each is indicated by bold text within a box outline.[4] The floor number values FLR in the planning forecast panel are also part of the design specification and may be changed to evaluate additional height options.

Components and Categories

There are 35 development forecast models, and the design specifications in each do not necessarily contain the same number of specification components. These components are defined and described in the glossary at the end of this chapter.[5] Ten of these components are used in Exhibit 2.1,

Exhibit 2.1 Example of Panel Arrangement within a Forecast Model

Forecast Model CG1L

*Development capacity forecast for **NONRESIDENTIAL BUILDINGS** based on the use of an adjacent **GRADE PARKING LOT** located on the same premises. When s and a equal zero in the design specification below, the forecast pertains to conditions when **NO PARKING** is required.*

Given: *Gross land area.* ***To Find:*** *Maximum development capacity of the land area (gross building area potential) based on the design specification values entered below.* ***Premise:*** *All building floors considered equal in area.*

DESIGN SPECIFICATION

Enter values in boxed areas where text is bold and blue. Express all fractions as decimals.

Given:	**Gross land area**	**GLA=**	**5.882**	acres	256,220	SF
Land Variables:	Public/ private right-of-way & paved easements	W=	**0.150**	fraction of GLA	38,433	SF
	Net land area	NLA=	5.000	acres	217,787	SF
	Unbuildable and/or future expansion areas	U=	**0.000**	fraction of GLA	0	SF
	Gross land area reduction	X=	0.150	fraction of GLA	38,433	SF
	Buildable land area remaining	**BLA=**	5.000	acres	217,787	SF
Parking Variables:	Est. gross pkg. lot area per space in SF	s =	**375**	ENTER ZERO IF NO PARKING REQUIRED		
	Building SF permitted per parking space	a =	**250**	ENTER ZERO IF NO PARKING REQUIRED		
	No. of loading spaces	l=	**1**			
	Gross area per loading space	b =	**1,000**	SF	1,000	SF
Site Variables:	**Project open space as fraction of BLA**	**S=**	**0.300**	⇦	65,336	SF
	Private driveways as fraction of BLA	R=	**0.030**		6,534	SF
	Misc. pavement as fraction of BLA	M=	**0.020**		4,356	SF
	Loading area as fraction of BLA	L=	0.005		1,000	SF
	Total site support areas as a fraction of BLA	Su=	0.355		77,225	SF
Core:	**Core development area as fraction of BLA**	**C=**	0.645	C+Su must = 1	140,562	SF

PLANNING FORECAST

no. of floors **FLR**	CORE minimum land area for BCG & PLA	gross building area **GBA**	parking lot area PLA	pkg. spaces NPS	footprint **BCA**	bldg SF / acre SFAC function of BLA	flr area ratio FAR function of BLA
1.00	140,562	**56,225**	84,337	224.9	**56,225**	11,246	0.258
2.00		**70,281**	105,421	281.1	**35,140**	14,057	0.323
3.00		**76,670**	115,005	306.7	**25,557**	15,335	0.352
4.00		**80,321**	120,481	321.3	**20,080**	16,065	0.369
5.00		**82,683**	124,025	330.7	**16,537**	16,538	0.380
6.00	*NOTE: Be aware when BCA becomes too small to be feasible.*	**84,337**	126,505	337.3	**14,056**	16,868	0.387
7.00		**85,559**	128,339	342.2	**12,223**	17,113	0.393
8.00		**86,499**	129,749	346.0	**10,812**	17,301	0.397
9.00		**87,245**	130,868	349.0	**9,694**	17,450	0.401
10.00		**87,851**	131,776	351.4	**8,785**	17,571	0.403
11.00		**88,353**	132,529	353.4	**8,032**	17,672	0.406
12.00		**88,776**	133,164	355.1	**7,398**	17,756	0.408
13.00		**89,137**	133,705	356.5	**6,857**	17,828	0.409
14.00		**89,448**	134,172	357.8	**6,389**	17,891	0.411
15.00		**89,720**	134,580	358.9	**5,981**	17,945	0.412

WARNING: These are preliminary forecasts that must not be used to make final decisions.
1) These forecasts are not a substitute for the "due diligence" research that must be conducted to support the final definition of "unbuildable areas" above and the final decision to purchase land. This research includes, but is not limited to, verification of adequate subsurface soil, zoning, environmental clearance, access, title, utilities and water pressure, clearance from deed restriction, easement and right-of-way encumbrances, clearance from existing above and below ground facility conflicts, etc.
2) The most promising forecast(s) made on the basis of data entered in the design specification from "due diligence" research must be verified at the drawing board before funds are committed and land purchase decisions are made. Actual land shape ratios, dimensions and irregularities encountered may require adjustments to the general forecasts above.
3) The software licensee shall take responsibility for the design specification values entered and any advice given that is based on the forecast produced.

collected within sections titled Land Variables, Parking Variables, and Site Variables. A core value is also calculated in this section, even though it is a forecast, since it flows from the values entered in the specification. Whenever a model is opened, it is recommended that the user read the definitions that pertain to the specification components included, since the values are functions of these definitions.

Land Variables

This design specification panel in Exhibit 2.1 contains a *gross land area* (GLA) component that must receive a value, and a component that must receive an estimate for the percentage of gross land area that will be devoted to public rights-of-way, private rights-of-way, and/or paved easements W. The model calculates the *net land area* (NLA) available from this data with the following equation:

$$\text{NLA} = (1 - W)\text{GLA}$$

The user is also asked to enter an estimate for the amount of unbuildable area U present on the land area available as a percentage[6] of the gross land area present. When the land area is known, U is generally known as well; but if U is unknown, it must be found, measured, and entered as a fraction of the gross land available. The model then calculates the gross land reduction X in order to produce the buildable land area available. The *buildable land area* (BLA) remaining is simply the gross land area GLA minus all land area consumed by roadways[7] and unbuildable areas X.[8] This is the land available for improvement. The capacity of this buildable area[9] is a function of the remaining values entered in the design specification panel and the building floor values FLR entered as a range in the planning forecast panel.

Parking Variables[10]

Parking variables represent the second major category of design specification values in Exhibit 2.1. The user is asked to enter the parking ratio planned or required *a* and the gross area per parking space anticipated *s* since forecast model CG1L is based on a surface parking lot system.

The parking ratio planned or required is expressed in a number of different ways in the development community. It can be stated as the number of parking spaces required per 1000 sq ft of building area, the number of spaces required per dwelling unit, the building area permitted per parking space, or in a more specific way, such as the number required per hospital bed. The development capacity collection of forecast models uses two methods of specifying a parking ratio. Nonresidential parking requirements are expressed as the number of building square feet permitted per parking space *a*. For example, a restaurant might be limited to 50 sq ft of building area per parking space, but the same parking space could support 200 sq ft of office area because of different demand characteristics. Residential parking requirements are expressed as the number of spaces *u* required per dwelling unit. In the case of nonresidential buildings therefore, the *number of parking spaces* (NPS) planned or required is equal to the *gross building area* (GBA) divided by the building area permitted per parking space *a*. The *parking lot area* (PLA) needed to accommodate this number is equal to the number found NPS times the gross average parking lot area planned per parking space *s*.[11]

When loading areas are planned or required as part of the overall parking function, the user is asked to enter values for the number required *l* and the gross average area per space *b* anticipated.[12] This gross area per loading space should include all required circulation, maneuvering, parking, and miscellaneous support areas that are a part of the loading function.

Site Variables[13]

Site variables represent the third major category of design specification values in Exhibit 2.1. The user is asked to enter values that are percentages of the buildable land area available for the following components: project open space *S*, private driveways *R*, and miscellaneous pavement *M*. The model also calculates the buildable area percentage devoted to loading area *L* from the parking values *l* and *b* previously entered by the user. The values *S*, *R*, *M*, and *L* are then added to produce the site support value Su shown. The land area remaining for building cover and parking cover is called the core area *C*, which is also expressed as a fraction of the buildable area available. Figure 2.2 illustrates these generic features of a site plan.

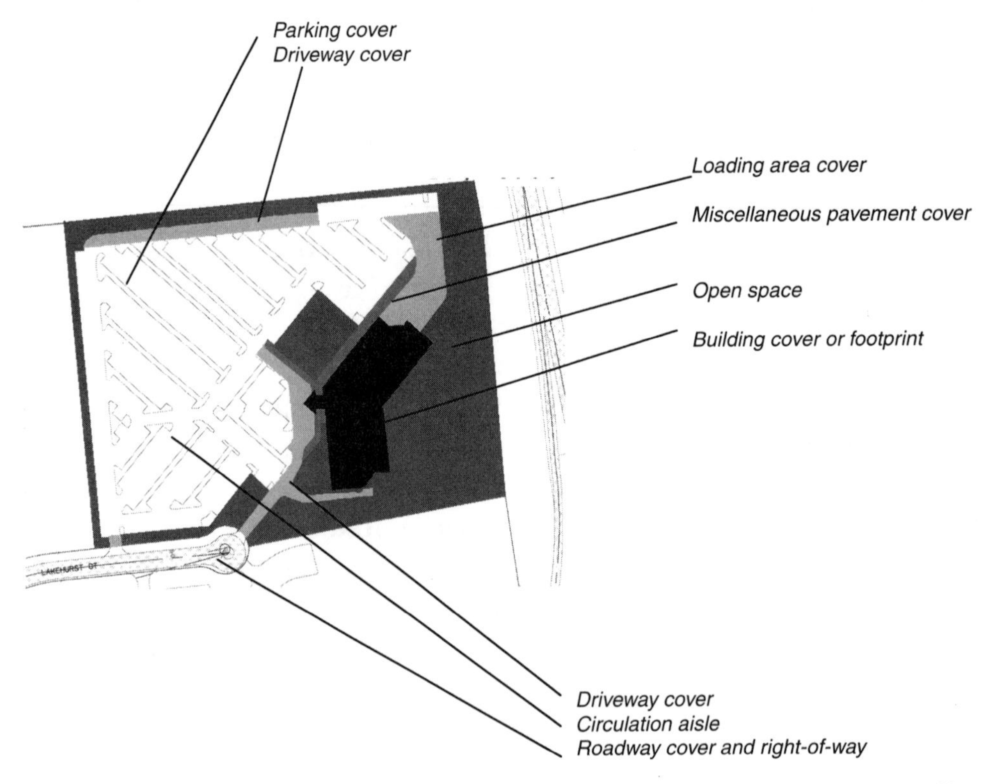

Figure 2.2 Elements of a site plan. The basic elements of a site plan include building cover, parking cover, loading cover, miscellaneous pavement cover, driveway cover, and roadway cover that combine to produce development cover. This is balanced by the project open space introduced.

Building cover combines with building height to produce building mass.

Building mass and development cover combine to produce development intensity that is offset by the project open space provided.

Core Area

The core area concept is at the heart of development capacity forecasting. The core area is the land area that remains after all site support elements are subtracted from the buildable area available.[14] The most influential site support element, in most cases, is project open space, since this allocation plays a major role in reducing the amount of land left for development cover. Building height and parking cover are the most influential core elements. Parking cover is influential because it consumes a great deal of land in relation to the building area it supports. Building height is influential because it reduces the land area needed for building cover and multiplies the building area that can be created. The mathematical expression of these relationships within the core area available is the foundation of development capacity analysis, and this expression varies with each forecast model included within the development capacity collection.

Design Specification When Building Area Is Known and Land Area Is to Be Forecast

When a gross building area objective is given instead of the gross land area available, the objective of the relevant forecast model is to predict the buildable land area needed to accommodate the objective based on values entered in its design specification. Exhibit 2.2 illustrates model CG1B, which is written to perform this function for nonresidential activities using a surface parking lot around but not under the building. A 50,000-sq-ft building is specified as the objective, and eight additional specification values are entered to produce the planning forecast in the lower half of the page. The BAC column in the forecast panel predicts that 4.45 net buildable acres will be needed with a 1-story building configuration, and 3.037 acres will be needed if the same building area is divided among 5 floors.

The lack of gross land and unbuildable land information in Exhibit 2.2 means that the focus of the forecast is on the buildable area needed and not the gross land area required. This focus is necessary since the unbuildable portion of an unknown land area cannot be estimated. Therefore, the BAC column in Exhibit 2.2 forecasts the net buildable acres required before the right-of-way *W* forecast is added. The GAC column forecasts the gross buildable acres required after the right-of-way forecast is added. This may or may not be enough land, however, depending on the amount of buildable land area found within the gross land area identified through a real estate search.

Design Specifications with Unique Features

Some forecast models contain specification panels that are relatively unique in the collection. The most prominent are found in two series: the residential apartment series and the single-family residential series.[15]

Dwelling Unit Mix Table

The dwelling unit mix table is found in all residential forecast models, except for single-family detached homes referred to as *suburb houses*. Exhibit 2.3 illustrates model RG1L,[16] which contains a dwelling unit mix table within its design specification panel. It asks the user to enter the names of the dwelling unit types DU being considered, the gross dwelling unit areas GDA planned, and the mix of dwelling unit types MIX under study. The user need not be concerned with the remaining values calculated by the model in this table, since they are used by the internal equations and are not useful in drawing-board analysis. The aggregate average dwelling unit area AGG calculated could be very useful to planners and zoning specialists as a regulatory index, however, and these readers are encouraged to review the definition and explanation of this value that is contained in the glossary at the end of this chapter.

Exhibit 2.2 Example of a Land Area Forecast Model

Forecast Model CG1B

*Development capacity forecast for **NONRESIDENTIAL BUILDINGS** based on the use of a **GRADE PARKING LOT** located on the same premises. When s and a equal zero in the design specification below, the forecast pertains to conditions when **NO PARKING** is required.*

***Given:** Gross bldg. area objective. **To Find:** Minimum buildable land area needed to achieve the gross building area objective given based on the design specification values entered below. **Premise:** All building floors considered equal in area.*

DESIGN SPECIFICATION

Enter values in boxed areas where text is bold and blue. Express all fractions as decimals.

Given:	**Gross building area objective**	**GBA=**	**50,000**	SF
Land Variables:	Public/ private right-of-way & paved easements	W=	**0.150**	fraction of GLA
	Future expansion and/or unbuildalbe areas	U=	**0.000**	
	Gross land area reduction	X=	0.150	
Parking Variables:	Estimated gross pkg. lot area per parking space in SF	s =	**375**	ENTER ZERO IF NO PARKING REQUIRED
	Building SF permitted per parking space	a =	**250**	ENTER ZERO IF NO PARKING REQUIRED
	No. of loading spaces	l=	**1**	
	Gross area per loading space	b=	**1,000**	SF
Site Variables: BLA= buildable land area	**Project open space as fraction of BLA**	**S=**	**0.300**	
	Private driveways as fraction of BLA	R=	**0.030**	
	Misc. pavement as fraction of BLA	M=	**0.020**	
	Total site support areas as a fraction of BLA	Su=	0.350	
Core:	**Core development area at grade as fraction of BLA:**	**C=**	0.650	C+Su must = 1

PLANNING FORECAST

NOTE: Be aware when BCG becomes too small to be feasible.

net pkg. area PLA	pkg. spaces NPS
75,000	200

NOTE:
Be aware when BCG becomes too small to be feasible.

no. of floors **FLR**	footprint BCA	minimum lot area CORE for BCG & PLA	buildable land BLA area forecast	buildable acre **BAC** forecast	BLA with expansion BLX forecast	BAC with expansion **BAX** forecast	bldg SF per SFAC BAX acre	flr area ratio FAR function of BLX
1.00	50,000	125,000	228,054	**5.235**	228,054	5.235	9,550	0.219
2.00	25,000	100,000	182,805	**4.197**	182,805	4.197	11,914	0.274
3.00	16,667	91,667	167,722	**3.850**	167,722	3.850	12,986	0.298
4.00	12,500	87,500	160,181	**3.677**	160,181	3.677	13,597	0.312
5.00	10,000	85,000	155,656	**3.573**	155,656	3.573	13,992	0.321
6.00	8,333	83,333	152,640	**3.504**	152,640	3.504	14,269	0.328
7.00	7,143	82,143	150,485	**3.455**	150,485	3.455	14,473	0.332
8.00	6,250	81,250	148,869	**3.418**	148,869	3.418	14,630	0.336
9.00	5,556	80,556	147,612	**3.389**	147,612	3.389	14,755	0.339
10.00	5,000	80,000	146,606	**3.366**	146,606	3.366	14,856	0.341
11.00	4,545	79,545	145,784	**3.347**	145,784	3.347	14,940	0.343
12.00	4,167	79,167	145,098	**3.331**	145,098	3.331	15,011	0.345
13.00	3,846	78,846	144,518	**3.318**	144,518	3.318	15,071	0.346
14.00	3,571	78,571	144,021	**3.306**	144,021	3.306	15,123	0.347
15.00	3,333	78,333	143,590	**3.296**	143,590	3.296	15,168	0.348

WARNING: These are preliminary forecasts that must not be used to make final decisions.

1) These forecasts are not a substitute for the "due diligence" research that must be conducted to support the final definition of "unbuildable areas" above and the final decision to purchase land. This research includes, but is not limited to, verification of adequate subsurface soil, zoning, environmental clearance, access, title, utilities and water pressure, clearance from deed restriction, easement and right-of-way encumbrances, clearance from existing above and below ground facility conflicts, etc.

2) The most promising forecast(s) made on the basis of data entered in the design specification from "due diligence" research must be verified at the drawing board before funds are committed and land purchase decisions are made. Actual land shape ratios, dimensions and irregularities encountered may require adjustments to the general forecasts above.

3) The software licensee shall take responsibility for the design specification values entered and any advice given that is based on the forecast produced.

Exhibit 2.3 Illustration of a Dwelling Unit Mix Table

Forecast Model RG1L

Development capacity forecast for ***APARTMENTS*** *based on the use of an adjacent* ***GRADE PARKING LOT*** *located on the same premises. When (s) and (u) equal zero in the design specification below, the forecast pertains to conditions when* ***NO PARKING*** *is required.*

Given: *Gross land area.* ***To Find:*** *Maximum dwelling unit capacity of the land area given based on the design specification values entered below.* ***Premise:*** *All building floors considered equal in area.*

DESIGN SPECIFICATION *Enter values in boxed areas where text is bold and blue. Express all fractions as decimals.*

Given:	**Gross land area**	GLA=	10.204	acres	444,486	SF
Land Variables:	Public/ private right-of-way & paved easements	W=	0.150	fraction of GLA	6,534	SF
	Net land area	NLA=	8.673	acres	377,813	SF
	Unbuildable and/or future expansion areas	U=	0.100	fraction of GLA	4,356	SF
	Gross land area reduction	X=	0.250	fraction of GLA	10,890	SF
	Buildable land area remaining	BLA=	7.653	acres	333,365	SF
Parking Variables:	Est. gross pkg. lot area per pkg. space in SF	s =	400	ENTER ZERO IF NO PARKING REQUIRED		
	Parking lot spaces planned or required per dwelling unit	u=	1.5	ENTER ZERO IF NO PARKING REQUIRED		
	Garage parking spaces planned or required per dwelling unit	Gn=	0.5	ENTER ZERO IF NO PARKING REQUIRED		
	Gross building area per garage space	Ga=	240	ENTER ZERO IF NO PARKING REQUIRED		
	No. of loading spaces	l =	1			
	Gross area per loading space	b =	1,000	SF	1,000	SF
Site Variables:	**Project open space as fraction of BLA**	S=	0.300		100,009	SF
	Private driveways as fraction of BLA	R=	0.050		16,668	SF
	Misc. pavement as fraction of BLA	M=	0.100		33,336	SF
	Loading area as fraction of BLA	L=	0.002		1,000	SF
	Total site support areas as a fraction of BLA	Su=	0.452		150,764	SF
Core:	**Core development area as fraction of BLA**	C=	0.548	C=Su must = 1	182,601	SF
Building Variables:	Building efficiency as percentage of GBA	Be=	0.600			
	Building support as fraction of GBA	Bu=	0.400	Be + Bu must = 1		

Dwelling Unit Mix Table:

DU dwelling unit type	GDA gross du area	CDA=GDA/Be comprehensive du area	MIX du mix	PDA = (CDA)MIX Prorated du area
EFF	500	833	10%	83
1 BR	750	1,250	20%	250
2 BR	1,200	2,000	40%	800
3 BR	1,500	2,500	20%	500
4 BR	1,800	3,000	10%	300
			Aggregate Avg. Dwelling Unit Area (AGG) =	1,933
			GBA sf per parking space a=	1,289

PLANNING FORECAST

no. of floors FLR	CORE minimum lot area for BCG & PLA	density per net acre dNA	dwelling units NDU	pkg. lot spaces NPS	parking lot area PLA	garage spaces GPS	garage area GAR	gross bldg area GBA no garages	footprint BCA	density per dBA bldable acre
1.00	182,601	7.93	68.8	103.2	41,292	34.4	8,258	133,051	133,051	8.99
2.00		12.48	108.3	162.4	64,957	54.1	12,991	209,305	104,653	14.15
3.00		15.43	133.8	200.7	80,297	66.9	16,059	258,734	86,245	17.49
4.00		17.50	151.7	227.6	91,047	75.9	18,209	293,375	73,344	19.83
5.00		19.02	165.0	247.5	99,000	82.5	19,800	319,001	63,800	21.56
6.00		20.20	175.2	262.8	105,122	87.6	21,024	338,726	56,454	22.89
7.00	*NOTE: Be aware when BCA becomes too small to be feasible.*	21.13	183.3	274.9	109,979	91.6	21,996	354,378	50,625	23.95
8.00		21.89	189.9	284.8	113,928	94.9	22,786	367,100	45,887	24.81
9.00		22.52	195.3	293.0	117,200	97.7	23,440	377,645	41,961	25.52
10.00		23.05	199.9	299.9	119,957	100.0	23,991	386,527	38,653	26.12
11.00		23.50	203.9	305.8	122,310	101.9	24,462	394,111	35,828	26.64
12.00		23.89	207.2	310.9	124,343	103.6	24,869	400,662	33,389	27.08
13.00		24.23	210.2	315.3	126,117	105.1	25,223	406,378	31,260	27.47
14.00		24.53	212.8	319.2	127,679	106.4	25,536	411,409	29,386	27.81

WARNING: These are preliminary forecasts that must not be used to make final decisions.
1) These forecasts are not a substitute for the "due diligence" research that must be conducted to support the final definition of "unbuildable areas" above and the final decision to purchase land. This research includes, but is not limited to, verification of adequate subsurface soil, zoning, environmental clearance, access, title, utilities and water pressure, clearance from deed restriction, easement and right-of-way encumbrances, clearance from existing above and below ground facility conflicts, etc.
2) The most promising forecast(s) made on the basis of data entered in the design specification from "due diligence" research must be verified at the drawing board before funds are committed and land purchase decisions are made. Actual land shape ratios, dimensions and irregularities encountered may require adjustments to the general forecasts above.
3) The software licensee shall take responsibility for the design specification values entered and any advice given that is based on the forecast produced.

Dwelling Unit Specification Panel

This panel is found in the RGT series of forecast models that address single-family *attached* homes referred to as grade-oriented *urban houses.* There are only two models in the series, so these models are included with the residential single-family (RSF) series. Exhibit 2.4 illustrates model RGTL, which pertains to attached single-family dwelling units when the gross land area is known.[17] This panel is an expanded version of the dwelling unit mix table and reflects the fact that land is directly associated with each dwelling unit in this building type category. The expanded panel asks for more detailed information regarding the types, sizes, and heights of the dwelling units intended; the mix under study; and the private yard area that will be associated with each unit. The sum of development cover and yard area per dwelling unit type produces a total land allocation footprint DYG per dwelling unit type and an aggregate average land area allocation AFP per dwelling unit that reflects the mix proposed. The AFP index is essential when forecasting dwelling unit capacity,[18] and could also be useful as a regulatory index, but is not particularly helpful in drawing-board analysis.

Common Open Space

Common open space is a specification component that is found in the grade-oriented series of single-family development capacity models. Exhibit 2.4 illustrates urban house model RGTL and contains an example of common open space specifications in the left-hand column of the dwelling unit forecast panel. *Common open space* in these models is distinct from *private open space* that is also provided for each dwelling unit. Both open space subsets function to balance development cover and reduce development intensity. Private open space, referred to by some as *defensible space,* is called *yard area* in these specifications. Project open space is called *common open space* since it creates a parklike

Exhibit 2.4 Illustration of a Dwelling Unit Specification Panel

Forecast Model RGTL

Development capacity forecast for ***TOWNHOUSES*** *based on the use of shared PARKING LOTS and / or PRIVATE GARAGES serving each dwelling unit.*
Given: *Gross land area.* ***To Find:*** *Maximum dwelling unit capacity of the land area given based on the design specification values entered below.*

DESIGN SPECIFICATION

Enter values in boxed areas where text is bold and blue. Express all fractions as decimals.

Given:	**Gross land area**	GLA=	**20.000**	acres	871,200	SF
Land Variables:	Public right-of-way as fraction of GLA	W=	**0.000**		0	SF
	Private right-of-way as fraction of GLA	Wp=	**0.200**		174,240	SF
	Net land area	NLA=	16.000	acres	696,960	SF
	Unbuildable and/or future expansion areas as fraction of GLA	U=	**0.000**		0	SF
	Total gross land area reduction as fraction of GLA	X=	0.200		8,712	SF
	Buildable land area remaining	BLA=	**16.000**	acres	696,960	SF
Parking Variables:	Pkg. lot spaces required or planned per DU	u=	**1.5**	enter zero if no shared parking lot		
	Est. gross pkg. area per pkg. lot space in SF	s =	**400**	enter zero if no shared parking lot		
	Roadway pkg. spaces per DU	Rn=	**0**	enter zero if no roadway parking		
	Est. net pkg. area per roadway space	Rs=	**0**	enter zero if no roadway parking		
	Est. gross area per garage parking space	Ga=	**250**	enter zero if no garage parking		
Support Variables:	Driveways as fraction of BLA	R=	**0.020**			
	Misc. pavement as fraction of BLA	M=	**0.010**		6,970	SF
	Outdoor social / entertainment area in SF	Sp=	**10,000**		0.014	fraction of BLA
	Social / service building(s) footprint in SF	Sf=	**5,000**		0.007	fraction of BLA
	Social / service building(s) gross area in SF	Sg=	**8,000**			
	Number of social / service parking spaces	Sn=	**20**	400 gross sf / space	0.011	fraction of BLA
	Total support areas as a fraction of BLA	Su=	0.063		43,909	SF
	Core + open space + roadway parking as fraction of BLA	CSR=	0.937	CS+Su must = 1	653,051	SF

NOTE:
Express all fractions as decimals

DWELLING UNIT SPECIFICATION

dwelling unit type DU	no. of flrs. FLR above grade	dwelling unit MIX mix	total HAB habitable area	habitable area FTP footprint	garage spaces GSD per dwelling	building and garage BCG cover per dwell unit	private yard area YRD with townhouse	bldg. garage and DYG yard per dwelling	du basement BSM area	du crawl CRW space area
EFF	**1**	**0.000**	**350**	**350**	**0.5**	475	**500**	975	**0**	**0**
1 BR	**1**	**0.100**	**500**	**500**	**0.5**	625	**500**	1,125	**0**	**0**
2 BR	**2**	**0.600**	**800**	**400**	**0.5**	525	**500**	1,025	**200**	**200**
3 BR	**2**	**0.300**	**1,000**	**500**	**0.5**	625	**500**	1,125	**200**	**300**
4 BR	**3**	**0.000**	**1,200**	**400**	**0.5**	525	**500**	1,025	**200**	**200**
Aggregate Avg. Areas / DU in SF given MIX		1.000	830	440	0.50	565	500		180	210

AFP = 1,065 aggregate average private area per du

DWELLING UNIT CAPACITY FORECAST

(based on common open space (COS) provided)

common COS open space	number of NDU dwelling units	EFF	1 BR	2 BR	3 BR	4 BR	density dNA per net acre	density per dBA buildable acre	parking lot NPS spaces	roadway RDS parking spaces
		dwelling unit type (DU) quantities given mix above								
0.000	392.2	0.0	39.2	235.3	117.7	0.0	24.51	24.51	588.3	0.0
0.100	350.4	0.0	35.0	210.2	105.1	0.0	21.90	21.90	525.5	0.0
0.200	308.5	0.0	30.9	185.1	92.6	0.0	19.28	19.28	462.8	0.0
0.300	266.6	0.0	26.7	160.0	80.0	0.0	16.67	16.67	400.0	0.0
0.400	224.8	0.0	22.5	134.9	67.4	0.0	14.05	14.05	337.2	0.0
0.500	182.9	0.0	18.3	109.8	54.9	0.0	11.43	11.43	274.4	0.0
0.600	141.1	0.0	14.1	84.6	42.3	0.0	8.82	8.82	211.6	0.0
0.700	99.2	0.0	9.9	59.5	29.8	0.0	6.20	6.20	148.8	0.0
0.800	57.3	0.0	5.7	34.4	17.2	0.0	3.58	3.58	86.0	0.0

DEFINITION of COMMON OPEN SPACE (COS): Open space provided for common benefit, massing separation and project setbacks in addition to that provided for private use adjacent to townhouses. On a given land area, private lot sizes or areas decrease as common open space increases when all other development factors are constant.
NOTE: Common open space values in bold & blue in the table above may be changed to examine other implications. When COS = 0.0 all open space is provided on private lots (OSA) in the form of yards and a typical subdivision is implied.

WARNING: These are preliminary forecasts that must not be used to make final decisions.
1) These forecasts are not a substitute for the "due diligence" research that must be conducted to support the final definition of "unbuildable areas" above and the final decision to purchase land. This research includes, but is not limited to, verification of adequate subsurface soil, zoning, environmental clearance, access, title, utilities and water pressure, clearance from deed restriction, easement and right-of-way encumbrances, clearance from existing above and below ground facility conflicts, etc.
2) The most promising forecast(s) made on the basis of data entered in the design specification from "due diligence" research must be verified at the drawing board before funds are committed and land purchase decisions are made. Actual land shape ratios, dimensions and irregularities encountered may require adjustments to the general forecasts above.
3) The software licensee shall take responsibility for the design specification values entered and any advice given that is based on the forecast produced.

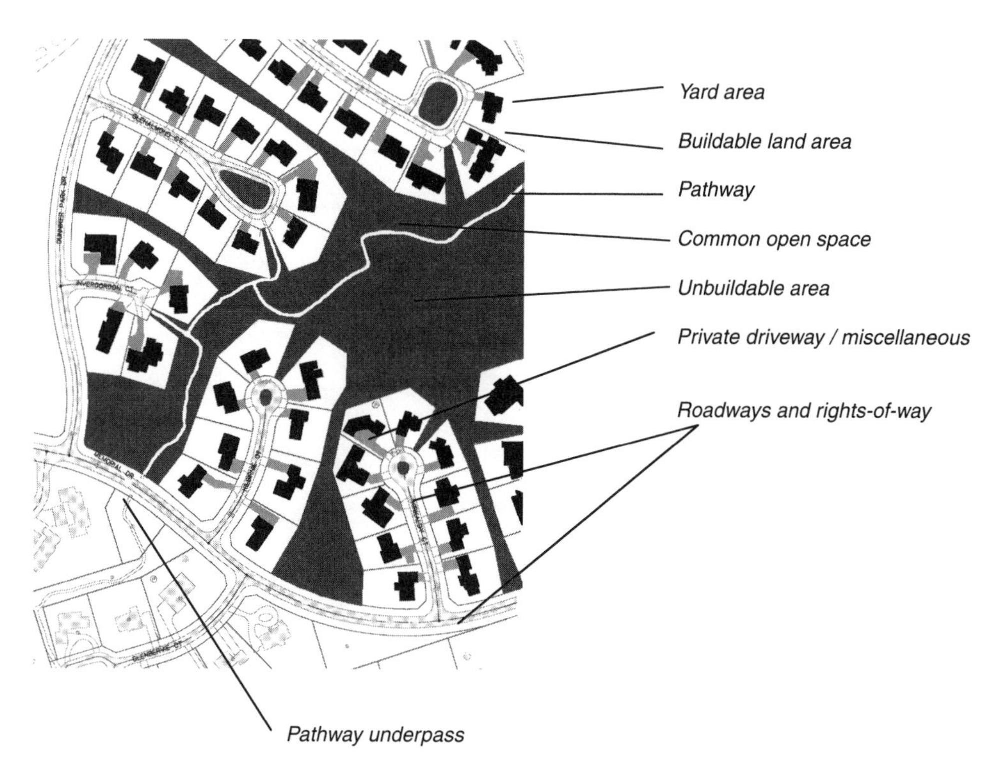

Figure 2.3 Buildable areas and common open space. Common open space is a variation of project open space that is shared within a development. It is typically associated with residential projects, but can be used as a planning technique within any land use area. Private land ownership is generally reduced to compensate for the shared land owned in common within a given land area.

setting that is shared by an entire development. These two project open space categories are separately identified to permit greater design flexibility within the urban house and suburb house design specification formats, and to allow the user to tailor these values to more clearly define the design approach intended. Figure 2.3 illustrates a typical suburb house example of common open space and yard areas.

Land Donation

Land donation is a unique design specification category that is included as a series of data-entry cells within the suburb house group of grade-oriented single-family *detached* residential forecast models. It is illustrated in Exhibit 2.5. The

Exhibit 2.5 Example of Land Donation

Forecast Model RSFL

Development capacity forecast for ***SINGLE- FAMILY DETACHED HOUSING*** *with no shared or common parking lots serving each dwelling unit.*
Given: *Gross land area.* ***To Find:*** *(1) Maximum number of lots that can be created from the land area given based on the design specification values entered below. (2) Maximum gross dwelling unit area that can be built on each lot created.*

DESIGN SPECIFICATION

Enter values in boxed areas where text is bold and blue. Express all fractions as decimals.

Given:	**Gross land area**	GLA=	**20.000**	acres		
Land Variables:	Est. right-of-way dedication as fraction of GLA	W=	**0.200**		8,712	SF
	Private roadway as fraction of GLA	R=	**0.000**			
	School land donation as fraction of GLA	Ds=	**0.000**			
	Park land donation as fraction of GLA	Dp=	**0.100**			
	Other land donation as fraction of GLA	Do=	**0.000**			
	Net land area	NLA=	14.000	acres	609,840	SF
	Unbuildable and/or future expansion areas as fraction of GLA	U=	**0.000**		0	SF
	Total gross land area reduction as fraction of GLA	X=	0.300		13,068	SF
	Buildable land area remaining	**BLA=**	**14.000**	acres	609,840	SF
Core:	**Core + open space available**	**CS=**	**14.000**	acres	609,840	SF

LOT SPECIFICATION

	total bldg footprint BCG with garage	garage GAR footprint	misc. development area Msf as multiple of BCG	total development DCA cover area	private yards YRD as multiple of DCA	total lot LOT area	lot LTX mix	aggregate avg. AAL lot areas in mix	total lot TLA areas in mix
Type 1 Interior Lot	**1,350**	**500**	**0.50**	2,025	**2.00**	6,075	**0.000**		
Type 2 Interior Lot	**1,500**	**500**	**0.60**	2,400	**2.00**	7,200	**0.000**		
Type 3 Interior Lot	**1,700**	**500**	**0.75**	2,975	**2.00**	8,925	**1.000**	8,925	8,925
Type 4 Interior Lot	**2,000**	**500**	**0.85**	3,700	**2.25**	12,025	**0.000**		
Type 5 Interior Lot	**2,450**	**750**	**0.75**	4,288	**2.50**	15,006	**0.000**		
Type 6 Interior Lot	**3,300**	**750**	**0.65**	5,445	**3.00**	21,780	**0.000**		
Type 7 Interior Lot	**5,600**	**750**	**0.55**	8,680	**4.00**	43,400	**0.000**		

NOTE: Zero lot line homes and townhouses are implied when YRD is less than 1.0 and approaches zero (0.0) in the YRD column above.

Mix Total (do not exceed 1.0)	1.00	AAL	
Aggregate Avg. Lot Size Based on Mix:		**8,925**	

LOT CAPACITY FORECAST

(based on common open space value (COS) entered below)

DEFINITION of COMMON OPEN SPACE (COS): Open space provided for common benefit and massing separation in addition to that provided for private use and setbacks on individual lots. On a given land area, private lot sizes decrease as common open space increases when all other development factors are constant.
NOTE: Common open space values in bold & blue in the table below may be changed to examine other implications. When COS = 0.0 all open space is provided on private lots (OSA) in the form of yards and a typical subdivision is implied.

COS:	**0.00**	**0.10**	**0.20**	**0.30**	**0.40**	**0.50**	**0.60**	**0.70**	**0.80**
No. of lots:	68.3	61.5	54.7	47.8	41.0	34.2	27.3	20.5	13.7
Net density (dNA):	4.88	4.39	3.90	3.42	2.93	2.44	1.95	1.46	0.98
BLA density (dBA):	4.88	4.39	3.90	3.42	2.93	2.44	1.95	1.46	0.98

GROSS BUILDING AREA per LOT FORECAST

	basement as fraction of BSM habitable footprint	basement BSA area	crawl space CRW area	NET POTENTIAL BUILDING AREA INCLUDING GARAGE but EXCLUDING BASEMENT and POTENTIAL EXPANSION number of building floors: 1	1.5	2	2.5	3	3.5
Type 1 Interior Lot	**1.000**								
Type 2 Interior Lot	**1.000**								
Type 3 Interior Lot	**1.000**	1,200	0	1,700	2,550	3,400	4,250	5,100	5,950
Type 4 Interior Lot	**1.000**								
Type 5 Interior Lot	**1.000**								
Type 6 Interior Lot	**1.000**								
Type 7 Interior Lot	**1.000**								

WARNING: These are preliminary forecasts that must not be used to make final decisions.
1) These forecasts are not a substitute for the "due diligence" research that must be conducted to support the final definition of "unbuildable areas" above and the final decision to purchase land. This research includes, but is not limited to, verification of adequate subsurface soil, zoning, environmental clearance, access, title, utilities and water pressure, clearance from deed restriction, easement and right-of-way encumbrances, clearance from existing above and below ground facility conflicts, etc.
2) The most promising forecast(s) made on the basis of data entered in the design specification from "due diligence" research must be verified at the drawing board before funds are committed and land purchase decisions are made. Actual land shape ratios, dimensions and irregularities encountered may require adjustments to the general forecasts above.
3) The software licensee shall take responsibility for the design specification values entered and any advice given that is based on the forecast produced.

land donation element in this specification includes components for school, park, and other requests. These data-entry cells are found only in suburb house forecast models (RSF series). If land donations or sales are required as a part of other development project proposals, the amount of land required should be included in the *unbuildable* specification category.

Summary

When a design specification is complete, it represents many facts, estimates, assumptions, and objectives regarding the final product desired and the project constraints present. This project profile is expressed in the specification through values assigned to its components. These values directly affect the range of gross building areas (GBAs) that can be produced on a given land area or lot. This GBA range in turn implies different levels of intensity, appearance, compatibility, construction cost, building population, traffic generation, investor return, and tax yield. By assigning different values to these specification components, hundreds of design alternatives and the environments they produce can be evaluated in the time it would take to draw one.

The following is a recap of the design specification values entered in the design specification of Exhibit 2.1. Values forecast are not noted. The values entered may be changed at will to forecast and compare alternate strategies and implications.

GLA	Gross lot area	5.882 acres
W	Public/private right-of-way and paved easements	15% of GLA
NLA	Net land area (calculated)	
U	Unbuildable areas such as ravines	0.0% of GLA
X	Gross land area reduction (calculated)	

BLA	Buildable land area remaining (calculated)	
s	Gross average parking lot area per parking space	375 sq ft
a	Building area permitted per parking space provided	250 sq ft
l	Number of loading spaces planned or required	1
b	Gross average area required per loading space	1000 sq ft
S	Project open space area	30% of BLA
R	Private driveways	3% of BLA
M	Miscellaneous pavement area	2% of BLA
L	Loading area (calculated)	
Su	Total site support area (calculated)	
C	Core development area (calculated)	
FLR	Number of building floors	Range from 1 to 15

Future Research

There is a broad range of values that can be assigned to each design component within a design specification, and each value implies a different result, not all of which are desirable. Each of the design specification components listed in the glossary at the end of this chapter is a candidate for additional research since the results produced from given values and sets of values are only vaguely and intuitively understood at the moment. Each of these components is an essential ingredient in a recipe, however, and each can play a major role in planning and forecasting our ability to shelter an expanding population in a way that is economically profitable, socially beneficial, and environmentally sustainable.

Notes

1. This does not mean that appearance controls are not needed. It simply means that they should not be confused with design specifications that forecast and define development capacity, produce balance, and limit intensity.

2. An *architectural program* is not a *design specification,* and the terms should not be used interchangeably. The elements of an architectural program are the rooms and relationships to be included within a building floor plan, and this room list varies with each client. The elements of a design specification are predetermined, and the values assigned establish the development capacity of land and the building intensity that results. An architectural program is primarily focused on areas within a building, and is an outline of internal space needs and potential costs that influence floor plan design. A design specification is primarily focused on land area allocations and building height relationships that create external environmental relationships among building mass, paved surfaces, and project open space. Building mass and paved surfaces produce development intensity. The two-dimensional component of intensity has been called *development cover,* which is counterbalanced by project open space. The relationship among development cover, building height, and project open space produces intensity and affects the amount of gross building area that can be placed on a given site or land area. In other words, the relationships among the components of development intensity affect the development capacity, environmental impact, and economic yield of the site.

3. The first panel is used for identification in the upper left-hand corner. It identifies the "call letters" or index of the model involved. The second is a customer/project identification panel that can be customized at will. The third is an explanation panel that provides a brief description of model applicability and objectives. The fifth is a forecast panel that is discussed in the next chapter, and the sixth explains the nature of the forecasts produced.

4. The bold text is also blue when viewed on the monitor.

5. This table includes definitions, descriptions, and explanations for each component.

6. When the term *percentage* is used, it means the decimal equivalent of the percentage.
7. Roadways are not parking lot circulation aisles or driveways. They include all public and private rights-of-way and paved easements within the project area that are part of the primary circulation system.
8. If you wish to include area for future expansion, include this area as part of the unbuildable area identified.
9. See glossary at end of this chapter for definition.
10. See glossary at end of this chapter for definitions and explanations of terms used in this section.
11. The value s should include all anticipated interior parking lot landscaping.
12. All model forecast equations are based on the premise that loading will not be in a parking structure. If it is, the average area per parking space s will need to be increased to accommodate this function, and the values l and b would be zero.
13. See glossary at end of chapter for definitions and explanations of terms used in this section.
14. This core area does not represent a location within a given gross land area, such as the area within zoning setback lines. It is simply the total land area available for building cover and parking cover. The total core area available can be comprised of several smaller areas as long as the sum of its parts do not exceed the total forecast. These areas do not need to be contiguous.
15. The apartment series includes the RG1, RG2, RS1, RS2, and RS3 groups of forecast models. The single-family series is represented by the RSF and RGT groups.
16. The explanation panel at the top of model RG1L explains the purpose of this particular forecast model. It is simply used here as an illustration.
17. See the explanation panel for a more complete definition.
18. There is no universally "typical" dwelling unit area. This area varies with each project proposal. The AFP value calculated represents the typical dwelling unit area within a specific project proposal.

Glossary

An expanded glossary of definitions, descriptions, and explanations for the indexes and terms used in the preceding specification follows. This glossary also includes many more specification terms, definitions, descriptions, and explanations that apply to the entire collection of development capacity models, but have not been used in this chapter. In order to hasten your familiarity with the concepts and use of the collection, it may be more helpful to read portions of this glossary as you use a particular model rather than attempt to read it all as a separate exercise. This glossary is designed to permit either approach, but in any event, it should not be overlooked since it provides much information that is not repeated elsewhere. A much shorter comprehensive glossary limited to terms and definitions is presented at the end of this book.

a Building area per parking space—the gross building area in square feet divided by the number of parking spaces provided.

These values can be found as requirements in most local zoning ordinances. They are expressed in relation to the land use activity anticipated and are also often expressed as the number of cars permitted per 1000 sq ft of gross building area. For instance, 4 parking spaces required per 1000 gross sq ft of building area would equal 250 sq ft of building area permitted per individual parking space (1000/4 = 250). Parking requirements generally range from 50 to 400 sq ft of building area permitted per parking space. This would equal 2.5 to 20 parking spaces required per 1000 sq ft of building area.

B Building plane percentage—the fraction of the core development area that is covered by a building footprint located over a gross parking structure footprint.

A parking structure above grade can occupy any percentage of the core area. If a building is located above the parking structure, it can also occupy any percentage of the core area below in terms of the building plane occupied. The value *B* permits the user to specify the percentage of the core area that will be occupied by the building floor plan. The value *P* permits the user to specify the percentage of the core area that will be occupied by the parking structure below the building floor plan. This permits greater planning and massing flexibility.

The design premise for the S3 parking system is based on a parking structure under a building but not underground. The building floor plan can be larger or smaller than the parking floor plan below, but neither can exceed the core area available. These relationships are expressed by the percentage of the core area that is devoted to the floor plan associated with each.

b Gross pavement area per loading space—The gross truck maneuvering, loading, and parking surface area in square feet per loading space provided (see Fig. 2.4).

Figure 2.4 Loading areas. Loading areas consume land area and are treated as a distinct category to avoid distorting the parking lot area calculated per parking space.

(Photo by W. M. Hosack)

This area can vary widely since many truck sizes and associated maneuvering areas can be involved. The gross area per loading space includes the space itself and a proportional share of all maneuvering areas, parking areas, and access drives required to serve the loading area. These statistics are not readily available, so the typical approach is to consult reference sources such as *Architectural Graphic Standards* and to measure similar areas in other existing facilities to determine the average needed.

BCA Building cover area, including garage—the area covered by the building floor plan, including attached or detached garages. Depending on the spreadsheet involved, this value is either calculated from other specification values entered or specified as a square-foot area.

This forecast defines the total footprint BCA to be included on a given lot type. The sum of the values entered for building cover BCA, miscellaneous pavement MSP, and yard area YRD produce a calculation of the total lot area LOT needed to satisfy the objectives defined. The total building cover proposed per dwelling unit is a major factor determining the development capacity of land.

Be Building efficiency—the fraction of the gross building area that is available for occupancy. This value is either specified or forecast as a fraction of the gross building area, depending on the spreadsheet involved. Most buildings fall within a 60 to 90% efficiency range.

Buildings contain mechanical areas, electrical closets, utility chases, corridors, wall thickness, plumbing facilities, elevators, etc. that reduce the net internal area available for occupancy. The difference between the gross building area forecast or desired and the net area available for occupancy is called *building efficiency*.

BLA Buildable land area—the total land area available for improvement in square feet. It does not include public and private rights-of-way, paved easements, unbuildable areas, and future expansion areas. This value is either specified or forecast in acres or square feet, depending on the spreadsheet involved. It is equal to the gross land area minus public and private rights-of-way, paved easements, and all unbuildable areas.

This term should not be confused with the more liberal zoning expression that often means the land area located within building setback lines on a given lot *plus* specified development encroachments permitted in required yard areas. When used in a zoning ordinance, this term does not necessarily mean the only buildable area on the lot. When used in this book, however, *buildable area* means the only land area available for development.

BSM Basement area—the gross basement area expressed as a fraction of the habitable footprint above. Depending on the spreadsheet involved, this value is either specified as a square-foot area that is less than or equal to the habitable area above or calculated from other specification values entered.

This definition is not intended to be legally precise and is often the subject of planning and zoning debate. Some interpretation will be required, but the intent is clear.

COS Common open space—net project open space that is shared by a number of homes within a private setting. Common open space is added to unbuildable areas such as ravines and lakes to produce the gross common open space available (see Fig. 2.3). This value is either specified as a fraction of the buildable land area available or forecast in square feet from other design specification values entered, depending on the spreadsheet in use. The range of values that can be specified produces a broad range of different housing environments.

A major portion of the character of a housing development is established by the way private yard areas and common open space are allocated in relation to the development cover provided. Large yard areas, private lots, and no common open space typify a standard residential subdivision. Small yard areas, small private lots, and large amounts of common open space within the same gross land area produce a larger parklike atmosphere with the housing clustered more compactly within. Common open space is net project open space. It is provided in addition to any unbuildable areas that may be present. When common open space is specified as a percentage of the buildable land area available, keep in mind that this is in addition to the unbuildable areas that may be present and that combine to produce the gross common open space available to the project.

CRW Crawl space—building area at least partially below grade that has headroom of less than 4 ft. Depending on the spreadsheet involved, this value is either specified as a square-foot area that is less than or equal to the habitable area above or calculated from other specification values entered.

These areas are intended as horizontal mechanical, electrical, and plumbing chases. They are often used for storage, but are not intended for human habitation.

d Dwelling units per net acre—the total number of dwelling units planned, permitted, or required per net acre of available land area. This value is either forecast from other design specification values entered or specified, depending on the forecast model in use. Density in the United States generally ranges from 0.2 to 200 dwelling units per net acre, depending on the building type and parking system chosen. Many definitions of low, medium, and high net density could be debated, particularly in relation to the urban environment being considered.

This forecast is a classic planning calculation, but all dwelling units are not equal in area. Therefore, when density is measured in dwelling units per net acre, the amount of building cover and development cover produced by the same density can vary significantly. This means that density regulations and calibrations are very imperfect measurements of the development balance and intensity that is planned, permitted, or produced by a specified density for a given land area.

Do Land donation—the amount of gross land area subtracted from the total for one or more public or semipublic purposes. This donation is specified as a fraction of the gross land area available. The legal issues associated with such donations are not within the scope of this book.

Land donations, or financial equivalents, have historically been required from developers by many local governments since the population introduced needs public services that imply additional public expense. These "donations" are called *exactions* by some. Those referring to these donations as exactions argue that payment should be made for the land area subtracted for this purpose.

Dp Land donation for parks—the amount of gross land area subtracted from the total for one or more future park sites. This donation is specified as a fraction of the gross land area available. The legal issues associated with such donations are not within the scope of this book.

Land donations for parks, or financial equivalents, have historically been required from developers by many local governments since the population introduced needs public services that imply additional public expense. These "donations" are called *exactions* by some. Those referring to these donations as exactions argue that payment should be made for the land area subtracted for this public purpose.

Ds Land donation for schools—the amount of gross land area subtracted from the total for one or more future school sites. This donation is specified as a fraction of the gross land area available. The legal issues associated with such donations are not within the scope of this book.

Land donations for schools, or financial equivalents, have historically been required from developers by many local governments since the population introduced needs public services that imply additional public expense. These "donations" are called *exactions* by some. Those referring to these donations as exactions argue that payment should be made for the land area subtracted for this public purpose.

FLR Building floors—the number of building floors planned, provided or required. Floor numbers are either specified in the planning panel of a forecast model as a beginning list or forecast as the result of a given gross building area objective. When floor values are specified, a beginning list from 1 to 15 floors has been entered in each affected forecast model, but these numbers can be changed at will.

The number of building floors contemplated is a big factor in forecasting the development capacity of a given land area. Increased building height produces increased building area, but not in the straight-line relationship one would assume when surface parking is involved. Since the building footprint must shrink as height increases to provide increased land area for the expanding surface-parking requirement, the increase in gross building area rapidly declines when all other design specification values are held constant.

FTP Footprint of habitable area—the gross area within the building foundation perimeter, minus the garage. In measuring these areas, measure to the centerline of common walls and express in square feet. This is the total land area covered by the habitable portion of a building floor plan.

G Underground parking structure area as fraction of BLA—a percentage of the buildable area that will be used for a parking structure floor plan below grade. This value is specified as a fraction of the buildable land area and can range anywhere up to 1.0 or 100%.

An underground parking structure does not need to occupy the entire underground area defined by the buildable area above. Any percentage of this buildable area can be chosen for evaluation.

Ga Residential garage area per parking space—the total garage area per parking space. This area is specified in square feet. Garage spaces demand more area than surface parking areas. These areas generally range from 200 to 300 sq. ft. per parking space, but this depends on the amount of storage area included.

This is a companion to the number of residential garage parking spaces provided and is needed to calculate the total land allocation devoted to this parking alternative.

GAR Garage footprint—the garage foundation plan area measured to the outside of all perimeter walls. This area is specified in square feet. In specifying these areas, measure to the centerline of common walls.

GBA Gross building area—the total building area as measured along the exterior perimeter, including exterior wall thicknesses. This value is expressed in square feet and is either specified or forecast from other specification values entered.

Gross building area is forecast based on the assumption that all building floors are equal in area. This is rarely the case in final architectural floor plans, but the purpose of this forecast model collection is not to predict the form and content of final floor plans and building appearance. The purpose is to forecast the relationships among building cover, parking cover, roadway cover, miscellaneous pavement, building height, and open space that can optimize development capacity based on the design specification values entered. Final floor plans, architectural sections, and building elevations can then be refined within the parameters established by the initial design specification values.

GDA Gross dwelling unit area—the gross building area provided per dwelling unit type. Specify gross dwelling unit area in square feet per dwelling unit type. This can also be expressed as the dwelling area measured to the outside of all perimeter walls, and to the centerline of shared walls.

Dwelling unit types are generally specified by the number of bedrooms provided and may also be differentiated by a special feature—for instance, "3-bedroom unit with den." These unit types and areas must be clearly differentiated in the dwelling unit mix table to produce an accurate development capacity forecast.

GLA Gross land area—the gross land area in acres as defined by recorded property lines. When the gross land area is known, it is specified in acres but converted to square feet by the spreadsheet.

The length, width, and shape of a given land area can affect the development capacity potential of that property. Excessively narrow and excessively irregular property shapes can make it difficult to efficiently arrange the rectilinear shapes typical of most building and parking plans. These land shapes cannot be anticipated by the forecast model collection, and the user must carefully consider the design values he or she specifies under these circumstances.

Gn Residential garage spaces per dwelling unit—the number of garage spaces to be provided per dwelling unit. The total number of parking spaces planned, provided, or required per dwelling unit generally ranges from 0.5 to 2.0 parking spaces per dwelling unit. The number of garage spaces provided is also specified in terms of the number provided per dwelling unit.

Garage spaces increase choices but also increase the development cover needed to provide the same number of surface parking spaces. Any portion of a parking requirement may be planned for this enhancement.

HAB Total habitable area—the gross dwelling unit area to be occupied. This does not include basement, crawl space, and garage. Specify the gross dwelling unit area in square feet minus basement, crawl space, and garage.

These areas can vary widely and are generally a function of the number of bedrooms and specialty rooms provided.

l Number of loading spaces—the number of loading spaces planned or required. Specify in whole numbers.

Most zoning ordinances specify loading space requirements in relation to the use and area of a building. These requirements should be checked in addition to owner preferences since they can demand large land area allocations.

LTX Lot type mix—the percentage of each lot area planned or provided. All percentages must be specified in their decimal equivalents.

Lot types are first specified in the lot mix table. The percentage of each type to be provided is then specified in the LTX column. Consult local zoning ordinances to ensure that the lot types specified conform with the requirements of the zoning district involved.

M Miscellaneous pavement—a fraction of the buildable land area that will be used for miscellaneous pavement (see Fig. 2.5). Specify as a fraction of the buildable land area. Express all fractions and percentages in their decimal equivalents. Most development projects have small amounts of miscellaneous pavement in the form of sidewalks, steps, terraces, and entrances that range from 1 to 15% of the buildable land area.

Miscellaneous pavement often represents a secondary land area allocation but it can become significant—for example, when formal plaza areas and large swimming pool decks are involved. It should not be included in the amount of project open space provided since it produces much greater stormwater runoff. The swimming pool

Figure 2.5 Miscellaneous pavement cover. Miscellaneous pavement is often a minor element in a site plan, but it represents impervious cover, stormwater runoff, and a pedestrian interface that can produce both pleasant and unpleasant transitions to building shelter.
(Photo by W. M. Hosack)

itself can be included as part of the open space provided if desired. This is often debated, since the pool can be considered to be similar to a pond that would be included in an open space provision or as a constructed active recreation area that does not offset development cover intensity as effectively as traditional project open space.

MIX Dwelling unit allocation—the percentage of each dwelling unit type planned or provided within the total. All percentages must be specified in their decimal equivalents.

Dwelling unit types are first identified in the dwelling unit mix table. The percentage of each type to be provided is then specified in the MIX column. Dwelling unit sizes, mixes, and building efficiencies vary widely and directly influence the density that can be achieved. Because of this, one development might find it very easy to meet a permitted density while another could find it exceedingly difficult.

Msf Miscellaneous building and pavement areas—the area covered by miscellaneous structures plus the area covered by sidewalks, terraces, swimming pools, swimming pool decks, and other miscellaneous single-family residential structures and pavement. Residential miscellaneous pavement MSP is specified as a percentage of the total residential building cover area BCA intended. It may exceed 1.0 or 100%.

This value can also be used to plan for and include future development expansion areas on a given lot or land area. These expansion area allowances are only prudent, since additions should be anticipated over time.

NLT Total lot number—the number of lots planned, permitted, or provided within a given land area. This value is either forecast from other design specification values entered or specified as a design objective, depending on the spreadhseet in use.

The number of lots that can be created within a larger land area does not always indicate the highest return on investment that can be achieved. The value of each lot should be carefully evaluated in relation to the number created. This is particularly true when shared common open space is involved.

P Parking structure cover—a fraction of the core development area that is covered by the gross parking structure floor plan (see Fig. 2.6). The core area is available for building cover and parking cover. A parking structure above grade can occupy any percentage of the core area. If a building is located above the parking structure, it can also occupy any percentage of the core area below in terms of the building plane occupied. The value *P* permits the user to specify the percentage of the core area that will be occupied by the parking structure floor plan. The value *B* permits the user to specify the percentage of the core area that will be occupied by the building floor plan above the parking structure. This permits greater planning and massing flexibility.

The design premise for this spreadsheet is based on a parking structure under a building but not underground. The building floor plan can be larger or smaller than the parking floor plan below, but neither can exceed the core area available. These relationships are expressed by the percentage of the core area that is devoted to the floor plan associated with each.

Figure 2.6 Parking-structure cover. Parking-structure cover multiplies the amount of parking that can be introduced to a given area and multiplies the gross building area that can be supported as a result. Development intensity is magnified by the increase in building mass, building population, and building cover that results. Surface parking lot comparisons are visible across the street.

(Photo by W. M. Hosack)

Pu Parking structure support areas—a fraction of the gross parking structure area that is not devoted to parking spaces or circulation aisles. This value is specified as a fraction of the gross parking structure area. This fraction would include areas devoted to stairways, elevators, walkways, columns, wall thicknesses, etc. Parking structure support can easily range from 10 to 30% of the gross structure area. These percentages should be verified with specialized parking consultants in relation to the internal circulation system being considered before commitments are made. Parking structure support can also be expressed as the gross parking structure area minus the parking structure area devoted to parking spaces and circulation aisles, which is then divided by the gross parking structure area planned or provided.

A more accurate forecast of the potential number of parking structure spaces possible is produced when the area devoted to parking structure support is subtracted from the gross area to produce the net parking structure area available for parking spaces and circulation aisles.

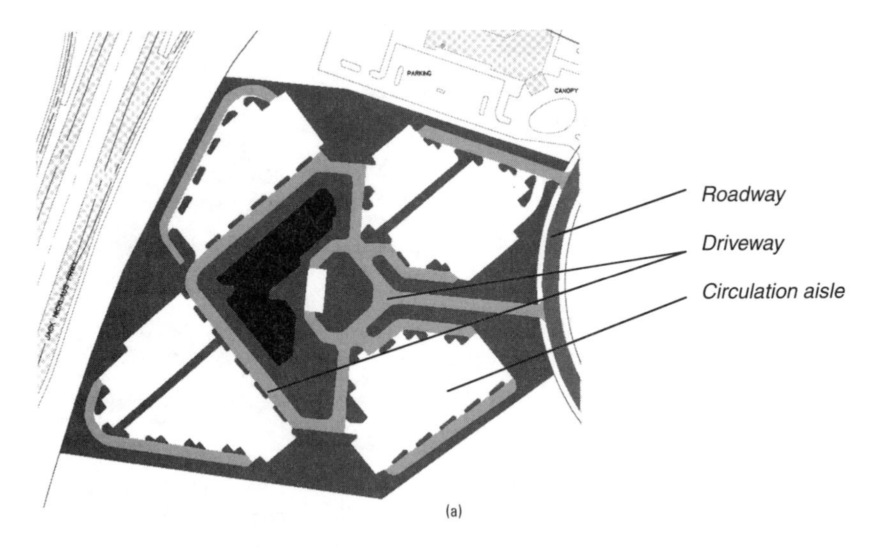

(a)

(b)

Figure 2.7 Roadways, driveways, and circulation aisles. Driveways separate traffic movement from the traffic maneuvering that takes place in parking lot circulation aisles. They are not always present to any significant degree, but are treated as a distinct category to avoid distorting the parking lot area calculated per parking space. (*a*) Site plan; (*b*) street view. (Photo by W. M. Hosack)

R Private driveways—a fraction of the buildable land area that will be used for driveway areas (see Fig. 2.7). Specify as a fraction of the buildable land area BLA. Express all percentages in their decimal equivalents.

Driveway cover includes all miscellaneous private roadway surfaces that are separate from parking lot circulation aisles serving marked parking spaces. They are not part of the primary movement system defined by rights-of-way and/or private easements. This category is intended to account for miscellaneous roadway that does not directly serve parking spaces and that connects parking lots and garages to primary circulation systems.

Rn Roadway parking spaces per dwelling unit—the number of roadway parking spaces planned or provided per dwelling unit. These spaces are served by a public or private right-of-way or easement, as opposed to parking lot spaces served by a parking lot circulation aisle (see Fig. 2.8).

Parking spaces are not always collected within parking lots. If a roadway parking design approach is anticipated, the number of parking spaces to be provided in this configuration can be specified per dwelling. The total parking spaces provided per dwelling unit generally range from 0.5 to 2.0, depending on the occupants anticipated. Roadway spaces can be introduced to provide some or all of this requirement.

Roadway parking is generally not desired since a parked car must back into oncoming traffic, and this traffic may not be expecting to encounter conflicting movement. Traffic along a parking lot circulation aisle is more alert to this potential conflict. Some lower-density multifamily residential arrangements wish to avoid the mass parking appearance of common parking lots, however, and introduce this parking arrangement along low-volume internal private circulation roadways.

Rs Net area per roadway parking space—the net parking area pavement excluding roadway pavement. The right-of-way is calculated elsewhere, so the user is only asked to specify the net parking area per space. A conservative estimate per space would range from 190 to 220 sq ft per space. This is a companion to the number of roadway parking spaces provided and is needed to calculate the

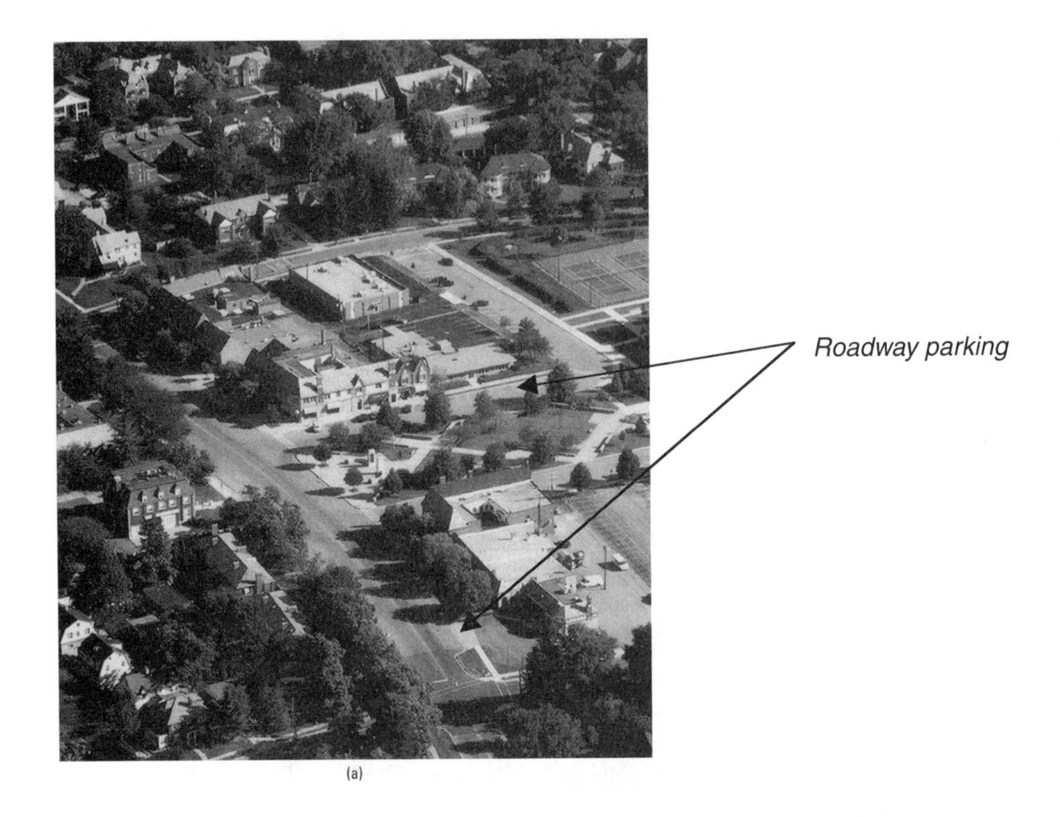

(a)

(b)

Figure 2.8 Roadway parking. Roadway parking has also been called *hitching-post parking*. It can be used in residential applications to reduce the impact of large parking lots, but introduces conflicting movement patterns and surprises as a result. (*a*) Overview; (*b*) street view.

(Photos by W. M. Hosack)

total land allocation devoted to this parking subset of the surface parking system.

Roadway parking spaces can often distribute and mitigate the impact of a surface parking lot. This design can improve the appearance of the environment created, but potential movement conflicts demand that it be given careful attention.

S Project open space—unpaved land within the buildable land area. It includes setback areas, mounds, gardens, and other unpaved landscape features, and is expressed as a percentage of the buildable land area. It can also be referred to as *common open space* in single-family and townhouse residential developments.

Specify project open space as a percentage of the buildable land area BLA available. Express all fractions and percentages in their decimal equivalents. Most suburban office development projects provide 20 to 40% net project open space around their buildings and parking lots.

Open space is a key factor when considering the balance, compatibility, character, and intensity of a project that will be produced. Miscellaneous pavement provided within project open space is separately specified.

Note: Required zoning setbacks must be included in this percentage, and the percentage must be large enough to accommodate these requirements. Zoning setback requirements often permit building and pavement intrusion, thus confusing the distinction between development cover and project open space relationships. This makes it difficult to focus on the actual amount of open space intended to offset, or counterbalance, the total amount of development cover and stormwater runoff permitted.

s Gross parking lot area per space—The gross parking lot area, including circulation aisles and internal landscaped areas, divided by the number of parking spaces present.

Specify in square feet. Many parking lot areas provide from 350 to 450 sq ft of total area per parking space, excluding loading space areas. This quantity is a function of the amount of internal parking lot landscaping, lot shape, and topographical features involved, in addition to the parking space areas and circulation aisle widths provided.

The parking lot area provided per space can imply severe asphalt surfaces or extensive landscape relief, depending on the gross area planned per parking space, since these values include the area devoted to internal parking lot landscape treatment.

Sb Social center gross building area—the total building area devoted to gathering and shared entertainment in a multifamily housing development. Specify the total area devoted to this function. It can vary widely and should be carefully considered during the land allocation planning process.

Buildings with entertainment centers, workout rooms, and kitchens are common attractions within multifamily development social center areas.

Sf Social center footprint—the building footprint area devoted to gathering and shared entertainment in a multifamily housing development. Specify this area in square feet. This footprint may be only part of the total building area involved.

Social centers, when included in a housing development, reduce the core development area available for building cover and parking cover. This reduces the potential development capacity of the land to support dwelling units and must be adequately anticipated to prevent unrealistic forecasts.

Sn Number of parking spaces for social center—the number of parking spaces allocated to serve the functions of a multifamily housing social center. The local zoning ordinance may specify the number of spaces required. It can also be used as a planning reference if these requirements are not specifically mentioned.

Additional spaces are required at a social center since guests may be involved and residents may be inclined to drive.

Sp Social center gross outdoor area—the common or shared outdoor area devoted to social activities and entertainment in a multifamily housing development. Specify the total land area devoted to these functions in square feet. These areas can vary widely and should be carefully considered during the land allocation process.

This area can include an extensive array of amenities depending on the project involved. Swimming pools and terraces are two of the more common provisions.

u Parking Spaces per dwelling unit—the number of parking spaces planned, provided, or required per dwelling unit. Zoning requirements generally range from 0.5 to 2.0 spaces per dwelling unit. Parking for senior housing is sometimes provided at 0.25 spaces per dwelling unit or joint occupancy room. Forecast model parking requirements must either be expressed as the number of spaces required per dwelling unit or the building square feet permitted per parking space in the case of nonresidential construction. Parking requirements expressed in such terms as the number of spaces per bed or the number of spaces per classroom may be more accurate expressions but are too numerous and diverse to be accommodated in a generic spreadsheet format. All such requirements must be converted to one of the two generic expressions mentioned in order to use a forecast model. Since parking requirements are not an exact science and are often debated, this is a concession that will have little effect on precision.

The parking space requirement is a major factor influencing development capacity since each space, whether provided on a surface parking lot or in a parking structure, consumes a great deal of land area in relation to the building square feet it can support. This is particularly true in the nonresidential land use group. Parking space requirements and project open space provisions are two major design components, often referred to as *design elements,* that reduce the amount of land available for building cover and thus influence the amount of gross building area that can be used.

U Unbuildable areas—a fraction of the gross land area that cannot be used for development cover or project open space improvements. Specify as a percentage of the gross land area available.

Examples of unbuildable areas include, but are not limited to, ecologically fragile areas, unstable soil, wetlands, ravines, ponds, marshes, lakes, severe topography, etc.

W Public and private right-of-way—the fraction of the gross land area allocated for vehicular circulation through public and private right-of-way dedication and/or paved easements. When right-of-way and/or paved easements are provided, they often consume 10 to 25% of the gross land area. Right-of-way is specified as a fraction of the gross land area available.

Right-of-way estimates should be immediately subtracted from the gross land available to begin defining the buildable land area that can actually be used for improvements.

Wp Private residential right-of-way—the fraction of the gross land area allocated for private residential roadway easements and circulation rights-of-way that are not dedicated to public use. These values can easily range from 10 to 25% of the gross land area available. The desire to build private rights-of-way generally implies a desire to keep this percentage at the low end of this range.

Roadways may be permitted within exclusive easements that do not require as much public right-of-way dedication. This approach is often based on a desire for exclusivity, a desire to increase the buildable area available, or both.

YRD Private yard area—lawn, courtyard, or other outdoor space that is owned or privately allocated to an individual dwelling unit. It is referred to by some as *defensible space*. Depending on the forecast model involved, this value is either specified as a square-foot area or as a fraction of the development cover planned, provided, or permitted. When forecast as a fraction, it may exceed 1.0 or 100%. As a frame of reference, the total yard areas in many single-family subdivisions often exceed 100% of the total development cover present. The courtyards in many Georgetown-type rowhouse developments are often less than 100% of the total development cover present.

Forecasting Development Capacity

The development forecast collection is capable of making a virtually infinite number of predictions regarding the capacity of land to accommodate development. These forecasts are not self-limiting and many are unacceptable from a number of physical, social, economic, and environmental perspectives. This book does not make judgments, however. It produces the ability to quickly and efficiently forecast options so that many may assess the implications. The forecast collection predicts the capacity of land to accommodate shelter. It also represents a language capable of building a foundation for future environmental evaluation and effective development regulation that can protect a scarce resource and contribute to a sustainable future.

Design Premises

The generic design premises of most building construction are illustrated in Figs. 1.4 to 1.9. Each premise is based on a parking system and not on a building style or land use activity, since many land use activities and building envelopes use the same parking system.[1] Table 3.1 identifies the design premise of each forecast series, which is also

Table 3.1 Forecast Models Organized by Design Premise

G1 Premise and Series

Premise:
A surface parking lot and building " footprint " adjacent to each other on the same premise. The total footprint area may not exceed the core development area available. no portion of the building is located above the parking lot. All building floors are considered equal in area. This series is also used when no parking is planned or required. The surface parking values requested by the design specification are entered as zero under these circumstances. (Photo by W. M. Hosack)

	Model	Given	Forecast
Land Use Family:		**NONRESIDENTIAL**	
Building Type:		Nonresidential	
	CG1B	Gross Building Area Objective	Minimum Buildable Land Area Needed
	CG1L	Gross Land Area	Gross Building Area Potential
Land Use Family:		**RESIDENTIAL**	
Building Type:		Apartment	
	RG1L	Gross Land Area	Maximum Dwelling Unit Capacity
	RG1B	Gross Building Area Objective	Minimum Buildable Land Area Needed
	RG1D	Net Density Objective and Gross Land Area	Minimum Number of Building Floors Needed

G2 Premise and Series

Premise:
Air rights above a surface parking lot are used by the building footprint to increase the number of parking spaces at grade. The parking lot area and the building plane area above grade do not exceed the core development area available. All building floors are considered equal in area. (Photo by W. M. Hosack)

	Model	Given	Forecast
Land Use Family:		**NONRESIDENTIAL**	
Building Type:		Nonresidential	
	CG2B	Gross Building Area Objective	Minimum Buildable Land Area Needed
	CG2L	Gross Land Area	Gross Building Area Potential
Land Use Family:		**RESIDENTIAL**	
Building Type:		Apartment	
	RG2L	Gross Land Area	Maximum Dwelling Unit Capacity
	RG2B	Gross Building Area Objective	Minimum Buildable Land Area and Height Needed
	RG2D	Net Density Objective and Gross Land Area	Minimum Number of Building Floors Needed

S1 Premise and Series

Premise:
A building footprint is adjacent to a parking garage footprint on the same premise. The total footprint area may not exceed the core development area available. All building floors are considered equal in area and all parking floors are considered equal in area. (Photo by W. M. Hosack)

	Model	Given	Forecast
Land Use Family:		**NONRESIDENTIAL**	
Building Type:		Nonresidential	
	CS1B	Gross Building Area Objective	Minimum Buildable Land Area Needed
	CS1L	Gross Land Area	Gross Building Area Potential
Land Use Family:		**RESIDENTIAL**	
Building Type:		Apartment	
	RS1L	Gross Land Area	Maximum Dwelling Unit Capacity
	RS1B	Gross Building Area Objective	Minimum Buildable Land Area Needed
	RS1D	Net Density Objective and Gross Land Area	Minimum Number of Parking Levels Needed

S2 Premise and Series

Premise:
An underground parking garage that may be less than or equal to the buildable lot area above and may be larger, smaller, or equal in area to the building footprint above. The building footprint area may not exceed the core development area available. All building floors are considered equal in area and all parking floors are considered equal in area. (Photo by W. M. Hosack)

	Model	Given	Forecast
Land Use Family:		**NONRESIDENTIAL**	
Building Type:		Nonresidential	
	CS2L	Gross Land Area	Gross Building Area Potential
	CS2B	Gross Building Area Objective	Minimum Buildable Land Area Needed
Land Use Family:		**RESIDENTIAL**	
Building Type:		Apartment	
	RS2L	Gross Land Area	Maximum Dwelling Unit Capacity
	RS2B	Gross Building Area Objective	Minimum Buildable Land Area Needed
	RS2D	Net Density Objective and Gross Land Area	Minimum Number of Building Floors and Parking Levels Needed

Table 3.1 Forecast Models Organized by Design Premise *(continued)*

S3 Premise and Series

Premise:
A parking garage footprint beneath a building footprint. These footprints may equal or exceed each other in area, but neither may exceed the total core development area available. The parking garage is partially or completely above grade. All building floors are considered equal in area and all parking floors are considered equal in area. (Photo by W. M. Hosack)

	Model	Given	Forecast
Land Use Family:		**NON-RESIDENTIAL**	
Building Type:		Nonresidential	
	CS3L	Gross Land Area	Gross Building Area Potential
	CS3B	Gross Building Area Objective	Minimum Buildable Land Area Needed
Land Use Family:		**RESIDENTIAL**	
Building Type:		Apartment	
	RS3L	Gross Land Area	Maximum Dwelling Unit Capacity
	RS3B	Gross Building Area Objective	Minimum Buildable Land Area Needed
	RS3D	Net Density Objective and Gross Land Area	Minimum Number of Building Floors and Parking Levels Needed

SF Group Premise Within SF Series -- Suburb houses

Premise:
Single-family detached housing. Shared parking rarely exists except for the public right-of-way, individual lots are provided for each dwelling unit, yards are prominent, all dwelling units are separated, foundations are unique to each unit, and there is a great emphasis on private garages and carports. All units are grade-oriented with main entries that do not require travel through intervening interior spaces under separate or common ownership. All floors are considered equal in area. (Photo by W. M. Hosack)

	Model	Given	Forecast
Land Use Family:		**RESIDENTIAL**	
Building Type:		Single- Family **Detached**	
	RSFD	Net Density Objective and Gross Land Area	Maximum Number of Lots Possible
	RSFL	Gross Land Area	Max. No.. of Lots and Max. DU Area Potential
	RSFN	Number of Lots Desired	Min. Buildable Land Area and Max. DU Areas

GT Group Premise Within SF Series -- Urban houses

Premise:
Single-family attached housing. Shared parking can be exclusive, combined with, or excluded in favor of private garage opportunities. Dwelling units are attached and foundations are connected. All units are grade-oriented with main entries that do not require travel through intervening interior spaces under separate or common ownership. All floors are considered equal in area. (Photo by W. M. Hosack)

	Model	Given	Forecast
Land Use Family:		**RESIDENTIAL**	
Building Type:		Single-Family **Attached**	
	RGTD	Net Density Objective and Gross Land Area	Design Specification Relationships Needed
	RGTL	Gross Land Area	Maximum Dwelling Unit Capacity

summarized in the explanation panel of each forecast model. Table 3.1 also describes the models that belong to each series, the information that must be known in order to use a particular model in the series, and a brief indication of the forecast content that will be produced by the model.

Some may find it easier to locate a model using Table 3.1 instead of Figs. 1.1 and 1.2 when they know the parking system or systems they plan to evaluate, since the universe of forecast model possibilities is quickly reduced when the design premise is known. For instance, assume that a developer wishes to predict the maximum office area that can be built on a given land area using a surface parking lot. The developer can turn to Table 3.1 and locate the forecast series whose design premise is a surface parking lot around but not under the building. This profile fits the G1 series. Since an office building is in a nonresidential land use group and is a nonresidential building type, only two forecast models can apply in the G1 series, models CG1B and CG1L. Since the gross land area is given, the choice narrows to model CG1L. At this point, the forecast model can be called to the screen and values can be entered in the design specification to evaluate the results produced.

Development Capacity Comparisons

If you wish to compare the gross building area results found in model CG1L with the results that could be constructed using an underground parking garage, you could turn to the S2 forecast series and locate model CS2L as the tool needed for a comparative analysis. The same process of entering alternate values in the design specification could be followed to evaluate the gross building area potential of this parking garage scenario. At an intuitive level, most would agree that more building area can be constructed when using an underground parking structure than when using a surface parking lot. The models, however, can quickly forecast how much more and under what conditions.

Forecast Categories

There are 49 forecast topics listed, defined, described, and explained in Fig. 3.1.[2] They appear throughout the forecast model collection, but not all appear in every model. These forecast topics fall into the following general categories:

1. Land area forecasts (when the gross building area objective is given)
2. Building cover area and height forecasts (when the gross land area is given)
3. Parking area and number forecasts (based on the design premise selected)
4. Structure parking and parking level forecasts (based on the design premise selected)
5. Open space area forecasts (when not specified as a design objective)
6. Dwelling unit density and number forecasts (when the land area is given)

Figure 3.1 Sample land purchase review.

	Information provided	Information missing	Notes
Location	X		
Land area	X		
Intended use		X	Information missing, but assume county garage.
Description of need and purpose		X	
Owner identification	X		
Jurisdiction identification	X		
Jurisdiction contact		X	No contact name provided and no contact apparently made.
Zoning district	X		H-C.
Contiguous zoning	X		H-C and L-C.
Zoning adequacy attachment		X	Not a clearly listed permitted use, but may be considered similar. Need statement from zoning inspector.
Pending actions		X	
Zoning map	X		Land is surrounded by highway commercial district and local commercial district zoning.
Zoning regulations	X		Land use section provided, but there may be related sections. No discussions with zoning official noted.
Professional survey		X	
Professional Geotechnical analysis		X	Natural hazards map indicates this area has poor natural drainage. Need more detailed explanation of potential problems.
Utilities available	X		All appear available, but survey not complete.
Utility information	X		
Stormwater detention	X		Required.
Clear title		X	Deed review of potential encumbrances and restrictions not provided.
Accessibility		X	Map provided indicates no problem with accessibility, but it is not a survey.
Clear title certification		X	Not verified.
Environment		X	UNOCAL required to do cleanup. Monitoring wells on site. Will future construction be delayed? Refer issue to environmental. How does this affect purchase price?
Asking price	X		$40,413.11 per acre for 44.54 acres out of the ROW. 22.635 acres in the interstate ROW not included in per acre calculation. Note environmental remediation under way and potential construction delay.
Appraised value		X	
Formal appraisal attachment		X	
Summary of request		X	
Option purchase data		X	It appears that options are not being considered, but the "no" box is not checked. An option to purchase might be useful while the information above is being assembled if there is other interest in the property.

Notes: Central water and sewer required by zoning. It appears to be available from form submitted.
Cannot tell from map if minimum area and yard requirements are met.
Note buffer strip requirements for site. Ensure that adequate land area is being considered.
Site plan review by zoning inspector required.
Land purchase area is directly south of 100-year floodplain on a natural hazard map in an area designated as "poor natural drainage." Further investigation by a professional geotechnical engineer appears warranted.
Conclusion: Concluding notes are also provided in the table above. Further research and attachments are required. The site may have a natural hazard problem. Environmental cleanup issues require further examination by the environmental group. The asking price should be evaluated in relation to these issues by a professional appraiser.

7. Residential lot area and number forecasts (when either the land area or the land area and net density objective are given)

The seven forecast topics above are noted where they apply in forecast models that intend to predict either the building area capacity of a given land area or the land needed for a given gross building area objective.

Decision Guides

The forecast models can be arranged in several different formats, depending on the search criteria. Figures 1.1 and 1.2 and Table 3.1 present alternate decision guide arrangements. Table 3.2 presents the fourth basic guide, and is based on the fact that all forecast models are written to produce either building capacity forecasts or land area fore-

Table 3.2 Forecast Models Organized by Category

	Building area forecast category			Land area forecast category	
		Residential land use family			
Design premise	Nonresidential land use family	Land area given	Land area and density given	Nonresidential land use family	Residential land use family
G1: Grade parking around but not under the building	CG1L	RG1L	RG1D	CG1B	RG1B
G2: Grade parking around and under the building	CG2L	RG2L	RG2D	CG2B	RG2B
S1: Adjacent parking structure	CS1L	RS1L	RS1D	CS1B	FS1B
S2: Underground parking structure	CS2L	RS2L	RS2D	CS2B	RS2B
S3: Parking beneath building	CS3L	RS3L	RS3D	CS3B	RS3B
SF: Single-family detached housing with surface parking	—	RSFL	RSFD	—	RSFN
GT: Single-family attached housing with surface parking	—	RGTL	RGTD	—	—

casts. These forecasts are based on a controlling design premise and the values entered in a design specification related to the premise. Table 3.2 arranges the forecast models in the collection based on these relationships. When reading this table, remember that the four-place name of a forecast model begins with an indication of the land use family involved. This is either *R* for residential or *C* for nonresidential.[3] The next two places indicate the parking system involved. For instance, *S2* indicates an underground parking structure. The last place in this name indicates the information given. For instance, *L* indicates that the gross land area is given, *D* indicates that a net residential density objective is given in addition to the gross land area, and *B* indicates that a gross building area objective is given. The result produces a name, such as forecast model *CS2B*.

Sample Forecast Panel

After a forecast model is selected and values are entered in its design specification panel, results are presented in its lower forecast panel. Model CG1L, Exhibit 3.1,[4] is a building area forecast model that contains 11 predictions, 7 of which are in this forecast panel.[5] The left-hand column in this panel is a specification column that is essential to the forecast produced. In this case, the column contains increasing floor heights FLR that are used to forecast increasing gross building area GBA. Any range of floors can be entered, but the range illustrated extends from 1 to 15 floors.

Core Area

The core development area calculated in a typical development forecast model is a major factor in determining the development capacity of any given land area. This value is forecast in the second column of the planning panel for forecast model CG1L (Exhibit 3.1). This core area is a statement of available land for building cover and parking cover. It

Exhibit 3.1 Forecast Model CG1L

Development capacity forecast for ***NONRESIDENTIAL BUILDINGS*** *based on the use of an adjacent* ***GRADE PARKING LOT*** *located on the same premises. When s and a equal zero in the design specification below, the forecast pertains to conditions when* ***NO PARKING*** *is required.*

Given: *Gross land area.* ***To Find:*** *Maximum development capacity of the land area (gross building area potential) based on the design specification values entered below.* ***Premise:*** *All building floors considered equal in area.*

DESIGN SPECIFICATION

Enter values in boxed areas where text is bold and blue. Express all fractions as decimals.

Given:	**Gross land area**	GLA=	**5.882**	acres	256,220	SF
Land Variables:	Public/ private right-of-way & paved easements	W=	**0.150**	fraction of GLA	38,433	SF
	Net land area	NLA=	5.000	acres	217,787	SF
	Unbuildable and/or future expansion areas	U=	**0.000**	fraction of GLA	0	SF
	Gross land area reduction	X=	0.150	fraction of GLA	38,433	SF
	Buildable land area remaining	**BLA=**	5.000	acres	217,787	SF
Parking Variables:	Est. gross pkg. lot area per space in SF	s =	**375**	ENTER ZERO IF NO PARKING REQUIRED		
	Building SF permitted per parking space	a =	**250**	ENTER ZERO IF NO PARKING REQUIRED		
	No. of loading spaces	l=	**1**			
	Gross area per loading space	b =	**1,000**	SF	1,000	SF
Site Variables:	**Project open space as fraction of BLA**	**S=**	**0.300**	⇦	65,336	SF
	Private driveways as fraction of BLA	R=	**0.030**		6,534	SF
	Misc. pavement as fraction of BLA	M=	**0.020**		4,356	SF
	Loading area as fraction of BLA	L=	0.005		1,000	SF
	Total site support areas as a fraction of BLA	Su=	0.355		77,225	SF
Core:	**Core development area as fraction of BLA**	**C=**	0.645	C+Su must = 1	140,562	SF

PLANNING FORECAST

no. of floors **FLR**	CORE (minimum land area for BCG & PLA)	gross building area **GBA**	parking lot area PLA	pkg. spaces NPS	footprint **BCA**	bldg SF / acre SFAC (function of BLA)	flr area ratio FAR (function of BLA)
1.00	140,562	**56,225**	84,337	224.9	**56,225**	11,246	0.258
2.00		**70,281**	105,421	281.1	**35,140**	14,057	0.323
3.00		**76,670**	115,005	306.7	**25,557**	15,335	0.352
4.00		**80,321**	120,481	321.3	**20,080**	16,065	0.369
5.00		**82,683**	124,025	330.7	**16,537**	16,538	0.380
6.00	*NOTE:*	**84,337**	126,505	337.3	**14,056**	16,868	0.387
7.00	*Be aware when BCA becomes*	**85,559**	128,339	342.2	**12,223**	17,113	0.393
8.00	*too small to be feasible.*	**86,499**	129,749	346.0	**10,812**	17,301	0.397
9.00		**87,245**	130,868	349.0	**9,694**	17,450	0.401
10.00		**87,851**	131,776	351.4	**8,785**	17,571	0.403
11.00		**88,353**	132,529	353.4	**8,032**	17,672	0.406
12.00		**88,776**	133,164	355.1	**7,398**	17,756	0.408
13.00		**89,137**	133,705	356.5	**6,857**	17,828	0.409
14.00		**89,448**	134,172	357.8	**6,389**	17,891	0.411
15.00		**89,720**	134,580	358.9	**5,981**	17,945	0.412

WARNING: These are preliminary forecasts that must not be used to make final decisions.
1) These forecasts are not a substitute for the "due diligence" research that must be conducted to support the final definition of "unbuildable areas" above and the final decision to purchase land. This research includes, but is not limited to, verification of adequate subsurface soil, zoning, environmental clearance, access, title, utilities and water pressure, clearance from deed restriction, easement and right-of-way encumbrances, clearance from existing above and below ground facility conflicts, etc.
2) The most promising forecast(s) made on the basis of data entered in the design specification from "due diligence" research must be verified at the drawing board before funds are committed and land purchase decisions are made. Actual land shape ratios, dimensions and irregularities encountered may require adjustments to the general forecasts above.
3) The software licensee shall take responsibility for the design specification values entered and any advice given that is based on the forecast produced.

does not, however, represent a location, such as the area within zoning setback limits.[6] It is simply the total land area available for building cover and parking cover. This core area can be composed of many separate land areas, but the sum of these areas should not exceed the total core area available.

Gross Building Area

The gross building area GBA forecast is a mathematical function of the design specification values entered and can be seen to increase with building height in Exhibit 3.1. This increase is not linear, however, and the optimum relationship between surface parking lot PLA and building cover BCA varies with each floor of building height involved. Exhibit 3.1 forecasts these relationships in the building cover area BCA column and the parking lot area PLA column.[7]

Number of Parking Spaces

The NPS column simply translates the gross parking lot area forecast into the number of parking spaces that can be accommodated based on the design specification value *s* entered previously.

Indexes of Intensity

The amount of building area that can be accommodated per buildable acre is an index of building intensity.[8] The SFAC column presents these values as they increase with building height FLR.

The floor area ratio FAR is simply the gross building area GBA forecast divided by the buildable land area BLA forecast. The FAR values do not regulate building height and open space, however, so two entirely different environmental results can be produced by the same FAR value.[9] These equivalent FAR values therefore are an interesting correlation but will not necessarily produce the same environmental results.

Companion Forecasts

The ability to forecast the gross building area capacity of a given land area unlocks the ability to forecast and evaluate a number of financial, social, economic, and environmental values that are a function of this gross building area potential. These forecasts can include but are certainly not limited to the following:

- The construction costs that may be involved
- The population, revenue, tax abatement, and traffic that may be introduced
- The operating cost, lease rate, and return on investment that may be possible
- The economic development opportunities present
- The environmental impact represented

Several forecast panels have been created to illustrate this opportunity. These panels have been linked to the CG1L forecast model in Exhibit 3.1 and demonstrate the parent–child relationship between an independent development capacity forecast represented by model CG1L and the additional forecasts that can be linked to this primary information. Exhibit 3.2 is the first of these illustrations, and is designed to create a link between development capacity and construction cost forecasts.

Construction Cost

Construction cost obviously affects project feasibility and is affected by a number of building and site-planning choices that are discretionary. Construction cost should not be confused with project cost. Project costs include construction costs and a number of related research, administrative, and consulting fees that are essential to the construction process but are generally excluded from construction cost estimates because of the wide variations involved. Project costs may include, but are not limited to, those listed in Fig. 3.2. Rec-

Exhibit 3.2 Construction Cost Forecast Panel Linked to Forecast Model CG1L

Project Open Space: 30.00%

BUILDING EFFICIENCY OBJECTIVES

Building skin	Bs=	0.05
Net building area	Bn=	0.95
Building core	Bc=	0.05
Building mechanical area	Bm=	0.02
Building circulation area	Bc=	0.08
BUILDING SUPPORT	Bs=	0.20
BUILDING EFFICIENCY	Be=	0.80

NOTE: Tenant finish costs generally apply to speculative office buildings. Interior finish and equipment costs generally apply to custom buildings. Both interior and tenant finish costs may not be needed on the same project.

COST SPECIFICATION

Land cost per acre in $/AC	$100,000	$2.30	0.380	=Tf	Tenant finishes if applicable as % of shell cost
Structure / surface parking in $/SF	Pc=	$2.50	0.120	=Df	Preconstruction fees as % of shell cost
Loading area in $/SF	Lc=	$5.00	0.085	=Mf	Marketing fees as % of shell cost
Open space grading & landscaping in $/SF	Sc=	$3.00	0.100	=Lf	Legal fees as % of shell cost
Public / private rights-of-way in $/SF	Rc=	$7.50	0.005	=Zf	Zoning, and permit fees as % of shell cost
Driveways in $/SF	Dc=	$2.50	0.000	=Ff	Financing costs as % of shell cost
Miscellaneous pavement in $/SF	Mc=	$2.50	0.000	=Cf	Contingency as % of shell cost
Interior finishes if applicable in $/SF	Ic=	$0.00	0.690	=Tc	TOTAL ADMINISTRATION & OVERHEAD
Furnishing, fixtures & equipment if applicable in $/SF	Ffe=	$0.00			
Demolition costs (lump sum)	Dc=	$100,000			
Replacement costs (lump sum)	Rc=	$100,000			

Links used in calculation

CONSTRUCTION COST FORECAST

Site acquisition, demolition, preparation, replacement, roadway and landscape improvement

$1,290,071 included in TCF below

no. of floors FLR	gross building area GBA	shell cost forecast SHC per sq. ft.	total shell TSC cost forecast	parking & loading PLD cost forecast	interior finish INF cost forecast	tenant finishes TNF cost forecast	fees, demolition ADM replacement & admin.	total cost **TCF** forecast	cost forecast **$SF** per sq. ft. of GBA
1.00	56,225	**$45.00**	$2,530,107	$215,842	see TNF	$961,441	$1,745,774	**$6,743,235**	**$119.93**
2.00	70,281	**$45.00**	3,162,634	268,553	see TNF	1,201,801	2,182,217	**$8,105,276**	**115.33**
3.00	76,670	**$50.00**	3,833,496	292,512	see TNF	1,456,728	2,645,112	**$9,517,919**	**124.14**
4.00	80,321	**$50.00**	4,016,043	306,203	see TNF	1,526,096	2,771,070	**$9,909,483**	**123.37**
5.00	82,683	**$55.00**	4,547,578	315,062	see TNF	1,728,080	3,137,829	**$11,018,620**	**133.26**
6.00	84,337	**$55.00**	4,638,530	321,263	see TNF	1,762,641	3,200,585	**$11,213,091**	**132.96**
7.00	85,559	**$60.00**	5,133,551	325,847	see TNF	1,950,749	3,542,150	**$12,242,368**	**143.09**
8.00	86,499	**$60.00**	5,189,963	329,373	see TNF	1,972,186	3,581,075	**$12,362,667**	**142.92**
9.00	87,245	**$65.00**	5,670,930	332,169	see TNF	2,154,953	3,912,942	**$13,361,064**	**153.14**
10.00	87,851	**$65.00**	5,710,311	334,441	see TNF	2,169,918	3,940,115	**$13,444,856**	**153.04**
11.00	88,353	**$70.00**	6,184,706	336,324	see TNF	2,350,188	4,267,447	**$14,428,736**	**163.31**
12.00	88,776	**$70.00**	6,214,298	337,909	see TNF	2,361,433	4,287,866	**$14,491,577**	**163.24**
13.00	89,137	**$75.00**	6,685,242	339,262	see TNF	2,540,392	4,612,817	**$15,467,785**	**173.53**
14.00	89,448	**$75.00**	6,708,617	340,431	see TNF	2,549,275	4,628,946	**$15,517,339**	**173.48**
15.00	89,720	**$80.00**	$7,177,609	$341,450	see TNF	$2,727,491	$4,952,550	**$16,489,171**	**$183.78**

Figure 3.2 Examples of project costs.

	Lump sum owner costs	*Construction costs subject to design fee*	
Due diligence research costs			
Property appraisal	________		
Land survey	________		
Geotechnical analysis	________		
Development capacity analysis	________		
Environmental impact	________		
Historic impact	________		
Zoning adequacy	________		
Existing structures evaluation	________		
Title research	________		
Other	________		
Administrative costs			
Option to purchase land	________	State amount and duration	
Land acquisition	________		
Legal fees	________		
Accounting fees	________		
Project programming	________		
Causes of loss—special form insurance	________		
Site security	________		
Loan fees	________		
Closing costs	________		
Other	________		
Construction costs			
Demolition		________	
Site preparation		________	
Building construction		________	
Landscape		________	
Roadways		________	
Parking		________	
Exterior lighting		________	
Miscellaneous pavement		________	
Exterior signage	________	________	If designed and specified by architect
Furniture and furnishings	________	________	If designed and specified by architect
Equipment (loose and fixed)	________	________	If designed and specified by architect
Telecommunications	________	________	If designed and specified by architect
Special inspections		________	If designed and specified by architect

Other ______
Contingencies ______
Bid contingency ______ %
Field cost contingency ______ %
Midpoint inflation contingency ______ %
Subtotal subject to design fee: $______

Consultant team costs (design fees)—check those that apply
Architect ___
Structural engineer ___
Mechanical engineer ___
Electrical engineer ___
Civil engineer ___
Acoustic engineer ___
Graphic designer ___
Interior designer ___
Kitchen consultant ___
Landscape architect ___
Other ___
Consultant team fee: ______%
Consultant team amount: $______

Subtotals: $______ $______

Grand total: $______

ognizing that construction cost is only part of total project cost can avoid surprises at a later date.

Exhibit 3.2 is a linked model that contains a panel of building efficiency objectives and construction cost specifications in the top half of the page. As usual, these values are variables[10] that may be changed to evaluate alternate possibilities. They are multiplied by the gross building area forecast GBA linked from Exhibit 3.1. All linked data from Exhibit 3.1 is shown in the first two columns of the construction cost forecast panel. The third column contains a series of building shell[11] cost specifications that can vary and that are entered by the user. The remainder of the panel forecasts various design element costs that add up to the total cost forecast in the TCF column. These costs vary with building height and are translated into costs per building square foot in the $SF column. Site improvement costs are noted above the TCF column since they do not increase with building height, but do increase with open space area. This site improvement cost is included in each TCF value forecast below it.

Linking this cost forecast panel allows the user to change any specification value in Exhibit 3.1, or any building efficiency objective or cost variable in Exhibit 3.2, in order to test the implications of various design alternatives. For instance, assume the project open space value in Exhibit 3.1 is increased to 60% because a corporate office park environment is desired. Exhibit 3.3 results, and a comparison with Exhibit 3.2 shows that the gross building area GBA capacity of the site declines and the total cost forecast TCF also declines, even though the site improvement cost[12] increases because of the expanded open space. It also shows that the building cost per square foot $SF increases because construction cost does not decline as rapidly as development capacity when the open space allocation is adjusted. This is an example of the interrelationships that can be evaluated when forecast model links are established.

Population, Revenue, and Tax Abatement

Building population is a function of development capacity. It drives the demand for services, affects traffic impact on surrounding areas, and produces revenue for local government reinvestment in the community.[13] Building value produces real estate tax revenue and requests for tax abatement. There may also be secondary sources of revenue, such as water and sewer surcharges and stormwater utility fees. These secondary sources vary by community and are not considered in this example for the sake of brevity. They can easily be added to any linked forecast model, however, to fine-tune revenue estimates to the unique revenue characteristics of a given community.

Exhibit 3.4 addresses revenue potential and is linked to the gross building area forecasts of Exhibit 3.1 and the

Exhibit 3.3 Construction Cost Forecast Panel Linked to Forecast Model CG1L

Project Open Space: 60.00%

BUILDING EFFICIENCY OBJECTIVES

Building skin	Bs=	**0.05**
Net building area	Bn=	0.95
Building core	Bc=	**0.05**
Building mechanical area	Bm=	**0.02**
Building circulation area	Bc=	**0.08**
BUILDING SUPPORT	Bs=	0.20
BUILDING EFFICIENCY	Be=	0.80

NOTE: Tenant finish costs generally apply to speculative office buildings. Interior finish and equipment costs generally apply to custom buildings. Both interior and tenant finish costs may not be needed on the same project.

COST SPECIFICATION

Land cost per acre in $/AC	**$100,000**	$2.30
Structure / surface parking in $/SF	Pc=	**$2.50**
Loading area in $/SF	Lc=	**$5.00**
Open space grading & landscaping in $/SF	Sc=	**$3.00**
Public / private rights-of-way in $/SF	Rc=	**$7.50**
Driveways in $/SF	Dc=	**$2.50**
Miscellaneous pavement in $/SF	Mc=	**$2.50**
Interior finishes if applicable in $/SF	Ic=	**$0.00**
Furnishing, fixtures & equipment if applicable in $/SF	Ffe=	**$0.00**
Demolition costs (lump sum)	Dc=	**$100,000**
Replacement costs (lump sum)	Rc=	**$100,000**

Value	Code	Item
0.380	=Tf	Tenant finishes if applicable as % of shell cost
0.120	=Df	Preconstruction fees as % of shell cost
0.085	=Mf	Marketing fees as % of shell cost
0.100	=Lf	Legal fees as % of shell cost
0.005	=Zf	Zoning and permit fees as % of shell cost
0.000	=Ff	Financing costs as % of shell cost
0.000	=Cf	Contingency as % of shell cost
0.690	=Tc	TOTAL ADMINISTRATION & OVERHEAD

Links used in calculation

CONSTRUCTION COST FORECAST

Site acquisition, demolition, preparation, replacement, roadway and landscape improvement

$1,616,751 included in TCF below

no. of floors FLR	gross building area GBA	shell cost forecast SHC per sq. ft.	total shell TSC cost forecast	parking & loading PLD cost forecast	interior finish INF cost forecast	tenant finishes TNF cost forecast	fees, demolition ADM replacement & admin.	total cost **TCF** forecast	cost forecast **$SF** per sq. ft. of GBA
1.00	30,090	**$45.00**	$1,354,058	$117,838	see TNF	$514,542	$934,300	**$4,537,489**	**$150.80**
2.00	37,613	**$45.00**	1,692,572	146,048	see TNF	643,177	1,167,875	**$5,266,423**	**140.02**
3.00	41,032	**$50.00**	2,051,603	158,870	see TNF	779,609	1,415,606	**$6,022,439**	**146.77**
4.00	42,986	**$50.00**	2,149,298	166,197	see TNF	816,733	1,483,016	**$6,231,995**	**144.98**
5.00	44,250	**$55.00**	2,433,764	170,938	see TNF	924,830	1,679,297	**$6,825,581**	**154.25**
6.00	45,135	**$55.00**	2,482,439	174,257	see TNF	943,327	1,712,883	**$6,929,657**	**153.53**
7.00	45,789	**$60.00**	2,747,363	176,710	see TNF	1,043,998	1,895,681	**$7,480,504**	**163.37**
8.00	46,293	**$60.00**	2,777,554	178,597	see TNF	1,055,471	1,916,512	**$7,544,885**	**162.98**
9.00	46,692	**$65.00**	3,034,957	180,094	see TNF	1,153,284	2,094,120	**$8,079,206**	**173.03**
10.00	47,016	**$65.00**	3,056,033	181,310	see TNF	1,161,293	2,108,663	**$8,124,049**	**172.79**
11.00	47,285	**$70.00**	3,309,919	182,317	see TNF	1,257,769	2,283,844	**$8,650,600**	**182.95**
12.00	47,511	**$70.00**	3,325,756	183,165	see TNF	1,263,787	2,294,771	**$8,684,231**	**182.78**
13.00	47,704	**$75.00**	3,577,795	183,890	see TNF	1,359,562	2,468,678	**$9,206,676**	**193.00**
14.00	47,871	**$75.00**	3,590,304	184,515	see TNF	1,364,316	2,477,310	**$9,233,197**	**192.88**
15.00	48,016	**$80.00**	$3,841,298	$185,061	see TNF	$1,459,693	$2,650,496	**$9,753,300**	**$203.13**

Exhibit 3.4 Revenue Forecast Panel Linked to Forecast Model CG1L

Project Open Space: 30.00%

BUILDING EFFICIENCY OBJECTIVES

Building skin	Bs=	0.05
Net building area	Bn=	0.95
Building core	Bc=	0.05
Building mechanical area	Bm=	0.02
Building circulation area	Bc=	0.08
BUILDING SUPPORT	Bs=	0.20
BUILDING EFFICIENCY	Be=	0.80
Avg. pop. per 1000 SF	Pop=	5.5

Linked from Exhibit 3.2

REVENUE SPECIFICATION

effective commercial real estate tax millage rates

Item	Symbol	Value	Millage	Fraction	Description
Avg. income per person	Inc=	**$30,000**	**6.838642**	10.2%	Fraction of REtax to local government
Avg. daily trips per person	Adt=	**5.0**	**45.490353**	67.8%	Fraction of REtax to local schools
Income tax rate		**0.02**	**13.754687**	20.5%	Fraction of REtax to county
Real estate tax millage		**67.082324**	**0.998642**	1.5%	Fraction of REtax to library
Fraction of real estate value assessed		**0.35**	**0.000000**	0.0%	Fraction of REtax to other
Personal property millage		**100.92**	67.082324	100.0%	Total effective millage rate
Personal property as % of building value		**0.10**			
Fraction of personal property value assessed		**0.25**			

ABATEMENT SPECIFICATION

First year real estate tax abatement request	**1.00**
Fraction of local school tax abated	**0.50**
Effective first year property tax abatement	56.5%

REVENUE FORECAST

no. of floors FLR	gross building area GBA	total project TCF cost forecast	total building POP population	income tax to ITX local government	total real estate REX tax	real estate tax to RGX local government after abatement	real estate tax to RSX local schools after abatement	personal property PPX tax to local gov.	yield to YTG local government after abatement	max. first year FYA tax abatement all sources
1.00	56,225	$6,743,235	247	$148,433	$158,323	$0	$53,682	$17,013	$165,446	$69,822
2.00	70,281	8,105,276	309	185,541	190,302	$0	$64,525	$20,450	$205,991	$125,778
3.00	76,670	9,517,919	337	202,409	223,469	$0	$75,770	$24,014	$226,422	$147,699
4.00	80,321	9,909,483	353	212,047	232,663	$0	$78,888	$25,002	$237,049	$153,775
5.00	82,683	11,018,620	364	218,284	258,704	$0	$87,717	$27,800	$246,084	$170,987
6.00	84,337	11,213,091	371	222,649	263,270	$0	$89,265	$28,291	$250,940	$174,005
7.00	85,559	12,242,368	376	225,876	287,436	$0	$97,459	$30,887	$256,764	$189,977
8.00	86,499	12,362,667	381	228,358	290,261	$0	$98,417	$31,191	$259,549	$191,844
9.00	87,245	13,361,064	384	230,327	313,702	$0	$106,365	$33,710	$264,037	$207,337
10.00	87,851	13,444,856	387	231,926	315,669	$0	$107,032	$33,921	$265,848	$208,637
11.00	88,353	14,428,736	389	233,252	338,770	$0	$114,864	$36,404	$269,655	$223,905
12.00	88,776	14,491,577	391	234,368	340,245	$0	$115,365	$36,562	$270,930	$224,880
13.00	89,137	15,467,785	392	235,321	363,165	$0	$123,136	$39,025	$274,346	$240,029
14.00	89,448	15,517,339	394	236,143	364,329	$0	$123,531	$39,150	$275,294	$240,798
15.00	89,720	$16,489,171	395	$236,861	$387,146	$0	$131,267	$41,602	$278,463	$255,879

building efficiency and cost information of Exhibit 3.2. It also contains original population, revenue, and tax abatement specifications in the top half of the model. As usual, the population, revenue, and abatement specifications may be changed or expanded to evaluate alternate possibilities. Once entered, these variables produce a population forecast that drives the income tax ITX calculation. These values are also used to forecast real estate tax REX and personal property tax PPX potential when linked to the building efficiency specifications and building cost forecasts of Exhibit 3.2. These revenue forecasts are displayed at the bottom of the page in the revenue forecast panel.[14]

A change to any of the specifications at the top of the model, or to any specification value in a linked model, obviously affects the yield to local government YTG forecast and the first year tax abatement FYA possible.[15] This example does not attempt to anticipate unique local taxing structures that will affect final revenue projections. Its purpose is to show that the development capacity of land is a parent forecast that can be linked to unique community characteristics and taxing structures to produce revenue forecasts related to this capacity.

Operating Cost, Lease Rate, and Return on Investment

The development capacity of land has first cost and continuing cost implications that affect a building owner's operating expense and return on investment. Development capacity is forecast in Exhibit 3.1. The first cost implications of this development are forecast in Exhibit 3.2. Exhibit 3.5 is linked to the gross building area forecasts of Exhibit 3.1 and the building efficiency objective information of Exhibit 3.2 in order to produce operating cost and return-on-investment forecasts. It also contains original operating and financing cost specifications in the top half of the model. As usual, the operating and financing specification

Exhibit 3.5 Return-on-Investment Forecast Panel Linked to Forecast Model CG1L

Project Open Space: 30.00%

BUILDING EFFICIENCY OBJECTIVES

Building skin	Bs=	0.05	Linked from Exhibit 3.2
Net building area	Bn=	0.95	Linked from Exhibit 3.2
Building core	Bc=	0.05	Linked from Exhibit 3.2
Building mechanical area	Bm=	0.02	Linked from Exhibit 3.2
Building circulation area	Bc=	0.08	Linked from Exhibit 3.2
Building support	Bs=	0.20	
Building efficiency	Be=	0.80	

OPERATION SPECIFICATION

Utilities	$1.50	per net sq. ft.
Maintenance	$2.50	per net sq. ft.
Management	$2.50	per net sq. ft.
Land lease/ year	$0.00	per net sq. ft.
Other/ year	$0.00	per net sq. ft.
Total not included in owner cost per net sq. ft.	$6.50	per net sq. ft.

FINANCING SPECIFICATION

Percent down	25.0%
Annual interest	7.5%
Term in years	$30.00
Cost markup	60.0%
Building occupancy rate	80.0%

RETURN on INVESTMENT FORECAST 17.5%

no. of floors FLR	gross GBA building area	total project TCF cost forecast	net leaseable NBA building area	debt service DBT per nsf	real estate tax RXA per nsf with abate	owner cost CST per nsf	average lease ALR rate no operating	risk adjusted RAL lease rate	gross GPR annual profit
1.00	56,225	$6,743,235	53,413	7.62	1.57	9.19	14.70	18.38	294,530
2.00	70,281	8,105,276	66,767	7.32	0.92	8.24	13.19	16.48	330,165
3.00	76,670	9,517,919	72,836	7.88	0.99	8.87	14.19	17.74	387,709
4.00	80,321	9,909,483	76,305	7.83	0.98	8.82	14.11	17.63	403,659
5.00	82,683	11,018,620	78,549	8.46	1.06	9.52	15.24	19.05	448,839
6.00	84,337	11,213,091	80,120	8.44	1.06	9.50	15.20	19.00	456,761
7.00	85,559	12,242,368	81,281	9.09	1.14	10.23	16.36	20.45	498,688
8.00	86,499	12,362,667	82,174	9.08	1.14	10.21	16.34	20.43	503,589
9.00	87,245	13,361,064	82,883	9.73	1.22	10.94	17.51	21.89	544,258
10.00	87,851	13,444,856	83,458	9.72	1.22	10.94	17.50	21.87	547,671
11.00	88,353	14,428,736	83,935	10.37	1.30	11.67	18.67	23.34	587,749
12.00	88,776	14,491,577	84,337	10.37	1.30	11.67	18.67	23.33	590,309
13.00	89,137	15,467,785	84,680	11.02	1.38	12.40	19.84	24.80	630,074
14.00	89,448	15,517,339	84,976	11.02	1.38	12.40	19.84	24.80	632,093
15.00	89,720	$16,489,171	85,234	11.67	1.46	13.13	21.01	26.27	671,680

variables may be changed to evaluate alternate possibilities. Once entered, these values and the specification data linked from Exhibits 3.1 and 3.2 produce debt service, tax, and building income predictions. These are used to predict owner costs per net square foot, average lease rates per net square foot, risk-adjusted lease rates, and gross annual profit forecasts. These forecasts are then used to predict the potential return on investment that is located in a box above the return-on-investment forecast panel at the bottom of the page.[16]

The premise of this example is a speculative office building. The distinction between a resident building owner and a building investor can be a major factor in the approach to financial evaluation, however.[17] For instance, a resident building owner might only be interested in the annual cost per net square foot CST forecast in Exhibit 3.5. An investor would be interested in the competitiveness of the average lease rate forecast ALR, the risk-adjusted lease rate RAL,[18] the gross annual profit from the net leaseable building area, and the return on the investment made.[19]

Since this forecast is based on a speculative office building premise, operating costs are not deducted from the gross annual profit forecast GPR, because they are assumed to be a tenant responsibility. Personal property tax is also not deducted, because it is assumed to be a tenant obligation. If this were a resident building owner premise, operating cost would be part of the owner cost projection in the CST column. Personal property tax might be a cost factor depending on the business, and the lease rate, profit, and return-on-investment data would generally be irrelevant.[20]

Financial Model

The construction of a financial model can become quite complex, and the alternatives available are beyond the scope of this effort. The point is, however, that financial analysis is only as good as the development capacity forecast on which it is based. The capacity forecast in this example is produced in Exhibit 3.1 and linked to the financial forecast. Some data are also linked from Exhibits 3.2 and 3.4. This makes the entire assembly quite powerful, since development capacity, cost, revenue, and investment alternatives can be evaluated with a very few keystrokes once the parent–child relationships are established. Since we are creatures of habit as well, these dependent models will probably be reused for many projects once created, since we tend to repeat financing and cost estimation techniques once established.[21]

Economic Development Capacity

The development forecast collection is relevant to economic development because the development capacity of land influences the economic yield produced. The basic point is that a larger building on the same land area will produce a greater yield per acre when the building occupancy and income rate per net square foot is constant. This is simply common sense, and often leads to overdevelopment, but the range of possibilities and implications has been difficult to predict in any reasonable time frame. Exhibit 3.4 shows that revenue forecasts can be easily calculated when tax rates and population rates are multiplied by the gross building area options and construction value forecasts produced by Exhibits 3.1 and 3.2, since these building areas and values drive the economic yield to both investor and community.

The ability to forecast the economic yield of a land use plan or development proposal can improve a city's ability to allocate the land within its jurisdiction in relation to the cash flow it needs to support the quality of life expected. Planning for land use allocation, balance, and intensity in relation to its economic yield can help a city conserve its land, stabilize its economy, increase its income, and improve its lifestyle. This increased yield can then be reinvested in the community to sustain its rate of property value appreciation.

Many cities currently enhance their yield with sprawl. Sprawl results from requests for annexation of agricultural land that is then converted to urban use. This annexation and conversion process changes a city's land use allocation ratios. For instance, if the change expands low-density residential land use, a city traditionally evaluates its current ability to serve this land with municipal services and may assess a development impact fee. It rarely evaluates its ability to afford the operating, maintenance, and improvement expense of this development over an extended period of time. Since the development is new, little expense is involved, and the additional revenue appears to be an improvement. It could represent a long-term liability, however, as age increases expense and residential revenue does not keep pace.

If the change expands office land use, a city's annual revenue may be enhanced by the change in its land use allocation ratios, but we are all beginning to realize that we cannot continue to compensate for past land use allocation mistakes by continuing to consume land with sprawl. This topic is covered in greater detail in later chapters, and the development forecast collection will help to quantify the discussion.

Due Diligence

The development forecast collection produces preliminary forecasts that must not be the sole basis for making final decisions. The word *forecast* is used throughout this book, and the word *prediction* is also used for this reason. The forecast collection is not a substitute for the *due diligence* research that must be conducted to support the final definition of *unbuildable areas,* the final decision to purchase land, and the final design specification that is written. As an example, due diligence research can include the information listed in the sample checklist format of Fig. 3.3,[22] but is not

Figure 3.3 Sample land purchase evaluation.

GENERAL PROPERTY INFORMATION

Location:

County:

Township:

Address of property or geographical reference:

Land area in acres to three decimal places:

Intended use of property:

Description of need and purpose of property request:

Name of property owner:

Address of property owner:

Telephone number of property owner:

LAND USE REGULATIONS

Jurisdiction having authority over property:

Jurisdiction contact name and phone number:

Zoning classification of property:

Zoning classification(s) for all contiguous property:

ATTACH: Letter or certificate from jurisdiction stating zoning adequacy for intended use or steps that must be taken to receive local approval of the intended use.

ATTACH: Letter of certificate from jurisdiction having authority that there are no known or pending inquiries, actions, or initiatives affecting the property under study *or* an explanation of actions pending.

ATTACH: Copy of zoning map covering property under study and all property within at least 500 feet of any property line.

ATTACH: Copy of zoning regulations applicable to property (not the entire ordinance).

LAND CHARACTERISTICS

ATTACH: Survey from a surveyor licensed to practice that includes all information listed in Exhibit A. Include definition of 100-year floodplain.

ATTACH: Soil boring and geotechnical report including analysis and recommendations from an engineer licensed to perform geotechnical analysis.

ATTACH: Map of soil boring locations and core boring logs.

UTILITIES

Are public utilities available? Yes No

If public utilities are not shown as available on survey, explain the method that will be used to provide or accommodate the following:

Type of heating energy source:

Is cooling to be provided? Yes No

Type of cooling energy source:

Is potable water available on site? Yes No

Note source of potable water and estimate gpm and pressure that will be available.

Type of sanitary disposal system contemplated:

Electrical energy source:

Is stormwater detention required? Yes No

Stormwater discharge design contemplated:

CURRENT DEED

ATTACH: Complete copy of current deed that enumerates any deed restrictions, easements, or other encumbrances that may pertain to the property under consideration and a summary of these encumbrances.

CLEAR LAND TITLE AVAILABLE

ATTACH: Title research from a title agency licensed to practice in the State of Ohio that identifies all deed restrictions, easements, liens, and other encumbrances recorded against the property under study.

ACCESSIBILITY

ATTACH: If the property is accessible only via easement or deed restriction, explain any restrictive provisions that pertain to these easements or deed restrictions. If the information is not available on the survey, define the width of the access available and the width of pavement within the access way. (This may be included in the title research)

REAL ESTATE CERTIFICATION

ATTACH: Certification of clear title and accessible property.

ENVIRONMENT

ATTACH: Copy of CE, EA, or EIS form.

PROPERTY VALUE

Asking price for land:

Asking price per acre:

Appraised value of land:

Appraised value per ace:

ATTACH: Appraisal of property value prepared by a member of the Real Estate Appraisal Institute.

Figure 3.3 Sample land purchase evaluation *(continued).*

SUMMARY OF REQUEST

ATTACH: Summary of conclusions from all research and attachments and explain need for property.

OPTION TO PURCHASE

Is an option to purchase being contemplated? Yes No

Option time period:

Does the option specify the purchaser's right to investigate subsurface soil conditions?

Option amount:

Reviewed and recommended by:

Authorized by:

FEASIBILITY EVALUATION

Comments will be attached after submittal and review.

PURCHASE APPROVAL

Purchase price offer:

Reviewed and recommended by:

Authorized by:

limited to this list since regional characteristics will affect the content, emphasis, and definition of due diligence. For instance, earthquake hazard research might be emphasized in one geographic area, seashore preservation in another, and abandoned underground mines in a third. None of these features are emphasized in Fig. 3.3 beyond the requests for deed research, surveys, and geotechnical analysis.

After data is entered in a design specification from due diligence research, these variables can also be adjusted to produce alternate forecasts for comparison. The most promising forecasts must also be verified at the drawing board before funds are committed and land purchase decisions are made. The forecast models cannot anticipate actual land shape ratios, dimensions, and field irregularities that may be encountered and that may require adjustments to the design specification values entered. For these reasons, the meaning of the words *forecast* and *prediction* should be clearly understood, and the practitioner must take responsibility for the design specification values entered in the software attached to this book and for any advice given that is based on the forecast produced.

It should also be understood that the forecast models do not make judgments. In some cases, forecast notes are included to remind the user to check for unrealistic extremes. One such note advises, "Be aware when BCA becomes too small to be feasible." This means that the model will not recognize when it is dividing a gross building area by such a large number of building floors that the resulting building area per floor, or the building footprint, becomes too small to be useful as a building floor plan. The minimum floor plan area per floor is a matter of professional judgment and will vary with the specific project encountered. For this reason, arbitrary cutoffs have been deleted and design limits are left to the interpretation and judgment of the user.[23]

Drawing-Board Evaluation

The specification and planning forecast panels in the development forecast collection are arranged to assist in drawing-board sketch evaluation of the most promising predictions. For this reason, the design specification panel converts the percentage values entered to square-foot values in its right-hand column. The forecast panel enumerates site plan areas based on the percentages entered previously and the building height being considered. All of these areas are needed when the forecast is taken to the drawing board for an evaluation of these predictions in relation to actual site conditions that may affect the utility of the buildable land area remaining.[24]

The most prominent site plan areas involved are the building footprint and the parking footprint and project open space. Exhibit 3.1 identifies parking footprint areas in the PLA column of the planning forecast panel and building footprint areas in the BCA column. These areas are noted in relation to the building height under consideration and the project open space percentage entered in the design specification panel above. They can be any shape, in any permitted location on the site, and in any relationship to each other, but should not exceed the total quantity noted for each category.

Project open space specifications limit the amount of development cover that can be introduced and the massing[25] relationships that can be formed. This in turn limits the development intensity that can be constructed and establishes the environmental design[26] concept for the project. The area required for driveways and miscellaneous pavement to serve these relationships is integrated with the final development plan and becomes part of the development cover introduced. Each solution can then be compared to other options that have also survived the initial development capacity analysis.

Final Decisions

In development there is rarely a "right" answer. There is generally a comparison of alternatives and a final choice. This is often confusing to the uninitiated. There are many potential solutions to a given problem and most have merit. Those searching for the "right" answer can become prematurely frustrated. The strength of a development solution is partially based on the quality of the comparisons that have led to its choice. With the advent of the computer, this comparative approach can be improved by contrasting hundreds of approaches in the time it would take to draw one. This can help us marry the intuitive knowledge of centuries with the power of design specifications to produce environments without dictating appearance.

Notes

1. This makes it possible to consolidate an infinite number of design possibilities into a manageable series of forecast groups.
2. These definitions and explanations can be read when using a forecast model, or they can be read in their entirety as a separate effort. In any case, they should not be overlooked, since most of the information is not included elsewhere.
3. The mixed-use M series is discussed later.
4. This is the same forecast model that was included as Exhibit 2.1 to illustrate a typical design specification format. It is repeated here as Exhibit 3.1 to complete the discussion regarding the basic panels of information contained in a forecast model.
5. The forecasts provided are arranged in the forecast section of each model to help design professionals when proceeding to the next level of site plan layout and drawing-board analysis.
6. Zoning setbacks can affect location and must be accounted for by the amount of project open space specified.
7. *Optimum* in this sense means the relationship that produces the largest gross building area GBA given the building height involved.
8. Buildable acres are used rather than net or gross acres because they represent the land modified to receive the improvement. The statistics that result are a more accurate indication of the design relationship proposed between the acres available and the gross building area introduced.
9. From an urban design, environmental design, and development capacity forecasting standpoint, this ratio may be useful as an index but is too vague to be effective in specifying the development results that are actually desired without additional amplification. A simpler index that could be used as an urban or environmental design specification is outlined in Chap. 1.

10. These variables are identified by the characteristic bold text within a box outline.
11. A building *shell* does not include interior tenant walls, equipment, and finishes. Interior public facilities, public safety requirements, means of egress, and vertical shafts are finished as part of the shell facility.
12. Not shown on the table, but included in the TCF value
13. In addition to a host of other demands that are not relevant to this example
14. The linked data from model CG1L (Exhibit 3.1) are presented in the first three columns of the forecast panel. The building objective data at the top of the model are linked from Exhibit 3.2.
15. Every community will have its own formula, and this is simply the author's equation based on the revenue potential and school-sharing format established in this equation.
16. The linked data from Exhibit 3.1 are presented in the first three columns of the forecast panel. The building efficiency objective data at the top of the model are also linked.
17. The numbers and approach have been simplified to illustrate the concept of linking dependent models to a parent development capacity forecast in order to expand the forecasting and evaluation potential of the model collection attached. For instance, construction financing is a world composed of many strategies that depend on the development capacity of land. In this example, a financing strategy has been used that is familiar to every homeowner, but would probably not be used in the world of development investors and balloon payments. It is simply based on an amount invested, a fixed annual interest rate, and a defined term of debt service.
18. Risk adjustment is calculated to anticipate an occupancy rate less than 100%.
19. In this example, a simplified financing approach is used as an example, and the return on investment is calculated based on the percentage of the total building cost used as collateral for the building loan.
20. This is not always the case, however, since some corporations prorate the cost of office space to their various department and division budgets.

21. In fact, different financing strategies can be linked and then saved under separate model collection names to avoid reinventing the wheel.
22. A sample review of information submitted based on Fig. 3.3 is included as Fig. 3.1.
23. There may be a future time when design limits become similar to "ultimate strength" and design stress values in steel handbooks, but for the moment these limits are a matter of intuition and professional judgment. They are an example of what is meant when the physical design disciplines are described as a blend of art and science.
24. Model forecasting is not a substitute for this analysis.
25. *Massing* is an environmental design term that means the relationship between three-dimensional building volume and the open space that surrounds it. Building style and appearance is generally ignored in favor of the building form produced. Open space often includes parking lots that become mass within the pedestrian environment as soon as cars are parked. Landscape design evaluation at the drawing board attempts to return parking lots to the pedestrian open space environment by using changes in grade and vertical shields to conceal these parked cars.
26. Often referred to as *urban design*.

Glossary

An expanded glossary of forecasting terms, symbols, definitions, descriptions, and explanations follows. It is intended as a supplement and extension of this discussion and is not simply a dictionary. It should not be overlooked for this reason, since much of the information is not repeated elsewhere.

AAL Aggregate average lot area—a single lot area calculated to represent the average lot areas and mix proposed. This value is forecast in square feet from other design specification values entered.

The AAL is an adjusted average lot area that accounts for the mix of lot areas specified. This value is compared to the average lot area potential of the site ALT given the common open space objective COS defined. If AAL is greater than ALT, adjustments must be made in the design specification to make AAL less than or equal to ALT.

AAL is calculated from the data entered in the lot mix table and is significantly affected by the common open space and net density objectives given.

AFP Aggregate average land allocation per dwelling unit type—the average amount of development cover and private yard area allocated per dwelling unit based on the dwelling unit mix proposed. This area includes development cover and private, personal, or "defensible" yard areas that are allocated per dwelling unit and are not shared among dwelling units. It does not include common open space that creates a parklike environment shared among dwelling units.

The distinction between private, personal, or "defensible" open space unique to each dwelling unit and common open space shared by all dwelling units allows everyone to focus on both the total environment created by these building type compositions and the privacy needed by each dwelling unit within these compositions.

AGG Aggregate average dwelling unit area—a single dwelling unit area calculated to represent the average dwelling unit areas and mix proposed. This value is forecast in square feet from other design specification values entered. The final value calculated is equal to the gross building area divided by the number of dwelling units present.

This is the theoretical dwelling unit referred to in all zoning ordinances, but its area and the land it consumes are not constant. It varies by the dwelling unit areas and mix planned, provided, or required in a specific housing project. A reference to a "dwelling unit" in a zoning ordinance, therefore, is not often specific enough to anticipate the environment that will result when the dwelling unit areas and mix are not known, or when the aggregate average dwelling unit area is not specified.

ALT Average lot area—the average lot area that can be provided within a given subdivision area. This value is forecast in square feet from other design specification values entered.

The potential ALT of the land area involved, based on the design specification values entered, is compared to the aggregate average lot area AAL needed based on the lot areas and mix desired. The ALT value must be greater than or equal to the AAL needed if the project is to appear feasible and worthy of further evaluation. The ALT value is significantly affected by the buildable land area calculated and the common open space objective defined.

BAC Buildable acres of land area—the total land area available for improvement in acres. It does not include public and private rights-of-way, paved easements, and unbuildable land areas. This value is forecast from other design specification values entered.

This term should not be confused with the more common but contradictory zoning expression that often means the land area located within the building setback lines on a given lot. (It is contradictory since this is not the only buildable area of the lot. Much is permitted outside of the buildable area in most, if not all, zoning ordinances.)

In the context of this work, *buildable area* means the total area available for improvement on a given land area and excludes only unbuildable areas of the site such as, but not limited to, public or private rights-of-way, ravines, lakes, unstable soil, wetland, etc.

BCA Building cover area—the area enclosed by the building foundation perimeter. This value is forecast from other design specification values entered. It is often referred to as the building *footprint*. In residential construction, when it includes the garage area it is referred to as BCG.

Building cover is one of the primary elements of construction and combines with parking cover, roadway cover, loading areas and miscellaneous pavement to produce development cover. The forecast models consider all building floors equal in area to avoid floor plan complexities that are impossible to forecast.

This forecast is limited to an expression of the geometric area and volume that can be accommodated on the site. Detailed sculpturing of this area and volume to fit specific floor plan, building form, and appearance objectives cannot be anticipated. These efforts are a function of later, more detailed phases of project design that use the envelope forecast as an expression of the outer limits that may be involved.

BCG Building cover area, including garage—the area covered by the building floor plan, including attached or detached garages. Depending on the forecast model involved, this value is either calculated from other specification values entered or specified as a square-foot area.

This forecast defines the total footprint BCG to be included on a given lot type. The sum of the values entered for building cover BCG, miscellaneous pavement MSP, and yard area YRD produce a calculation of the total lot or land area LOT needed to satisfy the objectives defined. The total building cover BCG proposed per dwelling unit is a major factor influencing the lot or land area required.

BLA Buildable land area—the total land area available for improvement in square feet. It does not include public and private rights-of-way, paved easements, and unbuildable land areas. This value is forecast in acres and square feet from other design specification values entered. It is equal to the gross land area minus public and private rights-of-way, paved easements, and all unbuildable areas.

This term should not be confused with the more common zoning expression that often means the land area located within the building setback lines on a given lot.

BPA Building plane area—the imaginary horizontal area above grade within a given lot or land area that is planned, permitted, or provided for an elevated building floor plan. This value is forecast in square feet from other design specification values entered.

Some buildings are built above grade with surface parking beneath. The total area encompassed by the building floor plan above this parking area is referred to as the *building plane area.*

When this configuration is adopted, careful attention should be paid to the edges of contact with surrounding land uses, since the temptation to pave to the property line can be significant. Perimeter project open space percentages and vertical screens can conceal this parking lot and promote compatibility.

BSA Basement area—the gross basement area expressed in square feet. A floor level with a floor elevation that is 3 ft or more below grade at any point along its exterior perimeter. This value is forecast as a fraction of the habitable area footprint above from other design specification values entered. The total building footprint BCA intended, including garage, and the total garage area GAR intended are entered in square feet by the lot type proposed in the lot mix table. The habitable footprint area HAB intended is equal to BCA minus GAR. The basement area BSM intended is entered as a fraction of HAB. The basement area BSA forecast is a product of BSM times HAB.

The basement is a separate construction cost consideration that is separately identified for this reason. It does not have additional land consumption implications, but it does have cost and quality implications.

Bu Building support—a fraction of the gross building area devoted to support functions such as circulation, mechanical, and plumbing services. This value is forecast from other design specification values entered. Building support areas generally range from 10 to 40% of the gross building area.

Buildings contain mechanical areas, electrical closets, utility chases, corridors, wall thicknesses, plumbing facilities, elevators, etc. that reduce the net internal area available for occupancy. The area devoted to these functions divided by the gross building area forecast or desired is the amount of building area devoted to building support.

C Core development area percentage—a percentage of the buildable land area that is exclusively available for building cover and parking cover. The core development area value is forecast as a percentage of the buildable land area available. All percentages are expressed in their decimal equivalents.

This is the area remaining for building cover and parking cover after all other land area demands have been either specified or forecast. The core area does not represent a location within a given gross land area, such as the area within zoning setback lines. It is simply a statement of the land available for building cover and parking cover. This area can be the sum of its parts, and these parts can be separated from each other or linked as long as the sum does not exceed the total available.

CDA Comprehensive dwelling unit area—the total building area constructed per dwelling unit. This value is forecast when the gross dwelling unit area specified in square feet for each dwelling unit type is divided by the building efficiency planned or present. This produces a total prorated gross building area per dwelling unit for use in other spreadsheet calculations.

This is not a traditional measurement, but is used to forecast the aggregate average dwelling unit area AGG implied by the dwelling unit areas and mix data entered.

CORE Core development area—the land area exclusively available, or used, for building cover and parking cover. This value is forecast from other design specification values entered, and is the area remaining for building cover and parking cover after all other land area demands have been either specified or forecast.

Identifying the core development area that is available for building cover and parking cover makes it easier to focus on the optimum relationships between these two primary ingredients of development cover.

COS Common open space—net project open space that is shared by a number of homes within a private setting. Common open space is added to private yard areas and unbuildable areas such as ravines and lakes to produce the gross project open space available. This value is either specified as a percentage of the buildable land area available or forecast in square feet from other design specification values entered, depending on the forecast model in use. The range of values that can be specified produces a broad range of different housing environments.

A major portion of the character of a housing development is established by the way private yard areas and common open space are allocated in relation to the development cover constructed. Large yard areas and private lots typify a suburban subdivision with no common open space. Smaller yard areas and lots with large amounts of common open space collected within the same gross land area produce a larger parklike atmosphere, with the housing clustered more compactly within.

Common open space is net project open space. It is provided in addition to any unbuildable areas that may be present. When common open space is specified as a percentage of the buildable land area available, keep in mind that this is in addition to the unbuildable areas that may be present and that combine to produce the gross project open space available to the project.

CS Core + open space area—an area including the core development area and the project open space provided. This value is forecast as a fraction of the buildable lot area from other design specification values entered.

This is an interim calculation that has no direct utility and is simply a step in the mathematical forecast produced.

CSL Core + open space + loading area—an area including the core development area, the total loading area, and the total project open space provided. This value is forecast as a percentage of the buildable lot area available from other design specification values entered. All percentages should be expressed in their decimal equivalents.

This is an interim calculation that has no direct utility and is simply a step in the mathematical forecast produced.

dBA Dwelling units per buildable acre—the total number of dwelling units provided per buildable acre available. This value is forecast from other design specification values entered and is a function of the buildable acres forecast.

This is an index that more accurately reflects the design configuration involved. The density recorded relates the number of dwelling units planned or provided to the actual number of buildable acres that will, or have, received them. It does not include the number of unbuildable acres that surround or interrelate with these dwelling units. These unbuildable acres may relieve the sense of intensity produced by the entire composition, but they serve to confuse the nature and intensity of the development massing and cover introduced. A separate expression for the amount of unbuildable open space helps to state and visualize the balance between the development cover introduced and the net project open space interspersed among the building elements proposed, or produced, on a given site. When both net and gross project open space are expressed with the development cover and building heights proposed, the expression helps to state and visualize the actual development cover intensity involved in relation to the offsetting open space provided within and around the project.

DCA Total development cover area—the sum of all impervious cover introduced including, but not limited to, building cover, parking cover, driveway cover, loading area, and miscellaneous pavement within the buildable area of a given lot or land area. Within the gross land area available, development cover also includes all public and private roadway cover. This value is forecast in square feet from other design specification values entered.

Development cover is counterbalanced by net and gross project open space and combines with building height to produce development intensity.

dNA Dwelling units per net acre—the total number of dwelling units provided per net acre available. This value is forecast from other design specification values entered, and it can be misleading.

Net density can understate the development intensity and design character present when the net area includes unbuildable ravines, wetlands, ponds, etc. that decrease the density calculated but actually increase the compact arrangement required and the physical presence created.

DUF Dwelling units per floor—the total number of dwelling units that can be placed on a single building floor. This value is forecast from other design specification values entered.

This forecast has two functions. It provides a general idea of the design parameters required to meet the forecast. It is also a feasibility indicator, since it is possible for these spreadsheets to produce forecasts that are unrealistic. For instance, when increased numbers of building floors are considered, it is possible for the forecast model to produce an unrealistically low number of potential dwelling units per floor. The user must take this into account in evaluating the feasibility of the forecasts produced.

DYG Dwelling unit footprint with yard and garage—the total private land area allocation per dwelling unit. This value is forecast as a square-foot area from other design specification values entered.

A major portion of the character of a housing development is established by the way private yard areas and common open space are allocated in relation to the dwelling units provided. In more compact developments, these private land areas have been referred to as *defensible space*. Large yard areas, private lots and no common open space often represent a typical suburban subdivision. Small yard areas, small private lots, and large amounts of common open space within the same gross land area produce a larger parklike atmosphere, with the housing clustered more compactly within. The total private land area allocation per dwelling unit is one indicator of the site planning approach adopted.

F Floor total—the total number of floors provided above grade, including parking structure levels. This value is forecast from other design specification values entered.

Building floor-to-floor heights and parking structure floor-to-floor heights can vary significantly. This value simply gives a rough indication of the total height involved.

FAR Floor area ratio—as used and calculated in this book, the ratio between the gross building area planned, permitted, or produced in square feet and the gross buildable area available in square feet. This value is forecast from other design specification values entered, and is found by dividing the total building area in square feet by the gross buildable area in square feet.

The floor area ratio specifies the amount of gross building area that may be constructed in relation to the gross land area involved. It is intended as a flexible development control, since it leaves the two components that control development intensity—project open space and building height—unaddressed. It is, however, presented in many forecast models as another method of illustrating the development capacity forecast, since many are familiar with this historic benchmark.

FLR Building floors—the number of building floors planned, provided, or required. Floor numbers are either specified in the forecast panel of a model or forecast as the result of a given gross building area objective. When floor values are specified, a beginning list from 1 to 15 floors has been entered in each forecast model, but these numbers may be changed at will.

The number of building floors contemplated is a big factor in forecasting the development capacity of a given land area. Increased building height produces increased building area, but not in the straight-line relationship one would assume when surface parking is involved. Since the building footprint must shrink as height increases to provide increased land area for the expanding surface-parking requirement, the increase in gross building area rapidly declines when all other design specification values are held constant.

GAC Gross acres of land area—the total land area given or forecast in acres. This value is forecast from other design specification values entered. Gross land area is also represented by the index GLA in some circumstances.

GBA Gross building area—the total building area as measured along the exterior perimeter, including exterior wall thicknesses. This value is expressed in square feet and is either specified or forecast from other specification values entered.

Gross building area is forecast based on the assumption that all building floors are equal in area. This is rarely the case in final architectural floor plans, but the purpose of this spreadsheet collection is not to predict the form and content of final floor plans and building appearance. The purpose is to forecast the relationships among building cover, parking cover, roadway cover, miscellaneous pavement, building height, and open space that can optimize development capacity based on the design specification values entered. Final floor plans, architectural sections, and building elevations can then be refined within the parameters established by the initial design specification values.

GPA Gross parking lot area—the entire parking lot area, including areas for building structure, service cores, and remote exit stairs when a building or structure is present above the parking lot. This value is forecast in square feet from other design specification values entered.

In a typical surface parking lot, it is not necessary to distinguish between net and gross parking areas. This becomes more important when a surface parking lot is beneath a building above, or when parking structures are being considered.

GPL Gross parking structure area per level—total parking structure area per level, including, but not limited to, building structure, wall thickness, service cores, remote exit stairs, parking spaces, and circulation aisles. This value is forecast in square feet from other design specification values entered.

The gross area per parking structure level contains many features not devoted to parking spaces and circulation aisles. Distinctions are made between gross and net parking structure areas in order to more effectively forecast the number of parking spaces that can be provided in the net area available.

L Loading area percentage—a fraction of the buildable land area devoted to truck maneuvering, loading, and parking. This value is forecast from other design specification values entered as a fraction of the buildable land area. All fractions and percentages are expressed in their decimal equivalents.

Loading areas must be carefully considered when required, since they can consume large amounts of land and can produce unacceptable land use relationships at edges of contact when not properly separated for compatibility.

LDA Gross loading area—the total exterior surface area devoted to truck loading, maneuvering, and parking. This value is forecast in square feet from other design specification values entered.

These areas, when present, can consume large portions of a given land area and should be carefully considered when required, since they can significantly reduce the development capacity of a given land area.

LOT Total lot area—the land area devoted to a single dwelling unit and defined by recorded property lines. This value is forecast in square feet from other design specification values entered.

Minimum lot areas vary widely and are often regulated in local zoning district ordinances. The lot sizes forecast as feasible based on the values entered in a design specification should be checked against the zoning district land areas required per lot.

The minimum lot sizes forecast should also be checked against the buildable areas that result and the gross building areas that can be placed on these buildable areas. This can ensure that the lot sizes forecast are compatible with the building construction objectives of the project.

NDU Number of dwelling units—the number of dwelling units that can be placed on a given land area. This value is forecast from other design specification values entered. The numbers forecast are a function of the dwelling unit areas and mix specified in addition to other values.

NGS Number of garage spaces—the number of parking spaces that will be included within one or more totally enclosed garage buildings. This value is forecast from other design specification values entered.

Do not confuse this with a carport that may not require more land area allocation than a typical parking space.

NLA Net land area—the fraction of the gross land area that is present after rights-of-way and paved easements are subtracted. This value is forecast in acres and square feet from other design specification values entered.

The net lot area includes unbuildable areas. This statistic is provided for correlation of development capacity forecasts with this traditional residential density benchmark.

NLT Total number of lots—the number of lots planned, permitted, or provided within a given land area. This value is either forecast from other design specification values entered or specified as a design objective, depending on the spreadsheet in use.

The number of lots that can be created within a larger land area does not always indicate the highest return on investment that can be achieved. The value of each lot should be carefully evaluated in relation to the number created. This is particularly true when shared common open space is involved.

NPA Net parking lot area—that portion of a parking lot area devoted to parking spaces and circulation aisles. This value is forecast in square feet from other design specification values entered.

When a building is above a surface parking lot, parking lot area can be consumed by the need for building structure, fire stairs, and service cores. (Service cores can include elevators, lobbies, chases, etc.) These areas and others like them are subtracted from the gross parking lot area to produce the net parking area available for parking spaces and circulation aisles.

When roadway parking is involved, net parking area does not include public and private roadway pavement serving the adjacent parking spaces.

NPL Net parking structure area per level—that portion of a parking structure area devoted to parking spaces and circulation aisles. This value is forecast in square feet from other design specification values entered.

The gross area per parking structure level contains many features not devoted to parking spaces and circulation aisles. Distinctions are made between gross and net parking structure areas in order to more effectively forecast the number of parking spaces that can be provided within the net area available.

NPS Number of parking spaces—the number of parking spaces planned, provided, or required. This value is forecast from other design specification values entered.

The number of parking spaces planned, provided, or required is critical and intensely debated since its determination is not an exact science and the need varies by the time of day and year. The need also varies by the land use activity involved. In fact, a permitted land use activity in a given zone may replace another permitted land use activity in a given building and require more parking than is available. If the new owner stumbles into this situation, it is very difficult if not impossible to resolve. These situations do occur and only serve to make the point that while parking demand is debatable, it is possible to provide too few parking spaces to support present and future activities. This is not desirable, since the competition for inadequate parking spaces reduces the quality and desirability of the environment created.

Requests for parking variances are generally based on the nature of a specific activity proposed. Since a zoning district permits many activities within a given zone, a parking space reduction tailored to a specific use in a given building may not effectively serve the next occupant of that building. This is a very difficult issue that can affect the long-term desirability of any given area.

OSAC Open space in acres—project open space expressed in acres. This value is forecast in acres from other design specification values entered.

It is sometimes helpful to review the project open space provided in terms of acres rather than square feet. This is a calculated field provided for that purpose.

PDA Prorated dwelling unit area—a comprehensive dwelling unit area prorated according to the dwelling unit mix anticipated. The comprehensive area of a given dwelling unit type CDA multiplied by the fraction of the dwelling unit mix devoted to this type. This value is calculated from other design specification values entered.

The sum of prorated dwelling unit areas for a given dwelling unit mix produces the aggregate average dwelling unit area for a given building. The aggregate average dwelling unit area is the theoretical dwelling unit referred to in all zoning ordinances.

Pe Parking structure efficiency—the fraction of the gross parking structure area devoted exclusively to parking spaces and circulation aisles. Calculated by dividing the parking structure area devoted to parking spaces and circulation aisles by the gross parking structure area, this value is forecast from other design specification values entered and is expressed as a percentage of the buildable land area.

Parking structure efficiency can easily range from 70 to 90% of the gross structure area. These percentages should be verified with specialized parking consultants in relation to the internal circulation system being considered before commitments are made, since some structure parking systems are more efficient than others. Efficiency multiplied by the gross parking structure area produces the net parking structure area available for parking spaces, circulation aisles, and miscellaneous transition space within the circulation plan.

PLA Total parking lot area—the total parking lot area, including all circulation aisles and internal landscape islands when no building is located above the parking lot. This value is forecast from other design specification values entered.

The parking lot area forecast includes all internal landscaping proposed. The amount of landscaping planned is a function of the specification value assigned to the gross area planned per parking space *s*. Parking lot appearance can vary widely depending on the amount of internal and perimeter landscaping provided. Area allocations over 400 sq ft begin to have extra surface area available for internal landscaping.

Parking lot perimeter landscaping is not included in the gross parking area planned per parking space *s*. The project open space percentage *S* specified must be large enough to accommodate any perimeter parking lot mounding and landscaping desired.

RDS Number of roadway parking spaces—the number of parking spaces served by a public and/or private right-of-way. This value is forecast from other design specification values entered.

The distinction between a parking lot circulation aisle serving parking spaces and a private roadway serving parking spaces may be difficult to make at times. It is not as important to precisely distinguish these features as to ensure that they are not double-counted.

SFAC Gross square feet of building area per buildable acre—the total number of building square feet provided per buildable acre of land available. This value is forecast from other design specification values entered.

This is an accurate reflection of the intensity of development introduced since it relates the gross building area produced to the total buildable area on which it is located. A separate net open space statement OSAC, expressed in square feet per buildable acre, gives an indication of the balance provided when contrasted with this index. When unbuildable areas are present, a gross open space statement OSGA, expressed in square feet per gross acre, gives an indication of the environment created when contrasted with the building area per gross acre SFGA provided.

Sg Gross open space—project open space, yard areas, and unbuildable areas such as, but not limited to, ecologically fragile areas, unstable soil areas, ravines, ponds, marshes, and *existing improved or expansion areas that are to remain within the gross land area available.*

Gross project open space is expressed as a fraction of the gross land available. It is not calculated by the forecast models but can be found by totaling all unbuildable area *U,* net project open space area *S* and yard area YRD allocations in square feet. This square-foot area can be divided by the gross land area available in square feet to find the percentage allocated to gross project open space. (Do not attempt to add percentages since they are not all expressed as fractions of the gross land area available.)

In many cases, unbuildable areas and yard areas will not be involved, and the net project open space planned will be equal to the gross project open space planned. These four types of project level open space are identified, however, because their unique characteristics and functions serve different social and environmental purposes.

Su Support areas of site—the buildable land area minus all areas devoted to building cover and parking cover. Support areas can include, but are not limited to, project open space and miscellaneous pavement, driveways, and loading areas. This value is forecast as a percentage of the buildable land area BLA.

Areas of the site not devoted to building cover or parking cover are considered support areas. These areas can include, but are not limited to, project open space and miscellaneous pavement, roadways, driveways, and loading areas. Separating all supporting elements of the site plan makes it easier to focus on the optimum relationships between building cover and parking cover.

TLA Tabular lot area—a lot area calculated from data entered in a given spreadsheet. This value is forecast in square feet from other design specification values entered.

A number of lot area options may be entered in a design specification, but not all lot areas entered must be used in each evaluation of the development capacity of a given land area. The column identified as LOT is a calculated column and identifies all lot areas that may be under consideration. If the MIX column includes 0%, or 0.000 for a given lot size in the LOT column, the TLA column will not include this lot area in the current forecast calculations.

X Gross reduction in land area—the fraction of the gross land area that is either unbuildable or devoted to public/private rights of way and/or paved easements. This value is forecast as a fraction of the gross land area available from other design specification values entered. Subtracting this value in square feet from the gross land area available in square feet produces the buildable land available.

Single-Family Housing Forecasts

Single-family housing is grade-oriented. It is distinguished by private yards that are related to each dwelling unit and main entries that do not require travel through intervening interior spaces under separate or common ownership. In the case of single-family *detached* housing—*suburb houses*—individual lots for each dwelling unit are provided, yards are prominent, all dwelling units are separated, foundations are unique to each unit, shared parking rarely exists except for the public right-of-way, and there is a great emphasis on private garages and carports. Several common methods of arranging these suburb houses are shown in Figs. 4.1 through 4.3.

In the case of single-family homes such as twin-singles and townhouses—*urban houses*—dwelling units are *attached,* foundations are connected, and increasing densities decrease the yard areas allocated per dwelling unit, the private garage spaces available to each unit, and the extent of separate lots available.[1] Increasing densities also increase the extent of shared parking, which may include roadway parking.[2] This shared parking can be exclusive, or it can be combined with, or excluded in favor of, private garage opportunities. As urban house density increases, private yard

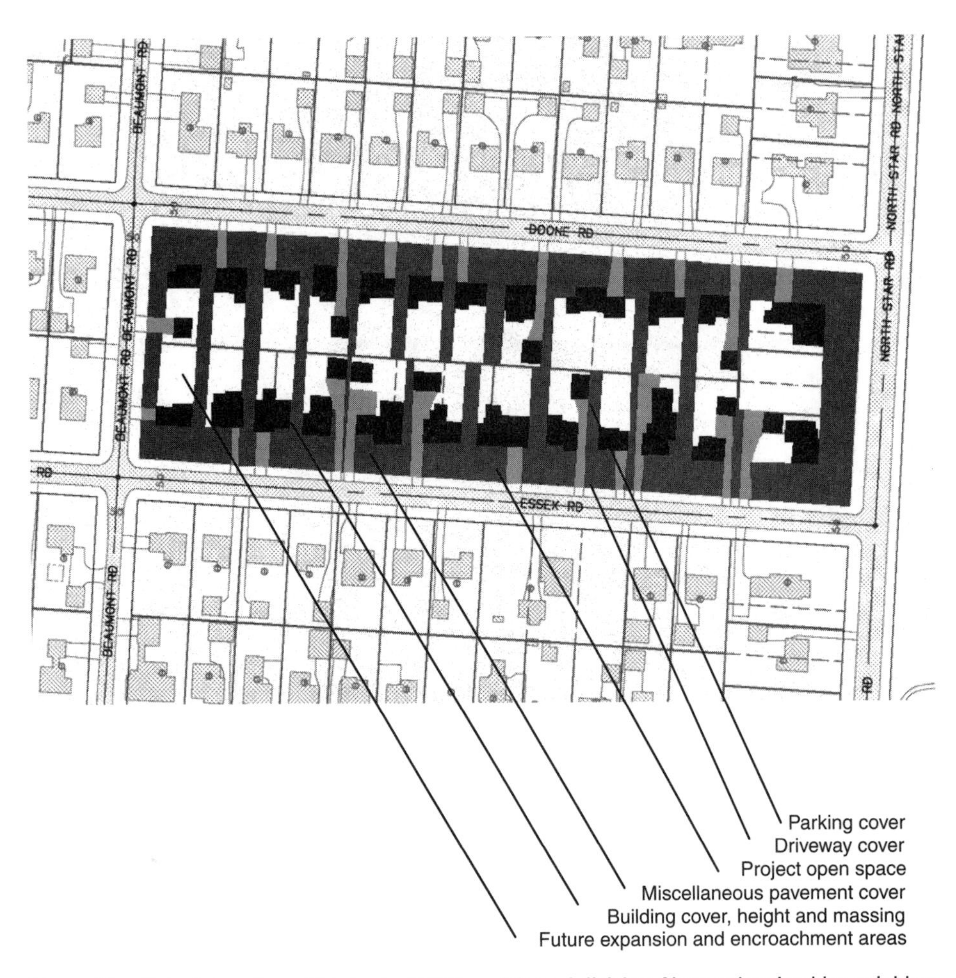

Figure 4.1 Suburb housing within typical 9000-sq-ft lot subdivision. Narrow lots in older neighborhoods often sacrifice rear yard areas for detached garages, fences, room additions, and other accessory structures that wall off neighbors, increase development cover and stormwater runoff, and produce side yard separation and rear yard isolation.

areas generally decrease and common open space often substitutes, as shown in Fig. 4.4.

The grade-oriented single-family characteristics of detached suburb houses and attached urban houses typically produce low housing densities that are forecast by five choices in two groups within the single-family series of development capacity forecasting models.[3] Three of these models are referred to as the *SF group* and two are called the

Figure 4.2 Suburb housing with common open space.

GT group. The SF group addresses suburb houses, and the GT group addresses urban houses. In some cases, you may choose to use them interchangeably.

The SF Group: Suburb House Development Forecast Models

The lot specification panel in Exhibit 4.1 is the first step in a brief tour of the SF group. It is a primary feature of the

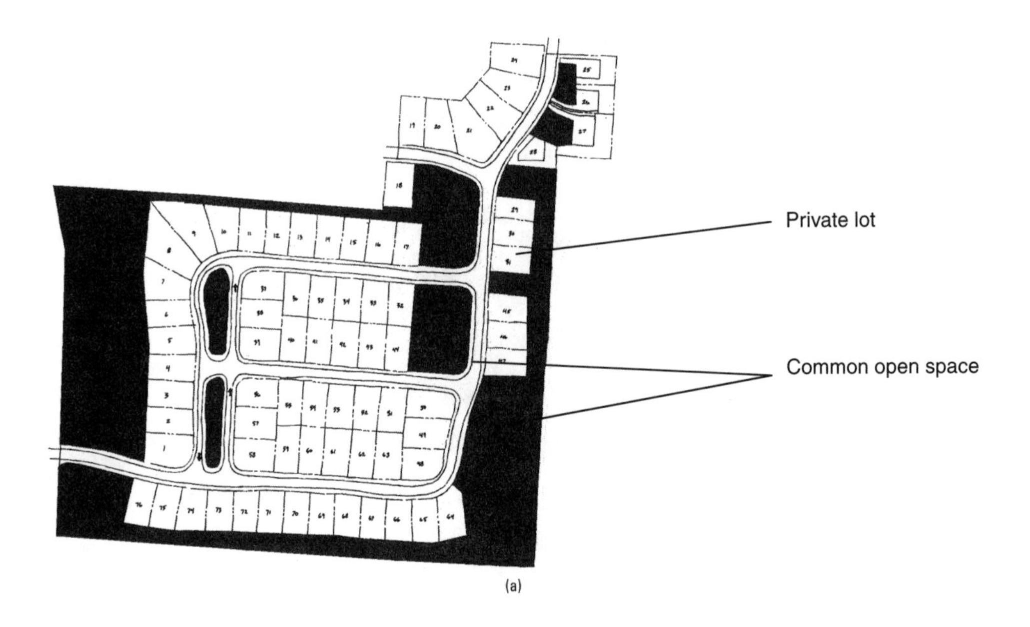

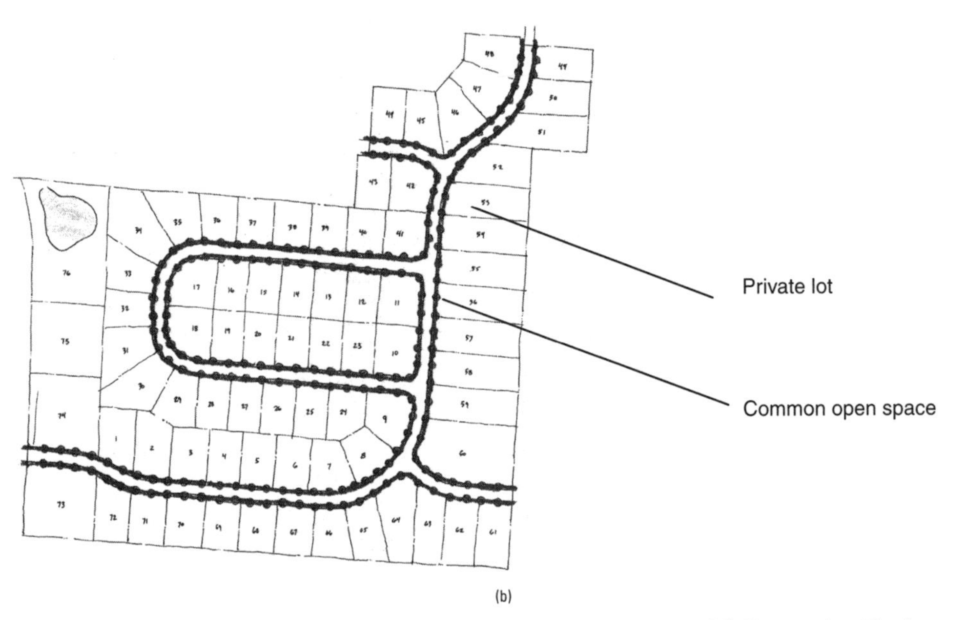

Figure 4.3 Suburb housing cluster plan and standard plan comparison. (*a*) Cluster plan: Total acres, 41.73; open space acres, 14.13 (34%); density, 1.82 DU/ac. (b) Standard plan: Total acres, 41.73; open space acres, 0; density, 1.82 DU/ac.

(Courtesy Jefferson Township, Ohio, and Lane Kendig, Inc.)

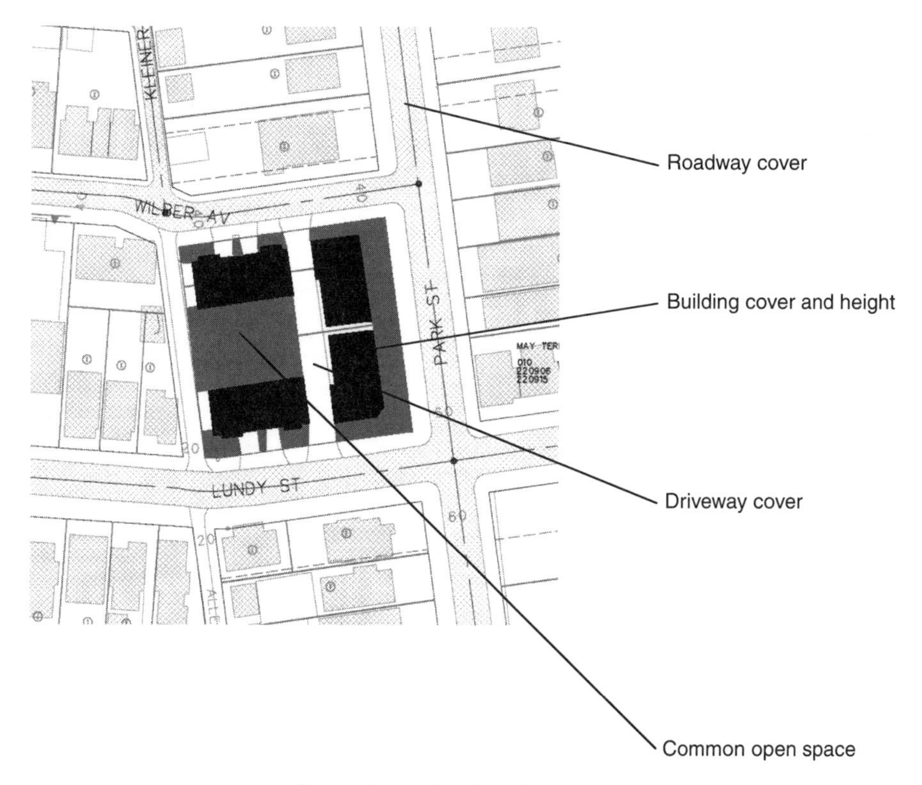

Figure 4.4 Urban housing cluster.

group, and allows the user to specify a lot area, or areas, for development capacity evaluation. Seven typical lot areas are shown in Exhibit 4.1 as a palette of possibilities, but any or all may be changed at will. The lot mix LTX column within the lot specification panel allows one or more lot areas to be chosen from the palette defined for analysis by entering the preferred mix percentages. For instance, if seven lot areas are defined but only one is to be evaluated, the mix percentage values entered in the LTX column would equal zero for all but the selected lot area, and its allocation value would equal 100%.

The current fashion is to plan neighborhoods around one typical lot size, with some developers making adjustments for corner lot parity within the neighborhood lot mix. The examples to follow focus on one lot size for simplicity and

Exhibit 4.1 Typical Detached Single-Family Forecast Format

Forecast Model RSFD

Development capacity forecast for ***SINGLE-FAMILY DETACHED HOUSING*** *with no shared or common parking lots serving each dwelling unit.*
Given: *Gross land area and net density objective.* ***To Find:*** *Maximum number of lots that can be created from the land area given based on the design specification values entered below.*

DESIGN SPECIFICATION

Enter values in boxed areas where text is bold and blue. Express all fractions as decimals.

Given:	**Gross land area**	GLA=	20.230	acres		
	Net density objective	d=	4.000	dwelling units per net acre		
Land Variables:	Est. right-of-way dedication as fraction of GLA	W=	0.000			
	Private roadway as fraction of GLA	R=	0.150			
	School land donation as fraction of GLA	Ds=	0.000			
	Park land donation as fraction of GLA	Dp=	0.000			
	Other land donation as fraction of GLA	Do=	0.000			
	Net land area	NLA=	17.196	acres	749,036	SF
	Unbuildable and/or future expansion areas as fraction of GLA	U=	0.100		4,356	SF
	Total gross land area reduction as fraction of GLA	X=	0.250		10,890	SF
	Buildable land area remaining	BLA=	15.173	acres	660,914	SF
Core:	**Core + open space available**	CS=	15.173	acres	660,914	SF
Lots:	Number of lots required by net density objective	NLT=	68.8	lots		

COMMON OPEN SPACE SPECIFICATION:

DEFINITION of COMMON OPEN SPACE: Open space provided for common benefit and massing separation in addition to that provided for private use and setbacks on individual lots. On a given land area, private lot sizes decrease as common open space increases when all other development factors are constant.
NOTE: Common open space (COS) values in bold & blue in the table below may be changed to examine other average lot area (ALT) implications. None of these values are used in spreadsheet calculations unless entered below. This table is provided for evaluation only. When COS = 0.0 all open space is provided on private lots (OSA) in the form of yards and a typical subdivision is implied. COS is expressed as a fraction of BLA.

COS:	**0.00**	**0.10**	**0.20**	**0.30**	**0.40**	**0.50**	**0.60**	**0.70**	**0.80**
Average lot area (ALT):	9,609	8,648	7,687	6,726	5,765	4,804	3,844	2,883	1,922

Common Open Space (COS) Specification:

Enter a common open space objective in the COS box below and adjust all bold & blue values in the table to express project design objectives. When zero (0) is entered in the COS box, a typical subdivision with no common open space is indicated. The average lot area ALT that can be created is calculated based on the COS value entered. Adjust the lot characteristics and lot mix in the LOT FORECAST table until the aggregate average lot area AAL is equal to or less than the ALT value produced here. The ALT value is duplicated next to the AAL value below for convenience. Read the number and combination of lots in the NLT box that can meet the design objective represented by the COS value entered.

COS = **0.00** | 9,609 | **= ALT**

LOT FORECAST

LOT SPECIFICATION	total footprint area BCG with garage	garage GAR footprint	misc. development area Msf as multiple of BCG	total development DCA cover area	private yards YRD as multiple of DCA	total lot **LOT** area	lot LTX mix	aggregate avg. **AAL** lot area	no. of lots **LBT** by type
Type 1 Interior Lot	**1,350**	**500**	**0.50**	2,025	**1.9630**	6,000	**0.000**		
Type 2 Interior Lot	**1,500**	**500**	**0.60**	2,400	**2.0000**	7,200	**0.000**		
Type 3 Interior Lot	**1,700**	**500**	**0.75**	2,975	**2.0252**	9,000	**1.000**	9,000	**68.8**
Type 4 Interior Lot	**2,000**	**500**	**0.85**	3,700	**2.2433**	12,000	**0.000**		
Type 5 Interior Lot	**2,450**	**750**	**0.75**	4,288	**2.4985**	15,000	**0.000**		
Type 6 Interior Lot	**3,300**	**750**	**0.65**	5,445	**3.0000**	21,780	**0.000**		
Type 7 Interior Lot	**5,600**	**750**	**0.55**	8,680	**4.0184**	43,560	**0.000**		
Total No. of Lots	*NOTE: Zero lot line homes and townhouses are implied when YRD is less than 1.0 and approaches zero (0.0) in the YRD column above.*								68.8
Mix Total (do not exceed 1.0)							1.00	**AAL**	ALT
Aggregate Avg. Lot Size Based on Mix:								**9,000**	9,609
							density =	4.84	

BUILDING AREA FORECAST per LOT

NOTE: When portions of this table are blank, the lot type referred to in the blank portion has been excluded from the mix under consideration above.

	basement as fraction of BSM habitable footprint	basement BSA area	crawl space CRW area	NET POTENTIAL BUILDING AREA INCLUDING GARAGE but EXCLUDING BASEMENT and POTENTIAL EXPANSION — number of building floors: 1	1.5	2	2.5	3	3.5
Type 1 Interior Lot	**1.000**								
Type 2 Interior Lot	**1.000**								
Type 3 Interior Lot	**1.000**	1,200	0	1,700	2,550	3,400	4,250	5,100	5,950
Type 4 Interior Lot	**1.000**								
Type 5 Interior Lot	**1.000**								
Type 6 Interior Lot	**1.000**								
Type 7 Interior Lot	**1.000**								

WARNING: These are preliminary forecasts that must not be used to make final decisions.
1) These forecasts are not a substitute for the "due diligence" research that must be conducted to support the final definition of "unbuildable areas" above and the final decision to purchase land. This research includes, but is not limited to, verification of adequate subsurface soil, zoning, environmental clearance, access, title, utilities and water pressure, clearance from deed restriction, easement and right-of-way encumbrances, clearance from existing above and below ground facility conflicts, etc.
2) The most promising forecast(s) made on the basis of data entered in the design specification from "due diligence" research must be verified at the drawing board before funds are committed and land purchase decisions are made. Actual land shape ratios, dimensions and irregularities encountered may require adjustments to the general forecasts above.
3) The software licensee shall take responsibility for the design specification values entered and any advice given that is based on the forecast produced.

do not compensate for corner lots, although an example of corner lot compensation is included in this chapter. While conformity may be the current trend, the panel also makes it possible to investigate a broader mix of lot size alternatives, or to establish a palette of options, by including seven lines within the table.

The goal of the SF group is to forecast the capacity of land to accommodate single-family detached housing. The objective of each model within the group is briefly described as follows:

- *Model RSFD:* To forecast the number and size of lots that can be created when a gross land area and net density objective is given
- *Model RSFL:* To forecast the maximum number of lots of a given size that can be placed on a given land area
- *Model RSFN:* To forecast the buildable land area needed to accommodate a given number of lots of a given size

Model RSFD: Single-Family Detached Homes

Given: Gross land area and net density objective

To find: Lot sizes, lot mix, number of lots, and home size potential that can meet the net density objective given

Model RSFD model (Exhibit 4.1) is used to forecast the number and size of single-family lots that can be placed on a given land area when a net density objective is given. Land data is entered in the design specification panel of the model where requested and lot data[4] is entered in the lot specification panel. Land information includes the gross land area involved, land to be donated, and unbuildable areas present. The buildable land area available for development is calculated from the data entered. This buildable area is further reduced by the common open space (COS) objective[5] entered at the top of the lot specification table.

The lot areas to be subdivided from the land area given in the design specification panel are assembled by entering values for the following parameters:

1. The building footprint to be accommodated, in the BCG column
2. The garage footprint to be accommodated, in the GAR column
3. The miscellaneous pavement and expansion area to be accommodated (as a multiple of the footprint), in the Msf column
4. The yard area to be accommodated (as a multiple of the development cover area [DCA] calculated), in the YRD column

The sum of these values produces a total lot area calculation in the LOT column. More than one lot area can be defined, and seven rows labeled Type 1 to 7 are provided for this information. The mix of lot types to be used in the subdivision is specified in the LTX column. The number of lots that can be subdivided is then forecast in the LBT column. The aggregate average lot area AAL produced by the lot areas and mix defined may exceed the target value ALT produced by the common open space value chosen. In this event, a warning will appear on the screen and the lot specification should be adjusted until the AAL value is less than or equal to the ALT target value listed next to it.

A typical subdivision provides little or no common open space that is shared among landowners. It allocates all buildable area to private lots and public services such as roads and easements.[6] It is possible, however, to build the same size and number of private homes on the same land area with common open space among them when private lot sizes are reduced. This is not a new idea, but it has not been a popular trend. A common open space table is provided in Exhibit 4.1 to assist in evaluating this option.[7] Exhibit 4.1

forecasts that 68.8 lots can be subdivided when common open space COS equals zero. When COS equals 40%, this table shows that the area per lot for 68.8 lots declines from 9609 sq ft (approximately 80 × 120 ft) to 5765 sq ft (approximately 50 × 120 ft), and the difference is assembled as common open space.

The lot series defined in the lot specification panel of Exhibit 4.1 is a typical subdivision series, and a COS value of zero has been entered in the data-entry box at the top of the panel. The lot mix LTX column in this table shows that only Type 3 lots are being considered. The building area capacity of the Type 3 lot specified is automatically presented in the building forecast panel below. This panel predicts that the 9000-sq-ft lot can support a range of gross building areas extending from a 1700-sq-ft ranch house to a 5100-sq-ft 3-story home.[8] Exhibit 4.2 shows that the same range of home areas can be accommodated on smaller lots when 40% COS is shared within the subdivision. When this value is entered in the COS box, however, the yard area multiple YRD per lot must be reduced from 2.02 times DCA to 0.937 times DCA to meet this objective. This yard area reduction implies less private, personal open space per dwelling unit, a different neighborhood environment, and a different floor plan arrangement to accommodate the smaller yard areas. This environment may not be for everyone, but the capacity to evaluate this option is included in the model format established.

The previous example should not leave the impression that a given lot area can accommodate unlimited building floor plan areas and volumes, however. The building area forecast panel is included in this series to predict these limits based on the number of building floors contemplated. If the gross building area forecast does not meet the present needs and future expansion capacity desired, then adjustments must be made to the specification values entered in order to increase the lot and building areas forecast.

Exhibit 4.2 Smaller Private Yard Areas and increased Common Open Space

Forecast Model RSFD

Development capacity forecast for ***SINGLE-FAMILY DETACHED HOUSING*** *with no shared or common parking lots serving each dwelling unit.*
Given: *Gross land area and net density objective.* ***To Find:*** *Maximum number of lots that can be created from the land area given based on the design specification values entered below.*

DESIGN SPECIFICATION

Enter values in boxed areas where text is bold and blue. Express all fractions as decimals.

Given:	**Gross land area**	GLA=	**20.230**	acres	
	Net density objective	d=	**4.000**	dwelling units per net acre	
Land Variables:	Est. right-of-way dedication as fraction of GLA	W=	**0.000**		
	Private roadway as fraction of GLA	R=	**0.150**		
	School land donation as fraction of GLA	Ds=	**0.000**		
	Park land donation as fraction of GLA	Dp=	**0.000**		
	Other land donation as fraction of GLA	Do=	**0.000**		
	Net land area	NLA=	17.196	acres	749,036 SF
	Unbuildable and/or future expansion areas as fraction of GLA	U=	**0.100**		4,356 SF
	Total gross land area reduction as fraction of GLA	X=	0.250		10,890 SF
	Buildable land area remaining	BLA=	**15.173**	acres	660,914 SF
Core:	**Core + open space available**	CS=	**15.173**	acres	660,914 SF
Lots:	Number of lots required by net density objective	NLT=	**68.8**	lots	

COMMON OPEN SPACE SPECIFICATION:

DEFINITION of COMMON OPEN SPACE: Open space provided for common benefit and massing separation in addition to that provided for private use and setbacks on individual lots. On a given land area, private lot sizes decrease as common open space increases when all other development factors are constant.
NOTE: Common open space (COS) values in bold & blue in the table below may be changed to examine other average lot area (ALT) implications. None of these values are used in spreadsheet calculations unless entered below. This table is provided for evaluation only. When COS = 0.0 all open space is provided on private lots (OSA) in the form of yards and a typical subdivision is implied. COS is expressed as a fraction of BLA.

COS:	**0.00**	**0.10**	**0.20**	**0.30**	**0.40**	**0.50**	**0.60**	**0.70**	**0.80**
Average lot area (ALT):	9,609	8,648	7,687	6,726	5,765	4,804	3,844	2,883	1,922

Common Open Space (COS) Specification:
Enter a common open space objective in the COS box below and adjust all bold & blue values in the table to express project design objectives. When zero (0) is entered in the COS box, a typical subdivision with no common open space is indicated. The average lot area (ALT) that can be created is calculated based on the COS value entered. Adjust the lot characteristics and lot mix in the LOT FORECAST table until the aggregate average lot area (AAL) is equal to or less than the ALT value produced here. The ALT value is duplicated next to the AAL value below for convenience. Read the number and combination of lots in the NLT box that can meet the design objective represented by the COS value entered.

COS = **0.40** | 5,765 = ALT

LOT FORECAST

LOT SPECIFICATION	total footprint area BCG with garage	garage GAR footprint	misc. development area Msf as multiple of BCG	total development DCA cover area	private yards YRD as multiple of DCA	total lot LOT area	lot LTX mix	aggregate avg. AAL lot area	no. of lots LBT by type
Type 1 Interior Lot	**1,350**	**500**	**0.50**	2,025	**1.963**	6,000	**0.000**		
Type 2 Interior Lot	**1,500**	**500**	**0.60**	2,400	**2.000**	7,200	**0.000**		
Type 3 Interior Lot	**1,700**	**500**	**0.75**	2,975	**0.937**	5,763	**1.000**	5,763	**68.8**
Type 4 Interior Lot	**2,000**	**500**	**0.85**	3,700	**2.243**	12,000	**0.000**		
Type 5 Interior Lot	**2,450**	**750**	**0.75**	4,288	**2.499**	15,000	**0.000**		
Type 6 Interior Lot	**3,300**	**750**	**0.65**	5,445	**3.000**	21,780	**0.000**		
Type 7 Interior Lot	**5,600**	**750**	**0.55**	8,680	**4.018**	43,560	**0.000**		
Total No. of Lots									68.8
Mix Total (do not exceed 1.0)							1.00	**AAL**	ALT
Aggregate Avg. Lot Size Based on Mix:								**5,763**	5,765
							density =	7.56	

NOTE: Zero lot line homes and townhouses are implied when YRD is less than 1.0 and approaches zero (0.0) in the YRD column above.

BUILDING AREA FORECAST per LOT

NOTE: When portions of this table are blank, the lot type referred to in the blank portion has been excluded from the mix under consideration above.

NET POTENTIAL BUILDING AREA INCLUDING GARAGE but EXCLUDING BASEMENT and POTENTIAL EXPANSION — number of building floors

	basement as fraction of BSM habitable footprint	basement BSA area	crawl space CRW area	1	1.5	2	2.5	3	3.5
Type 1 Interior Lot	**1.000**								
Type 2 Interior Lot	**1.000**								
Type 3 Interior Lot	**1.000**	1,200	0	1,700	2,550	3,400	4,250	5,100	5,950
Type 4 Interior Lot	**1.000**								
Type 5 Interior Lot	**1.000**								
Type 6 Interior Lot	**1.000**								
Type 7 Interior Lot	**1.000**								

WARNING: These are preliminary forecasts that must not be used to make final decisions.
1) These forecasts are not a substitute for the "due diligence" research that must be conducted to support the final definition of "unbuildable areas" above and the final decision to purchase land. This research includes, but is not limited to, verification of adequate subsurface soil, zoning, environmental clearance, access, title, utilities and water pressure, clearance from deed restriction, easement and right-of-way encumbrances, clearance from existing above and below ground facility conflicts, etc.
2) The most promising forecast(s) made on the basis of data entered in the design specification from "due diligence" research must be verified at the drawing board before funds are committed and land purchase decisions are made. Actual land shape ratios, dimensions and irregularities encountered may require adjustments to the general forecasts above.
3) The software licensee shall take responsibility for the design specification values entered and any advice given that is based on the forecast produced.

Model RSFL: Single-Family Detached Homes

Given: Gross land area

To find: Lot sizes, lot mix, number of lots, net density, and home size potential

The RSFL model (Exhibit 4.3) is similar to the RSFD model (Exhibit 4.1), but it does not specify a density objective. The intent of this model is to forecast the number of lots that can be subdivided from a larger land area when the lot areas are defined in the lot specification panel. Since the lot areas are held constant, the gross building area potential of each lot also remains constant while the number of potential lots declines with increasing amounts of common open space. This represents a classic set of relationships that is intuitively understood by most and has simply been mathematically expressed to expedite future evaluation.

Model RSFN: Single-Family Detached Homes

Given: Number of lots, lot sizes, and lot mix

To find: Total buildable area required and net density implied

A developer generally knows the land area involved, or the land area and density desired, when formulating a subdivision plan. A city planner, however, may want to examine the land area needed to accommodate a given number of dwelling units using a variety of development alternatives. The RSFN model (Exhibit 4.4) forecasts the buildable land area needed to accommodate a specified number of lots (NLT = 50) when each lot is equal to a defined area in the lot specification panel (12,000 sq ft; a 100- × 120-ft lot).

The buildable land area forecast panel predicts that 13.77 buildable acres are needed for the combination specified when no common open space is provided, and that 22.96 acres are needed when 40% common open space is provided. The gross building area per lot panel in this model

Exhibit 4.3 Development Capacity Based on Lot Sizes Without Density Objective

Forecast Model RSFL

Development capacity forecast for ***SINGLE-FAMILY DETACHED HOUSING*** *with no shared or common parking lots serving each dwelling unit.*

Given: *Gross land area.* ***To Find:*** *(1) Maximum number of lots that can be created from the land area given based on the design specification values entered below. (2) Maximum gross dwelling unit area that can be built on each lot created.*

DESIGN SPECIFICATION

Enter values in boxed areas where text is bold and blue. Express all fractions as decimals.

Given:	Gross land area	GLA=	**20.000**	acres		
Land Variables:	Est. right-of-way dedication as fraction of GLA	W=	**0.200**		8,712	SF
	Private roadway as fraction of GLA	R=	**0.000**			
	School land donation as fraction of GLA	Ds=	**0.000**			
	Park land donation as fraction of GLA	Dp=	**0.000**			
	Other land donation as fraction of GLA	Do=	**0.000**			
	Net land area	NLA=	16.000	acres	696,960	SF
	Unbuildable and/or future expansion areas as fraction of GLA	U=	**0.000**		0	SF
	Total gross land area reduction as fraction of GLA	X=	0.200		8,712	SF
	Buildable land area remaining	**BLA=**	**16.000**	acres	696,960	SF
Core:	**Core + open space available**	**CS=**	**16.000**	acres	696,960	SF

LOT SPECIFICATION

	total bldg footprint BCG with garage	garage GAR footprint	misc. development area Msf as multiple of BCG	total development DCA cover area	private yards YRD as multiple of DCA	total lot LOT area	lot LTX mix	aggregate avg. AAL lot areas in mix	total lot TLA areas in mix
Type 1 Interior Lot	**1,350**	**500**	**0.50**	2,025	**1.9630**	6,000	**0.000**		
Type 2 Interior Lot	**1,500**	**500**	**0.60**	2,400	**2.0000**	7,200	**0.000**		
Type 3 Interior Lot	**1,700**	**500**	**0.75**	2,975	**2.0252**	9,000	**1.000**	9,000	9,000
Type 4 Interior Lot	**2,000**	**500**	**0.85**	3,700	**2.2433**	12,000	**0.000**		
Type 5 Interior Lot	**2,450**	**750**	**0.75**	4,288	**2.4985**	15,000	**0.000**		
Type 6 Interior Lot	**3,300**	**750**	**0.65**	5,445	**3.0000**	21,780	**0.000**		
Type 7 Interior Lot	**5,600**	**750**	**0.55**	8,680	**4.0184**	43,560	**0.000**		

NOTE: Zero lot line homes and townhouses are implied when YRD is less than 1.0 and approaches zero (0.0) in the YRD column above.

Mix Total (do not exceed 1.0)	1.00	AAL	
Aggregate Avg. Lot Size Based on Mix:		**9,000**	

LOT CAPACITY FORECAST

(based on common open space value (COS) entered below)

DEFINITION of COMMON OPEN SPACE (COS): Open space provided for common benefit and massing separation in addition to that provided for private use and setbacks on individual lots. On a given land area, private lot sizes decrease as common open space increases when all other development factors are constant.
NOTE: Common open space values in bold & blue in the table below may be changed to examine other implications. When COS = 0.0 all open space is provided on private lots (OSA) in the form of yards and a typical subdivision is implied.

COS:	**0.00**	**0.10**	**0.20**	**0.30**	**0.40**	**0.50**	**0.60**	**0.70**	**0.80**
No. of lots:	77.4	69.7	62.0	54.2	46.5	38.7	31.0	23.2	15.5
Net density (dNA):	4.84	4.36	3.87	3.39	2.90	2.42	1.94	1.45	0.97
BLA density (dBA):	4.84	4.36	3.87	3.39	2.90	2.42	1.94	1.45	0.97

GROSS BUILDING AREA per LOT FORECAST

	basement as fraction of BSM habitable footprint	basement BSA area	crawl space CRW area	NET POTENTIAL BUILDING AREA INCLUDING GARAGE but EXCLUDING BASEMENT and POTENTIAL EXPANSION — number of building floors: 1	1.5	2	2.5	3	3.5
Type 1 Interior Lot	**1.000**								
Type 2 Interior Lot	**1.000**								
Type 3 Interior Lot	**1.000**	1,200	0	1,700	2,550	3,400	4,250	5,100	5,950
Type 4 Interior Lot	**1.000**								
Type 5 Interior Lot	**1.000**								
Type 6 Interior Lot	**1.000**								
Type 7 Interior Lot	**1.000**								

WARNING: These are preliminary forecasts that must not be used to make final decisions.
1) These forecasts are not a substitute for the "due diligence" research that must be conducted to support the final definition of "unbuildable areas" above and the final decision to purchase land. This research includes, but is not limited to, verification of adequate subsurface soil, zoning, environmental clearance, access, title, utilities and water pressure, clearance from deed restriction, easement and right-of-way encumbrances, clearance from existing above and below ground facility conflicts, etc.
2) The most promising forecast(s) made on the basis of data entered in the design specification from "due diligence" research must be verified at the drawing board before funds are committed and land purchase decisions are made. Actual land shape ratios, dimensions and irregularities encountered may require adjustments to the general forecasts above.
3) The software licensee shall take responsibility for the design specification values entered and any advice given that is based on the forecast produced.

Exhibit 4.4 Land Needed to Accommodate a Given Number and Size of Lots

Forecast Model RSFN

Development capacity forecast for ***SINGLE-FAMILY DETACHED HOUSING*** *with no shared or common parking lots serving each dwelling unit.*

Given: *Number of lots desired.* ***To Find:*** *(1) Minimum buildable acres required to accommodate the number of lots desired based on the design specification values entered below. (2) Maximum gross dwelling unit area that can be built on each lot created.*

DESIGN SPECIFICATION

Enter values in boxed areas where text is bold and blue. Express all fractions as decimals.

Number of lots desired. **50.00** = NLT

LOT SPECIFICATION

	total bldg footprint BCG with garage	garage GAR footprint	misc. development area Msf as multiple of BCG	total development DCA cover area	private yards YRD as multiple of DCA	total lot LOT area	lot LTX mix	aggregate avg. AAL lot areas in mix
Type 1 Interior Lot	**1,350**	**500**	**0.50**	2,025	**1.9630**	6,000	**0.000**	
Type 2 Interior Lot	**1,500**	**500**	**0.60**	2,400	**2.0000**	7,200	**0.000**	
Type 3 Interior Lot	**1,700**	**500**	**0.75**	2,975	**2.0252**	9,000	**0.000**	
Type 4 Interior Lot	**2,000**	**500**	**0.85**	3,700	**2.2433**	12,000	**1.000**	12,000
Type 5 Interior Lot	**2,450**	**750**	**0.75**	4,288	**2.4985**	15,000	**0.000**	
Type 6 Interior Lot	**3,300**	**750**	**0.65**	5,445	**3.0000**	21,780	**0.000**	
Type 7 Interior Lot	**5,600**	**750**	**0.55**	8,680	**4.0184**	43,560	**0.000**	

NOTE: Zero lot line homes and townhouses are implied when (YRD) is less than 1.0 and approaches zero (0.0) in the YRD column above.

Mix Total (do not exceed 1.0) 1.00

Aggregate Avg. Lot Size Based on Mix: **12,000**

BUILDABLE LAND AREA FORECAST

DEFINITION of COMMON OPEN SPACE (COS): Open space provided for common benefit and massing separation in addition to that provided for private use and setbacks on individual lots. On a given land area, private lot sizes decrease as common open space increases when all other development factors are constant.
NOTE: Common open space values in bold & blue in the table below may be changed to examine other implications. When COS = 0.0 all open space is provided on private lots (OSA) in the form of yards and a typical subdivision is implied.

CORE= 13.77 acres — Note: (BLA) below means the buildable land area expressed in acres.

(COS) = Common Open Space as % of BLA

COS:	**0.00**	**0.10**	**0.20**	**0.30**	**0.40**	**0.50**	**0.60**	**0.70**	**0.80**
BLA FORECAST:	13.77	15.30	17.22	19.68	22.96	27.55	34.44	45.91	68.87
BLA density (dBA):	3.63	3.27	2.90	2.54	2.18	1.81	1.45	1.09	0.73

GROSS LAND AREA FORECAST

The buildable land area (BLA) in acres forecast above does not include land area needed for public right-of-way (W), private roadways (R), school land donations (Ds), park land donations (Dp), other land donations (Do) and unbuildable areas (U). The potential gross lot area (GLA) may be forecast by entering values for these design variables in the table below:

Est. right-of-way dedication as fraction of GLA	W=	**0.200**	
Private roadway as fraction of GLA	R=	**0.000**	
School land donation as fraction of GLA	Ds=	**0.100**	
Park land donation as fraction of GLA	Dp=	**0.100**	
Other donation, unbuildable and/or expansion as fraction of GLA	Do=	**0.000**	
Total buildable land area reduction as fraction of GLA	X=	0.400	
Enter buildable area objective from above	BLA=	**22.960**	acres
Gross land area forecast	**GLA=**	**38.267**	acres

NOTE: Express all fractions as decimals

Identify COS design objective in table above and enter associated buildable acres (BLA) here.

GROSS BUILDING AREA per LOT FORECAST

NET POTENTIAL BUILDING AREA INCLUDING GARAGE but EXCLUDING BASEMENT and POTENTIAL EXPANSION (number of building floors)

	basement as fraction of BSM habitable footprint	basement BSA area	crawl space CRW area	1	1.5	2	2.5	3	3.5
Type 1 Interior Lot	**1.000**								
Type 2 Interior Lot	**1.000**								
Type 3 Interior Lot	**1.000**								
Type 4 Interior Lot	**1.000**	1,500	0	2,000	3,000	4,000	5,000	6,000	7,000
Type 5 Interior Lot	**1.000**								
Type 6 Interior Lot	**1.000**								
Type 7 Interior Lot	**1.000**								

WARNING: These are preliminary forecasts that must not be used to make final decisions.
1) These forecasts are not a substitute for the "due diligence" research that must be conducted to support the final definition of "unbuildable areas" above and the final decision to purchase land. This research includes, but is not limited to, verification of adequate subsurface soil, zoning, environmental clearance, access, title, utilities and water pressure, clearance from deed restriction, easement and right-of-way encumbrances, clearance from existing above and below ground facility conflicts, etc.
2) The most promising forecast(s) made on the basis of data entered in the design specification from "due diligence" research must be verified at the drawing board before funds are committed and land purchase decisions are made. Actual land shape ratios, dimensions and irregularities encountered may require adjustments to the general forecasts above.
3) The software licensee shall take responsibility for the design specification values entered and any advice given that is based on the forecast produced.

forecasts that each lot can accommodate a 2000-sq-ft ranch house, or a 6000-sq-ft 3-story house, depending on the building height involved.

The gross land area forecast panel is unique in this single-family series and is extremely theoretical since the user must enter a number of values that are unknown when the site location is not given. However, these guesses can be used to produce a rough estimate of real estate search parameters, and a city planner can use them to assess the implications of alternate policies and strategies.

Corner Lot Compensation for Model RSFL

Given: Gross land area

To find: Lot sizes, lot mix, number of lots, net density, and home size potential

If a corner lot in a subdivision is equal to the area of an interior lot, the corner lot will have less room for building construction and expansion if it is required to have two front yards. In such cases, these front yards can make the lot look bigger from the street, but the buildable area[9] within the setback lines is less than its interior lot neighbor. This is more of a problem for older homes on older lots that attempt to expand to meet current floor plan expectations, but all lots age and all must preserve their ability to adapt. Corner lots can be unfairly penalized in this respect if the subdivision plan does not increase these lot areas to compensate when such front yard requirements exist.

Exhibit 4.5 illustrates a simple method of compensating for such requirements in the lot specification panel. A Type 1 interior lot has been defined in the first row, and it has been estimated in the LTX column that 90% of the subdivision lots to be created will be represented by this lot type. The yard area YRD for this lot is specified as equal to twice the development cover DCA planned. A Type 1 corner lot is defined below it with a 10% subdivision allocation in the

Exhibit 4.5 Compensating for Corner Lot Yard Areas

Forecast Model RSFL

Development capacity forecast for ***SINGLE-FAMILY DETACHED HOUSING*** *with no shared or common parking lots serving each dwelling unit.*

Given: *Gross land area.* ***To Find:*** *(1) Maximum number of lots that can be created from the land area given based on the design specification values entered below. (2) Maximum gross dwelling unit area that can be built on each lot created.*

DESIGN SPECIFICATION

Enter values in boxed areas where text is bold and blue. Express all fractions as decimals.

Given:	Gross land area	GLA=	**20.230**	acres	
Land Variables:	Est. right-of-way dedication as fraction of GLA	W=	**0.000**		
	Private roadway as fraction of GLA	R=	**0.150**		
	School land donation as fraction of GLA	Ds=	**0.000**		
	Park land donation as fraction of GLA	Dp=	**0.000**		
	Other land donation as fraction of GLA	Do=	**0.000**		
	Net land area	NLA=	17.196	acres	749,036 SF
	Unbuildable and/or future expansion areas as fraction of GLA	U=	**0.100**		4,356 SF
	Total gross land area reduction as fraction of GLA	X=	0.250		10,890 SF
	Buildable land area remaining	**BLA=**	**15.173**	acres	660,914 SF
Core:	**Core + open space available**	**CS=**	**15.173**	acres	660,914 SF

LOT SPECIFICATION

	total bldg footprint BCG with garage	garage GAR footprint	misc. development area Msf as multiple of BCG	total development DCA cover area	private yards YRD as multiple of DCA	total lot LOT area	lot LTX mix	aggregate avg. AAL lot areas in mix	total lot TLA areas in mix
Type 1 Interior Lot	**1,350**	**500**	**0.50**	2,025	**2.00**	6,075	**0.900**	5,468	6,075
Type 1 CORNER LOT	**1,350**	**500**	**0.50**	2,025	**2.50**	7,088	**0.100**	709	7,088
Type 3 Interior Lot	**1,700**	**500**	**0.75**	2,975	**2.00**	8,925	**0.000**		
Type 4 Interior Lot	**2,000**	**500**	**0.85**	3,700	**2.25**	12,025	**0.000**		
Type 5 Interior Lot	**2,450**	**750**	**0.75**	4,288	**2.50**	15,006	**0.000**		
Type 6 Interior Lot	**3,300**	**750**	**0.65**	5,445	**3.00**	21,780	**0.000**		
Type 7 Interior Lot	**5,600**	**750**	**0.55**	8,680	**4.00**	43,400	**0.000**		

NOTE: Zero lot line homes and townhouses are implied when YRD is less than 1.0 and approaches zero (0.0) in the YRD column above.

Mix Total (do not exceed 1.0) 1.00

Aggregate Avg. Lot Size Based on Mix: **6,176**

LOT CAPACITY FORECAST

(based on common open space value (COS) entered below)

DEFINITION of COMMON OPEN SPACE (COS): Open space provided for common benefit and massing separation in addition to that provided for private use and setbacks on individual lots. On a given land area, private lot sizes decrease as common open space increases when all other development factors are constant.
NOTE: Common open space values in bold & blue in the table below may be changed to examine other implications. When COS = 0.0 all open space is provided on private lots (OSA) in the form of yards and a typical subdivision is implied.

COS:	**0.00**	**0.10**	**0.20**	**0.30**	**0.40**	**0.50**	**0.60**	**0.70**	**0.80**
No. of lots:	107.0	96.3	85.6	74.9	64.2	53.5	42.8	32.1	21.4
Net density (dNA):	6.22	5.60	4.98	4.36	3.73	3.11	2.49	1.87	1.24
BLA density (dBA):	7.05	6.35	5.64	4.94	4.23	3.53	2.82	2.12	1.41

GROSS BUILDING AREA per LOT FORECAST

	basement as fraction of BSM habitable footprint	basement BSA area	crawl space CRW area	NET POTENTIAL BUILDING AREA INCLUDING GARAGE but EXCLUDING BASEMENT and POTENTIAL EXPANSION — number of building floors: 1	1.5	2	2.5	3	3.5
Type 1 Interior Lot	**1.000**	850	0	1,350	2,025	2,700	3,375	4,050	4,725
Type 1 CORNER LOT	**1.000**	850	0	1,350	2,025	2,700	3,375	4,050	4,725
Type 3 Interior Lot	**1.000**								
Type 4 Interior Lot	**1.000**								
Type 5 Interior Lot	**1.000**								
Type 6 Interior Lot	**1.000**								
Type 7 Interior Lot	**1.000**								

WARNING: These are preliminary forecasts that must not be used to make final decisions.
1) These forecasts are not a substitute for the "due diligence" research that must be conducted to support the final definition of "unbuildable areas" above and the final decision to purchase land. This research includes, but is not limited to, verification of adequate subsurface soil, zoning, environmental clearance, access, title, utilities and water pressure, clearance from deed restriction, easement and right-of-way encumbrances, clearance from existing above and below ground facility conflicts, etc.
2) The most promising forecast(s) made on the basis of data entered in the design specification from "due diligence" research must be verified at the drawing board before funds are committed and land purchase decisions are made. Actual land shape ratios, dimensions and irregularities encountered may require adjustments to the general forecasts above.
3) The software licensee shall take responsibility for the design specification values entered and any advice given that is based on the forecast produced.

LTX column. The yard area for this lot is 2.5 times the development cover planned. This increase compensates for the expanded front yard requirements of the corner lot. All other buildable area factors remain constant for these two lots. This produces an aggregate average lot size of 6176 sq ft per lot and a total forecast of 107 lots when no common open space COS is provided.

Corner lot compensation reduces the number of lots that can be subdivided. Exhibit 4.6 removes the corner lot compensation provided in Exhibit 4.5 to show that 108.8 lots could be created when all lots are equal in area and no corner lot compensation is provided. This increase in lots however, is often nominal and must be weighed against the lot sizes involved and the impact on future corner lot building expansion and remodeling potential.

The GT Group: Urban House Development Forecast Models

Single-family *attached* homes include twin-singles, three-families, four-families, rowhouses, townhouses, garden apartments, etc. and all are referred to as *urban houses* for brevity. The transition from single-family detached home to single-family attached home begins with a configuration variously called a *twin-single,* a *duplex,* or a *double.* It proceeds through three-family and four-family attached homes with various regional nicknames to the typical townhouse configuration that generally appears when four or more 2-story single-family homes are attached with common wall separations.[10] All of these dwelling units share the same grade-oriented single-family characteristics, however. The appearance may vary. The parking system may vary. The ownership technique may vary, and the private open space provided per dwelling unit may vary, but all dwelling units in this category are entered from grade without travel through intervening interior spaces under separate or common ownership.

Exhibit 4.6 Removing Corner Lot Compensation

Forecast Model RSFL

Development capacity forecast for ***SINGLE-FAMILY DETACHED HOUSING*** *with no shared or common parking lots serving each dwelling unit.*

Given: *Gross land area.* ***To Find:*** *(1) Maximum number of lots that can be created from the land area given based on the design specification values entered below. (2) Maximum gross dwelling unit area that can be built on each lot created.*

DESIGN SPECIFICATION

Enter values in boxed areas where text is bold and blue. Express all fractions as decimals.

Given:	Gross land area	GLA=	**20.230**	acres		
Land Variables:	Est. right-of-way dedication as fraction of GLA	W=	**0.000**			
	Private roadway as fraction of GLA	R=	**0.150**			
	School land donation as fraction of GLA	Ds=	**0.000**			
	Park land donation as fraction of GLA	Dp=	**0.000**			
	Other land donation as fraction of GLA	Do=	**0.000**			
	Net land area	NLA=	17.196	acres	749,036	SF
	Unbuildable and/or future expansion areas as fraction of GLA	U=	**0.100**		4,356	SF
	Total gross land area reduction as fraction of GLA	X=	0.250		10,890	SF
	Buildable Land Area Remaining	BLA=	15.173	acres	660,914	SF
Core:	**Core + open space available**	CS=	15.173	acres	660,914	SF

LOT SPECIFICATION

	total bldg footprint BCG with garage	garage GAR footprint	misc. development area Msf as multiple of BCG	total development DCA cover area	private yards YRD as multiple of DCA	total lot LOT area	lot LTX mix	aggregate avg. AAL lot areas in mix	total lot TLA areas in mix
Type 1 Interior Lot	**1,350**	**500**	**0.50**	2,025	**2.00**	6,075	**1.000**	6,075	6,075
Type 1 CORNER LOT	**1,350**	**500**	**0.50**	2,025	**2.50**	7,088	**0.000**		
Type 3 Interior Lot	**1,700**	**500**	**0.75**	2,975	**2.00**	8,925	**0.000**		
Type 4 Interior Lot	**2,000**	**500**	**0.85**	3,700	**2.25**	12,025	**0.000**		
Type 5 Interior Lot	**2,450**	**750**	**0.75**	4,288	**2.50**	15,006	**0.000**		
Type 6 Interior Lot	**3,300**	**750**	**0.65**	5,445	**3.00**	21,780	**0.000**		
Type 7 Interior Lot	**5,600**	**750**	**0.55**	8,680	**4.00**	43,400	**0.000**		

NOTE: Zero lot line homes and townhouses are implied when YRD is less than 1.0 and approaches zero (0.0) in the YRD column above.

Mix Total (do not exceed 1.0) 1.00

Aggregate Avg. Lot Size Based on Mix: **6,075**

LOT CAPACITY FORECAST

(based on common open space value (COS) entered below)

DEFINITION of COMMON OPEN SPACE (COS): Open space provided for common benefit and massing separation in addition to that provided for private use and setbacks on individual lots. On a given land area, private lot sizes decrease as common open space increases when all other development factors are constant.
NOTE: Common open space values in bold & blue in the table below may be changed to examine other implications. When COS = 0.0 all open space is provided on private lots (OSA) in the form of yards and a typical subdivision is implied.

COS:	**0.00**	**0.10**	**0.20**	**0.30**	**0.40**	**0.50**	**0.60**	**0.70**	**0.80**
No. of lots:	108.8	97.9	87.0	76.2	65.3	54.4	43.5	32.6	21.8
Net density (dNA):	6.33	5.69	5.06	4.43	3.80	3.16	2.53	1.90	1.27
BLA density (dBA):	7.17	6.45	5.74	5.02	4.30	3.59	2.87	2.15	1.43

GROSS BUILDING AREA per LOT FORECAST

	basement as fraction of BSM habitable footprint	basement BSA area	crawl space CRW area	NET POTENTIAL BUILDING AREA INCLUDING GARAGE but EXCLUDING BASEMENT and POTENTIAL EXPANSION — number of building floors: 1	1.5	2	2.5	3	3.5
Type 1 Interior Lot	**1.000**	850	0	1,350	2,025	2,700	3,375	4,050	4,725
Type 1 CORNER LOT	**1.000**								
Type 3 Interior Lot	**1.000**								
Type 4 Interior Lot	**1.000**								
Type 5 Interior Lot	**1.000**								
Type 6 Interior Lot	**1.000**								
Type 7 Interior Lot	**1.000**								

WARNING: These are preliminary forecasts that must not be used to make final decisions.

1) These forecasts are not a substitute for the "due diligence" research that must be conducted to support the final definition of "unbuildable areas" above and the final decision to purchase land. This research includes, but is not limited to, verification of adequate subsurface soil, zoning, environmental clearance, access, title, utilities and water pressure, clearance from deed restriction, easement and right-of-way encumbrances, clearance from existing above and below ground facility conflicts, etc.

2) The most promising forecast(s) made on the basis of data entered in the design specification from "due diligence" research must be verified at the drawing board before funds are committed and land purchase decisions are made. Actual land shape ratios, dimensions and irregularities encountered may require adjustments to the general forecasts above.

3) The software licensee shall take responsibility for the design specification values entered and any advice given that is based on the forecast produced.

Urban houses may be owned as condominiums within one large parcel where all land is shared, even though each house may be personally identified. They may be separately owned with common party walls and private garden areas, and they may be rented. The ownership technique, however, should not confuse the fact that both private and shared open space at grade is associated with each dwelling unit in the urban house category. This private or personal open space might be as limited as the entry steps to a brownstone, and the shared or common open space might be as limited as the public right-of-way in front of the brownstone. Private and shared open space might also be as grand as the personal garden areas and shared parks of exclusive townhouse neighborhoods. In fact, the semantic distinction between a *rowhouse* and a *townhouse* is an indication of the differences in land ownership, private open space, and common open space allocation that exist within this building type category.[11]

The building ownership format and land allocation quantities associated with an urban house do not alter the fundamental characteristics that make this building type a variation of the single-family detached home, however. This has led to adjustments within the SF group in order to produce the GT group of model forecasting tools for urban houses.

The single-family SF group contains a lot specification panel and a building forecast panel. The urban house GT group illustrated by Exhibit 4.7 contains a dwelling unit specification panel and a dwelling unit capacity forecast panel. In both cases, however, personal land area is directly associated with each and every dwelling unit. This land area is not called a *lot* in the GT group, but it is a land area allocation that is similar and directly associated with each unit. In the GT group this land area allocation has two components. The first accommodates the building footprint, which includes both the building floor plan and the garage. This

Exhibit 4.7 Dwelling Unit Specification Required to Meet a Given Net Density Objective

Forecast Model RGTD

*Development capacity forecast for **TOWNHOUSES** based on the use of shared PARKING LOTS and / or PRIVATE GARAGES serving each dwelling unit.*

Given: *Gross land area and net density objective.* ***To Find:*** *A design specification to meet the net density objective, fit the land area given and produce desirable housing areas.*

DESIGN SPECIFICATION

Enter values in boxed areas where text is bold and blue. Express all fractions as decimals.

Given:	**Gross land area**	GLA=	10.204	acres	444,486	SF
	Net density objective	d=	10.000	du per net acre		
Land Variables:	Public right-of-way as fraction of GLA	W=	0.000		0	SF
	Private right-of-way as fraction of GLA	Wp=	0.150		66,673	SF
	Net Land Area	NLA=	8.673	acres	377,813	SF
	Unbuildable and/or future expansion areas as fraction of GLA	U=	0.100		4,356	SF
	Total gross land area reduction as fraction of GLA	X=	0.250		10,890	SF
	Buildable land area remaining	BLA=	7.653	acres	333,365	SF
Parking Variables:	Parking lot spaces required or planned per DU	u=	1	enter zero if no shared parking lot		
	Est. gross pkg. area per pkg. lot space in SF	s =	400	enter zero if no shared parking lot		
	Roadway pkg. spaces per DU	Rn=	1	enter zero if no roadway parking		
	Est. net pkg area per roadway space	Rs=	171	enter zero if no roadway parking	0.044	=Ra (fraction of BLA)
	Est. gross area per garage parking space	Ga=	250	enter zero if no garage parking		*NOTE: Express all fractions as decimals*
Support Variables:	Driveways as fraction of BLA	R=	0.030			
	Misc. pavement as fraction of BLA	M=	0.020		6,667	SF
	Outdoor social / entertainment area in SF	Sp=	10,000		0.030	fraction of BLA
	Social / service building(s) footprint in SF	Sf=	5,000		0.015	fraction of BLA
	Social / service building(s) gross area in SF	Sg=	8,000			
	Number of social / service parking spaces	Sn=	20	400 gross sf / space	0.024	=Sk (fraction of BLA)
	Total support areas as a fraction of BLA	Su=	0.163		54,500	SF
Core:	**Core + open space as fraction of BLA**	CS=	0.837	CS+Su must = 1	278,865	SF

DWELLING UNIT SPECIFICATION

dwelling unit type DU	no. of flrs. FLR above grade	dwelling unit MIX mix	total HAB habitable area	habitable area FTP footprint	garage spaces GSD per dwelling	building and garage BCG cover per dwell unit	private yard area YRD with townhouse	bldg, garage and DYG yard per dwelling	du basement BSM area	du crawl CRW space area
EFF	1	0.025	350	350	1.00	600	500	1,100	0	0
1 BR	1	0.050	500	500	1.00	750	500	1,250	0	0
2 BR	2	0.600	800	400	2.00	900	500	1,400	200	200
3 BR	2	0.300	1,000	500	2.00	1,000	500	1,500	200	300
4 BR	3	0.025	1,200	400	2.00	900	500	1,400	200	200
Aggregate Avg. Areas / DU in SF given MIX		1.000	844	434	1.93	915	500	1,415	185	215

DWELLING UNIT CAPACITY FORECAST

potential development capacity when common open space provided in the amounts shown in the left-hand column

common COS open space	CORE	NDU by TYPE (values produced by density objective)		CORE	NDU dwelling units	dNA density / net acre	dBA density / bldable acre	NPS parking lot	NGS garage pkg.	RDS roadway pkg.
0.000	**157,422**			278,865	153.6	17.71	20.08	153.6	295.8	153.6
0.100		**EFF**	**2.2**	245,528	135.3	15.60	17.68	135.3	260.4	135.3
0.200	no. dwelling units	**1 BR**	**4.3**	212,192	116.9	13.48	15.28	116.9	225.1	116.9
0.300	NDU	**2 BR**	**52.0**	178,856	98.5	11.36	12.88	98.5	189.7	98.5
0.400	**86.7**	**3 BR**	**26.0**							
0.500	open space %	**4 BR**	**2.2**							
0.600	COS									
0.700	**0.364**									
0.800										

NOTE (1): If the COS value in column 2 is negative the net density objective does not appear feasible and further drawing board investigation may be required.
NOTE (2): If rows in the potential capacity portion of the table above are blank the buildable land available cannot accommodate the open space value in the COS column in addition to the design specification values entered above. Consider adjusting the design specification if the common open space potential is inadequate.

DEFINITION of COMMON OPEN SPACE (COS): Open space provided for common benefit, massing separation and project setbacks in addition to that provided for private use adjacent to townhouse. On a given land area, private lot sizes or areas decrease as common open space increases when all other development factors are constant.
NOTE: Common open space values in bold & blue in the table above may be changed to examine other implications. When COS = 0.0 all open space is provided on private lots (OSA) in the form of yards and a typical subdivision is implied.

WARNING: These are preliminary forecasts that must not be used to make final decisions.
1) These forecasts are not a substitute for the "due diligence" research that must be conducted to support the final definition of "unbuildable areas" above and the final decision to purchase land. This research includes, but is not limited to, verification of adequate subsurface soil, zoning, environmental clearance, access, title, utilities and water pressure, clearance from deed restriction, easement and right-of-way encumbrances, clearance from existing above and below ground facility conflicts, etc.
2) The most promising forecast(s) made on the basis of data entered in the design specification from "due diligence" research must be verified at the drawing board before funds are committed and land purchase decisions are made. Actual land shape ratios, dimensions and irregularities encountered may require adjustments to the general forecasts above.
3) The software licensee shall take responsibility for the design specification values entered and any advice given that is based on the forecast produced.

footprint area is identified in the BCG column of the dwelling unit specification panel. The second component is personal open space that is directly associated with the privacy, identity, and personal enjoyment of each urban house dwelling unit environment. This is referred to in the dwelling unit specification table as the private yard area YRD associated with the urban house. The combined land area allocation per dwelling unit is referred to as the building, garage, and yard area allocation DYG per dwelling unit in the dwelling unit specification panel. This is a key quantity in forecasting urban house development capacity, since achievable density is directly influenced by the amount of land allocated to each private urban house environment.

Two models have been created to produce urban house forecasts. The first is written to evaluate the specifications needed to reach a net density objective on a given land area, and the other is written to forecast the maximum dwelling unit capacity of a given land area.

Model RGTD: Single-Family Attached Housing

Given: Gross land area and net density objective

To find: Dwelling unit capacity

The RGTD model (Exhibit 4.7) is written to evaluate the dwelling unit specification required to meet a given net density objective on a given land area. This model includes a design specification panel that is extensive but similar in format to all others, and a dwelling unit specification panel that assembles values entered into a total land allocation per dwelling unit forecast.

When a dwelling unit density is permitted by a zoning ordinance for a given land area, it does not mean that the density can be achieved by every developer.[12] In order to meet a given density objective, land area allocations per dwelling unit type must be reconciled with a number of other influential design specification values. Exhibit 4.7 is included to illustrate how the RGTD model functions as a forecasting and reconciliation tool in this context. It begins by asking the user to enter values in its design specification and dwelling unit specification panels. After completion, a dwelling unit forecast panel at the bottom of the model uses these values in two tables. The left-hand table responds to the implied question: "How many dwelling units can be placed on the gross land area available?" The table forecasts answers to this question using the specification values entered. The forecast includes a dwelling unit quantity NDU, the common open space COS remaining, and the number of each dwelling unit type that can be provided within the total quantity forecast. The right-hand side of the panel forecasts the number of dwelling units that can be built if common open space is provided in the smaller amounts shown in the COS column.[13] This table illustrates the classic tension that always exists between the amount of project open space[14] provided and the reduction in development capacity[15] that results. The panel shows that a developer can increase the potential number of dwelling units that can be built by 77% if the common open space provision is reduced from 36.4 to 0.0%. This reduction might be unacceptable and unrealistic in some urban settings and acceptable in others, but at the present time we can only rely on our intuition to give us the proper response.[16]

Exhibit 4.7 is based on a density objective of 10 dwelling units per net acre. Exhibit 4.8 is identical to exhibit 4.7, except that the density objective has been increased to 18. The left-hand table within the dwelling unit capacity panel of Exhibit 4.8 shows the dwelling unit statistics that meet the density objective entered, but the COS value is negative and the right-hand table within this panel is blank. This indicates that the density objective does not appear feasible, given the design and dwelling unit specifications that were entered.

Exhibit 4.8 Unachievable Net Density Objective

Forecast Model RGTD

Development capacity forecast for ***TOWNHOUSES*** *based on the use of shared PARKING LOTS and / or PRIVATE GARAGES serving each dwelling unit.*

Given: *Gross land area and net density objective.* ***To Find:*** *A design specification to meet the net density objective, fit the land area given and produce desirable housing areas.*

DESIGN SPECIFICATION

Enter values in boxed areas where text is bold and blue. Express all fractions as decimals.

Given:	**Gross land area**	GLA=	**10.204**	acres	444,486	SF
	Net density objective	d=	**18.000**	du per net acre		
Land Variables:	Public right-of-way as fraction of GLA	W=	**0.000**		0	SF
	Private right-of-way as fraction of GLA	Wp=	**0.150**		66,673	SF
	Net land area	NLA=	8.673	acres	377,813	SF
	Unbuildable and/or future expansion areas as fraction of GLA	U=	**0.100**		4,356	SF
	Total gross land area reduction as fraction of GLA	X=	0.250		10,890	SF
	Buildable land area remaining	BLA=	**7.653**	acres	333,365	SF
Parking Variables:	Parking lot spaces required or planned per DU	u=	**1**	enter zero if no shared parking lot		
	Est. gross pkg. area per pkg. lot space in SF	s =	**400**	enter zero if no shared parking lot		
	Roadway pkg. spaces per DU	Rn=	**1**	enter zero if no roadway parking		
	Est. net pkg. area per roadway space	Rs=	**171**	enter zero if no roadway parking	0.080	=Ra (fraction of BLA)
	Est. gross area per garage parking space	Ga=	**250**	enter zero if no garage parking		*NOTE: Express all fractions as decimals*
Support Variables:	Driveways as fraction of BLA	R=	**0.030**			
	Misc. pavement as fraction of BLA	M=	**0.020**		6,667	SF
	Outdoor social / entertainment area in SF	Sp=	**10,000**		0.030	fraction of BLA
	Social / service building(s) footprint in SF	Sf=	**5,000**		0.015	fraction of BLA
	Social / service building(s) gross area in SF	Sg=	**8,000**			
	Number of social / service parking spaces	Sn=	**20**	**400** gross sf / space	0.024	=Sk (fraction of BLA)
	Total support areas as a fraction of BLA	Su=	0.199		66,365	SF
Core:	**Core + open space as fraction of BLA**	CS=	0.801	CS+Su must = 1	267,000	SF

DWELLING UNIT SPECIFICATION

dwelling unit type DU	no. of flrs. FLR above grade	dwelling unit MIX mix	total HAB habitable area	habitable area FTP footprint	garage spaces GSD per dwelling	building and garage BCG cover per dwell unit	private yard area YRD with townhouse	bldg, garage and DYG yard per dwelling	du basement BSM area	du crawl CRW space area
EFF	**1**	**0.025**	**350**	**350**	**1.00**	600	**500**	1,100	**0**	**0**
1 BR	**1**	**0.050**	**500**	**500**	**1.00**	750	**500**	1,250	**0**	**0**
2 BR	**2**	**0.600**	**800**	**400**	**2.00**	900	**500**	1,400	**200**	**200**
3 BR	**2**	**0.300**	**1,000**	**500**	**2.00**	1,000	**500**	1,500	**200**	**300**
4 BR	**3**	**0.025**	**1,200**	**400**	**2.00**	900	**500**	1,400	**200**	**200**
Aggregate Avg. Areas / DU in SF given MIX		1.000	844	434	1.93	915	500	1,415	185	215

DWELLING UNIT CAPACITY FORECAST

potential development capacity when common open space provided in the amounts shown in the left-hand column

common COS open space	CORE	NDU by TYPE (values produced by density objective)		CORE	NDU dwelling units	dNA density / net acre	dBA density / bldable acre	NPS parking lot	NGS garage pkg.	RDS roadway pkg.
0.000	**283,360**									
0.100		**EFF**	**3.9**							
0.200	no. dwelling units	**1 BR**	**7.8**							
0.300	NDU	**2 BR**	**93.7**							
0.400	**156.1**	**3 BR**	**46.8**							
0.500	open space %	**4 BR**	**3.9**							
0.600	COS									
0.700	**-0.049**									
0.800										

NOTE (1): If the COS value in column 2 is negative the net density objective does not appear feasible and further drawing board investigation may be required.
NOTE (2): If rows in the potential capacity portion of the table above are blank the buildable land available cannot accommodate the open space value in the COS column in addition to the design specification values entered above. Consider adjusting the design specification if the common open space potential is inadequate.

DEFINITION of COMMON OPEN SPACE (COS): Open space provided for common benefit, massing separation and project setbacks in addition to that provided for private use adjacent to townhouse. On a given land area, private lot sizes or areas decrease as common open space increases when all other development factors are constant.
NOTE: Common open space values in bold & blue in the table above may be changed to examine other implications. When COS = 0.0 all open space is provided on private lots (OSA) in the form of yards and a typical subdivision is implied.

WARNING: These are preliminary forecasts that must not be used to make final decisions.
1) These forecasts are not a substitute for the "due diligence" research that must be conducted to support the final definition of "unbuildable areas" above and the final decision to purchase land. This research includes, but is not limited to, verification of adequate subsurface soil, zoning, environmental clearance, access, title, utilities and water pressure, clearance from deed restriction, easement and right-of-way encumbrances, clearance from existing above and below ground facility conflicts, etc.
2) The most promising forecast(s) made on the basis of data entered in the design specification from "due diligence" research must be verified at the drawing board before funds are committed and land purchase decisions are made. Actual land shape ratios, dimensions and irregularities encountered may require adjustments to the general forecasts above.
3) The software licensee shall take responsibility for the design specification values entered and any advice given that is based on the forecast produced.

Exhibit 4.9 illustrates how changes can be made to the design and dwelling unit specification panels of Exhibit 4.8 to reconcile the net density objective of 18 with the specification parameters needed to achieve this objective. Arrows point to changes that have been made. These statistical changes define the following planning strategy revisions. The private roadway objective Wp has been reduced to 10% from 15% of the gross land area.[17] Surface parking *u* has increased from 1.0 to 1.5 spaces per dwelling unit. Roadway parking Rn has declined from 1.0 to 0.0 spaces per dwelling unit,[18] and garage parking GSD has declined from 1.93 to 0.5 spaces per dwelling unit.[19] The planned area per parking space *s* has been reduced from 400 to 350 sq ft per space.[20] All outdoor entertainment, social centers, and social center parking have been eliminated,[21] and the dwelling unit mix MIX has been reduced to include only one- and two-bedroom units at a 15:85 ratio.[22] No changes have been made to habitable areas HAB, heights FLR, or yard areas YRD allocated per dwelling unit. This basket of changes, or reconciliation, represents an entirely different approach to the development of the site and the environment that will be created. It can make the net density objective of 18 achievable, however. In fact, this higher density objective ensures that changes must be made to the specification that produced a net density of 10. The result of these adjustments is a new design specification that represents a revised development strategy. Part of this strategy is represented by the changes that have been entered, and part is represented by the COS allocation that has fallen from 36.4 to 22.2% in response.[23] It is important to keep track of this allocation at these densities, since the private yard areas provided per dwelling unit are too small to provide adequate separation among the larger dwelling unit structures. Inadequate common open space, therefore, can easily produce a sense of overcrowding among urban houses when it is too small to adequately separate the buildings and private yard areas involved.

Exhibit 4.9 Dwelling Unit Reconciliation to Meet a Given Net Density Objective

Forecast Model RGTD

Development capacity forecast for ***TOWNHOUSES*** *based on the use of shared PARKING LOTS and / or PRIVATE GARAGES serving each dwelling unit.*

Given: *Gross land area and net density objective.* ***To Find:*** *A design specification to meet the net density objective, fit the land area given and produce desirable housing areas.*

DESIGN SPECIFICATION *Enter values in boxed areas where text is bold and blue. Express all fractions as decimals.*

Given:	**Gross land area**	GLA=	10.204	acres	444,486	SF
	Net density objective	d=	18.000	du per net acre		
Land Variables:	Public right-of-way as fraction of GLA	W=	0.000		0	SF
	Private right-of-way as fraction of GLA	Wp=	0.100		44,449	SF
	Net land area	NLA=	9.184	acres	400,038	SF
	Unbuildable and/or future expansion areas as fraction of GLA	U=	0.100		4,356	SF
	Total Gross Land Area Reduction as fraction of GLA	X=	0.200		8,712	SF
	Buildable land area remaining	BLA=	8.163	acres	355,589	SF
Parking Variables:	Pkg lot spaces required or planned per DU	u=	1.5	enter zero if no shared parking lot		
	Est. gross pkg. area per pkg. lot space in SF	s =	350	enter zero if no shared parking lot		
	Roadway pkg. spaces per DU	Rn=	0	enter zero if no roadway parking		
	Est. net pkg. area per roadway space	Rs=	171	enter zero if no roadway parking	0.000	=Ra (fraction of BLA)
	Est. gross area per garage parking space	Ga=	250	enter zero if no garage parking		*NOTE: Express all fractions as decimals*
Support Variables:	Driveways as fraction of BLA	R=	0.030			
	Misc. pavement as fraction of BLA	M=	0.020		7,112	SF
	Outdoor social / entertainment area in SF	Sp=	0		0.000	fraction of BLA
	Social / service building(s) footprint in SF	Sf=	0		0.000	fraction of BLA
	Social / service building(s) gross area in SF	Sg=	0			
	Number of social / service parking spaces	Sn=	0	400 gross sf / space	0.000	=Sk (fraction of BLA)
	Total support Areas as a fraction of BLA	Su=	0.050		17,779	SF
Core:	**Core + open space as fraction of BLA**	CS=	0.950	CS+Su must = 1	337,810	SF

DWELLING UNIT SPECIFICATION

dwelling unit type DU	no. of flrs. FLR above grade	dwelling unit MIX mix	total HAB habitable area	habitable area FTP footprint	garage spaces GSD per dwelling	building and garage BCG cover per dwell unit	private yard area YRD with townhouse	bldg, garage and DYG yard per dwelling	du basement BSM area	du crawl CRW space area
EFF	1	0.000	350	350	0.00	350	500	850	0	0
1 BR	1	0.150	500	500	0.50	625	500	1,125	0	0
2 BR	2	0.850	800	400	0.50	525	500	1,025	200	200
3 BR	2	0.000	1,000	500	0.00	500	500	1,000	200	300
4 BR	3	0.000	1,200	400	0.00	400	500	900	200	200
Aggregate Avg. Areas / DU in SF given MIX		1.000	755	415	0.50	540	500	1,040	170	170

DWELLING UNIT CAPACITY FORECAST

common COS open space	CORE	NDU by TYPE (values produced by density objective)	
0.000	258,702		
0.100		EFF	0.0
0.200	no. dwelling units	1 BR	24.8
0.300	NDU	2 BR	140.5
0.400	165.3	3 BR	0.0
0.500	open space %	4 BR	0.0
0.600	COS		
0.700	0.222		
0.800			

potential development capacity when common open space provided in the amounts shown in the left hand column

CORE	NDU dwelling units	dNA density / net acre	dBA density / bldable acre	NPS parking lot	NGS garage pkg.	RDS roadway pkg.
337,810	215.9	23.50	26.44	323.8	107.9	0.0
302,251	193.1	21.03	23.66	289.7	96.6	0.0
266,692	170.4	18.56	20.88	255.6	85.2	0.0

NOTE (1): If the COS value in column 2 is negative the net density objective does not appear feasible and further drawing board investigation may be required.
NOTE (2): If rows in the potential capacity portion of the table above are blank the buildable land available cannot accommodate the open space value in the COS column in addition to the design specification values entered above. Consider adjusting the design specification if the common open space potential is inadequate.

DEFINITION of COMMON OPEN SPACE (COS): Open space provided for common benefit, massing separation and project setbacks in addition to that provided for private use adjacent to townhouse. On a given land area, private lot sizes or areas decrease as common open space increases when all other development factors are constant.
NOTE: Common open space values in bold & blue in the table above may be changed to examine other implications. When COS = 0.0 all open space is provided on private lots (OSA) in the form of yards and a typical subdivision is implied.

WARNING: These are preliminary forecasts that must not be used to make final decisions.
1) These forecasts are not a substitute for the "due diligence" research that must be conducted to support the final definition of "unbuildable areas" above and the final decision to purchase land. This research includes, but is not limited to, verification of adequate subsurface soil, zoning, environmental clearance, access, title, utilities and water pressure, clearance from deed restriction, easement and right-of-way encumbrances, clearance from existing above and below ground facility conflicts, etc.
2) The most promising forecast(s) made on the basis of data entered in the design specification from "due diligence" research must be verified at the drawing board before funds are committed and land purchase decisions are made. Actual land shape ratios, dimensions and irregularities encountered may require adjustments to the general forecasts above.
3) The software licensee shall take responsibility for the design specification values entered and any advice given that is based on the forecast produced.

Model RGTL Single-Family Attached Housing

Given: Gross land area

To find: Dwelling unit capacity

Model RGTL (Exhibit 4.10) is designed to forecast the dwelling unit capacity of a given land area based on values assigned to elements within the design specification and dwelling unit specification panels of the model. This capacity is predicted in the NDU column of the dwelling unit forecast panel, and is based on values entered in the specification panels. The NDU column is adjacent to a left-hand column of COS alternatives that directly affect the forecast produced, and that can be changed at will.[24]

Four columns to the right of the NDU column are labeled EFF, 1 BR, 2 BR, 3 BR, and 4 BR. These abbreviations refer to the number of bedrooms that distinguish each dwelling unit type, and the data below each label itemizes the total number of units that can be built as common open space increases. For instance, the RGTL model forecasts that 224.8 dwelling units can be built on the land area given when 40% common open space is provided. Based on the unit mix specified, none would be efficiency units (EFF), 22.5 would be one-bedroom units (1 BR), 134.9 would be two-bedroom units (2 BR), 67.4 would be three-bedroom units (3 BR), and none would be four-bedroom units (4 BR).[25]

The right-hand section of the dwelling unit forecast panel predicts the densities and parking that will result from the specification values entered. Garage spaces are not separately forecast due to a lack of space, and are in addition to the parking lot spaces forecast.[26] All forecasts in the panel are made by row in relation to the common open space values in the COS column, and all can be seen to decrease as common open space increases, which is what would be expected when all other design specification values remain constant.

Exhibit 4.10 Dwelling Unit Capacity of a Given Land Area

Forecast Model RGTL

*Development capacity forecast for **TOWNHOUSES** based on the use of shared PARKING LOTS and / or PRIVATE GARAGES serving each dwelling unit.*
***Given:** Gross land area. **To Find:** Maximum dwelling unit capacity of the land area given based on the design specification values entered below.*

DESIGN SPECIFICATION

Enter values in boxed areas where text is bold and blue. Express all fractions as decimals.

Given:	**Gross land area**	**GLA=**	**20.000**	acres	871,200	SF
Land Variables:	Public right-of-way as fraction of GLA	W=	**0.000**		0	SF
	Private right-of-way as fraction of GLA	Wp=	**0.200**		174,240	SF
	Net land area	NLA=	16.000	acres	696,960	SF
	Unbuildable and/or future expansion areas as fraction of GLA	U=	**0.000**		0	SF
	Total Gross Land Area Reduction as fraction of GLA	X=	0.200		8,712	SF
	Buildable land area remaining	**BLA=**	**16.000**	acres	696,960	SF
Parking Variables:	Pkg lot spaces required or planned per DU	u=	**1.5**	enter zero if no shared parking lot		
	Est. gross pkg. area per pkg. lot space in SF	s =	**400**	enter zero if no shared parking lot		
	Roadway pkg. spaces per DU	Rn=	**0**	enter zero if no roadway parking	*NOTE: Express all fractions as decimals*	
	Est. net pkg. area per roadway space	Rs=	**0**	enter zero if no roadway parking		
	Est. gross area per garage parking space	Ga=	**250**	enter zero if no garage parking		
Support Variables:	Driveways as fraction of BLA	R=	**0.020**			
	Misc. pavement as fraction of BLA	M=	**0.010**		6,970	SF
	Outdoor social / entertainment area in SF	Sp=	**10,000**		0.014	fraction of BLA
	Social / service building(s) footprint in SF	Sf=	**5,000**		0.007	fraction of BLA
	Social / service building(s) gross area in SF	Sg=	**8,000**			
	Number of social / service parking spaces	Sn=	**20**	**400** gross sf / space	0.011	fraction of BLA
	Total support areas as a fraction of BLA	Su=	0.063		43,909	SF
	Core + open space + roadway parking as fraction of BLA	**CSR=**	0.937	CS+Su must = 1	653,051	SF

DWELLING UNIT SPECIFICATION

dwelling unit type DU	no. of flrs. FLR above grade	dwelling unit MIX mix	total HAB habitable area	habitable area FTP footprint	garage spaces GSD per dwelling	building and garage BCG cover per dwell unit	private yard area YRD with townhouse	bldg. garage and DYG yard per dwelling	du basement BSM area	du crawl CRW space area
EFF	**1**	**0.000**	**350**	**350**	**0.5**	475	**500**	975	**0**	**0**
1 BR	**1**	**0.100**	**500**	**500**	**0.5**	625	**500**	1,125	**0**	**0**
2 BR	**2**	**0.600**	**800**	**400**	**0.5**	525	**500**	1,025	**200**	**200**
3 BR	**2**	**0.300**	**1,000**	**500**	**0.5**	625	**500**	1,125	**200**	**300**
4 BR	**3**	**0.000**	**1,200**	**400**	**0.5**	525	**500**	1,025	**200**	**200**
Aggregate Avg. Areas / DU in SF given MIX		1.000	830	440	0.50	565	500		180	210
							AFP =	1,065	aggregate average private area per du	

DWELLING UNIT CAPACITY FORECAST

(based on common open space (COS) provided)

common COS open space	number of NDU dwelling units	EFF	1 BR	2 BR	3 BR	4 BR	density dNA per net acre	density per dBA buildable acre	parking lot NPS spaces	roadway RDS parking spaces
		dwelling unit type (DU) quantities given mix above								
0.000	392.2	0.0	39.2	235.3	117.7	0.0	24.51	24.51	588.3	0.0
0.100	350.4	0.0	35.0	210.2	105.1	0.0	21.90	21.90	525.5	0.0
0.200	308.5	0.0	30.9	185.1	92.6	0.0	19.28	19.28	462.8	0.0
0.300	266.6	0.0	26.7	160.0	80.0	0.0	16.67	16.67	400.0	0.0
0.400	224.8	0.0	22.5	134.9	67.4	0.0	14.05	14.05	337.2	0.0
0.500	182.9	0.0	18.3	109.8	54.9	0.0	11.43	11.43	274.4	0.0
0.600	141.1	0.0	14.1	84.6	42.3	0.0	8.82	8.82	211.6	0.0
0.700	99.2	0.0	9.9	59.5	29.8	0.0	6.20	6.20	148.8	0.0
0.800	57.3	0.0	5.7	34.4	17.2	0.0	3.58	3.58	86.0	0.0

DEFINITION of COMMON OPEN SPACE (COS): Open space provided for common benefit, massing separation and project setbacks in addition to that provided for private use adjacent to townhouses. On a given land area, private lot sizes or areas decrease as common open space increases when all other development factors are constant.
NOTE: Common open space values in bold & blue in the table above may be changed to examine other implications. When COS = 0.0 all open space is provided on private lots (OSA) in the form of yards and a typical subdivision is implied.

WARNING: These are preliminary forecasts that must not be used to make final decisions.
1) These forecasts are not a substitute for the "due diligence" research that must be conducted to support the final definition of "unbuildable areas" above and the final decision to purchase land. This research includes, but is not limited to, verification of adequate subsurface soil, zoning, environmental clearance, access, title, utilities and water pressure, clearance from deed restriction, easement and right-of-way encumbrances, clearance from existing above and below ground facility conflicts, etc.
2) The most promising forecast(s) made on the basis of data entered in the design specification from "due diligence" research must be verified at the drawing board before funds are committed and land purchase decisions are made. Actual land shape ratios, dimensions and irregularities encountered may require adjustments to the general forecasts above.
3) The software licensee shall take responsibility for the design specification values entered and any advice given that is based on the forecast produced.

Conclusion

A lot specification panel distinguishes the SF group of development forecasting models. A dwelling unit specification panel distinguishes the GT group. Both groups also share a number of common characteristics. Both use a common open space panel to specify shared open space. Both use yard area values to specify private open space per lot or dwelling unit. Both use a universal design specification panel. Both are designed to forecast the housing capacity of a given land area. Both are designed to forecast the development capacity of land to provide housing, and both assume that surface parking will be the parking system of choice. The fundamental difference is that the SF group forecasts the number of lots that can be created and the home area that can be placed on each lot. The GT group forecasts the number of dwelling units that can be accommodated based on given areas per dwelling unit.

The single-family forecasting series discussed in this chapter and the apartment forecasting collection discussed earlier introduce a quick and comprehensive ability to predict the development capacity of land to accommodate the housing needs of an increasing population. These models can also be used with panel attachments to forecast the cost, yield, and return on investment associated with these predictions as discussed in Chap. 3. Chapters 12, 13, and 14 present case studies to demonstrate the practical application of these tools.

Notes

1. The extent of separate lots available declines in most, but not all, cases. In fact, separate townhouse lot subdivisions are discouraged by most zoning ordinances since side yards are not present. However, separate lot townhouses can encourage property maintenance, neighborhood pride, and community association. The common walls associated with this building type were an historic problem when they were not built

with fire ratings in mind. The evolution from this building type that produced extremely narrow side yards in the 6-in to 5-ft range is still a problem that should be avoided. The separate-lot urban house, however, represents an opportunity to learn to live within limits by borrowing from the past while avoiding its technological weaknesses.

2. Roadway parking seems to be more prevalent when these townhouse units are owned in a condominium arrangement, but this is simply an empirical observation.
3. Eliminating the grade orientation provided for all dwelling units and reducing the parking requirement per dwelling unit is one way to significantly increase housing densities. Unfortunately, project open space is also reduced or nonexistent as well, and little attempt has been made to study the inverse relationships involved.
4. In this context, land area is a larger parcel that may be subdivided into smaller lots. Both areas are defined by property lines and both are owned through recorded deeds. One is simply larger than the other and represents the raw material to be subdivided.
5. This is an emerging planning approach for single-family homes that has historic roots. The concept is to reduce the size of single-family lots, cluster them, and surround them with the open space that would remain if the same number of dwelling units were placed on larger lots in a typical subdivision arrangement. Clustering can provide construction economies, environmental preservation, and the appearance of a village among open fields—if enough common open space is provided. Please refer to the definition and explanation of this open space category in Table 3.2.
6. This is often the result of a zoning and subdivision ordinance that specifies minimum lot areas.
7. Do not confuse common open space with land donation for other public purposes.
8. These areas include the garage but do not include the basement.
9. Area available for initial construction and future expansion on the lot.
10. Single-story and 3-story units can also be attached to the assembly.
11. The social implications of rowhouse and townhouse lifestyles have had the attention of some of the world's greatest authors and social reformers for generations.

12. The level of expectation produced by some of these ordinances can be a factor that stimulates overdevelopment. The ability to meet a density limit is a function of the building footprint and private open space allocated per dwelling unit. These allocations vary widely and are often a function of perceived market demand, attraction, and property location.
13. Remember that common open space is *shared* open space and private open space is called *yard area.*
14. A generic term that means common open space in this context.
15. A generic term that means dwelling unit capacity in this context.
16. For instance, public rights-of-way in some classic brownstone settings might be considered an acceptable level of common open space. In this setting, entry steps and basement light wells might also be considered acceptable levels of private open space.
17. This implies that great efficiency must be achieved during the drawing-board phase.
18. This is an attempt to consolidate and minimize the land allocated to parking.
19. This implies less visitor and personal parking available.
20. This implies no landscaping in the lot.
21. This reduction in amenity implies less personal appeal and a potential reduction in rental range.
22. This implies less family orientation.
23. This value can be seen in the left-hand table of the dwelling unit forecast panel.
24. The NDU column confirms the expectation that the number of potential dwelling units will decline as the amount of common open space increases, when all other design specification values remain constant.
25. The models do not round numbers because they are guidelines and not final answers, but uneven numbers will obviously become whole numbers in a final design and can be rounded up or down depending on the judgment and professional opinion of the user. If the user wishes to ensure adequate provision of the open space intended, however, the user would round these forecasts down to the next lowest whole number.
26. This number can easily be found by multiplying the GSD average value by the NDU value that corresponds to the amount of common open space under consideration in the COS column.

Mixed-Use Forecasting

Mixed-use development plans combine land use activities that have historically been segregated by zoning district. The term *mixed use* can refer to either horizontal or vertical relationships among these land use activities. When used in a horizontal context, it refers to the arrangement of these uses within a larger plan that permits the association of these mixed activities. When used in a vertical context, it refers to a combination of these uses within the same building envelope.[1] For instance, assembling first-floor retail business with apartments above is a vertical mixture that combines activities often listed and segregated by zoning district. Figure 5.1 illustrates an early vertical mixture that contains first-floor retail activity and second-floor office activity.

Land use separation was originally introduced in response to public health, safety, and welfare risks that were, and are, produced when some land use activities come into close contact. It has evolved under the broad concept of general welfare to include regulations that protect property value,[2] privacy, and appearance but has had little success protecting the public from excessive land consumption.[3]

Years of experience with land use separation has convinced many planners that some zoning distinctions are more aca-

(a)

(b)

(c)

Figure 5.1 Early mixed use. The concept of mixed use is as old as history and represents a problem only when the edges of contact produce incompatible activity relationships or association with processes, products, and appearances that are either life threatening or degrading to the surrounding area. Mixed-use conflicts increased as the processes, products, and activities of the Industrial Revolution required shelter, and zoning was introduced at the beginning of the twentieth century in an attempt to arrange these activities in nonthreatening patterns of association. Zoning ordinances, however, defined land use activities with such precision that many nonthreatening mixed-use relationships were prohibited because each activity required a separate zoning category. Many are now searching for new tools and language that can reintroduce diversity while also preventing the conflict and degradation of the past.

The historic buildings in these photographs represent another typical mixed-use problem. The buildings contain neighborhood retail activities at the first level and growing businesses at the second. The mixed use is compatible with the surrounding area, but the development suffers because it does not adequately accommodate the automobile. In fact, the center has no parking area of its own. The "hitching-post" parking provided is in the public right-of-way and provides too few spaces for both retail customers and office visitors. It also introduces movement conflicts and surprises as vehicles attempt to maneuver into spaces while others attempt to pass through. These historic patterns often require adjustment to remain competitive. Resolving the parking problem, however, will increase traffic activity to levels that were not anticipated when the project was constructed to serve both horses and the emerging automobile.

New mixed-use construction can avoid these problems of compatibility, but historic mixed use represents a significant community asset when these issues can be resolved.

(*a*) Overview; (*b*) and (*c*) parking areas.

(Photos by W. M. Hosack)

demic than essential, and have resulted in the isolation of complementary nonconforming activities that can bring variety and spontaneity to the urban environment. This has led them to suggest that these land use activities can be combined, or "mixed," to reduce land consumption and reintroduce variety. Not all agree, however, and the debate intensifies when the issue becomes site specific, since nonconformity can produce fear of incompatibility and real estate depreciation.[4]

This chapter does not debate the virtues of mixing land use activities, but it does demonstrate how to calculate the development capacity of land when considering mixed-use possibilities. This should help to objectively evaluate the merits of these concepts in relation to the social and economic benefits they imply, the intensity they propose, and the environments they produce.

The forecast collection discussed in Chap. 1 has been expanded to include five models that predict the development capacity of land when vertically integrated mixed-use proposals are under consideration.[5] Four of the five address new construction, and the fifth, MG1B, is written to forecast the capacity of an existing facility when remodeled to accommodate mixed-use activities. Table 5.1 is included as a decision guide to help in selecting one of the five. The models themselves can be found on the CD-ROM within a workbook entitled MX Design Premise.

There is no land use decision to be made in Table 5.1 since the design premise of all M-series forecast models is that residential apartments and nonresidential land use activities will be combined within the same building or buildings. All five models are also based on the assumption that the land area is known. In the case of MG1B, the building footprint to remain must also be known, but the value entered in the design specification can be modified to test alternate footprint possibilities.

A separate M series of forecast models is required because both nonresidential building area and residential dwelling

Table 5.1 The M-Series Decision Guide for Vertically Integrated Mixed-Use Forecast Models

	Parking System			
To Find	**Adjacent Parking Lot on Same Premise**	**Adjacent Parking Structure on Same Premise**	**Underground Parking Structure**	**Parking Structure Partially or Completely Above Grade Under Building**
Maximum commercial building area and apartment dwelling unit capacity of a given land area when the allocation between residential and non-residential activities varies, and when building height varies.	MG1L	MS1L	MS2L	MS3L
Maximum commercial building area and apartment dwelling unit capacity of an *existing building footprint* within an existing land area when the allocation between residential and nonresidential activities varies, and when building height varies.	MG1B			

NOTE: All M-series models are included in the MX Design Premise workbook on the CD.

unit capacity must be predicted in the same model when vertically mixed land uses are considered. The M-series models have two parent spreadsheets that are not visible in order to accommodate this requirement. The relationship of these parents is defined by the residential allocation percentage (RAP) entered in the planning forecast panel of each M-series model. This percentage establishes the amount of buildable land area that will be used to support residential land use activities, and is specified as a percentage of the total buildable land area available.[6] The nonresidential allocation percentage CAP is simply found by subtracting the RAP value from 100%.[7] The model and its parents then use the entire set of design specification values entered to produce the predictions displayed in the planning forecast panel of each model.

Before continuing with an explanation of the forecast models listed in Table 5.1, however, the term *mixed-use* should be clarified, because the forecast models in Table 5.1 focus on vertically integrated activities. The forecast models discussed in previous chapters are capable of accommodating horizontal applications. The following is included as a brief explanation.

Horizontally Integrated Mixed-Use Development Capacity

Horizontally integrated mixed-use development is simply a function of the land area allocated to each land use category within a larger project area. For instance, if a developer owns 50 acres and allocates 10 acres to office use, 15 acres to apartment use, and 25 acres to single-family detached homes, that developer is proposing a mixed-use development. In order to evaluate the development capacity of the combined proposal, the developer would evaluate the capacity of each subarea with the appropriate forecast model listed in either Fig. 1.1 or 1.2. In this example, assuming a surface lot parking system, the forecast model for the office use would be CG1L, the model for the apartment use would be RG1L, and the model for the single family detached use would be RSFL. The ensuing evaluation would simply be a function of the design specification values entered in each respective forecast model as described in previous chapters. The maximum development capacity of each subarea is then determined by the way the internal equations of the forecast model use these design specification values to optimize the allocation of land within the core development area. This produces a series of development capacity forecasts for each subarea based on increasing building height. The process is the same as that described in previous chapters. It is simply applied to more than one subarea, and may involve more than one forecast model when a mixed-use proposal is under study.

In a horizontally integrated mixed-use application, therefore, identify the subareas allocated to each land use family and building type, since each family and building type involves a separate forecast series of models (see Table 5.2). If there are separate lots within a subarea, treat each lot within the subarea as a separate parcel for development capacity evaluation (even though that lot area might not be identified by separate deed or physical demarcation), and choose the forecast model that applies from either Fig. 1.1 or 1.2. For instance, a nonresidential subarea could involve many lots devoted to different nonresidential land use activities, such as offices, banks, restaurants, etc. The development capacity of each nonresidential lot could be evaluated with the C series of forecast models. If all nonresidential activities planned to use surface parking and the land area for each was known, they could all be evaluated with the CG1L forecast model. The development capacity results produced for each lot would depend on the values entered in the CG1L specification panel for each lot.

In summary, when evaluating a lot within a subarea of a mixed-use development plan, choose the forecast model that relates to the land use family involved, the building type planned, and the parking system anticipated (with the help of Fig. 1.1 or 1.2). After selection, proceed with an evaluation of the lot by entering appropriate values in the design specification panel of the forecast model chosen. Repeat this process for each subarea or lot defined and summarize the results produced to predict the development capacity of the entire horizontally integrated mixed-use proposal under consideration. This process could apply to a 50-acre tract, a 500-acre annexation area, a district, a city, or a region since they all involve horizontally mixed land use areas.

Table 5.2 Mixed-Use Forecast Models Arranged by Land Use and Building Type

Land Use	Building Type	Forecast Model Series	Forecast Models		Decision Guide
Nonresidential	All	C series	CG1L CG1B CG2L CG2B CS1L	CS1B CS2L CS2B CS3L CS3B	Figure 1.2
Residential	Apartments	R series	RG1L RG1B RG1D RG2L RG2B RG2D RS1L RS1B	RS1D RS2L RS2B RS2D RS3L RS3B RS3D	Figure 1.1
Residential	Single-family detached suburb houses	SF series	RSFL RSFD RSFN		Figure 1.1
Residential	Single-family attached urban houses	GT series	RGTL RGTD		Figure 1.1
Mixed use	Vertically integrated mixed uses	M series	MG1L MG1B MS1L MS2L MS3L		Table 5.1

Key to Forecast Model Designations

Land Use Series	Parking System	Information Given
C: Nonresidential	G1: Grade parking around but not under building(s)	L: Land area
R: Residential	G2: Grade parking around and under building(s)	B: Building area objective
M: Mixed use	S1: Parking structure adjacent to building on same premise	D: Dwelling unit density objective
	S2: Parking structure underground	N: Number of lots desired
	S3: Parking structure below building and at least partially above grade.	
	SF: Single-family detached grade parking	
	GT: Single-family attached grade parking	

Vertically Integrated Mixed-Use Development Capacity

The second category of mixed land use activities involves vertical integration. The most common form of vertically integrated mixed-use development involves nonresidential activity at lower floor levels and residential apartments above, with both using surface pavement for parking. This works with increasing effectiveness as the residential parking demand per dwelling unit decreases. Parking structures are also used with this approach and can dramatically increase both the capacity and intensity of the project under consideration.

Surface Parking

Two conditions are often encountered using a surface parking approach. The first involves new construction on vacant land or land that is to be cleared and made available. The second involves existing buildings that are to be renovated and expanded for mixed-use activities by rearranging the use and allocation of their parking lots (such as shopping centers).

NEW CONSTRUCTION

Exhibit 5.1 uses forecast model MG1L and a 25-acre land area to illustrate mixed-use forecasting when new construction is involved. Additional values have been entered in the design specification of this exhibit to complete the illustration. These values assume that no public right-of-way *W* is involved and no unbuildable areas *U* exist. A parking space is estimated to require 375 gross sq ft of pavement *s,* and 1.5 parking spaces are proposed for each dwelling unit provided *u*. No garage spaces are planned (Gn and Ga equal 0), and every nonresidential parking space will permit the construction of 200 sq ft of nonresidential building area *a*. Three loading spaces *l* will be provided, at an average area *b* of 1000 sq ft per space. Miscellaneous driveways *R* and

Exhibit 5.1 Mixed-Use Development Capacity with 30% Project Open Space

Forecast Model MG1L

Development capacity forecast for ***MIXED USE*** *based on an adjacent* ***GRADE PARKING LOT*** *located on the same premises.*
Given: *Gross land area.* ***To Find:*** *Maximum commercial building area and apartment dwelling unit capacity of the land area given when the residential land use allocation varies.* ***Premise:*** *All building floors considered equal in area.*

DESIGN SPECIFICATION *Enter values in boxed areas where text is bold and blue. Express all fractions as decimals.*

Given:	**Gross land area**	GLA=	25.000	acres	1,089,000	SF
Land Variables:	Public/ private right-of-way & paved easements	W=	0.000	fraction of GLA	0	SF
	Net land area	NLA=	25.000	acres	1,089,000	SF
	Unbuildable and/or future expansion areas	U=	0.000	fraction of GLA	0	SF
	Gross Land Area Reduction	X=	0.000	fraction of GLA	0	SF
	Buildable land area remaining	BLA=	25.000	acres	1,089,000	SF
Parking Variables:	Est. gross pkg. lot area per pkg. space in SF	s =	375			
	Parking lot spaces planned or required per dwelling unit	u=	1.5			
	Garage parking spaces planned or required per dwelling unit	Gn=	0.00			
	Gross building area per garage space	Ga=	0			
	Nonresidential building SF permitted per parking space	a=	200			
	No. of loading spaces	l =	3			
	Gross area per loading space	b =	1,000	SF	3,000	SF
Site Variables:	**Project open space as fraction of BLA**	S=	0.300		326,700	SF
	Private driveways as fraction of BLA	R=	0.010		10,890	SF
	Misc.pavement as fraction of BLA	M=	0.050		54,450	SF
	Loading area as fraction of BLA	L=	0.003		3,000	SF
	Total site support areas as a fraction of BLA	Su=	0.363		395,040	SF
Core:	**Core development area as fraction of BLA**	C=	0.637	C=Su must = 1	693,960	SF
Building Variables:	Res. bldg. efficiency as percentage of GBA	Be=	0.600			
	Building support as fraction of GBA	Bu=	0.400	Be + Bu must = 1		

Dwelling Unit Mix Table:

DU dwelling unit type	GDA gross du area	CDA=GDA/Be comprehensive du area	MIX du mix	PDA = (CDA)MIX Prorated du area
EFF	500	833	5%	42
1 BR	750	1,250	25%	313
2 BR	1,200	2,000	50%	1,000
3 BR	1,500	2,500	20%	500
4 BR	1,800	3,000	0%	0
			Aggregate avg. dwelling unit area (AGG) =	1,854
			GBA sf per parking space a=	1,236

MIXED-USE PLANNING FORECAST

50.00% =(RAP): residential land use allocation percentage
50.00% =(CAP): nonresidential land use allocation percentage

total floors FLR	CGBA nonres GBA	RGBA res GBA	total bldg MBCA cover area	nonres CFLR floors	total parking MPLA lot area	total parking MNPS spaces	total dwelling MNDU units	density per dBA bldable acre
1.00	120,689	266,793	387,482	0.31	307,228	819	143.9	5.8
2.00	146,097	432,839	289,468	0.50	405,242	1,081	233.4	9.3
3.00	157,123	546,141	234,421	0.67	460,289	1,227	294.5	11.8
4.50	165,447	661,596	183,787	0.90	510,923	1,362	356.8	14.3
5.00	167,219	690,803	171,604	0.97	523,106	1,395	372.6	14.9
6.00	169,949	739,792	151,624	1.12	543,086	1,448	399.0	16.0
7.00	171,955	779,265	135,889	1.27	558,821	1,490	420.3	16.8
8.00	173,490	811,750	123,155	1.41	571,555	1,524	437.8	17.5
9.00	174,703	838,951	112,628	1.55	582,082	1,552	452.5	18.1
10.00	175,686	862,060	103,775	1.69	590,935	1,576	464.9	18.6
11.00	176,498	881,937	96,221	1.83	598,489	1,596	475.7	19.0
12.00	177,181	899,215	89,700	1.98	605,010	1,613	485.0	19.4
13.00	177,763	914,372	84,010	2.12	610,700	1,629	493.1	19.7
14.00	178,265	927,776	79,003	2.26	615,707	1,642	500.4	20.0
15.00	178,702	939,716	74,561	2.40	620,149	1,654	506.8	20.3

WARNING: These are preliminary forecasts that must not be used to make final decisions.
1) These forecasts are not a substitute for the "due diligence" research that must be conducted to support the final definition of "unbuildable areas" above and the final decision to purchase land. This research includes, but is not limited to, verification of adequate subsurface soil, zoning, environmental clearance, access, title, utilities and water pressure, clearance from deed restriction, easement and right-of-way encumbrances, clearance from existing above and below ground facility conflicts, etc.
2) The most promising forecast(s) made on the basis of data entered in the design specification from "due diligence" research must be verified at the drawing board before funds are committed and land purchase decisions are made. Actual land shape ratios, dimensions and irregularities encountered may require adjustments to the general forecasts above.
3) The software licensee shall take responsibility for the design specification values entered and any advice given that is based on the forecast produced.

pavement *M* will consume 6% of the buildable area, and 30% project open space *S* will be provided.

The dwelling unit mix specification in Exhibit 5.1 also contains entries that affect the forecast produced. These entries specify that 5% of the total dwelling unit capacity will be efficiency apartments with a gross area of 500 sq ft. 25% will be one-bedroom apartments with a gross area of 750 sq ft. 50% will be two-bedroom apartments with a gross area of 1200 sq ft, and 20% will be three-bedroom apartments with a gross area of 1500 sq ft.

The design specification entries are complete when a residential allocation percentage (RAP) is entered in the upper left-hand corner of the planning forecast panel. The ratio for the example in Exhibit 5.1 proposes that 50% of the development capacity be used to support residential land use activity and 50% be used to support nonresidential activity. Based on the entire set of values entered throughout the design specification described, the planning forecast panel at the bottom of Exhibit 5.1 predicts that the development capacity of the 25-acre land area can accommodate a 5-story building when the building footprint MBCA is no greater than 171,604 sq ft in area and the parking lot MPLA equals 523,106 sq ft. This 5-story building would contain 162,219 sq ft of nonresidential area CGBA, equaling 97% of the total first-floor area CFLR, and 690,803 sq ft of residential area. The total number of parking spaces MNPS would be 1395. The total number of dwelling units would be 372.6, and the density would be 14.9 dwelling units per buildable acre.[8] The total project open space remaining to offset the development intensity proposed would equal 30% of the buildable area available.

If the 30% project open space *S* specification in Exhibit 5.1 were increased by 20% to enhance the residential character of the proposal, the development capacity impact could easily be evaluated by entering 50% in the design specification cell labeled *S* in Exhibit 5.1.

If senior housing were anticipated, the development capacity implications could be assessed by lowering the parking requirement u in Exhibit 5.1 to 1.0 or even to 0.5 spaces per dwelling unit.

Exhibit 5.2 illustrates the impact of a 50% open space requirement when all other values in Exhibit 5.1 remain constant. Exhibit 5.3 illustrates the development capacity impact of a parking reduction to 1.0 space per dwelling unit. The dwelling unit capacity results are summarized in Table 5.3 to reveal the general principle that development capacity decreases when project open space increases and

Table 5.3 Development Capacity Evaluation

Principle:

When all other values remain constant, development capacity decreases as project open space increases.

Principle:

When all other values remain constant, development capacity increases as parking requirements decrease.

Dwelling Unit Capacity Forecasts from Exhibits 5.1, 5.2, and 5.3

Exhibits	Number of Dwelling Units Floors 1	5	10	15	Notes
E5.1	143.9	372.6	464.9	506.8	Baseline data from Exhibit 5.1
E5.2	98.8	255.9	319.3	348.1	Project open space increased to 50% in Exhibit 5.2
E5.3	156	466.2	620.5	697.4	Parking reduced from 1.5 to 1.0 spaces per dwelling unit in Exhibit 5.3

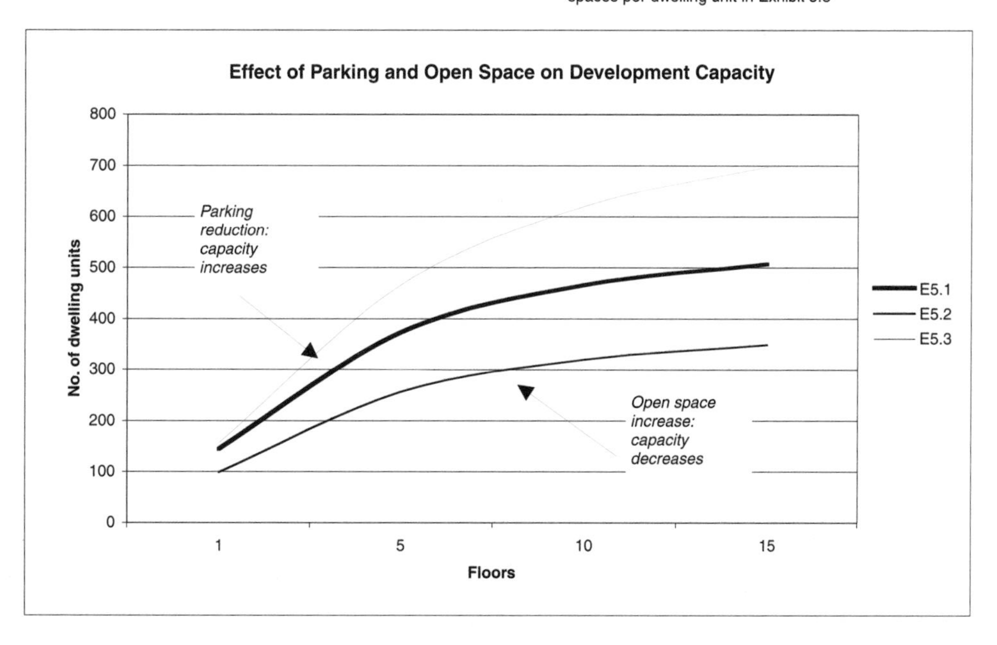

Exhibit 5.2 Mixed-Use Development Capacity with 50% Project Open Space

Forecast Model MG1L

*Development capacity forecast for **MIXED USE** based on an adjacent **GRADE PARKING LOT** located on the same premises.*
***Given:** Gross land area. **To Find:** Maximum commercial building area and apartment dwelling unit capacity of the land area given when the residential land use allocation varies. **Premise:** All building floors considered equal in area.*

DESIGN SPECIFICATION — *Enter values in boxed areas where text is bold and blue. Express all fractions as decimals.*

Given:	**Gross land area**	GLA=	**25.000**	acres	1,089,000	SF
Land Variables:	Public/ private right-of-way & paved easements	W=	**0.000**	fraction of GLA	0	SF
	Net land area	NLA=	25.000	acres	1,089,000	SF
	Unbuildable and/or future expansion areas	U=	**0.000**	fraction of GLA	0	SF
	Gross land area reduction	X=	0.000	fraction of GLA	0	SF
	Buildable land area remaining	BLA=	**25.000**	acres	1,089,000	SF
Parking Variables:	Est. gross pkg. lot area per pkg. space in SF	s =	**375**			
	Parking lot spaces planned or required per dwelling unit	u=	**1.5**			
	Garage parking spaces planned or required per dwelling unit	Gn=	**0.00**			
	Gross building area per garage space	Ga=	**0**			
	Nonresidential building SF permitted per parking space	a=	**200**			
	No. of loading spaces	l =	**3**			
	Gross area per loading space	b =	**1,000**	SF	3,000	SF
Site Variables:	**Project open space as fraction of BLA**	S=	**0.500**		544,500	SF
	Private driveways as fraction of BLA	R=	**0.010**		10,890	SF
	Misc. pavement as fraction of BLA	M=	**0.050**		54,450	SF
	Loading area as fraction of BLA	L=	0.003		3,000	SF
	Total site support areas as a fraction of BLA	Su=	0.563		612,840	SF
Core:	**Core development area as fraction of BLA**	C=	0.437	C=Su must = 1	476,160	SF
Building Variables:	Res. bldg. efficiency as percentage of GBA	Be=	**0.600**			
	Building support as fraction of GBA	Bu=	0.400	Be + Bu must = 1		

Dwelling Unit Mix Table:

DU dwelling unit type	GDA gross du area	CDA=GDA/Be comprehensive du area	MIX du mix	PDA = (CDA)MIX Pro-rated du area
EFF	**500**	833	**5%**	42
1 BR	**750**	1,250	**25%**	313
2 BR	**1,200**	2,000	**50%**	1,000
3 BR	**1,500**	2,500	**20%**	500
4 BR	**1,800**	3,000	**0%**	0
			Aggregate avg. dwelling unit area (AGG) =	**1,854**
			GBA sf per parking space a=	1,236

MIXED-USE PLANNING FORECAST

50.00% =(RAP): residential land use allocation percentage
50.00% =(CAP): nonresidential land use allocation percentage

total floors FLR	CGBA nonres GBA	RGBA res GBA	total bldg MBCA cover area	nonres CFLR floors	total parking MPLA lot area	total parking MNPS spaces	total dwelling MNDU units	density per dBA bldable acre
1.00	82,810	183,240	266,051	0.31	210,859	562	98.8	4.0
2.00	100,244	297,285	198,765	0.50	278,145	742	160.3	6.4
3.00	107,810	375,104	160,971	0.67	315,939	843	202.3	8.1
4.50	113,522	454,401	126,205	0.90	350,705	935	245.1	9.8
5.00	114,737	474,461	117,840	0.97	359,070	958	255.9	10.2
6.00	116,611	508,108	104,120	1.12	372,790	994	274.0	11.0
7.00	117,987	535,220	93,315	1.26	383,595	1,023	288.7	11.5
8.00	119,040	557,531	84,571	1.41	392,339	1,046	300.7	12.0
9.00	119,872	576,213	77,343	1.55	399,567	1,066	310.8	12.4
10.00	120,547	592,086	71,263	1.69	405,647	1,082	319.3	12.8
11.00	121,104	605,737	66,077	1.83	410,833	1,096	326.7	13.1
12.00	121,573	617,604	61,598	1.97	415,312	1,107	333.1	13.3
13.00	121,972	628,014	57,691	2.11	419,219	1,118	338.7	13.5
14.00	122,316	637,221	54,253	2.25	422,657	1,127	343.7	13.7
15.00	122,616	645,421	51,202	2.39	425,708	1,135	348.1	13.9

WARNING: These are preliminary forecasts that must not be used to make final decisions.
1) These forecasts are not a substitute for the "due diligence" research that must be conducted to support the final definition of "unbuildable areas" above and the final decision to purchase land. This research includes, but is not limited to, verification of adequate subsurface soil, zoning, environmental clearance, access, title, utilities and water pressure, clearance from deed restriction, easement and right-of-way encumbrances, clearance from existing above and below ground facility conflicts, etc.
2) The most promising forecast(s) made on the basis of data entered in the design specification from "due diligence" research must be verified at the drawing board before funds are committed and land purchase decisions are made. Actual land shape ratios, dimensions and irregularities encountered may require adjustments to the general forecasts above.
3) The software licensee shall take responsibility for the design specification values entered and any advice given that is based on the forecast produced.

Exhibit 5.3 Mixed-Use Development Capacity with 50% Project Open Space and Reduced Parking Requirement

Forecast Model MG1L

*Development capacity forecast for **MIXED USE** based on an adjacent **GRADE PARKING LOT** located on the same premises.*
***Given:** Gross land area. **To Find:** Maximum commercial building area and apartment dwelling unit capacity of the land area given when the residential land use allocation varies. **Premise:** All building floors considered equal in area.*

DESIGN SPECIFICATION *Enter values in boxed areas where text is bold and blue. Express all fractions as decimals.*

Given:	**Gross land area**	GLA=	**25.000**	acres	1,089,000	SF
Land Variables:	Public/ private right-of-way & paved easements	W=	**0.000**	fraction of GLA	0	SF
	Net land area	NLA=	25.000	acres	1,089,000	SF
	Unbuildable and/or future expansion areas	U=	**0.000**	fraction of GLA	0	SF
	Gross land area reduction	X=	0.000	fraction of GLA	0	SF
	Buildable land area remaining	BLA=	**25.000**	acres	1,089,000	SF
Parking Variables:	Est. gross pkg. lot area per pkg. space in SF	s =	**375**			
	Parking lot spaces planned or required per dwelling unit	u=	**1.00**			
	Garage parking spaces planned or required per dwelling unit	Gn=	**0.00**			
	Gross building area per garage space	Ga=	**0**			
	Nonresidential building SF permitted per parking space	a=	**200**			
	No. of loading spaces	l =	**3**			
	Gross area per loading space	b =	**1,000**	SF	3,000	SF
Site Variables:	**Project open space as fraction of BLA**	S=	**0.300**		326,700	SF
	Private driveways as fraction of BLA	R=	**0.010**		10,890	SF
	Misc. pavement as fraction of BLA	M=	**0.050**		54,450	SF
	Loading area as fraction of BLA	L=	0.003		3,000	SF
	Total site support areas as a fraction of BLA	Su=	0.363		395,040	SF
Core:	**Core development area as fraction of BLA**	C=	0.637	C+Su must = 1	693,960	SF
Building Variables:	Res. bldg. efficiency as percentage of GBA	Be=	**0.600**			
	Building support as fraction of GBA	Bu=	0.400	Be + Bu must = 1		

Dwelling Unit Mix Table:

DU dwelling unit type	GDA gross du area	CDA=GDA/Be comprehensive du area	MIX du mix	PDA = (CDA)MIX Prorated du area
EFF	**500**	833	**5%**	42
1 BR	**750**	1,250	**25%**	313
2 BR	**1,200**	2,000	**50%**	1,000
3 BR	**1,500**	2,500	**20%**	500
4 BR	**1,800**	3,000	**0%**	0
			Aggregate Avg. Dwelling Unit Area (AGG) =	**1,854**
			GBA sf per parking space a=	1,854

MIXED-USE PLANNING FORECAST

50.00% =(RAP): residential land use allocation percentage
50.00% =(CAP): nonresidential land use allocation percentage

total floors FLR	CGBA nonres GBA	RGBA res GBA	total bldg MBCA cover area	nonres CFLR floors	total parking MPLA lot area	total parking MNPS spaces	total dwelling MNDU units	density per dBA bldable acre
1.00	120,689	289,233	409,922	0.29	284,788	759	156.0	6.2
2.00	146,097	495,168	320,632	0.46	374,078	998	267.1	10.7
3.00	157,123	649,258	268,794	0.58	425,916	1,136	350.2	14.0
4.50	165,447	819,211	218,813	0.76	475,897	1,269	441.8	17.7
5.00	167,219	864,468	206,338	0.81	488,372	1,302	466.2	18.6
6.00	169,949	942,578	185,421	0.92	509,289	1,358	508.4	20.3
7.00	171,955	1,007,608	168,509	1.02	526,201	1,403	543.4	21.7
8.00	173,490	1,062,591	154,510	1.12	540,200	1,441	573.1	22.9
9.00	174,703	1,109,688	142,710	1.22	552,000	1,472	598.5	23.9
10.00	175,686	1,150,482	132,617	1.32	562,093	1,499	620.5	24.8
11.00	176,498	1,186,159	123,878	1.42	570,832	1,522	639.7	25.6
12.00	177,181	1,217,625	116,234	1.52	578,476	1,543	656.7	26.3
13.00	177,763	1,245,584	109,488	1.62	585,222	1,561	671.8	26.9
14.00	178,265	1,270,591	103,490	1.72	591,220	1,577	685.3	27.4
15.00	178,702	1,293,091	98,120	1.82	596,590	1,591	697.4	27.9

WARNING: These are preliminary forecasts that must not be used to make final decisions.
1) These forecasts are not a substitute for the "due diligence" research that must be conducted to support the final definition of "unbuildable areas" above and the final decision to purchase land. This research includes, but is not limited to, verification of adequate subsurface soil, zoning, environmental clearance, access, title, utilities and water pressure, clearance from deed restriction, easement and right-of-way encumbrances, clearance from existing above and below ground facility conflicts, etc.
2) The most promising forecast(s) made on the basis of data entered in the design specification from "due diligence" research must be verified at the drawing board before funds are committed and land purchase decisions are made. Actual land shape ratios, dimensions and irregularities encountered may require adjustments to the general forecasts above.
3) The software licensee shall take responsibility for the design specification values entered and any advice given that is based on the forecast produced.

all other values remain constant. The table also illustrates the principle that development capacity increases when parking requirements decrease and all other values remain constant. The same principles apply to nonresidential development.

Chapter 16 includes case studies entitled Kingsdale Shopping Center, The Expansion Model, The Barley Block, and Barley Block Office/Residential Mixed Use that address the issue of mixed use in more detail.

RENOVATION AND EXPANSION

Exhibit 5.4 illustrates forecast model MG1B and is written to predict the mixed-use development capacity of existing facilities. The design specification in this exhibit is similar to that in Exhibit 5.3, but also includes two values located in the planning forecast panel. These values define the total building footprint BCA in square feet to remain and the percentage of the total parking area that will be allocated to support residential activity (shown as 150,000 sq ft and 30%, respectively, in this example). Additional values have been added to the design specification to produce this exhibit. Based on these values, the gross building area table in the planning forecast panel predicts that the gross building area potential of the site is 740,994 sq ft, with 203,078 sq ft allocated to nonresidential building area. Since the building footprint to remain is only 150,000 sq ft, and since all floors are considered equal in area, the upper right-hand corner of the planning forecast panel predicts that 1.35 nonresidential floors and 3.59 residential floors are required to accommodate the capacity forecast. This means that a 5-story 740,994-sq ft building could be accommodated by the site when the building footprint is no more than 150,000 sq ft, when no more than 207,078 sq ft is devoted to nonresidential activity, and when all other specification values are met in plan development at the drawing board. This would produce 290.1 dwelling units in addition to the

Exhibit 5.4 Mixed-Use Renovation and Expansion Forecast

Forecast Model MG1B

Development capacity forecast for ***MIXED-USE*** *based on an adjacent* ***GRADE PARKING LOT*** *located on the same premises.*
Given: *Gross land area and building cover footprint.* ***To Find:*** *Maximum commercial building area and apartment dwelling unit capacity of the land area given when the residential parking lot area allocation varies.* ***Premise:*** *All building floors considered equal in area.*

DESIGN SPECIFICATION

Enter values in boxed areas where text is bold and blue. Express all fractions as decimals.

Given:	**Gross land area**	**GLA=**	**25.000**	acres	1,089,000	SF
Land Variables:	Public/ private right-of-way & paved easements	W=	**0.000**	fraction of GLA	0	SF
	Net land area	NLA=	25.000	acres	1,089,000	SF
	Unbuildable and/or future expansion areas	U=	**0.000**	fraction of GLA	0	SF
	Gross land area reduction	X=	0.000	fraction of GLA	0	SF
	Buildable land area remaining	**BLA=**	**25.000**	acres	1,089,000	SF
Parking Variables:	Est. gross pkg. lot area per pkg. space in SF	s =	**375**			
	Parking lot spaces planned or required per dwelling unit	u=	**1.5**			
	Garage parking spaces planned or required per dwelling unit	Gn=	**0**			
	Gross building area per garage space	Ga=	**0**			
	Nonresidential building SF permitted per parking space	a=	**200**			
	No. of loading spaces	l =	**3**			
	Gross area per loading space	b =	**1,000**	SF	3,000	SF
Site Variables:	**Project open space as fraction of BLA**	**S=**	**0.300**		326,700	SF
	Private driveways as fraction of BLA	R=	**0.010**		10,890	SF
	Misc. pavement as fraction of BLA	M=	**0.050**		54,450	SF
	Loading area as fraction of BLA	L=	0.003		3,000	SF
	Total site support areas as a fraction of BLA	Su=	0.363		395,040	SF
Core:	**Core development area as fraction of BLA**	**C=**	0.637	C+Su must = 1	693,960	SF
Building Variables:	Res. bldg. efficiency as percentage of GBA	Be=	**0.600**			
	Building support as fraction of GBA	Bu=	0.400	Be + Bu must = 1		

Dwelling Unit Mix Table:

DU dwelling unit type	GDA gross du area	CDA=GDA/Be comprehensive du area	MIX du mix		PDA = (CDA)MIX Pro-rated du area
EFF	**500**	833	**5%**		42
1 BR	**750**	1,250	**25%**		313
2 BR	**1,200**	2,000	**50%**		1,000
3 BR	**1,500**	2,500	**20%**		500
4 BR	**1,800**	3,000	**0%**		0
			Aggregate avg. dwelling unit area	(AGG) =	**1,854**
			GBA sf per parking space	a=	1,236

MIXED USE PLANNING FORECAST

	Building cover footprint desired or to remain	BCA=	**150,000**	SF	**1.35**	Nonres. floors needed
					3.59	Res. floors needed.
Parking allocation percentages	Residential parking area		**0.300**	% of total available	**4.94**	Total floors needed
	Nonresidential parking area		0.700	% of total available		
	Core development area	CORE=	693,960	SF		
	Parking area available	PLA=	543,960	SF		
	Garage access as multiple of bldg. area		no garages			
	Garage adjusted area per pkg. space	s=	375	SF	**290.1**	**No. res. dwell units**
	Parking capacity	NPS=	**1,450.6**	total spaces	**14.5**	EFF
					72.5	1 BR
Gross building area potential					**145.1**	2 BR
	Nonres. bldg. area		**203,078**	SF	**58.0**	3 BR
	Res. bldg. area		537,916	SF	**0.0**	4 BR
	Gross building area potential		**740,994**	SF	290.1	11.6 dBA

WARNING: These are preliminary forecasts that must not be used to make final decisions.
1) These forecasts are not a substitute for the "due diligence" research that must be conducted to support the final definition of "unbuildable areas" above and the final decision to purchase land. This research includes, but is not limited to, verification of adequate subsurface soil, zoning, environmental clearance, access, title, utilities and water pressure, clearance from deed restriction, easement and right-of-way encumbrances, clearance from existing above and below ground facility conflicts, etc.
2) The most promising forecast(s) made on the basis of data entered in the design specification from "due diligence" research must be verified at the drawing board before funds are committed and land purchase decisions are made. Actual land shape ratios, dimensions and irregularities encountered may require adjustments to the general forecasts above.
3) The software licensee shall take responsibility for the design specification values entered and any advice given that is based on the forecast produced.

nonresidential area forecast. The lower right-hand corner of the planning forecast panel shows that the dwelling unit mix would include 14.5 efficiency apartments, 72.5 one-bedroom units, 145.1 two-bedroom units, and 58.0 three-bedroom units, based on the values entered in the dwelling unit mix table of the specification. The dwelling unit density would equal 11.6 dwelling units per buildable acre in addition to the nonresidential area produced.[9]

All of these predictions can be found in the planning forecast panel of Exhibit 5.4, and can be used by an architect or engineer at the drawing board to begin schematic plan preparation. They can also be used by developers to evaluate investment opportunity; by planners to evaluate economic development potential; by elected officials to evaluate public policy alternatives; and by residents to evaluate the neighborhood impact of a mixed-use proposal. They should not be used to make final decisions, however, until the options that survive this analysis are tested at the drawing board. Exhibit 5.4 demonstrates the untapped potential of an underutilized shopping center parking lot. Figure 5.2 is not the basis for this particular exhibit, but it illustrates one of these parking lots. The possibilities are extensive and can be further expanded with the use of parking structures.

Parking Structures

Three generic parking structure systems have been identified in previous chapters. These can also be used by mixed-use development proposals to expand the potential development capacity of a given land area. The scope of this expanded potential is predicted by forecast model MS1L when using adjacent parking structure systems. Expanded development capacity produced by underground parking structure systems is forecast by model MS2L, and that produced by parking structures partially or fully above grade under a building is forecast by model MS3L.[10]

Figure 5.2 Typical strip shopping center. The parking lot in a strip shopping center represents a hidden asset that can be leveraged to produce mixed-use development when it is not required by the customer volume at the center.

(Photo by W. M. Hosack)

Any Ms-series model can be chosen to evaluate the expanded development capacity potential of a given parking structure system when considering a mixed-use development approach. The process of evaluation is similar to that used with all other models described. Values are entered where requested in a predefined design specification, and results are predicted in the planning forecast panel of each forecast model based on the values entered.[11]

ADJACENT PARKING STRUCTURES

Exhibit 5.5 illustrates forecast model MS1L and is based upon the format just described. However, there is one unique specification in MS1L that is found directly below the planning forecast panel. This value specifies the development capacity percentage allocated to support residential apartment land use within the mixed-use equation and is called the *residential allocation percentage* (RAP). In the case of this example, the value entered is 50%. This is an abstract value, however, and does not mean that the land area will be divided in half.[12] In a vertically integrated mixed-use context, this value means that 50% of the buildable land area will be used to calculate the residential development capacity of the site. The parking area and building

Exhibit 5.5 Mixed Use Development Capacity with Adjacent Parking Structure

Forecast Model MS1L

Development capacity forecast for ***APARTMENTS*** *using an* ***ADJACENT PARKING STRUCTURE*** *on the same premise.*
Given: *Gross land area available.* ***To Find:*** *Maximum dwelling unit capacity of the buildable land area given based on the design specification values entered below.*
Design Premise: *Building footprint adjacent to parking garage footprint within the core development area. All similar floors considered equal in area.*

DESIGN SPECIFICATION *Enter values in boxed areas where text is bold and blue. Express all fractions as decimals.*

Given:	**Gross land area**	**GLA=**	**25.000**	acres	1,089,000	SF
Land Variables:	Public/ private right-of-way & paved easements	W=	**0.000**	fraction of GLA	0	SF
	Net land area	NLA=	25.000	acres	1,089,000	SF
	Unbuildable and/or future expansion areas	U=	**0.000**	fraction of GLA	0	SF
	Gross land area reduction	X=	0.000	fraction of GLA	0	SF
	Buildable land area remaining	**BLA=**	**25.000**	acres	1,089,000	SF
Parking Variables:	Estimated net pkg. structure area per parking space	s =	**375**	SF		
	Parking spaces required per dwelling unit	u =	**1.5**			
	Nonresidential building SF permitted per parking space	a=	**200**			
	No. of parking levels contemplated	**p=**	**2.00**			
	Pkg. support as fraction of gross pkg. structure area (GPA)	Pu=	**0.200**			
	Net area for parking & circulation as fraction of GPA	Pe=	0.800	Pe+Pu must = 1		
	No. of loading spaces	l=	**3**			
	Gross area per loading space	b =	**1,000**	SF		
Site Variables:	**Project open space as fraction of BLA**	**S=**	**0.300**		326,700	SF
	Private driveways as fraction of BLA	R=	**0.010**		10,890	SF
	Misc. pavement as fraction of BLA	M=	**0.050**		54,450	SF
	Loading area as fraction of BLA	L=	0.003		3,000	SF
	Total site support areas as a fraction of BLA	Su=	0.363		395,040	SF
Core:	**Core development area as fraction of BLA**	**C=**	0.637	C+Su must = 1	693,960	SF
Building Variables:	Building efficiency as fraction of GBA	Be=	**0.600**			
	Building support as fraction of GBA	Bu=	0.400	Be+Bu must = 1		

NOTE: p=1 is a grade parking lot based on design premise. Increase the number of parking levels to increase the capacity forecast below. Other variables in blue, box and bold may also be changed.

Dwelling Unit Mix Table

DU dwelling unit type	GDA gross du area	CDA=GDA/Be comprehensive du area	MIX du mix	PDA = (CDA)MIX Prorated du area
EFF	**500**	833	**5%**	42
1 BR	**750**	1,250	**25%**	313
2 BR	**1,200**	2,000	**50%**	1,000
3 BR	**1,500**	2,500	**20%**	500
4 BR	**1,800**	3,000	**0%**	0
			Aggregate Avg. Dwelling Unit Area (AGG) =	**1,854**
			Building area planned, provided or required per parking space a =	1,236

MIXED-USE PLANNING FORECAST

50.00% =(RAP): residential land use allocation percentage
50.00% =(CAP): nonresidential land use allocation percentage

FLR total floors	CGBA nonres GBA	RGBA res GBA	gross MGBA bldg area	total bldg MBCA cover area	nonres CFLR floors	total parking MPLA cover	total parking MNPS spaces	total dwelling MNDU units	density per dBA bldable acre
1.00	159,761	291,676	451,437	451,437	0.4	242,523	1,034.8	157.3	6.29
2.00	207,539	503,156	710,696	355,348	0.6	338,612	1,444.7	271.4	10.85
3.00	230,520	663,518	894,037	298,012	0.8	395,948	1,689.4	357.9	14.31
4.00	244,030	789,296	1,033,326	258,332	0.9	435,628	1,858.7	425.7	17.03
5.00	252,924	890,590	1,143,514	228,703	1.1	465,257	1,985.1	480.3	19.21
6.00	259,222	973,915	1,233,137	205,523	1.3	488,437	2,084.0	525.3	21.01
7.00	263,917	1,043,662	1,307,579	186,797	1.4	507,163	2,163.9	562.9	22.51
8.00	267,551	1,102,901	1,370,452	171,306	1.6	522,654	2,230.0	594.8	23.79
9.00	270,447	1,153,839	1,424,286	158,254	1.7	535,706	2,285.7	622.3	24.89
10.00	272,810	1,198,107	1,470,917	147,092	1.9	546,868	2,333.3	646.2	25.85
11.00	274,774	1,236,936	1,511,709	137,428	2.0	556,532	2,374.5	667.1	26.68
12.00	276,432	1,271,268	1,547,700	128,975	2.1	564,985	2,410.6	685.6	27.43
13.00	277,851	1,301,843	1,579,694	121,515	2.3	572,445	2,442.4	702.1	28.08
14.00	279,079	1,329,245	1,608,325	114,880	2.4	579,080	2,470.7	716.9	28.68
15.00	280,152	1,353,945	1,634,097	108,940	2.6	585,020	2,496.1	730.2	29.21

WARNING: These are preliminary forecasts that must not be used to make final decisions.
1) These forecasts are not a substitute for the "due diligence" research that must be conducted to support the final definition of "unbuildable areas" above and the final decision to purchase land. This research includes, but is not limited to, verification of adequate subsurface soil, zoning, environmental clearance, access, title, utilities and water pressure, clearance from deed restriction, easement and right-of-way encumbrances, clearance from existing above and below ground facility conflicts, etc.
2) The most promising forecast(s) made on the basis of data entered in the design specification from "due diligence" research must be verified at the drawing board before funds are committed and land purchase decisions are made. Actual land shape ratios, dimensions and irregularities encountered may require adjustments to the general forecasts above.
3) The software licensee shall take responsibility for the design specification values entered and any advice given or decisions made that are based on the forecast produced.

footprint produced by this allocation may or may not actually consume 50% of the buildable land area, however, given the building height and parking ratios being considered for the mixed-use combination. It is a complex set of relationships that is controlled by the RAP specified in the planning forecast panel. If the results produced in the forecast panel do not display the number of dwelling units MNDU and the commercial building area CGBA desired, the RAP can be adjusted. All other design specification values can also be adjusted to explore alternate strategies. This does not mean that every development intensity objective is achievable, but that specification values can be adjusted to produce the drawing-board recipe closest to the development goal established.

The planning forecast in Exhibit 5.5 is summarized in Table 5.4. This table shows that the number of dwelling units MNDU increases 4.64 times (from 157.3 to 730.2) as building height increases to 15 floors. This disparity exists because 1.5 parking spaces will permit a 1854-sq-ft average dwelling unit area AGG to be constructed but only permit 300 sq ft of commercial area *a* to be built, based on the parking values and dwelling unit values entered in the design specification. This does not mean that there is a problem with these parking requirements, since very differ-

Table 5.4 Summary of Exhibit 5.5 Forecast

Floors	CGBA	*r*	MBCA	*r*	MGBA	*r*	MNPS	*r*	MNDU	*r*
1	159,761		451,437		451,437		1035		157	
5	252,924	1.58	228,703	0.51	1,143,514	2.53	1985	1.92	480	3.05
10	272,810	1.71	147,092	0.33	1,470,917	3.26	2333	2.25	646	4.11
15	280,152	1.75	108,940	0.24	1,634,097	3.62	2496	2.41	730	4.64

NOTE: *r* = rate of increase
CGBA = nonresidential gross building area in sq ft
MBCA = total mixed-use building cover or footprint in sq ft
MGBA = total mixed-use building area in sq ft
MNPS = total number of parking spaces for all mixed uses in sq ft
MNDU = total number of dwelling units in mixed-use configuration in sq ft

ent demand characteristics are involved. It does mean that the RAP should be considered in relation to this knowledge, since land allocated to residential use produces many times the building area produced by commercial use, but also produces much less yield per square foot of building area. (In the case of this example, 6.18 times the building area was produced.) Table 5.4 also shows that the gross building area MGBA forecast increases 3.67 times and the parking count MNPS increases 2.41 times as building height increases to 15 floors.[13] Please note, however, that all of these forecasts are based on the design specification value *p*. This value stipulates that two parking levels are contemplated in the adjacent parking structure. If this value were changed to 3, 4, or 5 levels, all forecasts and relationships would change accordingly.

UNDERGROUND PARKING STRUCTURES

Exhibit 5.6 illustrates forecast model MS2L, which is based on an underground parking structure design premise. The design specification values in this model are identical to those in Exhibit 5.5, except where additional or unique information pertaining to an underground parking system is required. The method of mixed-use forecasting is different, however, since the number of parking levels is not specified but forecast. For example, Exhibit 5.6 predicts the gross building area potential MGBA of the land area given based on the design specification values entered. The residential allocation percentage RAP defines the amount of the gross *building* area to be used for residential purposes, and the parking level forecast *p* predicts the number of underground parking levels needed to support this mixed-use specification defined. Total dwelling unit capacity MNDU is calculated based on the residential square feet available and the aggregate average dwelling unit area AGG found in the dwelling unit mix table. The total number of parking spaces MNPS needed is calculated from the parking ratio

Exhibit 5.6 Mixed-Use Development Capacity with Underground Parking Structure

Forecast Model MS2L

Development capacity forecast for ***APARTMENTS*** *based on the use of an* ***UNDERGROUND PARKING STRUCTURE.***
Given: *Gross land area.* ***To Find:*** *Dwelling unit capacity of the land area given based on the design specification values entered below.*
Design Premise: *Underground parking footprint may be larger, smaller or equal to the building footprint above. All similar floors considered equal in area.*

DESIGN SPECIFICATION *Enter values in boxed areas where text is bold and blue. Express all fractions as decimals.*

Given:	**Gross land area**	GLA=	**25.000**	acres	1,089,000	SF
Land Variables:	Public/ private right-of-way & paved easements	W=	**0.000**	fraction of GLA	0	SF
	Net land area	NLA=	25.000	acres	1,089,000	SF
	Future expansion and/or unbuildalbe areas	U=	**0.000**	fraction of GLA	0	SF
	Gross land area reduction	X=	0.000	fraction of GLA	0	SF
	Buildable land area remaining	BLA=	**25.000**	acres	1,089,000	SF
Parking Variables:	Est. net parking structure area per parking space	s =	**375**	SF		
	Parking spaces required per dwelling unit	u =	**1.5**			
	Non-residential building SF permitted per parking space	a =	**200**			
	No. of loading spaces	l =	**3**			
	Gross area per loading space	b =	**1,000**		3,000	SF
Site Variables at Grade:	**Project open space as fraction of BLA**	S=	**0.300**		326,700	SF
	Private driveways as fraction of BLA	R=	**0.010**		10,890	SF
	Misc. pavement as fraction of BLA	M=	**0.050**		54,450	SF
	Loading area as fraction of BLA	L=	0.003		3,000	SF
	Total site support areas at grade as a fraction of BLA	Su=	0.363		395,040	SF
Core:	**Core development area at grade as fraction of BLA**	C=	0.637	C+Su must = 1	693,960	SF
Below Grade:	Gross underground parking area (UNG) as fraction of BLA	G=	**0.900**		980,100	SF
	Parking support within parking structure as fraction of UNG	Pu=	**0.200**			
	Net parking structure for pkg. & circulation as fraction of UNG	Pe=	0.800			
Building Variables:	Building efficiency as fraction of GBA	Be=	**0.600**			
	Building support as fraction of GBA	Bu=	0.400	Be+Bu must = 1		

Dwelling Unit Mix Table

DU dwelling unit type	GDA gross du area	CDA=GDA/Be comprehensive du area	MIX du mix	PDA = (CDA)MIX Prorated du area
EFF	**500**	833	**5%**	42
1 BR	**750**	1,250	**25%**	313
2 BR	**1,200**	2,000	**50%**	1,000
3 BR	**1,500**	2,500	**20%**	500
4 BR	**1,800**	3,000	**0%**	0
			Aggregate Avg. Dwelling Unit Area	(AGG) = **1,854**

MIXED-USE PLANNING FORECAST

50.00% =(RAP): building area allocated to residential land use
50.00% =(CAP): building area allocated to nonresidential land use

FLR total bldg floors	no. of p parking levels	nonres gross CGBA bldg area	residential gross RGBA bldg area	gross MGBA bldg area	nonres CFLR floors	total parking MPLA area per level	total parking MNPS spaces	total dwelling MNDU units	density per dBA bldable acre
1.00	**1.93**	346,980	346,980	693,960	0.5	980,100	2,016	187.1	7.5
2.00	**3.86**	693,960	693,960	1,387,920	1.0		4,031	374.3	15.0
3.00	**5.78**	1,040,940	1,040,940	2,081,880	1.5		6,047	561.4	22.5
4.00	**7.71**	1,387,920	1,387,920	2,775,840	2.0	total bldg	8,062	748.5	29.9
5.00	**9.64**	1,734,900	1,734,900	3,469,800	2.5	MBCA	10,078	935.7	37.4
6.00	**11.57**	2,081,880	2,081,880	4,163,760	3.0	cover area	12,094	1,122.8	44.9
7.00	**13.50**	2,428,860	2,428,860	4,857,720	3.5	693,960	14,109	1,309.9	52.4
8.00	**15.42**	2,775,840	2,775,840	5,551,680	4.0		16,125	1,497.1	59.9
9.00	**17.35**	3,122,820	3,122,820	6,245,640	4.5		18,140	1,684.2	67.4
10.00	**19.28**	3,469,800	3,469,800	6,939,600	5.0		20,156	1,871.4	74.9
11.00	**21.21**	3,816,780	3,816,780	7,633,560	5.5		22,172	2,058.5	82.3
12.00	**23.14**	4,163,760	4,163,760	8,327,520	6.0		24,187	2,245.6	89.8
13.00	**25.06**	4,510,740	4,510,740	9,021,480	6.5		26,203	2,432.8	97.3
14.00	**26.99**	4,857,720	4,857,720	9,715,440	7.0		28,218	2,619.9	104.8
15.00	**28.92**	5,204,700	5,204,700	10,409,400	7.5		30,234	2,807.0	112.3

WARNING: These are preliminary forecasts that must not be used to make final decisions.
1) These forecasts are not a substitute for the "due diligence" research that must be conducted to support the final definition of "unbuildable areas" above and the final decision to purchase land. This research includes, but is not limited to, verification of adequate subsurface soil, zoning, environmental clearance, access, title, utilities and water pressure, clearance from deed restriction, easement and right-of-way encumbrances, clearance from existing above and below ground facility conflicts, etc.
2) The most promising forecast(s) made on the basis of data entered in the design specification from "due diligence" research must be verified at the drawing board before funds are committed and land purchase decisions are made. Actual land shape ratios, dimensions and irregularities encountered may require adjustments to the general forecasts above.
3) The software licensee shall take responsibility for the design specification values entered and any advice given or decisions made that are based on the forecast produced.

requirements specified. The number of parking levels required is a function of the number of parking spaces calculated. The density per buildable acre dBA is a function of the dwelling unit quantities forecast. The results are summarized in Table 5.5. These results show, for instance, that a 1-story building, or buildings, on 70% of a 25-acre land area would require approximately 2 underground parking structure levels on 90% of the land area and produce only 30 more dwelling units than Exhibit 5.5. It would, however, produce more that twice as much commercial building area. If more dwelling units and less commercial building area were desired, the RAP would have to be increased in the design specification. This increase would produce a smaller underground parking level requirement, since square feet devoted to residential use produce a much lower parking demand than those devoted to commercial use, as mentioned earlier.

While the capacity of an underground parking structure appears dramatically greater when Table 5.5 is compared to Table 5.4, keep in mind that Table 5.4 is based on a two-level garage above grade on an adjacent portion of the buildable area. Table 5.5 is based on an unlimited number of underground parking levels covering 90% of the available underground land area.

Table 5.5 Summary of Exhibit 5.6 Forecast

Floors	*p*	CGBA	MBCA	MGBA	MNPS	MNDU
1	1.93	346,980	693,960	693,960	2,016	187
5	9.64	1,734,900	693,960	3,469,800	10,078	936
10	19.28	3,469,800	693,960	6,939,600	20,156	1871
15	28.92	5,204,700	693,960	10,409,400	30,234	2807

NOTE: p = number of underground parking levels
CGBA = nonresidential gross building area in sq ft
MBCA = total mixed-use building cover or footprint in sq ft
MGBA = total mixed-use building area in sq ft
MNPS = total number of parking spaces for all mixed uses in sq ft
MNDU = total number of dwelling units in mixed-use configuration in sq ft

Since the number of underground parking levels *p* needed to accommodate increasing building height and area is unlimited by the planning forecast panel in Exhibit 5.6, the user must exercise professional judgment to determine when the number of underground levels becomes unrealistic.[14]

Parking Structure Above Grade Under Building

Exhibit 5.7 illustrates mixed-use forecast model MS3L. This model is based on the premise that a building is located above a parking garage, that the parking garage is at least partially above grade, and that each may be larger or smaller in footprint area than the other. It also assumes that all floors within each are equal in area.

The format of the planning forecast panel within model MS3L begins with a parking column that ranges from 1 to 15 levels. Increasing numbers of parking levels *p* produce increasing commercial areas CGBA, increasing residential areas RGBA, and increasing parking counts MNPS. These in turn produce increasing parking spaces MNPS, increasing numbers of building floors FLR, increasing numbers of dwelling units MNDU, and increasing densities dBA.

None of this is unexpected. What is evident again in this exhibit is the great disparity between the amount of residential apartment area RGBA that can be supported by 50% of the land area allocation and the amount of commercial area CGBA that can be supported by the remaining 50%. This is a recurring theme that is actually a design principle based on the different parking demand characteristics of the two land use families. Activities within the residential land use family have less parking demand and produce greater building areas per parking space than most, if not all, activities within the nonresidential land use family. This principle can be very useful in the mixed-use conversion of old shopping centers, for instance, since a small reduction in the amount of commercial area can free up a large number of parking spaces to serve new residential apartments placed above.

Exhibit 5.7 Mixed-Use Development Capacity with Parking Structure Above Grade Under Building

Forecast Model MS3L

Development capacity forecast for ***APARTMENTS*** *based on the use of a* ***PARKING STRUCTURE*** *either partially or completely* ***ABOVE*** *grade* ***UNDER*** *the building or buildings.*

Given: *Gross land area.* ***To Find:*** *Maximum dwelling unit capacity of the gross land area given based on the design specification values entered below.*
Design Premise: *Parking footprint is beneath building footprint. The footprints may equal or exceed each other within the core development area. All similar floors considered equal in area.*

DESIGN SPECIFICATION

Enter values in boxed areas where text is bold and blue. Express all fractions as decimals.

Given:	**Gross land area**	GLA=	**5.000**	acres	217,800
Land Variables:	Public/ private right-of-way & paved easements	W=	**0.000**	fraction of GLA	0
	Net land area	NLA=	5.000	acres	217,800
	Future expansion and/or unbuildalbe areas	U=	**0.000**	fraction of GLA	0
	Gross land area reduction	X=	0.000	fraction of GLA	0
	Buildable land area remaining	BLA=	**5.000**	acres	217,800
Parking Variables:	Est. net parking structure area per parking space	s =	**375**	SF	
	Building SF permitted per parking space	a=	**200**		
	Parking spaces required per dwelling unit	u =	**1.5**		
	No. of loading spaces	l =	**3.00**		
	Gross area per loading space	b =	**1,000**	SF	3,000
Site Variables:	**Project open space as fraction of BLA**	S=	**0.300**		65,340
	Private driveways as fraction of BLA	R=	**0.010**		2,178
	Misc. pavement as fraction of BLA	M=	**0.050**		10,890
	Loading area as fraction of BLA	L=	0.014		3,000
	Total site support areas as a fraction of BLA	Su=	0.374		81,408
Core:	**Core development area as fraction of BLA**	C=	0.626	C+Su must = 1	136,392
Pkg. Structure Variables:	Gross pkg. structure cover as fraction of core area (C)	P=	**1.000**	DO NOT EXCEED 1.0	136,392
	Parking support as fraction of gross parking area (GPA)	Pu=	**0.200**		
	Net area for parking & circulation as fraction of GPA	Pe=	0.800	Pu+Pe must = 1	109,114
Building Variables	Bldg. footprint over pkg structure as fraction of C	B=	**1.000**	DO NOT EXCEED 1.0	
	Building efficiency as percentage of gross bldg. area (GBA)	Be=	**0.600**		
	Bldg. support as fraction of GBA	Bu=	0.400	Bu+Be must equal 1	

Dwelling Unit Mix	DU dwelling unit type	GDA gross du area	CDA/Be comprehensive du area	MIX du mix	PDA = (CDA)MIX Prorated du area
	EFF	**500**	833	**0%**	0
	1 BR	**750**	1,250	**25%**	313
	2 BR	**1,200**	2,000	**50%**	1,000
	3 BR	**1,500**	2,500	**20%**	500
	4 BR	**1,800**	3,000	**0%**	0
				Aggregate avg. dwelling unit area (AGG) =	**1,813**
				Building area planned, provided or required per parking space a =	1,208

MIXED-USE PLANNING FORECAST

50.00% =(RAP): residential land use allocation percentage
50.00% =(CAP): nonresidential land use allocation percentage

no. of pkg. levels p	CORE	CGBA nonres GBA	nonres CFLR floors	residential RGBA gross bldg area	gross bldg area MGBA	total parking MNPS spaces	no. of res + comm FLR bldg. floors	total floors F bldg + pkg	total dwelling MNDU units	density per dBA bldable acre
NOTE: p <= 1 = grade parking lot beneath building										
1.00	136,392	29,097	0.2	175,794	204,891	291.0	3.0	4.0	97.0	19.40
2.00		58,194	0.4	351,588	409,782	581.9	6.0	8.0	194.0	38.80
3.00	footprint	87,291	0.6	527,382	614,673	872.9	9.0	12.0	291.0	58.19
4.00	MBCA	116,388	0.9	703,177	819,564	1,163.9	12.0	16.0	388.0	77.59
5.00	136,392	145,485	1.1	878,971	1,024,455	1,454.8	15.0	20.0	484.9	96.99
6.00		174,582	1.3	1,054,765	1,229,347	1,745.8	18.0	24.0	581.9	116.39
7.00	gross pkg struc area	203,679	1.5	1,230,559	1,434,238	2,036.8	21.0	28.0	678.9	135.79
8.00	GPL	232,776	1.7	1,406,353	1,639,129	2,327.8	24.0	32.0	775.9	155.18
9.00	136,392 per level	261,873	1.9	1,582,147	1,844,020	2,618.7	27.0	36.0	872.9	174.58
10.00		290,970	2.1	1,757,941	2,048,911	2,909.7	30.0	40.0	969.9	193.98
11.00	net pkg area per level	320,067	2.3	1,933,735	2,253,802	3,200.7	33.0	44.0	1066.9	213.38
12.00	NPL	349,164	2.6	2,109,530	2,458,693	3,491.6	36.1	48.1	1163.9	232.78
13.00	109,114	378,260	2.8	2,285,324	2,663,584	3,782.6	39.1	52.1	1260.9	252.17
14.00		407,357	3.0	2,461,118	2,868,475	4,073.6	42.1	56.1	1357.9	271.57
15.00		436,454	3.2	2,636,912	3,073,366	4,364.5	45.1	60.1	1454.8	290.97

WARNING: These are preliminary forecasts that must not be used to make final decisions.
1) These forecasts are not a substitute for the "due diligence" research that must be conducted to support the final definition of "unbuildable areas" above and the final decision to purchase land. This research includes, but is not limited to, verification of adequate subsurface soil, zoning, environmental clearance, access, title, utilities and water pressure, clearance from deed restriction, easement and right-of-way encumbrances, clearance from existing above and below ground facility conflicts, etc.
2) The most promising forecast(s) made on the basis of data entered in the design specification from "due diligence" research must be verified at the drawing board before funds are committed and land purchase decisions are made. Actual land shape ratios, dimensions and irregularities encountered may require adjustments to the general forecasts above.
3) The software licensee shall take responsibility for the design specification values entered and any advice given or decisions made that are based on the forecast produced.

This point can be further illustrated by changing the values in Exhibit 5.7 to create Exhibit 5.8. The residential allocation percentage RAP has been increased to 90% and the parking ratio has been reduced to 0.75 spaces per dwelling unit to explore the implications of classic high-rise residential apartment configurations found in high-density urban areas. The remaining 10% commercial allocation percentage CAP establishes the mixed-use ratio represented by the exhibit. Project open space remains at 30% to compensate for the building height involved,[15] and three loading spaces are provided for nonresidential services. Dwelling unit data is defined in the dwelling unit mix table and the resulting forecast predicts that a small city could be created within one building on 5 buildable acres.

Exhibit 5.8 reveals that a parking garage with a p of 15 levels and a GPL of 136,392-sq-ft area per floor could provide an MNPS of 4,364.5 spaces and support a gross habitable building area MGBA of 9,580,174 sq ft based on the specification values entered.[16] Within a building footprint of equal area, 87,291 sq ft, or a CFLR of 0.64 floors, would be devoted to commercial use (CGBA). The parking provided would support a FLR of approximately 84 total floors of residential and commercial activity, and the total building height including garage F would equal approximately 99 floors. The resulting density on this 5 acres could be 1,047.49 dwelling units per buildable acre. While this building type and density is not for everyone, it does serve to illustrate one end of the development intensity spectrum that is used to shelter a growing population. It should also be apparent that the tools used to make these predictions can also be used to index the results and limit the impact of excessive intensity.

The impact of an increase in residential allocation RAP and a decrease in the residential parking requirement can be seen by comparing Exhibit 5.7 to 5.8. When the RAP is 50% in Exhibit 5.7 and the parking requirement is 1.5

Exhibit 5.8 Revised RAP, CAP, and u Values in Exhibit 5.7

Forecast Model MS3L

Development capacity forecast for ***APARTMENTS*** *based on the use of a* ***PARKING STRUCTURE*** *either partially or completely* ***ABOVE*** *grade* ***UNDER*** *the building or buildings.*

Given: *Gross land area.* ***To Find:*** *Maximum dwelling unit capacity of the gross land area given based on the design specification values entered below.*
Design Premise: *Parking footprint is beneath building footprint. The footprints may equal or exceed each other within the core development area. All similar floors considered equal in area.*

DESIGN SPECIFICATION — *Enter values in boxed areas where text is bold and blue. Express all fractions as decimals.*

Given:	**Gross land area**	GLA=	**5.000**	acres	217,800
Land Variables:	Public/ private right-of-way & paved easements	W=	**0.000**	fraction of GLA	0
	Net land area	NLA=	5.000	acres	217,800
	Future expansion and/or unbuildalbe areas	U=	**0.000**	fraction of GLA	0
	Gross land area reduction	X=	0.000	fraction of GLA	0
	Buildable land area remaining	BLA=	**5.000**	acres	217,800
Parking Variables:	Est. net parking structure area per parking space	s =	**375**	SF	
	Building SF permitted per parking space	a=	**200**		
	Parking spaces required per dwelling unit	u =	**0.75**		
	No. of loading spaces	l =	**3.00**		
	Gross area per loading space	b =	**1,000**	SF	3,000
Site Variables:	**Project open space as fraction of BLA**	S=	**0.300**		65,340
	Private driveways as fraction of BLA	R=	**0.010**		2,178
	Misc. pavement as fraction of BLA	M=	**0.050**		10,890
	Loading area as fraction of BLA	L=	0.014		3,000
	Total site support areas as a fraction of BLA	Su=	0.374		81,408
Core:	**Core development area as fraction of BLA**	C=	0.626	C+Su must = 1	136,392
Pkg. Structure Variables:	Gross pkg. structure cover as fraction of core area (C)	P=	**1.000**	DO NOT EXCEED 1.0	136,392
	Parking support as fraction of gross parking area (GPA)	Pu=	**0.200**		
	Net area for parking & circulation as fraction of GPA	Pe=	0.800	Pu+Pe must = 1	109,114
Building Variables	Bldg. footprint over pkg structure as fraction of C	B=	**1.000**	DO NOT EXCEED 1.0	
	Building efficiency as percentage of gross bldg. area (GBA)	Be=	**0.600**		
	Bldg. support as fraction of GBA	Bu=	0.400	Bu+Be must equal 1	

Dwelling Unit Mix	DU dwelling unit type	GDA gross du area	CDA/Be comprehensive du area	MIX du mix	PDA = (CDA)MIX Prorated du area
	EFF	**500**	833	**0%**	0
	1 BR	**750**	1,250	**25%**	313
	2 BR	**1,200**	2,000	**50%**	1,000
	3 BR	**1,500**	2,500	**20%**	500
	4 BR	**1,800**	3,000	**0%**	0
				Aggregate avg. dwelling unit area (AGG) =	**1,813**
				Building area planned, provided or required per parking space a =	2,417

MIXED USE PLANNING FORECAST

90.00% =(RAP): residential land use allocation percentage
10.00% =(CAP): nonresidential land use allocation percentage

no. of pkg. levels p	CORE	CGBA nonres GBA	nonres CFLR floors	residential RGBA gross bldg area	gross bldg area MGBA	total parking MNPS spaces	no. of res + comm FLR bldg. floors	total floors F bldg + pkg	total dwelling MNDU units	density per dBA bldable acre
NOTE: p <= 1 = grade parking lot beneath building										
1.00	136,392	5,819	0.04	632,859	638,678	291.0	5.6	6.6	349.2	69.83
2.00		11,639	0.09	1,265,718	1,277,357	581.9	11.2	13.2	698.3	139.67
3.00	footprint	17,458	0.13	1,898,577	1,916,035	872.9	16.7	19.7	1047.5	209.50
4.00	MBCA	23,278	0.17	2,531,436	2,554,713	1,163.9	22.3	26.3	1396.7	279.33
5.00	136,392	29,097	0.21	3,164,294	3,193,391	1,454.8	27.9	32.9	1745.8	349.16
6.00		34,916	0.26	3,797,153	3,832,070	1,745.8	33.5	39.5	2095.0	419.00
7.00	gross pkg struc area	40,736	0.30	4,430,012	4,470,748	2,036.8	39.1	46.1	2444.1	488.83
8.00	GPL	46,555	0.34	5,062,871	5,109,426	2,327.8	44.7	52.7	2793.3	558.66
9.00	136,392 per level	52,375	0.38	5,695,730	5,748,104	2,618.7	50.2	59.2	3142.5	628.49
10.00		58,194	0.43	6,328,589	6,386,783	2,909.7	55.8	65.8	3491.6	698.33
11.00	net pkg area per level	64,013	0.47	6,961,448	7,025,461	3,200.7	61.4	72.4	3840.8	768.16
12.00	NPL	69,833	0.51	7,594,307	7,664,139	3,491.6	67.0	79.0	4190.0	837.99
13.00	109,114	75,652	0.55	8,227,165	8,302,818	3,782.6	72.6	85.6	4539.1	907.83
14.00		81,471	0.60	8,860,024	8,941,496	4,073.6	78.2	92.2	4888.3	977.66
15.00		87,291	0.64	9,492,883	9,580,174	4,364.5	83.7	98.7	5237.5	1047.49

WARNING: These are preliminary forecasts that must not be used to make final decisions.
1) These forecasts are not a substitute for the "due diligence" research that must be conducted to support the final definition of "unbuildable areas" above and the final decision to purchase land. This research includes, but is not limited to, verification of adequate subsurface soil, zoning, environmental clearance, access, title, utilities and water pressure, clearance from deed restriction, easement and right-of-way encumbrances, clearance from existing above and below ground facility conflicts, etc.
2) The most promising forecast(s) made on the basis of data entered in the design specification from "due diligence" research must be verified at the drawing board before funds are committed and land purchase decisions are made. Actual land shape ratios, dimensions and irregularities encountered may require adjustments to the general forecasts above.
3) The software licensee shall take responsibility for the design specification values entered and any advice given or decisions made that are based on the forecast produced.

spaces per dwelling unit, 15 parking levels can support a gross building area MGBA of 3,073,366 sq ft and produce a dwelling unit density of 290.97 units per buildable acre. When the RAP value is increased to 90% in Exhibit 5.8 and the parking requirement is reduced to 0.75 spaces per dwelling unit, the same number of parking levels can support a MGBA of 9,492,883 sq ft. This produces a density of 1,047.49 units per buildable acre. The issue of environmental balance and necessary levels of project open space in relation to these intensity levels is beyond the scope of this effort and the scientific ability of the author. The power of a design specification to forecast and regulate these levels to shelter growing populations, however, should be clear.

Conclusion

Mixed use offers a more compact solution to the problem of providing shelter for diverse activities. The development components of these activities are shared, and a common footprint is used. Parking and miscellaneous pavement are shared, and roadways, when present, connect stacked activities within shorter distances. Unfortunately, these efficiencies are achieved with taller buildings that increase the intensity proposed. It is possible to increase project open space to offset this intensity and still provide less open space than would be provided in two separate projects, but the amount required is simply an intuitive judgment at the present time.

A mixed-use lifestyle will not appeal to everyone, since the intensity produced is greater than that found in suburban sprawl. We think of intensity in personal terms, however. In reality, low-intensity sprawl represents a high level of environmental intensity that places a great burden on our silent partner, and we are still debating the balance that is required.

Notes

1. It is possible that these relationships could be horizontal within the same building, but the term *vertical* has been adopted for convenience when referring to building construction. The term *horizontal* has been adopted to describe two-dimensional land use relationships.
2. Property value affects not only the welfare of the landowner but the economic base of the community.
3. It has also had limited success protecting property value in older neighborhoods where historically based random land use relationships establish a precedent that undermines newer zoning district objectives.
4. There is a difference, however, between a randomly evolving mixture that produces uncertainty and deterioration and a planned mixture that produces a new neighborhood.
5. The collection of 30 models discussed previously can address horizontal applications.
6. In the case of MS2L, the RAP value is expressed as a percentage of the gross building area.
7. This percentage does not mean that the land will be divided according to this value, since this would be a horizontal mixed-use application. It means that this fraction of the total development capacity will be allocated to residential land use. A more complete explanation will follow.
8. The breakdown among dwelling unit types can be calculated by multiplying the appropriate percentage in the MIX column by the total number of dwelling units forecast in the MNDU column. These breakdowns are calculated by the model but are not shown because of space limitations.
9. The maximum achievable density is less than the ideal since the footprint area does not represent the best relationship between parking and building area that will produce the maximum density possible. If this were important, the user could test a number of footprint areas to arrive near the maximum potential dwelling unit capacity of the site.
10. Hybrid parking structure systems may evolve during the plan preparation process, but are too complex to predict. For this reason, development capacity forecasts predict the initial parameters of potential

development capacity but not necessarily the final design solution produced at the drawing board.

11. Mixed-use forecast models have two parents that work behind the scene. For instance, the forecast model MS1L distributes the values entered in its design specification to its parent models CS1L and RS1L. Forecasts are produced in these parent models and sent back to the planning forecast panel of model MS1L. These values are displayed in MS1L, and some are modified based on the mixed-use information provided. The result is a mixed-use forecast that is driven by the data entered and the parent processing models associated with each. This format pertains to all mixed-use forecast models in Table 5.1.
12. This would be a horizontally integrated mixed-use approach. This example is based upon a vertically integrated mixed-use concept.
13. Building cover and parking cover footprints are adjusted in Exhibit 5.4 in relation to building height to achieve these increases.
14. Keep in mind with this forecast model that an increase in the residential allocation percentage (RAP) and a corresponding decrease in the commercial allocation percentage can reduce the number of parking spaces and parking levels required.
15. Research is needed to determine what levels of intensity are acceptable.
16. Keep in mind that this area does not need to be located in a single building.

Land Use Allocation and Economic Stability

Municipal land use allocation has been briefly discussed in the economic development section of Chap. 3. It is a simple approach that requires a somewhat different aggregation of existing financial data; an expanded view of the purpose of land within a city; and a simplified method of planning, regulating, and evaluating development proposals. A *land use allocation plan* is simply a land use plan whose areas and locations are not only based on compatible relationships and future growth areas, but on existing financial data and future economic development objectives as well. Land use allocation is meant to correlate the realistic development capacity of a city's land area with the annual yield required to produce a stable financial platform for all residents and landowners. The financial profile of a bedroom suburb surrounded by other incorporated areas is presented in Table 6.1 and is used as the basis for this discussion.

Yield

All municipal acres do not provide the same yield, but the combined average income per acre in a stable community should meet or exceed the city's total annual expense per

Table 6.1 Profile of a Bedroom Suburb

	Residential	Effective Rates Res %	Nonresidential
MILLAGES			
1 City: 6.29	6.138015	10.5%	6.290000
2 County: 14.57	11.830000	20.3%	13.329492
3 School: 74.66	39.549064	67.8%	43.042970
4 Library: 1.00	0.850000	1.5%	1.000000
Total millage	58.367079		63.662462
Personal prop. millage			96.52
VARIABLES			
Income tax %	0.02		
Real prop. valuation %	0.35		
Total income tax paid	$13,347,366		
Personal prop. valuation	0.25		
			avg. res. density
No. of dwelling units	13,956		2.61
RESIDENTIAL REVENUE			
Avg. real estate tax	$4,169,874		
Avg. income tax	$2,808,321		
OTHER RESIDENTIAL REVENUE			
Estate tax	$3,028,710		
Permits & fees	$399,468		
Other	$3,081,022		
Total other	$6,509,200		
NONRESIDENTIAL VARIABLES			
Income tax	$5,367,211		
Real estate tax	$605,893		
Other	$258,130		
AREAS			
Residential AC	5,353		
Institutional AC	589		
Res+Inst AC	5,942		
Nonresidential AC	296		
Total	6,238		
LAND VALUATION			personal prop.
Residential	$721,903,410		$93,051,224
Nonresidential	$88,434,130		
CITY EXPENSE			
Operations & maintenance	$20,171,314		$3,234
Depreciated assets	$838,813		$134
Old debt service	$2,336,870		$375
COST/ AC			$3,743
Capital improvements	$7,236,000	**NOTE	
CITY REVENUE			
Residential revenue	$13,487,395		
Commercial revenue	$6,231,234		
Institutional revenue	$843,239		
Other revenue	$2,772,071		
REV/ AC			$3,741

**NOTE

This is a five year capital improvement program funded with a 20 year debt issue that is not reflected in this expense and revenue example since it distorts the view of annually recurring income. A "stable financial platform" in this example is considered to be annually recurring income that does not require additional voter approval for continuation. Estate, or inheritance, tax revenue is included in this example, but it is a questionable component of "stable income" since it represents an annual "windfall" that is essentially unreliable even when consistent for a number of years. If inheritance tax were not included in this summary, however, the annual expense would obviously be greater than the city's "stable" annual income. The addition of five year debt service income would simply disguise this "stable annual income picture". Since five year debt income ends with fifteen years of remaining debt service on each resident's property tax bill, this income is not considered "stable annually recurring income" in this example since it must be renewed in year six with another debt issue requiring voter approval. This adds to a resident's tax bill and must be repeated in five year increments until the first issue is retired in twenty years. This approach increases the real estate tax burden, and voter approval is a question mark that has removed this income from the "stable" category. The debt service for this "income" has also been removed. If this debt service were added and the estate tax income removed, a very different picture of the city's "stable financial platform" would quickly emerge. It is beyond the scope of this book, but it would be interesting to study "financial stability benchmarking" by exploring results produced by dividing a city's stable annual income by its total annual income, and by dividing its annual operating expense by its total annual expense.

acre. When a city can record the yield it receives from each land use category, and knows the acres allocated to each category, it can evaluate the performance of those acres and the combined yield needed to support its current lifestyle. It can also evaluate economic development targets, adjustments needed, land use designations for future annexation areas, the capacity of specific land areas to meet its economic targets, and the extent of reasonable tax abatement. This was a much more difficult task before the advent of geographic information systems (GIS) linked to county real estate data. The land area measurement and tax information required are now within the realm of possibility for any community. In this case study, the first three columns of Table 6.2 define physical information and financial data pertaining to a community that will be called Center City. The block of data called *Areas* in Table 6.2 shows that approximately 4.7% (296 acres) of the city's land is devoted to nonresidential land use and 95.3% is devoted to residential and institutional land uses. The *Variable* block shows that 13,956 dwelling units are present, which is 61% of the total that would theoretically be permitted in the same area by the city's zoning ordinance. The *millage* block shows that the municipal real estate tax revenue received by the city from its residential land use is 10.5% of the total paid by its residents, with the majority going to the city's school system and county. The income tax revenue received by the city is 21% of that paid by its residents, since it is a highly paid bedroom community that exports its workforce to the surrounding area.

Allocation

Table 6.2, entitled "Economic Implications of Geographic Land Use Allocation," illustrates the role of land use in this community. The data labeled #1 illustrate the disparity between the real estate taxes paid to the schools and the

Table 6.2 Economic Implications of Land Use Allocation

	Effective Rates	
	Residential	Nonresidential
MILLAGES		
1 City: 6.29	6.138015	6.290000
2 County: 14.57	11.830000	13.329492
3 School: 74.66	39.549064	43.042970
4 Library: 1.00	0.850000	1.000000
Total millage	58.367079	63.662462
Personal prop. millage		96.52
VARIABLES		
Income tax %	0.02	
Real prop. valuation %	0.35	
Total income tax paid	$13,347,366	
Personal prop. valuation	0.25	
		avg. res. density
No. of dwelling units	13,956	2.61
RESIDENTIAL REVENUE		
Avg. real estate tax	$4,169,874	
Avg. income tax	$2,808,321	
OTHER RESIDENTIAL REVENUE		
Estate tax	$3,028,710	
Permits & fees	$399,468	
Other	$3,081,022	
Total Other	$6,509,200	
NONRESIDENTIAL VARIABLES		
Income tax	$5,367,211	
Real estate tax	$605,893	
Other	$258,130	
AREAS		
Residential AC	5353	
Institutional AC	589	
Res+Inst AC	5942	
Commercial AC	296	
Total	6238	
LAND VALUATION		personal prop.
Residential	$721,903,410	$93,051,224
Commercial	$88,434,130	
CITY EXPENSE		
Operations	$20,171,314	$3,234
Depreciated assets	$838,813	$134
Old debt service	$2,336,870	$375
1994 improvements		
Proj. 1994 Debt Service		
COST/ AC		$3,743
		Rev/ AC
CITY REVENUE		
Residential revenue	$13,487,395	
Commercial revenue	$6,231,234	
Institutional revenue	$843,239	
Other fevenue	$2,772,071	$3,741

Zone	RA	RB	RC	RD	R1	R2	R3		R4, R5, & R6			R7	R7	Non-residential	Totals
Land use %	3.37%	0.52%	4.58%	0.13%	1.62%	38.64%	37.45%	1.34%	3.56%	3.56%	0.45%	0.03%	0.00%		
Acres/DU required	3	2	1	0.5	0.34	0.27	0.21	0.21	0.10	0.09	0.07	0.05	0.03		
Est. road allocation %	0.2	0.2	0.2	0.15	0.15	0.15	0.1	0.1	0.1	0.1	0.1	0.1	0.1		
Net density	0.33	0.50	1.00	2.00	2.90	3.63	4.84	4.84	9.68	10.89	14.52	**21.78**	**34.85**		
Gross density	0.27	0.40	0.80	1.70	2.47	3.09	4.36	4.36	8.71	9.80	13.07	19.60	31.37		
Land use AC	210.46	32.36	285.81	8.24	101.14	2410.59	2336.10	83.59	222.07	222.07	28.07	1.62	0.00		5,942
Est. no. of DU	34	8	138	14	151	4498	6162	220	1171	1318	222	19	0		13,956
Avg REtax paid/ DU	**$9,035**	**$8,192**	**$5,421**	**$4,217**	**$3,614**	**$3,313**	**$3,012**	**$2,409**	**$1,807**	**$1,205**	**$964**	**$843**			$39,651,804
Income tax to city/ DU	**$585**	**$487**	**$390**	**$341**	**$292**	**$234**	**$195**	**$175**	**$156**	**$136**	**$117**	**$97**			$2,808,321
School property tax/ DU	$5,478	$4,967	$3,287	$2,557	$2,191	$2,009	$1,826	$1,461	$1,096	$730	$584	$511			$24,042,271
County property tax/ DU	$1,639	$1,486	$983	$765	$655	$601	$546	$437	$328	$218	$175	$153			$7,191,575
Library property tax/ DU	$118	$107	$71	$55	$47	$43	$39	$31	$24	$16	$13	$11			$516,723
City property tax/ DU	$950	$861	$570	$443	$380	$348	$317	$253	$190	$127	$101	$89		$2,047	$4,169,874
City other tax/ DU	$1,483	$1,345	$890	$692	$593	$544	$494	$396	$297	$198	$158	$138		$872	$6,509,200
City income tax/ DU	$585	$487	$390	$341	$292	$234	$195	$175	$156	$136	$117	$97		$18,132	$2,808,321
Institutional revenue/ DU	$192	$174	$115	$90	$77	$70	$64	$51	$38	$26	$20	$18			$843,239
TOTAL CITY REV/ DU	**$3,210**	**$2,868**	**$1,965**	**$1,566**	**$1,343**	**$1,197**	**$1,070**	**$876**	**$681**	**$486**	**$397**	**$342**			
TOTAL CITY REV/ AC	$518	$695	$952	$2,663	$2,004	$2,233	$2,822	$2,309	$3,592	$2,887	$3,141	$4,065	$2,412	$21,051	
SCHOOL, COUNTY, & LIBRARY REV per DU	$8,085	$7,331	$4,851	$3,773	$3,234	$2,965	$2,695	$2,156	$1,617	$1,078	$862	$755			
per AC	$1,305	$1,775	$2,350	$6,414	$4,827	$5,532	$7,108	$5,687	$8,530	$6,397	$6,824	$8,956	$5,971	$12,835	

\# 1 \# 2 \# 2 \# 3 \# 4 \# 5

Avg. City Revenue per Residential AC ($2,412)

Avg. City Revenue per Commercial AC ($21,051)

Avg. School, County, & Library Revenue per Residential AC ($5,971)

Avg. School, County & Library Revenue per Commercial AC ($12,835)

NOTE: 4.7% commercial land use
22.7% of total revenue provided by commercial land use

NOTE: Institutional = income tax & stormwater revenue from city, school, and other institutions

income taxes paid to the city. This disparity exists because of the daily workforce migration mentioned earlier. The data labeled #2 show that the low-density residential RA zone produced the highest municipal revenue per dwelling unit at $3210 per dwelling unit involved. However, this zone requires 3 acres per dwelling unit, and it produced the lowest revenue per acre, equal to $518. The R3 zone, by contrast, permits 4.84 dwelling units per net acre and produces much less city revenue per dwelling at $1070 per unit. However, it produces much more city revenue per acre at $2822 since the density has increased from 0.33 to 4.84 dwelling units per acre. The message is *not* that low-density housing is unacceptable, because it produces a very positive image for the city as a whole, and this translates into real estate appreciation for everyone. The message is that each acre of urban land produces a yield that contributes to the support of a community standard, and these acres must be balanced to produce the total revenue needed to maintain that community's lifestyle and rate of property value appreciation over time.

Balance

The data labeled #3 show that the average yield per residential acre was $2412 and the average yield per commercial acre was $21,051. The data labeled #4 show that the combined average yield from all acreage in the city was $3741, since only 4.7% of the city's land was devoted to the higher-yielding $21,000 acreage. It also illustrates that the city was in a break-even position since operating, maintenance, and improvement costs equaled $3743 per acre.

The data labeled #5 show that 4.7% of the city's land use allocation yielded 22.7% of the revenue, and that this revenue was reinvested in the city to maintain each shareholder's appreciation rate and quality of life. This 4.7% allocation is devoted to the nonresidential land use group, and the bulk of this group allocation is devoted to retail land use. Since this city does not share in the sales tax received by the county and state, retail land is still a relatively low-yielding use. It consumes a great deal of land area and houses few workers per 1000 sq ft of building area. When this population density is combined with the average salary of the workers involved, a relatively low yield is produced. Even with these lower yields, however, the lesson from this example is that every acre devoted to the nonresidential land use group can produce 10 times the yield of an average residential acre, and heavier concentrations of prime office land use could produce 20 times this yield.

Land Use Modeling

A city could engage in land use modeling to evaluate the yield of its current land use allocation portfolio and its future growth objectives. Table 6.3 presents a simple example of this approach. The table consists of two panels, both of which represent a city of 10 sq mi. Five square miles is currently developed in the lower panel and 10 sq mi is developed in the upper panel. Traditional land use categories are listed in the left-hand column of each. Land use allocation percentages and net acre translations are listed in the second and third columns, respectively, of each. The average yield currently received per acre for each land use category is listed in the fourth column, and the prorated revenue per acre is calculated in the fifth. The lower panel of Table 6.3 represents the implications of the city's current land use allocation; the upper panel represents the change that could take place if the remaining 5 sq mi were developed to alter the overall land use allocation as shown in its second column. The lower panel has 5% of its land allocated to office, institutional, and regional shopping land use; the upper panel has 18% devoted to these activities. A number of changes have also been made to the average revenue per acre targets in the upper panel since this example will

Table 6.3 Land Use Allocation Modeling Comparison—Different Land Use Allocation in Remaining Five Square Miles

Land Use Category	Land Use Allocation Percentage	Net Acres Excluding Rights-of-Way	Average Revenue per Acre	Prorated Revenue per Acre
Net Area (in sq mi) **10.0 Developed** **10.0 Total**				
Residential	5.0%	320.0	$550	$28
Low density				
Med density				
High density				
Educational	50.0%	3200.0	$2,200	$1,100
Office	3.0%	192.0	$3,000	$90
Institutional	10.0%	640.0	$3,500	$350
Manufacturing	8.0%	512.0	$35,000	$2,800
Commercial	3.0%	192.0	$5,500	$165
Neighborhood	1.0%	64.0	$75,000	$750
Community	1.0%	64.0	$3,000	$30
Regional	2.0%	128.0	$5,000	$100
Open space	3.0%	192.0	$12,000	$360
Public, no revenue	10.0%	640.0	$0	$0
Public, revenue	0.0%	0.0	$0	$0
Private, golf	2.0%	128.0	$500	$10
Private, other	2.0%	128.0	$500	$10
Other				$500
	100.0%	6400.0	Yield / acre	$6,293
			Total yield	$40,272,000
Net Area (in sq mi) **5.0 Developed** **10.0 Total**				
Residential				
Low density	5.0%	160.0	$550	$28
Med density	70.0%	2240.0	$2,200	$1,540
High density	0.0%	0.0	$3,500	$0
Educational	10.0%	320.0	$3,500	$350
Office	2.0%	64.0	$21,000	$420
Institutional	1.0%	32.0	$6,000	$60
Manufacturing	0.0%	0.0	$75,000	$0
Commercial				
Neighborhood	0.0%	0.0	$3,000	$0
Community	0.0%	0.0	$5,000	$0
Regional	2.0%	64.0	$12,000	$240
Open space				
Public, no revenue	3.0%	96.0	$0	$0
Public, revenue	0.0%	0.0	$0	$0
Private, golf	7.0%	224.0	$500	$35
Private, other	0.0%	0.0	$500	$0
Other				$500
	100.0%	3200.0	Yield / acre	$3,173
			Total yield	$10,152,000
			Difference in yield per acre	$3,120

demonstrate how these targets can be established more effectively. The difference in yield per acre from the reallocation in Table 6.3 is $3120 per acre. This is twice the original yield per acre and it produces four times the annual revenue when the city expands to 10 sq mi.

The difference in total yield after city completion can be seen in the upper panel of Table 6.4. If the city simply continued with its current land use allocation trend, the entire 10-sq-mi land area could be expected to yield $20,304,000 per year upon completion. If the city adjusted this allocation according to the second column in the upper panel of Table 6.3 while growing into the 10-sq-mi area, the difference in total anticipated yield at completion would be $19,968,000 annually.

If the city were already completed according to the allocation percentages in the second column of the upper panel in Table 6.4, calculations could be undertaken to identify redevelopment options. A very simplified example of this approach is provided in Table 6.5, which shows current land use allocation and yield data in the second column. If the city forecasted a need of $6 million in additional revenue and planned to raise $4 million of it with redevelopment, the third column shows that the city would need to convert 105 acres to commercial land use. This would increase its commercial land use allocation from 5.2% to 7.1% of the total city area. This may not be a politically realistic solution, so a more feasible approach is to establish land use yield targets while the city is still developing.

Land Use Yield Targets

The economic yield objective for office land use in Table 6.3 is $35,000 per net acre. The feasibility of this objective can be evaluated by looking at forecast model CG1L, which is included as Exhibit 6.1. A gross land area of 1 acre has been specified and no land area is allocated for public right-of-way. A parking requirement of one space for every 250 sq ft

Table 6.4 Land Use Allocation Pattern Continuation—Continuation of Same Land Use Allocation Pattern over Remaining Five Square Miles

Land Use Category	Land Use Allocation Percentage	Net Acres Excluding Rights-of-Way	Average Revenue per Acre	Prorated Revenue per Acre
Net Area (in sq mi)				
10.0 Developed				
10.0 Total				
Residential				
Low density	5.0%	320.0	$550	$28
Med density	70.0%	4480.0	$2,200	$1,540
High density	0.0%	0.0	$3,500	$0
Educational	10.0%	640.0	$3,500	$350
Office	2.0%	128.0	$21,000	$420
Institutional	1.0%	64.0	$6,000	$60
Manufacturing	0.0%	0.0	$75,000	$0
Commercial				
Neighborhood	0.0%	0.0	$3,000	$0
Community	0.0%	0.0	$5,000	$0
Regional	2.0%	128.0	$12,000	$240
Open space				
Public, no revenue	3.0%	192.0	$0	$0
Public, revenue	0.0%	0.0	$0	$0
Private, golf	7.0%	448.0	$500	$35
Private, other	0.0%	0.0	$500	$0
Other				$500
	100.0%	6400.0	Yield / acre	$3,173
			Total yield	$20,304,000
Net Area (in sq mi)				
5.0 Developed				
10.0 Total				
Residential				
Low density	5.0%	160.0	$550	$28
Med density	70.0%	2240.0	$2,200	$1,540
High density	0.0%	0.0	$3,500	$0
Educational	10.0%	320.0	$3,500	$350
Office	2.0%	64.0	$21,000	$420
Institutional	1.0%	32.0	$6,000	$60
Manufacturing	0.0%	0.0	$75,000	$0
Commercial				
Neighborhood	0.0%	0.0	$3,000	$0
Community	0.0%	0.0	$5,000	$0
Regional	2.0%	64.0	$12,000	$240
Open space				
Public, no revenue	3.0%	96.0	$0	$0
Public, revenue	0.0%	0.0	$0	$0
Private, golf	7.0%	224.0	$500	$35
Private, other	0.0%	0.0	$500	$0
Other				$500
	100.0%	3200.0	Yield / acre	$3,173
			Total yield	$10,152,000
			Difference in yield per acre	$0

Table 6.5 Redevelopment Evaluation—Land Use Reallocation to Achieve a Given Yield Objective*

	Current Data	Future Option	
Current data			
Gross residential acres	5,353	5,353	
Gross commercial acres	296	296	
Residential revenue	$12,119,531	$12,119,531	
Commercial revenue	$6,410,101	$6,410,101	
Total residential + commercial + institutional revenue	$18,529,632	$18,529,632	
Average yield per residential acre	$2,264.06	$2,264.06	
Total residential + commercial acres	5,649	5,649	
Average yield per commercial acre	$21,655.75	$21,655.75	
Yield objective			
Total additional residential + commercial revenue needed per year		$4,000,000	Revenue balance option
Target residential share of additional revenue needed		$2,000,000	
Resulting total residential yield per acre		$2,638	
Reallocation needed			
Total commercial acres needed	296	401	
Residential reallocation needed in acres	0	105	Adjustment needed
Commercial land as % of residential + commercial	5.2%	7.1%	Resulting land use balance needed among residential and commercial land uses
Residential land as % of residential + commercial	94.8%	92.9%	

*Note: Forecast based upon the following formulas:

$I = xc + R(a - x)$

$I = xc + Ra - Rx$

$I = x(c - R) + Ra$

$\boxed{x = (I - Ra) / (c - R)}$

When $I = (I + i)$ and $R = (R + r)$

$\boxed{x = ((I + i) - ((R + r)a)) / (c - (R + r))}$

where:

x = gross commercial acres needed
c = average yield per commercial acre
I = municipal income from residential and commercial land
i = additional municipal income needed per year
R = average yield per residential acre
r = additional residential yield needed
a = total residential + commercial acres

Exhibit 6.1 Forecast Model CG1L

*Development capacity forecast for **NONRESIDENTIAL BUILDINGS** based on the use of an adjacent **GRADE PARKING LOT** located on the same premises. When s and a equal zero in the design specification below, the forecast pertains to conditions when **NO PARKING** is required.*

***Given:** Gross land area. **To Find:** Maximum development capacity of the land area (gross building area potential) based on the design specification values entered below. **Premise:** All building floors considered equal in area.*

DESIGN SPECIFICATION

Enter values in boxed areas where text is bold and blue. Express all fractions as decimals.

Given:	**Gross land area**	**GLA=**	**1.000**	acres	43,560	SF
Land Variables:	Public/ private right-of-way & paved easements	W=	**0.000**	fraction of GLA	0	SF
	Net land area	NLA=	1.000	acres	43,560	SF
	Unbuildable and/or future expansion areas	U=	**0.000**	fraction of GLA	0	SF
	Gross land area reduction	X=	0.000	fraction of GLA	0	SF
	Buildable land area remaining	**BLA=**	1.000	acres	43,560	SF
Parking Variables:	Est. gross pkg. lot area per space in SF	s =	**350**	ENTER ZERO IF NO PARKING REQUIRED		
	Building SF permitted per parking space	a =	**250**	ENTER ZERO IF NO PARKING REQUIRED		
	No. of loading spaces	l=	**0**			
	Gross area per loading space	b =	**1,000**	SF	0	SF
Site Variables:	**Project open space as fraction of BLA**	**S=**	**0.300**		13,068	SF
	Private driveways as fraction of BLA	R=	**0.020**		871	SF
	Misc. pavement as fraction of BLA	M=	**0.010**		436	SF
	Loading area as fraction of BLA	L=	0.000		0	SF
	Total site support Areas as a fraction of BLA	Su=	0.330		14,375	SF
Core:	**Core development area as fraction of BLA**	**C=**	0.670	C+Su must = 1	29,185	SF

PLANNING FORECAST

no. of floors **FLR**	**CORE** minimum land area for BCG & PLA	gross building area **GBA**	parking lot area **PLA**	pkg. spaces **NPS**	footprint **BCA**	bldg SF / acre **SFAC** function of BLA	flr area ratio **FAR** function of BLA
1.00	29,185	**12,161**	17,025	48.6	**12,161**	12,161	0.279
2.00		**15,361**	21,505	61.4	**7,680**	15,361	0.353
3.00		**16,838**	23,573	67.4	**5,613**	16,838	0.387
4.00		**17,688**	24,763	70.8	**4,422**	17,688	0.406
5.00		**18,241**	25,537	73.0	**3,648**	18,241	0.419
6.00	*NOTE: Be aware when BCA becomes too small to be feasible.*	**18,629**	26,080	74.5	**3,105**	18,629	0.428
7.00		**18,916**	26,483	75.7	**2,702**	18,916	0.434
8.00		**19,138**	26,793	76.6	**2,392**	19,138	0.439
9.00		**19,314**	27,039	77.3	**2,146**	19,314	0.443
10.00		**19,457**	27,240	77.8	**1,946**	19,457	0.447
11.00		**19,575**	27,406	78.3	**1,780**	19,575	0.449
12.00		**19,675**	27,546	78.7	**1,640**	19,675	0.452
13.00		**19,761**	27,665	79.0	**1,520**	19,761	0.454
14.00		**19,835**	27,768	79.3	**1,417**	19,835	0.455
15.00		**19,899**	27,859	79.6	**1,327**	19,899	0.457

WARNING: These are preliminary forecasts that must not be used to make final decisions.
1) These forecasts are not a substitute for the "due diligence" research that must be conducted to support the final definition of "unbuildable areas" above and the final decision to purchase land. This research includes, but is not limited to, verification of adequate subsurface soil, zoning, environmental clearance, access, title, utilities and water pressure, clearance from deed restriction, easement and right-of-way encumbrances, clearance from existing above and below ground facility conflicts, etc.
2) The most promising forecast(s) made on the basis of data entered in the design specification from "due diligence" research must be verified at the drawing board before funds are committed and land purchase decisions are made. Actual land shape ratios, dimensions and irregularities encountered may require adjustments to the general forecasts above.
3) The software licensee shall take responsibility for the design specification values entered and any advice given that is based on the forecast produced.

of building area has been specified with a total parking area allocation of 350 sq ft per space. No loading is specified. Miscellaneous pavement is at a minimum 3%, and project open space is not underweighted at 30%. The model forecasts that this 1 acre could accommodate 12,161 to 16,838 sq ft of gross building area based on a building height range extending from 1 to 3 stories. The revenue potential of this capacity is forecasted in Exhibit 6.2 based on the building, efficiency, revenue, and tax abatement values entered in the respective specification panels.[1] The tax abatement is shown as equal to 0%, and the potential yield to local government, based on the construction costs and real estate values that were duplicated from Exhibit 3.2, ranges from $39,716 to $55,143 per acre. Therefore, since the area chosen for evaluation in the model was 1 acre, the $35,000 target appears feasible based on the specification values used.[2]

Tax Abatement

If 100% city real estate tax and 50% school real estate tax abatement were offered, the revenue forecast panel in Exhibit 6.3 shows that a 1-story office building would just meet the yield target in column YTG. Therefore, this example shows that abatement could be offered, and that it would be in the city's interest to encourage a 2- or 3-story building since the forecast shows that it could be easily accommodated while still providing a reasonable footprint area. In fact, not only could abatement be offered, but an incentive rebate could be discussed if the building were increased in area and the $35,000 target was exceeded.

Application

The 1-acre office example could produce a yield of $39,716 and meet the design specification values entered if:

Exhibit 6.2 Revenue Forecast Panel Linked to Forecast Model CG1L

Project Open Space: 30.00% Parking Area per Space: 350 No Tax Abatement

BUILDING EFFICIENCY OBJECTIVES

Building skin	Bs=	0.05
Net building area	Bn=	0.95
Building core	Bc=	0.05
Building mechanical area	Bm=	0.02
Building circulation area	Bc=	0.08
BUILDING SUPPORT	Bs=	0.20
BUILDING EFFICIENCY	Be=	0.80
Avg. pop. per 1000 SF	Pop=	**5.5**

REVENUE SPECIFICATION

			effective commercial real estate tax millage rates		
Avg. income per person	Inc=	**$30,000**	**6.838642**	10.2%	Fraction of REtax to local government
Avg. daily trips per person	Adt=	**5.0**	**45.490353**	67.8%	Fraction of REtax to local schools
Income tax rate		**0.02**	**13.754687**	20.5%	Fraction of REtax to county
Real estate tax millage		**67.082324**	**0.998642**	1.5%	Fraction of REtax to library
Fraction of real estate value assessed		**0.35**	**0.000000**	0.0%	Fraction of REtax to other
Personal property millage		**100.92**	67.082324	100.0%	Total effective millage rate
Personal property as % of building value		**0.10**			
Fraction of personal property value assessed		**0.25**			

ABATEMENT SPECIFICATION

Municipal tax abatement request	**0.00**
School tax abatement request	**0.00**
Total first year REtax abatement	0.0%

REVENUE FORECAST

no. of floors FLR	gross building area GBA	total project TCF cost forecast	total building POP population	income tax to ITX local government	total real estate REX tax	real estate tax to RGX local government after abatement	real estate tax to RSX local schools after abatement	personal property PPX tax to local gov.	yield to YTG local government after abatement	max. first year FYA tax abatement all sources
1.00	12,161	$1,548,275	54	$32,104	$36,352	$3,706	$24,651	$3,906	$39,716	$0
2.00	15,361	1,857,568	68	40,552	43,613	$4,446	$29,575	4,687	49,685	$0
3.00	16,838	2,174,588	74	44,451	51,057	$5,205	$34,623	5,486	55,143	$0
4.00	17,688	2,265,579	78	46,696	53,193	$5,423	$36,072	5,716	57,835	$0
5.00	18,241	2,513,515	80	48,156	59,014	$6,016	$40,019	6,342	60,513	$0
6.00	18,629	2,559,059	82	49,180	60,084	$6,125	$40,744	6,457	61,762	$0
7.00	18,916	2,788,579	83	49,939	65,473	$6,675	$44,399	7,036	63,649	$0
8.00	19,138	2,816,865	84	50,524	66,137	$6,742	$44,849	7,107	64,373	$0
9.00	19,314	3,039,224	85	50,988	71,357	$7,274	$48,389	7,668	65,931	$0
10.00	19,457	3,058,974	86	51,366	71,821	$7,322	$48,704	7,718	66,405	$0
11.00	19,575	3,277,958	86	51,679	76,963	$7,846	$52,190	8,270	67,795	$0
12.00	19,675	3,292,795	87	51,943	77,311	$7,881	$52,427	8,308	68,132	$0
13.00	19,761	3,509,992	87	52,169	82,410	$8,401	$55,885	8,856	69,426	$0
14.00	19,835	3,521,706	87	52,363	82,685	$8,429	$56,071	8,885	69,678	$0
15.00	19,899	$3,737,884	88	$52,533	$87,761	$8,947	$59,513	$9,431	$70,911	$0

Exhibit 6.3 Revenue Forecast Panel Linked to Forecast Model CG1L

Project Open Space: 30.00% Parking Area per Space: 350 With Tax Abatement

BUILDING EFFICIENCY OBJECTIVES

Building skin	Bs=	0.05
Net building area	Bn=	0.95
Building core	Bc=	0.05
Building mechanical area	Bm=	0.02
Building circulation area	Bc=	0.08
BUILDING SUPPORT	Bs=	0.20
BUILDING EFFICIENCY	Be=	0.80
Avg. pop. per 1000 SF	Pop=	**5.5**

REVENUE SPECIFICATION

			effective commercial real estate tax millage rates		
Avg. income per person	Inc=	**$30,000**	**6.838642**	10.2%	Fraction of REtax to local government
Avg. daily trips per person	Adt=	**5.0**	**45.490353**	67.8%	Fraction of REtax to local schools
Income tax rate		**0.02**	**13.754687**	20.5%	Fraction of REtax to county
Real estate tax millage		**67.082324**	**0.998642**	1.5%	Fraction of REtax to library
Fraction of real estate value assessed		**0.35**	**0.000000**	0.0%	Fraction of REtax to other
Personal property millage		**100.92**	67.082324	100.0%	Total effective millage rate
Personal property as % of building value		**0.10**			
Fraction of personal property value assessed		**0.25**			

ABATEMENT SPECIFICATION

Municipal tax abatement request	**100%**
School tax abatement request	**50%**
Total first year REtax abatement	56.5%

NOTE: All forecasts below are per acre forecasts when the gross land area in the parent model is equal to one acre.

REVENUE FORECAST

no. of floors FLR	gross building area GBA	total project TCF cost forecast	total building POP population	income tax to ITX local government	total real estate REX tax	real estate tax to RGX local government after abatement	real estate tax to RSX local schools after abatement	personal property PPX tax to local government	yield to YTG local government after abatement	max. first year FYA tax abatement all sources
1.00	12,161	$1,548,275	54	$32,104	$36,352	$0	$12,326	$3,906	$36,010	$16,031
2.00	15,361	1,857,568	68	40,552	43,613	$0	$14,788	4,687	45,239	$19,234
3.00	16,838	2,174,588	74	44,451	51,057	$0	$17,311	5,486	49,938	$22,516
4.00	17,688	2,265,579	78	46,696	53,193	$0	$18,036	5,716	52,412	$23,459
5.00	18,241	2,513,515	80	48,156	59,014	$0	$20,010	6,342	54,497	$26,026
6.00	18,629	2,559,059	82	49,180	60,084	$0	$20,372	6,457	55,637	$26,497
7.00	18,916	2,788,579	83	49,939	65,473	$0	$22,199	7,036	56,975	$28,874
8.00	19,138	2,816,865	84	50,524	66,137	$0	$22,425	7,107	57,631	$29,167
9.00	19,314	3,039,224	85	50,988	71,357	$0	$24,195	7,668	58,656	$31,469
10.00	19,457	3,058,974	86	51,366	71,821	$0	$24,352	7,718	59,084	$31,674
11.00	19,575	3,277,958	86	51,679	76,963	$0	$26,095	8,270	59,949	$33,941
12.00	19,675	3,292,795	87	51,943	77,311	$0	$26,213	8,308	60,251	$34,095
13.00	19,761	3,509,992	87	52,169	82,410	$0	$27,942	8,856	61,024	$36,344
14.00	19,835	3,521,706	87	52,363	82,685	$0	$28,036	8,885	61,249	$36,465
15.00	19,899	$3,737,884	88	$52,533	$87,761	$0	$29,757	$9,431	$61,964	$38,703

- 30% of the buildable area is provided as project open space[3]
- No abatement is offered
- Parking is provided at a ratio of four spaces per 1000 sq ft of building area[4]
- Parking areas require no more than 350 sq ft of paving and landscape area per space[5]

If the project open space value is changed from 30% to 50%, or if a number of other specification values are changed, the potential yield from this gross acre declines. Exhibit 6.4 predicts that the 1-acre, 1-story yield of the previous example would decline to $28,622 per gross acre if the project open space were increased from 30% to 50% of the buildable area, and if all other specification values remain constant. The message here is that a city can control both the environment and yield produced per acre of land available when it focuses on the pressure points of development capacity[6] and uses them to produce a land use allocation plan that includes economic development objectives.

For instance, assume 100 acres of land zoned for office development with a project open space requirement of 50%, a height limit of 10 stories,[7] and associated parking requirements. The potential yield from this specification could easily be forecast using any design premise and forecast model by constructing linked spreadsheets similar to those illustrated. Using a grade parking design solution, the aforementioned scenario would produce an intensity index INX equal to 10.5 and a parking index PLX that could equal[8] 250.400. If this intensity specification, which can be expressed as INX = 10.5 and[9] PLX = 250.400, were entered in Exhibit 6.1, Exhibit 6.5 results and forecasts that a 1-story, 1-acre yield expectation would be $26,597 as shown in the YTG column of the revenue forecast panel.[10] This value represents a realistic land use yield objective for this

Exhibit 6.4 Revenue Forecast With 50% Open Space and 350 SF per Parking Space

Project Open Space: 50.00% Parking Area per Space: 350

BUILDING DESIGN OBJECTIVES

Building skin	Bs=	0.05
Net building area	Bn=	0.95
Building core	Bc=	0.05
Building mechanical area	Bm=	0.02
Building circulation area	Bc=	0.08
BUILDING SUPPORT	Bs=	0.20
BUILDING EFFICIENCY	Be=	0.80
Avg. pop. per 1000 SF	Pop=	**5.5**

REVENUE SPECIFICATION

Item		Value	effective commercial real estate tax millage rates		
Avg. income per person	Inc=	**$30,000**	**6.838642**	10.2%	Fraction of REtax to local government
Avg. daily trips per person	Adt=	**5.0**	**45.490353**	67.8%	Fraction of REtax to local schools
Income tax rate		**0.02**	**13.754687**	20.5%	Fraction of REtax to county
Real estate tax millage		**67.082324**	**0.998642**	1.5%	Fraction of REtax to library
Fraction of real estate value assessed		**0.35**	**0.000000**	0.0%	Fraction of REtax to other
Personal property millage		**100.92**	67.082324	100.0%	Total effective millage rate
Personal property as % of building value		**0.10**			
Fraction of personal property value assessed		**0.25**			

ABATEMENT SPECIFICATION

Municipal tax abatement request	**0.00**
School tax abatement request	**0.00**
Total first year REtax abatement	0.0%

REVENUE FORECAST

no. of floors FLR	gross building area GBA	total project TCF cost forecast	total building POP population	income tax to ITX local government	total real estate REX tax	real estate tax to RGX local government after abatement	real estate tax to RSX local schools after abatement	personal property PPX tax to local. government	yield to YTG local government after abatement	max. first year FYA tax abatement all sources
1.00	8,531	$1,240,996	38	$22,521	$29,137	$2,970	$19,759	$3,131	$28,622	$0
2.00	10,775	1,457,962	47	28,447	34,231	$3,490	$23,213	3,678	35,615	$0
3.00	11,811	1,680,349	52	31,182	39,453	$4,022	$26,754	4,240	39,444	$0
4.00	12,408	1,744,179	55	32,757	40,951	$4,175	$27,770	4,401	41,332	$0
5.00	12,796	1,918,104	56	33,781	45,035	$4,591	$30,539	4,839	43,211	$0
6.00	13,068	1,950,053	57	34,500	45,785	$4,667	$31,048	4,920	44,087	$0
7.00	13,270	2,111,059	58	35,032	49,565	$5,053	$33,611	5,326	45,411	$0
8.00	13,425	2,130,902	59	35,442	50,031	$5,100	$33,927	5,376	45,919	$0
9.00	13,548	2,286,885	60	35,768	53,693	$5,474	$36,411	5,770	47,011	$0
10.00	13,649	2,300,740	60	36,033	54,019	$5,507	$36,632	5,805	47,344	$0
11.00	13,732	2,454,355	60	36,253	57,625	$5,875	$39,077	6,192	48,319	$0
12.00	13,802	2,464,763	61	36,438	57,870	$5,899	$39,243	6,219	48,556	$0
13.00	13,862	2,617,125	61	36,596	61,447	$6,264	$41,669	6,603	49,463	$0
14.00	13,914	2,625,343	61	36,732	61,640	$6,284	$41,800	6,624	49,640	$0
15.00	13,959	$2,776,990	61	$36,852	$65,200	$6,647	$44,214	$7,006	$50,505	$0

Exhibit 6.5 Revenue Forecast With 50% Open Space and 400 SF per Parking Space

Project Open Space: 50.00% Parking Area per Space: 400

BUILDING DESIGN OBJECTIVES

Building skin	Bs=	0.05
Net building area	Bn=	0.95
Building core	Bc=	0.05
Building mechanical area	Bm=	0.02
Building circulation area	Bc=	0.08
BUILDING SUPPORT	Bs=	0.20
BUILDING EFFICIENCY	Be=	0.80
Avg. pop. per 1000 SF	Pop=	**5.5**

REVENUE SPECIFICATION

			effective commercial real estate tax millage rates		
Avg. income per person	Inc=	**$30,000**	**6.838642**	10.2%	Fraction of REtax to local government
Avg. daily trips per person	Adt=	**5.0**	**45.490353**	67.8%	Fraction of REtax to local schools
Income tax rate		**0.02**	**13.754687**	20.5%	Fraction of REtax to county
Real estate tax millage		**67.082324**	**0.998642**	1.5%	Fraction of REtax to library
Fraction of real estate value assessed		**0.35**	**0.000000**	0.0%	Fraction of REtax to other
Personal property millage		**100.92**	67.082324	100.0%	Total effective millage rate
Personal property as % of building value		**0.10**			
Fraction of personal property value assessed		**0.25**			

ABATEMENT SPECIFICATION

Municipal tax abatement request	**0.00**
School tax abatement request	**0.00**
Total first year REtax abatement	0.0%

REVENUE FORECAST

no. of floors FLR	gross building area GBA	total project TCF cost forecast	total building POP population	income tax to ITX local government	total real estate REX tax	real estate tax to RGX local government after abatement	real estate tax to RSX local schools after abatement	personal property PPX tax to local. government	yield to YTG local government after abatement	max. first year FYA tax abatement all sources
1.00	7,874	$1,181,512	35	$20,788	$27,740	$2,828	$18,812	$2,981	$26,597	$0
2.00	9,749	1,363,652	43	25,738	32,017	$3,264	$21,712	3,440	32,442	$0
3.00	10,590	1,554,904	47	27,957	36,507	$3,722	$24,757	3,923	35,601	$0
4.00	11,067	1,606,182	49	29,216	37,711	$3,844	$25,573	4,052	37,113	$0
5.00	11,374	1,756,949	50	30,027	41,251	$4,205	$27,973	4,433	38,665	$0
6.00	11,589	1,782,240	51	30,594	41,845	$4,266	$28,376	4,497	39,356	$0
7.00	11,747	1,922,478	52	31,012	45,137	$4,601	$30,609	4,850	40,464	$0
8.00	11,869	1,938,067	52	31,333	45,504	$4,639	$30,857	4,890	40,861	$0
9.00	11,965	2,074,254	53	31,587	48,701	$4,965	$33,025	5,233	41,785	$0
10.00	12,043	2,085,089	53	31,794	48,955	$4,991	$33,198	5,261	42,045	$0
11.00	12,108	2,219,375	53	31,965	52,108	$5,312	$35,336	5,599	42,876	$0
12.00	12,162	2,227,489	54	32,108	52,299	$5,332	$35,465	5,620	43,060	$0
13.00	12,209	2,360,773	54	32,231	55,428	$5,651	$37,587	5,956	43,838	$0
14.00	12,249	2,367,164	54	32,337	55,578	$5,666	$37,689	5,972	43,975	$0
15.00	12,284	$2,499,876	54	$32,430	$58,694	$5,984	$39,802	$6,307	$44,720	$0

100-acre zoning area, and could be entered in the city's general ledger of land use allocation and yield expectations.[11] If $35,000 per acre were still the objective, then 3-story buildings using this design premise would be needed if no other specification values changed.

The land use specification for this 100-acre area could then begin with the land use activity permitted. In this case, the term *office park* has been used to indicate a list of permitted uses. The specification could then proceed with the intensity INX permitted, the parking and parking lot landscaping PLX required, and the minimum yield YLX objective stated in thousands of dollars for adequate overall economic development of the city. The complete definition could read as follows:

Office Park

AREA = 100

INX = 10.5

PLX = 250.400

YLX = 26

This is a land use *specification* that includes an economic development component. This text can combine with a land use allocation plan to define both an environmental and economic future that is compatible. It will also have a chance of meeting expectations because it is based on a realistic forecast of development capacity and a set of design specification values that produce this forecast. Since these specification values are in their infancy, however, much work is required to monitor and evaluate performance in order to refine and improve the results achieved.

Appearance and Compatibility

Nothing has been said about setbacks, fence requirements, dumpster locations, permitted yard encroachments, sign regulations, landscape design regulations, building materials, architectural style, swimming pools, antennas, and so on that are addressed in typical zoning ordinances. This is not because these issues are unimportant. It is because they are appearance issues and not land use, intensity, and yield issues that shape the primary urban context in which they exist. Think of these topics as if they were included in a separate appearance chapter in a zoning ordinance.[12] This appearance chapter would focus on ensuring that the context established by the land use, intensity, and economic development chapters was well executed in its final form. Expecting these appearance issues to produce adequate land use allocation, preservation, capacity, intensity, and yield is simply placing too much responsibility on the wrong set of tools. This reduces their credibility when it is simply their purpose and application that are misunderstood.

Conclusion

In conclusion, the process of evaluation used to establish economic yield objectives for the office park example can be used for any land use group or subset of a land use group. This approach can help to build a local or regional plan of land use allocation that conserves resources, increases capacity, and establishes economic stability since the plan can be based on a realistic appraisal of the development capacity of land. A statement of economic yield objectives for each land use area can also be included within a complete zoning ordinance to improve the discussion among all affected parties regarding the objectives of every stakeholder involved in the construction of our communities.

Notes

1. This exhibit is linked to the parent CG1L forecast model and is the same panel used for Exhibit 3.4.

2. This forecast is based on one design premise using a grade parking lot around but not under the building. There are five nonresidential design premises included in the collection, and the remaining four should be able to produce much higher yields per acre than the premise chosen.
3. These statistics can be read at the top of Exhibit 6.1.
4. This abbreviated method of stating the parking requirement is adopted here for brevity. Divide 1000 by 4 to produce the *a* value used in Exhibit 6.1. The resulting *a* value is the amount of building area provided per parking space. The ratio of 4 indicates the number of parking spaces provided per 1000 sq ft of building area.
5. This yield forecast also depends on the building efficiency and cost specification values that produced the total real estate value TCF forecast illustrated in Exhibit 6.3, since all real estate tax predictions are based on these total value forecasts. Exhibits 6.2 and 6.3 are linked to the values in Exhibit 6.1, and to each other. A change to any specification value in any of the three models will affect the yield forecast produced in Exhibit 6.2. This format is repeated here to lay the groundwork for a further extension of this example.
6. In fact, anyone engaged in development design can produce these results.
7. According to Chap. 1, the INX specification would be 10.5 for this land area.
8. This means 250 sq ft of building area permitted per parking space and 400 sq ft of parking lot area planned per parking space.
9. A parking structure index PSX might read 250.350.
10. Note that a 3-story building with 50% project open space would produce $35,601 per acre when built to the design specifications entered.
11. This ledger could track both current performance of developed land and future land use portfolio expectations. These expectations could then become a part of a city's development negotiations with the private sector.
12. In fact, it would be very helpful to any designer or developer if all of these items were collected by topic in a single chapter instead of often being scattered throughout the ordinance.

Case Studies and Context Records

The purpose of all case studies included in this book is to forge a relationship between actual development projects constructed and the development forecast collection, which is written to predict these development results before they are constructed. The context record system represents this link, and is the heart of the case study process. It is designed to record past experience through a standardized form of case study measurement and converts this information to design specification values that define the project under study. These values represent a recipe that can be used to reproduce the context created without duplicating form and appearance, and they can be adjusted within a forecast model to predict alternate outcomes. The objective is to learn from the past and forecast a future based on this knowledge. It would be ideal if one forecast model could do it all. Unfortunately, the number of development options and shelter types have been too extensive to reduce this collection to less than the 35 forecast models that are listed in Figs. 1.1 and 1.2 and Table 5.1. The case studies in the following chapters are each written to introduce a different facet of this development forecast collection.

The American Heritage Dictionary, Third Edition, says that *context* means, "that part of a text or statement that surrounds a particular word or passage and determines its meaning." In the language of city planning, urban design, and architecture, *context* means the built environment created by paved surfaces, parking areas, and open space that surround building mass and affect the meaning and quality of the lifestyle within. Pavement, parking, building mass, and open space are the raw materials of urban design, and they combine within projects and among projects to produce relative levels of intensity on site, in neighborhoods, throughout communities, and across regions. Context, therefore, begins with the allocation of these raw materials at the project level. These allocations are then formed, arranged, and styled in relation to the site characteristics present in order to produce final project appearance. These projects collect and combine over time to produce built environments, compositions, and lifestyle alternatives. A farmhouse surrounded by 5 sq mi of pasture sits at one end of the context spectrum and a Manhattan apartment on the fiftieth floor sits at the other. These lifestyles and intensities couldn't be further apart, and they are established by the context that surrounds the shelter provided. The context record system can measure and evaluate the environments and lifestyles produced by this spectrum of possibilities to produce a vocabulary and a library of knowledge. This vocabulary and knowledge can then be used with the development forecast collection to predict future development possibilities based on design specification values with known context results.

The case studies to follow use context record measurements to define existing development projects. These case studies have been selected to illustrate the use of this system to define existing projects and evaluate future possibilities with the development forecast collection. Each case is designed to look beneath the appearance of the building and site to the fundamental design components and quantities that were formed, arranged, and styled to produce the environment photographed. Each case study also represents a recurring development issue that can benefit from expanded context research and comparative evaluation of forecast possibilities.

A context record pertains to a single development project or section of a project. It contains a series of project measurements that are translated by equations within its spreadsheet format to produce context values. Three record formats are available, depending on the nature of the project under study. Form CR is displayed as Table 7.1 and is the workhorse of the set. It can be used to pertain to all but single-family subdivisions. Form GT is displayed as Table 7.2 and can be used as an alternative to form CR when urban house[1] building types are involved. Form SF is included as Table 7.3 and pertains to the residential land use family when suburb house[2] development is involved. All forms calculate context values based on the project measurements requested, and selected values within each summary represent design specification components capable of reproducing the context measured without plagiarizing appearance.

A design specification is also used in a development forecast model to predict development capacity options before a project is designed and constructed. After construction, however, the specification values chosen define the project context produced. When an existing project does not have a recorded design specification, it can be re-created by entering requested project measurements in a context record spreadsheet. These measurements are then used to calculate the context values and design specification that produced the environment encountered. Since neighborhoods, districts, cities, and regions are composed of individual projects that slowly multiply, our ability to effectively define, evaluate, plan, and guide this larger urban context toward a sustainable future will depend on our ability to achieve planned results at the project level.

Table 7.1 Context Record

Form CR

For Nonresidential and Apartment houses

Name ______________________

Address ______________________

Location ______________________

Class and Order ______________________

Family, Genus, Species ______________________

Parking System	check box		
Surface parking around but not under building	______	Structure parking adjacent to building	______
Surface parking around and under building	______	Structure parking underground	______
Surface and/or garage parking	______	Structure parking above grade under building	______
No Parking	______		

		Description	Project Measurements		Context Summary SF	Context Summary %	
Site							
		Gross land area in acres	☐	GLA	0		
		Public/ private ROW and paved easements in SF	☐	W			of GLA
		Net lot area		NLA	0		of GLA
		Total unbuildalble area in SF	☐	U			of GLA
		Water area within unbuildable area in SF	☐	WAT			of U
		Buildable land area		BLA	0		of GLA
Development Cover							
		Gross building area in SF	☐	GBA			
		Total building footprint in SF	☐	B		0.0%	of BLA
		Total bldg. support in SF (stairs, corridors, elevators, etc.)	☐	BSU			
		Building efficiency		Be			of GBA
	Residental only	Number of dwelling units	☐	DU			
		Aggregate average dwelling unit area in SF		AGG			
		Number of building floors	☐	FLR			
	G2 only	Unenclosed surface parking area under building in SF	☐	PUB			
		Total surface parking area in SF**	☐	P or G		0.0%	of BLA
		Number of surface parking spaces on premise	☐	NPSs			
		Building cover over surface parking area		AIR			of GPA
		Gross surface parking area per parking space in SF		ss			
		Gross building area per surface parking space in SF		as			
		Surface parking spaces per dwelling unit		us	0.00		
	Pkg. struc. only	Total parking structure footprint in SF	☐	P		0.0%	of BLA
	Pkg. struc. only	Parking structure area below grade in SF	☐	UNG			of BLA
	Pkg. struc. only	Parking structure area above grade under building in SF	☐	PSC			
	Pkg. struc. only	Number of parking structure floors	☐	p			
	Pkg. struc. only	Total parking structure area in SF	☐	GPA			
	Pkg. struc. only	Total parking structure spaces	☐	NPSg			
	Pkg. struc. only	Total pkg. struc. support % (stairs, elevators, etc.)	☐	PSS			
	Pkg. struc. only	Parking structure support SF		Pu			of GPA
	Pkg. struc. only	Gross parking structure area per space in SF		sg	0		
	Pkg. struc. only	Net parking structure area per space in SF		ag			
	Pkg. struc. only	Gross building area per structure parking space in SF			0		
	Pkg. struc. only	Parking structure spaces per dwelling unit		ug	0.00		
	Residential only	Total dwelling unit garage area in SF	☐	GPA			
	Residential only	Dwelling unit garage above grade under builiding in SF	☐	GUB			
	Residential only	Number of dwelling unit garage parking spaces	☐	GPS			
	Residential only	Garage parking spaces per dwelling		Gn	0.0		
	Residential only	Total dwelling unit garage area per space in SF		Ga			
	Residential only	Total dwelling unit garage cover in addition to building cover		Gc		0.0%	of BLA
		Total parking spaces		NPS	0.0		
	Residential only	Total parking spaces per dwelling unit		u			
		Gross building area per parking space total		a			
		Number of loading spaces	☐	l			
		Total loading area in SF	☐	LDA			
		Gross area per loading space in SF		b			
		Loading area percentage		L		0.0%	of BLA
		Driveway areas in SF	☐	R		0.0%	of BLA
		Misc. pavement and social center pvmt. in SF	☐	M		0.0%	of BLA
		TOTAL DEVELOPMENT COVER		D		**0.0%**	of BLA
Project Open Space							
		PROJECT OPEN SPACE		S		**0.0%**	of BLA
		Water area within project open space in SF	☐	wat			of S
			0.0%				
Summary							
		Development balance index		BNX			
		Development intensity index		INX			
		Floor area ratio per buildable acre		BFAR			
		Capacity index (gross building sq. ft. per buildable acre)		SFAC			
		Density per buildable acre if applicable		dBA			
		Density per net acre if applicable		dNA			
		Density per gross acre if applicable		dGA			

** Include internal parking lot landscaping, circulation aisles, and private roadway parking space areas. Express private roads separately as driveway areas.

Table 7.2 Urban House Context Record

Form GT

Name ______________________

Address or Location ______________________

Land Use ______________________

Parking System *Any combination of parking lot, roadway, and/or garage parking*

NOTE: An urban house is a single -family ATTACHED home that includes twin-singles, three-families, four-families, rowhouses, townhouses, garden apartments, etc. , OR a single-family detached home on a lot less than 40 feet wide. A townhouse implies separate ownership and defined private open space. A rowhouse implies rental status and less defined private open space.

	Description	Project Measurements		Context Summary	%	
Site						
	Gross land area in acres		GLA	0		
	Public/ private ROW and paved easements in SF		W			of GLA
	Net lot area		NLA	0		of GLA
	Total unbuildalbe and/or future expansion in SF		U			of GLA
	Water area within unbuildable area in SF		WAT			of U
	Buildable land area		BLA	0		of GLA
Development Cover						
	Gross habitable building area in SF		GBA			
	Total building footprint in SF		B		0.0%	of BLA
	Number of dwelling units		DU			
	Total private yard areas in SF					
	Avg. private yard area per dwelling unit			0		
	Aggregate average dwelling unit area in SF		AGG			
	Avg. habitable footprint plus yard in SF (no parking)		AFP	0		
	Number of building floors		FLR			
	Parking lot footprint in SF (include internal landscape)		P or G		0.0%	of BLA
	Parking lot footprint under building(s) in SF					
	Number of parking spaces		NPS			
	Parking lot spaces per dwelling unit		u	0.00		
	Gross parking lot area per parking space in SF		s			
	Private roadway parking spaces		RPS			
	Net roadway parking area in SF		RPA		0.0%	of BLA
	Roadway parking area under building(s) in SF					
	Private roadway parking spaces per dwelling unit		Rn	0.00		
	Net parking area per roadway space in SF		Rs			
	Total garage area in SF		GPA			
	Garage parking area under building(s) in SF		Gu			
	Number of garage parking spaces		GPS			
	Garage parking spaces per dwelling		Gn	0.00		
	Total garage area per space in SF		Ga			
	Total garage building cover percentage		Gc			of BLA
	Total parking spaces per dwelling unit			0.00		
	Social and misc. building footprint(s) in SF		Sf			
	No. of social bldg. parking spaces		Sn			
	Social parking footprint (include internal landscape) in SF		Spla			of BLA
	Driveway areas in SF		R			of BLA
	Misc. pavement and social center pvmt. in SF		M			of BLA
	TOTAL DEVELOPMENT COVER		D		**0.0%**	of BLA
Project Open Space						
	TOTAL PROJECT OPEN SPACE (OSA)		S	0	**0.0%**	of BLA
	Common open space area			0		of BLA
	Private yard area			0		of BLA
	Private yard area per dwelling unit			0		
	Private yard percentage of project open space					of OSA
	Water area within project open space in SF		wat	0		of OSA
Summary						
	Development balance index		BNX			
	Development intensity index		INX			
	Floor area ratio per buildable acre		BFAR			
	Capacity index (gross building sq. ft. per buildable acre)		SFAC			
	Density per buildable acre if applicable		dBA			
	Density per net acre if applicable		dNA			
	Density per gross acre if applicable		dGA			

Table 7.3 Suburb House Context Record

Form SF

Name ______
Address or Location ______
Land Use ______

Parking System *Interior parking in garage or carport or exterior parking on site*

NOTE: Suburb house means a single family DETACHED dwelling unit (home) intended for the use of only one family unit..

	Description	Project Measurements		Context Summary SF	%	
Site						
	Gross land area in acres	[ ]	GLA	0	____	
	Public/ private ROW and paved easements in SF	[ ]	W		____	of GLA
	Net lot area		NLA	0	____	of GLA
	Total unbuildable area in SF not common open space	[ ]	U		____	of GLA
	Water area within unbuildable area in SF	[ ]	WAT		____	of U
	Buildable land area		BLA	**0**	____	of GLA
Common Open Space						
	Common open space as a % of BLA	[ ]	COS	0		of BLA
	Water area within common open space in SF	[ ] 0.0%				
Subdivision Development						
	Total number of lots	[ ]	NLT			
	Average private lot area in SF		ALA	**0**		
	Avg. private lot area plus prorated common open space			**0**		
	Avg. Lot frontage in feet	[ ]	FRN			
	Avg. lot depth to width ratio					
	Total dev. cover planned per lot as a % of BLA	[ ]	D	0	____	
	Avg. total private yard area per lot in SF		YRD**	0	0.0%	of LOT
	Private yard as multiple of total dev. cover area		Y			
	Misc. pavement incl. driveways as a % of lot	[ ]	M	0	____	
	Total potential building cover area per lot		BCG	0	0.0%	of LOT
	Misc. pavement as % of total bldg. cover potential		Mbc		____	of BCG
	Total avg. building cover present per lot in SF	[ ]	FTP		____	
	Building footprint expansion capacity remaining		FXA	0	____	of FTP
	Current private and common open space including bldg. expansion areas		Sn	0	0.0%	of BCG
	Anticipated private and common open space minimum permanent provision		S	0	0.0%	of BCG
	No. of lots with 1 story homes	[ ]				
	No. of lots with 1.5 story homes	[ ]				
	No. of lots with 2 story homes	[ ]				
	No. of lots with 2.5 story homes	[ ]				
	No. of lots with 3.0 story homes	[ ]				
	No. of lots with 3.5 story homes	[ ]				
	Average building height in stories				0.00	
	Max. permitted building height in stories	[ ]				
Summary						
	Lots per buildable acre				____	
	Lots per net acre				____	
	Lots per gross acre				____	
	Current Development Intensity Index		CINX		____	
	Anticipated Max. Development Intensity Index		AINX		____	

Typical subdivision. Gross buildable area within dashed lines with emphasis on front yards. Subtract side yards to find net buildable area. Divide by number of lots to find average per lot.

Common open space. Separate lot ownership. Front yards de-emphasized. When there is no separate lot ownership, buildable areas may or not be defined within condominium arrangement.

***This can be a deceiving statistic since many zoning ordinances permit detached garages, swimming pools, sheds, fences ,etc. to be placed in side and rear yards as additional development cover. This displaces the open space required by setback lines and in the case of fences and garages disrupts the continuity of this rear yard amenity. In older neighborhoods where detached garages are common, the combination of fences, garages, and sheds reduces rear and side yards to postage stamp amenities that bear no resemblance to the open space implied by the setback statistic quoted.*

Figure 7.1 Bradenton office. Example of suburban office building with more than average open space allocation.

Figure 7.1 is a photograph of a suburban office building that could be found in many areas of the world. Its appearance may be unique, but its fundamental characteristics and site plan components are common to all surface parking design solutions. Figure 7.2 illustrates the site plan arrange-

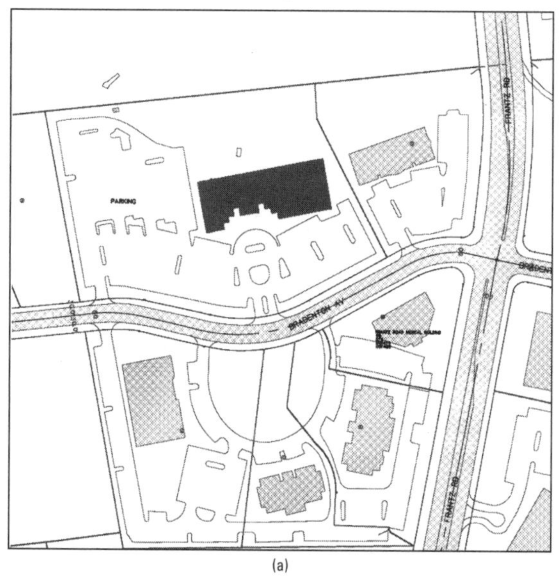

(a)

(b)

Figure 7.2 Bradenton site plan. This project has a high open space allocation that is not evident from the road because a large amount of this space is located behind the building and associated with a stream floodplain. (*a*) Site plan; (*b*) ground-level view.

(Photos by W. M. Hosack)

ment of these components and the relationship between development cover and project open space that produces project balance. Figure 7.1 also reveals the height of the building that combines with the two-dimensional balance of the site plan to produce project intensity.

Balance and *intensity* have been intuitive terms within the development community, but they can be measured, recorded, and indexed when their components are identified. Table 7.4 presents the measurements of Fig. 7.1 in the form of a context record. Sixteen values have been entered in the project measurement column of this record. Four boxes have *not* been completed because they pertain to residential projects, and seven have not been completed because they pertain to parking structure design solutions. The description column to the left of these boxes explains the nature of the information requested, and the context summary column to the right of these boxes converts the measurements entered to a series of context values that contain the elements of a design specification.

Selected context values calculated in Table 7.4 have been entered in Exhibit 7.1 as design specifications. Exhibit 7.1 has used these values to produce the predictions displayed in its forecast panel. The forecast actually chosen, built, and photographed in Fig. 7.1 is defined by the values in row three. Exhibit 7.1 demonstrates that a full table of options was present during the design of Fig. 7.1, and many other option tables could have been produced for evaluation by changing the design specification values used in Exhibit 7.1. It also shows that a 3-story building with 39.5% project open space was the approach adopted.[3]

Producing measured context records to define desirable and undesirable projects can help build a vocabulary and library of knowledge. Object lessons are all around us. We simply need to decipher their message so that a greater degree of success can be achieved by all. The highest priority from this perspective is not the appearance of pavement, parking,

Table 7.4 Bradenton Office Context Record

Name	**5000 Bradenton Avenue**
Address or Location	**5000 Bradenton Avenue**
Land Use	**Office CG1**

Parking System	*check box*
Surface parking around but not under building	x
Surface parking around and under building	
No Parking	
Notes	

		Description	Project Measurements		Context Summary SF	%	
Site							
		Gross land area in acres	5.230	GLA	227,819		
		Public/ private ROW and paved easements in SF	0	W		0.0%	of GLA
		Net lot area		NLA	227,819	100.0%	of GLA
		Total unbuildalbe area in SF	0	U		0.0%	of GLA
		Water area within unbuildable area in SF	0	WAT			of U
		Buildable land area		BLA	227,819	100.0%	of GLA
Development Cover							
		Gross building area in SF	65,000	GBA			
		Total building footprint in SF	21,667	B		9.5%	of BLA
		Total bldg. support in SF (stairs, corridors, elevators, etc.)	12,050	BSU			
		Building efficiency		Be		81.5%	of GBA
	Residental only	Number of dwelling units		DU			
		Aggregate average dwelling unit area in SF		AGG			
		Number of building floors	3.00	FLR			
	G2 only	Unenclosed surface parking area under building in SF	0	PUB			
		Total surface parking area in SF**	109,329	P or G		48.0%	of BLA
		Number of surface parking spaces on premise	260	NPSs			
		Building cover over surface parking area		AIR			of GPA
		Gross surface parking area per parking space in SF		ss	420		
		Gross building area per surface parking space in SF		as	250		
		Surface parking spaces per dwelling unit		us	0.00		
	Pkg. struc. only	Total parking structure footprint in SF		P		0.0%	of BLA
	Pkg. struc. only	Parking structure area below grade in SF					
	Pkg. struc. only	Parking structure area above grade under building in SF					
	Pkg. struc. only	Number of parking structure floors		p			
	Pkg. struc. only	Total parking structure area in SF		GPA			
	Pkg. struc. only	Total parking structure spaces		NPSg			
	Pkg. struc. only	Total pkg. struc. support in SF (stairs, elevators, etc.)		PSS			
	Pkg. struc. only	Parking structure support percentage		Pu			of GPA
	Pkg. struc. only	Gross parking structure area per space in SF		sg	0		
	Pkg. struc. only	Gross building area per structure parking space in SF		ag	0		
	Pkg. struc. only	Parking structure spaces per dwelling unit		ug	0.00		
	Residential only	Total dwelling unit garage area in SF		GPA			
	Residential only	Dwelling unit garage above grade under builiding in SF		GUB			
	Residential only	Number of dwelling unit garage parking spaces		GPS			
	Residential only	Garage parking spaces per dwelling		Gn	0.0		
	Residential only	Total dwelling unit garage area per space in SF		Ga			
	Residential only	Total dwelling unit garage cover in addition to building cover		Gc		0.0%	of BLA
		Total parking spaces			260.0		
	Residential only	Total parking spaces per dwelling unit		u			
		Gross building area per parking space total		a	250.0		
		Number of loading spaces	0	l			
		Total loading area in SF	0	LDA			
		Gross area per loading space in SF		b			
		Loading area percentage		L		0.0%	of BLA
		Driveway areas in SF	5,695	R		2.5%	of BLA
		Misc. pavement and social center pvmt. in SF	1,139	M		0.5%	of BLA
		TOTAL DEVELOPMENT COVER		D	137,830	**60.5%**	of BLA
Project Open Space							
		PROJECT OPEN SPACE		S	89,989	**39.5%**	of BLA
		Water area within project open space in SF	0.0%	wat			of S
Summary							
		Development balance index		BNX		0.605	
		Development intensity index		INX		3.605	
		Floor area ratio per buildable acre		BFAR		0.285	
		Capacity index (gross building sq. ft. per buildable acre)		SFAC		12,428	
		Density per buildable acre if applicable		dBA			
		Density per net acre if applicable		dNA			
		Density per gross acre if applicable		dGA			

** Include internal parking lot landscaping, circulation aisles, and private roadway parking space areas. Express private roads separately as driveway areas..

Exhibit 7.1 Bradenton Design Specification

Forecast Model CG1L

Development capacity forecast for ***NONRESIDENTIAL BUILDINGS*** *based on the use of an adjacent* ***GRADE PARKING LOT*** *located on the same premises. When s and a equal zero in the design specification below, the forecast pertains to conditions when* ***NO PARKING*** *is required.*

Given: *Gross land area.* ***To Find:*** *Maximum development capacity of the land area (gross building area potential) based on the design specification values entered below.* ***Premise:*** *All building floors considered equal in area.*

DESIGN SPECIFICATION

Enter values in boxed areas where text is bold and blue. Express all fractions as decimals.

Given:	**Gross land area**	**GLA=**	**5.230**	acres	227,819	SF
Land Variables:	Public/ private right-of-way & paved easements	W=	**0.000**	fraction of GLA	0	SF
	Net land area	NLA=	5.230	acres	227,819	SF
	Unbuildable and/or future expansion areas	U=	**0.000**	fraction of GLA	0	SF
	Gross land area reduction	X=	0.000	fraction of GLA	0	SF
	Buildable land area remaining	**BLA=**	5.230	acres	227,819	SF
Parking Variables:	Est. gross pkg. lot area per space in SF	s =	**420.5**	ENTER ZERO IF NO PARKING REQUIRED		
	Building SF permitted per parking space	a =	**250**	ENTER ZERO IF NO PARKING REQUIRED		
	No. of loading spaces	l=	**0**			
	Gross area per loading space	b =	**0**	SF	0	SF
Site Variables:	**Project open space as fraction of BLA**	**S=**	**0.395**		89,988	SF
	Private driveways as fraction of BLA	R=	**0.025**		5,695	SF
	Misc. pavement as fraction of BLA	M=	**0.005**		1,139	SF
	Loading area as fraction of BLA	L=	0.000		0	SF
	Total site support areas as a fraction of BLA	Su=	0.425		96,823	SF
Core:	**Core development area as fraction of BLA**	**C=**	0.575	C+Su must = 1	130,996	SF

PLANNING FORECAST

no. of floors **FLR**	**CORE** minimum land area for BCG & PLA	gross building area **GBA**	parking lot area **PLA**	pkg. spaces **NPS**	footprint **BCA**	bldg SF / acre **SFAC** function of BLA	flr area ratio **FAR** function of BLA
1.00	130,996	48,843	82,153	195.4	48,843	9,339	0.214
2.00		60,035	100,978	240.1	30,017	11,479	0.264
3.00		**65,000**	**109,329**	**260.0**	**21,667**	**12,428**	**0.285**
4.00		67,803	114,045	271.2	16,951	12,964	0.298
5.00		69,605	117,075	278.4	13,921	13,309	0.306
6.00	*NOTE: Be aware when BCA becomes too small to be feasible.*	70,860	119,186	283.4	11,810	13,549	0.311
7.00		71,784	120,741	287.1	10,255	13,725	0.315
8.00		72,494	121,934	290.0	9,062	13,861	0.318
9.00		73,055	122,879	292.2	8,117	13,968	0.321
10.00		73,511	123,645	294.0	7,351	14,056	0.323
11.00		73,887	124,279	295.5	6,717	14,128	0.324
12.00		74,205	124,812	296.8	6,184	14,188	0.326
13.00		74,475	125,267	297.9	5,729	14,240	0.327
14.00		74,708	125,659	298.8	5,336	14,285	0.328
15.00		74,912	126,002	299.6	4,994	14,323	0.329

WARNING: These are preliminary forecasts that must not be used to make final decisions.

1) These forecasts are not a substitute for the "due diligence" research that must be conducted to support the final definition of "unbuildable areas" above and the final decision to purchase land. This research includes, but is not limited to, verification of adequate subsurface soil, zoning, environmental clearance, access, title, utilities and water pressure, clearance from deed restriction, easement and right-of-way encumbrances, clearance from existing above and below ground facility conflicts, etc.

2) The most promising forecast(s) made on the basis of data entered in the design specification from "due diligence" research must be verified at the drawing board before funds are committed and land purchase decisions are made. Actual land shape ratios, dimensions and irregularities encountered may require adjustments to the general forecasts above.

3) The software licensee shall take responsibility for the design specification values entered and any advice given that is based on the forecast produced.

shelter, land form, and plant material selection. It is, rather, the quantities of each that are provided, since they represent essential elements of the urban environment that are often difficult or impossible to replace, remodel, or expand when missing.[4] These essential ingredients raise the following questions:

- What quantities of pavement, parking, mass, and open space[5] can be assembled, shaped, and styled to produce an acceptable quality of life?[6]
- What is the capacity of these quantities to accommodate an expanding population?
- What is the extent of our planet's environmental ability to sustain these quantities over time.

Pavement, parking, mass, and open space combine to produce the context in which we live. These contexts represent and support different lifestyles, but the lifestyles supported do not always represent an acceptable quality of life. Unfortunately, we have little idea how to relate context to a sustainable future that supports an acceptable quality of life as we continue to sprawl. We intuitively know, however, that there is a limit. We also intuitively know that the intensity of development context produces lifestyle choices. It also limits population capacity and influences sustainable environmental relationships. The context record system is designed to record project-level data regarding the quantities in use at specific project locations. These elements and quantities represent a project design specification that can be indexed and recorded in relation to the intensity created, the lifestyle represented, the quality of life produced, and the population served. The library of information created by these measurements can then be used as habitable standards in the forecast of future development possibilities when using the development forecast collection.

Each context record format (forms CR, GT, and SF) is included on the CD-ROM for those who wish to begin examining existing projects. These formats can be copied and collections can be assembled to produce a library of context values. This library can provide the knowledge we need to magnify the skills, abilities, and success of all when using the forecast collection. Context intuition is not enough since it rests in the hands of too few, the clients are too limited, and the environments created are always subject to future modification. Context knowledge[7] must be recorded so that it can be indexed, expanded, and improved over time by a much larger group of professionals that are equal to the scope of the populations involved.

A simple example may convince at least some that we take our urban environment for granted and that it is dangerous to continue this pattern. We use asphalt pavement every day as we drive to work and park our cars. It is neatly contained within precise edges, and we have used it so often that we rarely take notice. This asphalt, however, is a petroleum product that represents the greatest oil spill on the face of the planet. It has displaced and killed more wildlife than 1000 ruptured tankers, and it continues to spread each day under our complacent gaze. There is no effort to clean up and little effort to contain it because it is taken for granted, but the context it expands must be taken seriously. It is essential, therefore, that we create the ability to define the development capacity of land in relation to the population it can shelter and the context and quality of life produced. This will help us begin to define the relationships that can shelter an expanding population within an urban context that is livable and environmentally sustainable.

The context record system is designed to record the results produced by existing projects. This can produce the knowledge needed to define development capacity in relation to the context created. The development forecast collection is designed to use these definitions to predict the capacity of future land development to shelter populations within contexts that can produce similar quality for the lifestyles implied.

The following chapters will not produce conclusions; they will simply set the stage for further examination by illustrating the use of the context record system and the development forecast collection in relation to common development issues that are repeated many times each day. This does not mean that these systems are the only tools needed in this effort. It does mean that these tools can be used to translate knowledge into a development vocabulary that can coordinate and produce the results needed from a very broad and diverse group of participants.

Notes

1. An *urban house* is a single-family attached home that includes twin-singles, three-families, four-families, rowhouses, townhouses, garden apartments, and so on, or a single-family detached home on a lot that is less than or equal to 40 ft wide. A *townhouse* implies separate ownership and defined private open space. A *rowhouse* implies rental status and less defined private open space. Both have a similar appearance.
2. A *suburb house* is a detached dwelling unit (home) intended for the use of a single family unit.
3. This represents an intensity index of 3.605 and a balance index of 0.605.
4. This does not mean that appearance should be discounted, since quantities alone cannot produce an enjoyable urban environment. It does mean that quantities provide the raw materials available, and that appearance cannot substitute for an inadequate supply.
5. These quantities are frequently referred to as the *built environment*. It is an interesting term, since it consumes the natural environment and has its own "unnatural" biological classification system. (See Appendix A.)
6. These quantities, once provided, are then arranged, combined, shaped, and styled to produce final urban form and appearance. If adequate quantities do not exist, appearance attempts to substitute for the context needed.
7. Context knowledge cannot solve the problem of excessive population growth, but it can help to increase our ability to shelter a sustainable population level with a series of urban intensities and lifestyle alternatives that represents desirable quality-of-life choices.

Single-System Comparisons

A *single-system comparison* of development capacity[1] alternatives involves one parking system. Single-system comparisons alter values assigned to components within the design specification of a single-forecast model. The forecasts that result are arranged to evaluate the different development capacities predicted and the context environments implied in the planning forecast panel of each. Since surface parking lots are often foregone conclusions where land is readily available at a reasonable price, where cost is an issue, and where maximum development capacity is not an objective, this case study focuses on surface parking as the single system of choice. Five forecast models, listed in Table 3.1, can be used to predict development capacity and context when this system is chosen. Five models are required because both residential and nonresidential land use families use this parking approach, and because three givens affect the forecast equations involved. If the land area is known, either forecast model CG1L or RG1L pertains, depending on the land use family involved. If the gross building area objective is known, forecast model CG1B or RG1B pertains. If a dwelling unit density objective and gross land area are known, forecast model RG1D pertains. In reality, the information known will lead to one of the five models mentioned.[2] Development capacity options forecast by this model can then be explored by comparing the results produced with known context records and environments that pertain to the same parking system. Hence, the term *single-system comparison*.

Single systems can produce widely divergent results. Figure 8.1 is a photograph of a typical nonresidential office development using a G1 parking system (i.e., surface parking). Figure 8.2 illustrates another G1 system with a completely different appearance. While these projects appear unique, they use identical design components and are distinguished by the quantity of each introduced on site. The appearances applied to these components tend to distract from this underlying similarity. They also disguise the fact that a development capacity forecast can be evaluated by comparing the quantities predicted to existing projects with known values and results. Figures 8.3 and 8.4 illustrate this point within the residential land use family. The context pictures are quite different, but both use the G1 parking system and both use the same design specification components. Tables 8.1 and 8.2 present the context record measurements that make Figs. 8.1 and 8.2 unique. Tables 8.3 and 8.4 record the context measurements for Figs. 8.3 and 8.4. The design specification values that define each project have been calculated in these context records and are summarized in Tables 8.5 and 8.6.

Cramer Creek

The design specification values that define Fig. 8.1 are located in Table 8.5. They have been entered in the design specification panel of forecast model CG1L and displayed as Exhibit 8.1. This exhibit predicts not only the design specification photographed, but some of the development options that were available at the time this project was created. The planning forecast panel in this exhibit displays these alterna-

(a)

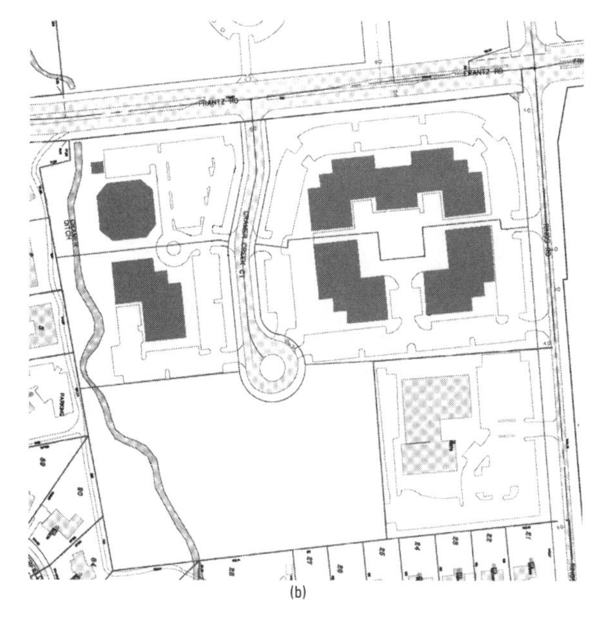

(b)

(c)

Figure 8.1 Cramer Creek Office. This is an example of suburban office lease space that has modestly increased its landscape budget in an affluent community. (*a*) Overview of Cramer Creek Office; (*b*) site plan; (*c*) ground-level frontal view.

(Photos by W. M. Hosack)

(a)

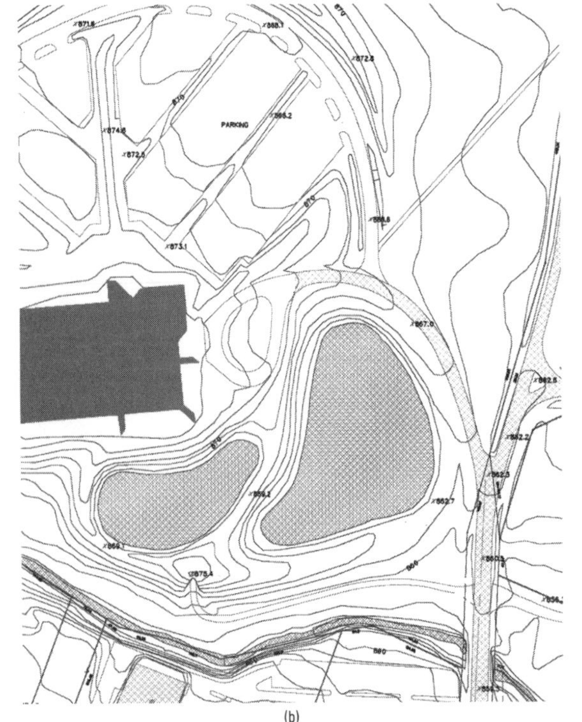

(b)

(c)

Figure 8.2 Post Road Office. This is an example of a corporate office park of significant quality and project open space dedication. (*a*) Overview; (*b*) site plan; (*c*) ground-level frontal view.

(Photos by W. M. Hosack)

tives, and the option constructed is illustrated by the row underlined. The point is that the options forecast in Exhibit 8.1 represent single-system alternatives that produce different development capacities and contexts for comparison. The easiest comparison is simply to contrast the results possible when different building heights are considered. For instance,

(a)

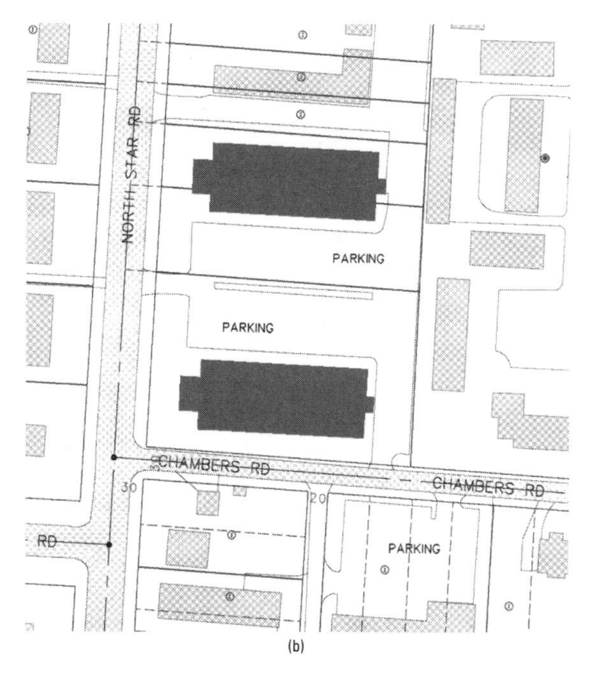

(b)

Figure 8.3 Jefferson Apartments. This is a common example of a high-density, parking lot–oriented apartment complex that is found in many variations in most cities. (*a*) Overview; (*b*) site plan.

(Photo by W. M. Hosack)

(a)

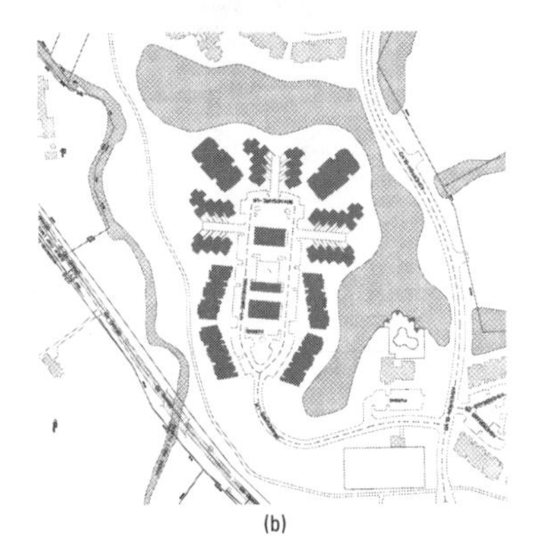

(b)

(c)

(d)

Figure 8.4 Asherton. This is one of five clusters within a water-oriented complex that conserves project open space by using compact parking cores on the interior of each cluster. (*a*) Overview; (*b*) site plan; (*c*) front yard; (*d*) service and parking interior of cluster.

(Photos by W. M. Hosack)

Table 8.1 Cramer Creek Context Record

Form CR

Name — **Cramer Creek**
Address or Location — **281 and 325 Cramer Creek**
Land Use — **Office**

Parking System	*check box*
Surface parking around but not under building	**x**
Surface parking around and under building	
No parking	
Notes	

	Description	Project Measurements		Context Summary SF	Context Summary %	
Site						
	Gross land area in acres	**4.200**	GLA	182,952		
	Public/ private ROW and paved easements in SF	**0**	W		0.0%	of GLA
	Net lot area		NLA	182,952	100.0%	of GLA
	Total unbuildable area in SF	**0**	U		0.0%	of GLA
	Water area within unbuildable area in SF	**0**	WAT			of U
	Buildable land area		BLA	182,952	100.0%	of GLA
Development Cover						
	Gross building area in SF	**39,008**	GBA			
	Total building footprint in SF	**39,008**	B		21.3%	of BLA
	Total bldg. support in SF (stairs, corridors, elevators, etc.)	**2,000**	BSU			
	Building efficiency		Be		94.9%	of GBA
Residential only	Number of dwelling units		DU			
	Aggregate average dwelling unit area in SF		AGG			
	Number of building floors	**2.00**	FLR			
G2 only	Unenclosed surface parking area under building in SF		PUB			
	Total surface parking area in SF**	**80,825**	P or G		44.2%	of BLA
	Number of surface parking spaces on premise	**195**	NPSs			
	Building cover over surface parking area		AIR			of GPA
	Gross surface parking area per parking space in SF		ss	414		
	Gross building area per surface parking space in SF		as	200		
	Surface parking spaces per dwelling unit		us	0.00		
Pkg. struc. only	Total parking structure footprint in SF		P		0.0%	of BLA
Pkg. struc. only	Parking structure area below grade in SF					
Pkg. struc. only	Parking structure area above grade under building in SF					
Pkg. struc. only	Number of parking structure floors		p			
Pkg. struc. only	Total parking structure area in SF		GPA			
Pkg. struc. only	Total parking structure spaces		NPSg			
Pkg. struc. only	Total pkg. struc. support in SF (stairs, elevators, etc.)		PSS			
Pkg. struc. only	Parking structure support percentage		Pu			of GPA
Pkg. struc. only	Gross parking structure area per space in SF		sg	0		
Pkg. struc. only	Gross building area per structure parking space in SF		ag	0		
Pkg. struc. only	Parking structure spaces per dwelling unit		ug	0.00		
Residential only	Total dwelling unit garage area in SF		GPA			
Residential only	Dwelling unit garage above grade under builiding in SF		GUB			
Residential only	Number of dwelling unit garage parking spaces		GPS			
Residential only	Garage parking spaces per dwelling		Gn	0.0		
Residential only	Total dwelling unit garage area per space in SF		Ga			
Residential only	Total dwelling unit garage cover in addition to building cover		Gc		0.0%	of BLA
	Total parking spaces			195.0		
Residential only	Total parking spaces per dwelling unit		u			
	Gross building area per parking space total		a	200.0		
	Number of loading spaces		l			
	Total loading area in SF		LDA			
	Gross area per loading space in SF		b			
	Loading area percentage		L		0.0%	of BLA
	Driveway areas in SF	**1,830**	R		1.0%	of BLA
	Misc. pavement and social center pvmt. in SF	**3,659**	M		2.0%	of BLA
	TOTAL DEVELOPMENT COVER		D	125,322	**68.5%**	of BLA
Project Open Space						
	PROJECT OPEN SPACE		S	57,630	**31.5%**	of BLA
	Water area within project open space in SF		wat			of S
Summary		0.0%				
	Development balance index		BNX		0.685	
	Development intensity index		INX		2.685	
	Floor area ratio per buildable acre		BFAR		0.213	
	Capacity index (gross building sq. ft. per buildable acre)		SFAC		9,288	
	Density per buildable acre if applicable		dBA			
	Density per net acre if applicable		dNA			
	Density per gross acre if applicable		dGA			

** Include internal parking lot landscaping, circulation aisles and private roadway parking space areas. Express private roads separately as driveway areas..

Table 8.2 Post Road Office Context Record

Form CR

Name	**Post Road**
Address or Location	**6565 Post Road**
Land Use	**Office**

Parking System	*check box*
Surface parking around but not under building	**x**
Surface parking around and under building	
No parking	
Notes	

		Description	Project Measurements		Context Summary SF	%	
Site					SF	%	
		Gross land area in acres	**42.260**	GLA	1,840,846		
		Public/ private ROW and paved easements in SF	**0**	W		0.0%	of GLA
		Net lot area		NLA	1,840,846	100.0%	of GLA
		Total unbuildable area in SF	**0**	U		0.0%	of GLA
		Water area within unbuildable area in SF	**0**	WAT			of U
		Buildable land area		BLA	1,840,846	100.0%	of GLA
Development Cover							
		Gross building area in SF	**293,062**	GBA			
		Total building footprint in SF	**73,266**	B		4.0%	of BLA
		Total bldg. support in SF (stairs, corridors, elevators, etc.)	**80,000**	BSU			
		Building efficiency		Be		72.7%	of GBA
	Residential only	Number of dwelling units		DU			
		Aggregate average dwelling unit area in SF		AGG			
		Number of building floors	**4.00**	FLR			
	G2 only	Unenclosed surface parking area under building in SF		PUB			
		Total surface parking area in SF**	**257,410**	P or G		14.0%	of BLA
		Number of surface parking spaces on premise	**713**	NPSs			
		Building cover over surface parking area		AIR			of GPA
		Gross surface parking area per parking space in SF		ss	361		
		Gross building area per surface parking space in SF		as	411		
		Surface parking spaces per dwelling unit		us	0.00		
	Pkg. struc. only	Total parking structure footprint in SF		P		0.0%	of BLA
	Pkg. struc. only	Parking structure area below grade in SF					
	Pkg. struc. only	Parking structure area above grade under building in SF					
	Pkg. struc. only	Number of parking structure floors		p			
	Pkg. struc. only	Total parking structure area in SF		GPA			
	Pkg. struc. only	Total parking structure spaces		NPSg			
	Pkg. struc. only	Total pkg. struc. support in SF (stairs, elevators, etc.)		PSS			
	Pkg. struc. only	Parking structure support percentage		Pu			of GPA
	Pkg. struc. only	Gross parking structure area per space in SF		sg	0		
	Pkg. struc. only	Gross building area per structure parking space in SF		ag	0		
	Pkg. struc. only	Parking structure spaces per dwelling unit		ug	0.00		
	Residential only	Total dwelling unit garage area in SF		GPA			
	Residential only	Dwelling unit garage above grade under builiding in SF		GUB			
	Residential only	Number of dwelling unit garage parking spaces		GPS			
	Residential only	Garage parking spaces per dwelling		Gn	0.0		
	Residential only	Total dwelling unit garage area per space in SF		Ga			
	Residential only	Total dwelling unit garage cover in addition to building cover		Gc		0.0%	of BLA
		Total parking spaces			713.0		
	Residential only	Total parking spaces per dwelling unit		u			
		Gross building area per parking space total		a	411.0		
		Number of loading spaces	**4**	l			
		Total loading area in SF	**8,040**	LDA			
		Gross area per loading space in SF		b	2,010		
		Loading area percentage		L		0.4%	of BLA
		Driveway areas in SF	**36,817**	R		2.0%	of BLA
		Misc. pavement and social center pvmt. in SF	**18,409**	M		1.0%	of BLA
		TOTAL DEVELOPMENT COVER		D	393,942	**21.4%**	of BLA
Project Open Space							
		PROJECT OPEN SPACE		S	1,446,904	**78.6%**	of BLA
		Water area within project open space in SF		wat			of S
Summary			0.0%				
		Development balance index		BNX		0.214	
		Development intensity index		INX		4.214	
		Floor area ratio per buildable acre		BFAR		0.159	
		Capacity index (gross building sq. ft. per buildable acre)		SFAC		6,935	
		Density per buildable acre if applicable		dBA			
		Density per net acre if applicable		dNA			
		Density per gross acre if applicable		dGA			

** Include internal parking lot landscaping, circulation aisles, and private roadway parking space areas. Express private roads separately as driveway areas..

Table 8.3 Jefferson Context Record

Form CR

Name	**Jefferson Apartments**
Address or Location	**1800 N. Star**
Land Use	**Apartment**

Parking System	*check box*
Surface parking around but not under building	x
Surface parking around and under building	
Notes	

		Description	Project Measurements		Context Summary SF	Context Summary %	
Site							
		Gross land area in acres	1.756	GLA	76,491		
		Public/ private ROW and paved easements in SF	0	W		0.0%	of GLA
		Net lot area		NLA	76,491	100.0%	of GLA
		Total unbuildable area in SF	0	U		0.0%	of GLA
		Water area within unbuildable area in SF	0	WAT			of U
		Buildable land area		BLA	76,491	100.0%	of GLA
Development Cover							
		Gross building area in SF	62,400	GBA			
		Total building footprint in SF	20,800	B		27.2%	of BLA
		Total bldg. support in SF (stairs, corridors, elevators, etc.)		BSU			
		Building efficiency		Be			of GBA
	Residential only	Number of dwelling units	72	DU			
		Aggregate average dwelling unit area in SF		AGG	867		
		Number of building floors	3.00	FLR			
	G2 only	Unenclosed surface parking area under building in SF		PUB			
		Total surface parking area in SF (include internal landscape)	31,725	P or G		41.5%	of BLA
		Number of surface parking spaces	102	NPSs			
		Building cover over surface parking area		AIR			of GPA
		Gross surface parking area per parking space in SF		ss	311		
		Gross building area per surface parking space in SF		as	612		
		Surface parking spaces per dwelling unit		us	1.42		
	Pkg. struc. only	Total parking structure footprint in SF		P		0.0%	of BLA
	Pkg. struc. only	Parking structure area under building in SF					
	Pkg. struc. only	Number of parking structure floors		p			
	Pkg. struc. only	Total parking structure area in SF		GPA			
	Pkg. struc. only	Total parking structure spaces		NPSg			
	Pkg. struc. only	Total pkg. struc. support in SF (stairs, elevators, etc.)		PSS			
	Pkg. struc. only	Parking structure support percentage		Pu			of GPA
	Pkg. struc. only	Gross parking structure area per space in SF		sg	0		
	Pkg. struc. only	Gross building area per structure parking space in SF		ag	0		
	Pkg. struc. only	Parking structure spaces per dwelling unit		ug	0.00		
	Pkg. struc. only	Total dwelling unit garage area in SF		GPA			
	Residential only	Dwelling unit garage under builiding in SF		GUB			
	Residential only	Number of dwelling unit garage parking spaces		GPS			
	Residential only	Garage parking spaces per dwelling		Gn	0.0		
	Residential only	Total dwelling unit garage area per space in SF		Ga			
	Residential only	Total dwelling unit garage cover in addition to building cover		Gc		0.0%	of BLA
	Residential only	Total parking spaces			102.0		
		Total parking spaces per dwelling unit		u	1.4		
	Residential only	Gross building area per parking space total		a	611.8		
		Number of loading spaces		l			
		Total loading area in SF		LDA			
		Gross area per loading space in SF		b			
		Loading area percentage		L		0.0%	of BLA
		Driveway areas in SF	640	R		0.8%	of BLA
		Misc. pavement and social center pvmt. in SF	100	M		0.1%	of BLA
		TOTAL DEVELOPMENT COVER		D	53,265	**69.6%**	of BLA
Project Open Space							
		PROJECT OPEN SPACE		S	23,226	**30.4%**	of BLA
		Water area within project open space in SF		wat			of S
Summary			0.0%				
		Development balance index		BNX		0.696	
		Development intensity index		INX		3.696	
		Floor area ratio per buildable acre		BFAR		0.816	
		Capacity index (gross building sq. ft. per buildable acre)		SFAC		35,535	
		Density per buildable acre if applicable		dBA		41.00	
		Density per net acre if applicable		dNA		41.00	
		Density per gross acre if applicable		dGA		41.00	

Table 8.4 Asherton Context Record

Form CR

Name	**Asherton**
Address or Location	
Land Use	**Urbanhouse**

Parking System	*check box*
Surface parking around but not under building	x
Surface parking around and under building	
No parking	
Notes	

	Description	Project Measurements		Context Summary SF	Context Summary %	
Site						
	Gross land area in acres	**21.330**	GLA	929,135		
	Public/ private ROW and paved easements in SF	**35,750**	W		3.8%	of GLA
	Net lot area		NLA	893,385	96.2%	of GLA
	Total unbuildable area in SF		U		0.0%	of GLA
	Water area within unbuildable area in SF		WAT			of U
	Buildable land area		BLA	893,385	96.2%	of GLA
Development Cover						
	Gross building area in SF	**178,700**	GBA			
	Total building footprint in SF	**89,350**	B		10.0%	of BLA
	Total bldg. support in SF (stairs, corridors, elevators, etc.)		BSU			
	Building efficiency		Be			of GBA
Residential only	Number of dwelling units	**103**	DU			
	Aggregate average dwelling unit area in SF		AGG	1,735		
	Number of building floors	**2.00**	FLR			
G2 only	Unenclosed surface parking area under building in SF		PUB			
	Total surface parking area in SF**	**60,780**	P or G		6.8%	of BLA
	Number of surface parking spaces on premise	**146**	NPSs			
	Building cover over surface parking area		AIR			of GPA
	Gross surface parking area per parking space in SF		ss	416		
	Gross building area per surface parking space in SF		as	1,224		
	Surface parking spaces per dwelling unit		us	1.42		
Pkg. struc. only	Total parking structure footprint in SF		P		0.0%	of BLA
Pkg. struc. only	Parking structure area below grade in SF		UNG			
Pkg. struc. only	Parking structure area above grade under building in SF		PSC			
Pkg. struc. only	Number of parking structure floors		p			
Pkg. struc. only	Total parking structure area in SF		GPA			
Pkg. struc. only	Total parking structure spaces		NPSg			
Pkg. struc. only	Total pkg. struc. support in SF (stairs, elevators, etc.)		PSS			
Pkg. struc. only	Parking structure support percentage		Pu			of GPA
Pkg. struc. only	Gross parking structure area per space in SF		sg	0		
Pkg. struc. only	Gross building area per structure parking space in SF		ag	0		
Pkg. struc. only	Parking structure spaces per dwelling unit		ug	0.00		
Residential only	Total dwelling unit garage area in SF	**10,000**	GPA			
Residential only	Dwelling unit garage above grade under builiding in SF	**0**	GUB			
Residential only	Number of dwelling unit garage parking spaces	**40**	GPS			
Residential only	Garage parking spaces per dwelling		Gn	0.4		
Residential only	Total dwelling unit garage area per space in SF		Ga	250.0		
Residential only	Total dwelling unit garage cover in addition to building cover		Gc		1.1%	of BLA
	Total parking spaces		NPS	186.0		
Residential only	Total parking spaces per dwelling unit		u	1.8		
	Gross building area per parking space total		a	960.8		
	Number of loading spaces	**0**	l			
	Total loading area in SF	**0**	LDA			
	Gross area per loading space in SF		b			
	Loading area percentage		L		0.0%	of BLA
	Driveway areas in SF	**6,500**	R		0.7%	of BLA
	Misc. pavement and social center pvmt. in SF	**17,200**	M		1.9%	of BLA
	TOTAL DEVELOPMENT COVER		D	183,830	**20.6%**	of BLA
Project Open Space						
	PROJECT OPEN SPACE		S	709,555	**79.4%**	of BLA
	Water area within project open space in SF	**179,500** 25.3%	wat			of S
Summary						
	Development balance index		BNX		0.206	
	Development intensity index		INX		2.206	
	Floor area ratio per buildable acre		BFAR		0.200	
	Capacity index (gross building sq. ft. per buildable acre)		SFAC		8,713	
	Density per buildable acre if applicable		dBA		5.02	
	Density per net acre if applicable		dNA		5.02	
	Density per gross acre if applicable		dGA		4.83	

** Include internal parking lot landscaping, circulation aisles, and private roadway parking space areas. Express private roads separately as driveway areas..

Table 8.5 Design Specifications for Cramer Creek and Post Road

Index	Specification Component	Land Use Family and Building Type Nonresidential Nonresidential **Cramer Creek**	 **Post Road**
GLA	Gross land area in acres	4.2	42.26
W	Public/ private right-of-way and paved areas in sq. ft.	0.0	0.0
U	Unbuildable and/or future expansion areas in sq. ft.	0.0	0.0
s	Gross parking lot area per parking space in sq. ft.	414	361
a	Gross nonresidential building sq. ft. per parking space	200	411
l	No. of loading spaces	0	4
b	Gross area per loading space in sq. ft.	0	2,010
S	Project open space as percentage of buildable lot area	31.4%	78.6%
R	Private driveways as percentage of buildable lot area	1.0%	2.0%
M	Miscellaneous pavement as percentage of buildable lot area	2.0%	1.0%

Development Capacity Indexes

Index	Specification Component	Cramer Creek	Post Road
BNX	Development balance index	0.685	0.214
INX	Development intensity index	1.685	4.214
SFAC	Gross building square feet constructed per buildable acre	9,288	6,935

WM Hosack

Cramer Creek

WM Hosack

Post Road

Table 8.6 Design Specifications for Jefferson and Asherton

Index	Specification Component	Land Use Family and Building Type	
		Residential	
		Apartment	
		Jefferson	**Asherton**
GLA	Gross land area in acres	1.756	21.33
W	Public/ private right-of-way and paved area	0	35,750
U	Unbuildable areas, future expansion areas, and social center areas	0	0
DU	Total number of dwelling units	72	103
NPS	Total parking spaces	102	146
PLA	Total parking lot area including internal landscaped areas	31,725	60,780
s	Gross parking lot area per parking space in sq. ft.	311	416
u	Parking lot spaces provided per dwelling unit	1.42	1.42
Gn	Garage parking spaces provided per dwelling unit	0	0.4
Ga	Gross garage building area per garage space	0	250
l	No. of loading spaces	0	0
b	Gross area per loading space	0	0
S	Project open space as percentage of buildable lot area	30.4%	79.4%
R	Private driveways as percentage of buildable lot area	0.8%	0.7%
M	Miscellaneous pavement as percentage of buildable lot area	0.1%	1.9%
Be	Residential building efficiency as percentage of the gross building area		
GBA	Gross building area	62,400	178,700
AGG	Aggregate average dwelling unit area	867	1,735

Development Capacity Indexes

Index	Specification Component	Jefferson	Asherton
BNX	Development balance index	.696	.206
INX	Development intensity index	3.696	2.206
SFAC	Gross building square feet constructed per buildable acre	35,535	8,713

B. Higgins

Jefferson

WM Hosack

Asherton

Exhibit 8.1 Cramer Creek Development Capacity Forecast (Fig. 8.1)

Forecast Model CG1L

Development capacity forecast for ***NONRESIDENTIAL BUILDINGS*** *based on the use of an adjacent* ***GRADE PARKING LOT*** *located on the same premises. When s and a equal zero in the design specification below, the forecast pertains to conditions when* ***NO PARKING*** *is required.*

Given: *Gross land area.* ***To Find:*** *Maximum development capacity of the land area (gross building area potential) based on the design specification values entered below.* ***Premise:*** *All building floors considered equal in area.*

DESIGN SPECIFICATION

Enter values in boxed areas where text is bold and blue. Express all fractions as decimals.

Given:	**Gross land area**	**GLA=**	**4.200**	acres	182,952	SF
Land Variables:	Public/ private right-of-way & paved easements	W=	**0.000**	fraction of GLA	0	SF
	Net land area	NLA=	4.200	acres	182,952	SF
	Unbuildable and/or future expansion areas	U=	**0.000**	fraction of GLA	0	SF
	Gross land area reduction	X=	0.000	fraction of GLA	0	SF
	Buildable land area remaining	**BLA=**	4.200	acres	182,952	SF
Parking Variables:	Est. gross pkg. lot area per space in SF	s =	**414.4**	ENTER ZERO IF NO PARKING REQUIRED		
	Building SF permitted per parking space	a =	**200**	ENTER ZERO IF NO PARKING REQUIRED		
	No. of loading spaces	l=	**0**			
	Gross area per loading space	b =	**0**	SF	0	SF
Site Variables:	**Project open space as fraction of BLA**	**S=**	**0.315**		57,630	SF
	Private driveways as fraction of BLA	R=	**0.010**		1,830	SF
	Misc. pavement as fraction of BLA	M=	**0.020**		3,659	SF
	Loading area as fraction of BLA	L=	0.000		0	SF
	Total site support areas as a fraction of BLA	Su=	0.345		63,118	SF
Core:	**Core development area as fraction of BLA**	**C=**	0.655	C+Su must = 1	119,834	SF

PLANNING FORECAST

no. of floors **FLR**	**CORE** minimum land area for BCG & PLA	gross building area **GBA**	parking lot area **PLA**	pkg. spaces **NPS**	footprint **BCA**	bldg SF / acre **SFAC** function of BLA	flr area ratio **FAR** function of BLA
1.00	**119,834**	**39,008**	**80,825**	**195.0**	**39,008**	**9,288**	**0.213**
2.00		46,592	96,538	233.0	23,296	11,093	0.255
3.00		49,820	103,227	249.1	16,607	11,862	0.272
4.00		51,608	106,932	258.0	12,902	12,288	0.282
5.00		52,744	109,285	263.7	10,549	12,558	0.288
6.00	*NOTE: Be aware when BCA becomes too small to be feasible.*	53,529	110,912	267.6	8,921	12,745	0.293
7.00		54,104	112,104	270.5	7,729	12,882	0.296
8.00		54,544	113,016	272.7	6,818	12,987	0.298
9.00		54,891	113,735	274.5	6,099	13,069	0.300
10.00		55,172	114,316	275.9	5,517	13,136	0.302
11.00		55,404	114,797	277.0	5,037	13,191	0.303
12.00		55,599	115,200	278.0	4,633	13,238	0.304
13.00		55,764	115,544	278.8	4,290	13,277	0.305
14.00		55,907	115,840	279.5	3,993	13,311	0.306
15.00		56,032	116,098	280.2	3,735	13,341	0.306

WARNING: These are preliminary forecasts that must not be used to make final decisions.

1) These forecasts are not a substitute for the "due diligence" research that must be conducted to support the final definition of "unbuildable areas" above and the final decision to purchase land. This research includes, but is not limited to, verification of adequate subsurface soil, zoning, environmental clearance, access, title, utilities and water pressure, clearance from deed restriction, easement and right-of-way encumbrances, clearance from existing above and below ground facility conflicts, etc.

2) The most promising forecast(s) made on the basis of data entered in the design specification from "due diligence" research must be verified at the drawing board before funds are committed and land purchase decisions are made. Actual land shape ratios, dimensions and irregularities encountered may require adjustments to the general forecasts above.

3) The software licensee shall take responsibility for the design specification values entered and any advice given that is based on the forecast produced.

this exhibit shows that a 5-story building supported by a surface parking lot could have produced 52,744 gross sq ft of building area if the building footprint (BCA) had been reduced to 10,549 sq ft per floor. This is an increase of 35%, or 13,736 sq ft over the 1-story project that was built. It is also closer to the maximum development capacity of the site under G1 parking conditions when the project open space allocation remains constant. If the project open space allocation is reduced, this maximum capacity could be increased at the potential expense of the context produced.

A second method of single-system comparison is to change the values assigned to components within the design specification of a forecast model. Three component values, in addition to building height (FLR), significantly affect the development capacity of land and the context produced, and all are common to both residential and nonresidential construction:[3] (1) the amount of project open space (*S*) provided, (2) the parking provision[4] (*a*) planned or required, and (3) the total amount of parking area (*s*) planned per parking space.[5] Changing one or more of these values within a forecast model can have a significant effect on the development capacity of land and the context predicted. Exhibit 8.2 illustrates the reduction in gross building area capacity that is forecast when the project open space value *S* in Exhibit 8.1 is increased to 80% of the buildable area. When Exhibit 8.2 is compared with Exhibit 8.1, it is clear that the original gross building area of 39,008 sq ft can no longer be reached, no matter what building height is chosen. In fact, a 3-story building could only reach 12,930 sq ft of gross building area under these circumstances, which is 33% of that actually built.

Post Road

The photograph in Fig. 8.2 illustrates a project with 78.6% project open space. It is a classic example of a prosperous suburban office that has created its own private park and

Exhibit 8.2 Cramer Creek Open Space Increase

Forecast Model CG1L

*Development capacity forecast for **NONRESIDENTIAL BUILDINGS** based on the use of an adjacent **GRADE PARKING LOT** located on the same premises. When s and a equal zero in the design specification below, the forecast pertains to conditions when **NO PARKING** is required.*

***Given:** Gross land area. **To Find:** Maximum development capacity of the land area (gross building area potential) based on the design specification values entered below. **Premise:** All building floors considered equal in area.*

DESIGN SPECIFICATION *Enter values in boxed areas where text is bold and blue. Express all fractions as decimals.*

Given:	**Gross land area**	**GLA=**	**4.200**	acres	182,952	SF
Land Variables:	Public/ private right-of-way & paved easements	W=	**0.000**	fraction of GLA	0	SF
	Net land area	NLA=	4.200	acres	182,952	SF
	Unbuildable and/or future expansion areas	U=	**0.000**	fraction of GLA	0	SF
	Gross land area reduction	X=	0.000	fraction of GLA	0	SF
	Buildable land area remaining	**BLA=**	4.200	acres	182,952	SF
Parking Variables:	Est. gross pkg. lot area per space in SF	s =	**414.4**	ENTER ZERO IF NO PARKING REQUIRED		
	Building SF permitted per parking space	a =	**200**	ENTER ZERO IF NO PARKING REQUIRED		
	No. of loading spaces	l=	**0**			
	Gross area per loading space	b =	**0**	SF	0	SF
Site Variables:	**Project open space as fraction of BLA**	**S=**	**0.800**	⬅	146,362	SF
	Private driveways as fraction of BLA	R=	**0.010**		1,830	SF
	Misc. pavement as fraction of BLA	M=	**0.020**		3,659	SF
	Loading area as fraction of BLA	L=	0.000		0	SF
	Total site support areas as a fraction of BLA	Su=	0.830		151,850	SF
Core:	**Core development area as fraction of BLA**	**C=**	0.170	C+Su must = 1	31,102	SF

PLANNING FORECAST

no. of floors **FLR**	**CORE** minimum land area for BCG & PLA	gross building area **GBA**	parking lot area **PLA**	pkg. spaces **NPS**	footprint **BCA**	bldg SF / acre **SFAC** function of BLA	flr area ratio **FAR** function of BLA
1.00	**31,102**	**10,124**	**20,978**	**50.6**	**10,124**	**2,411**	**0.055**
2.00		12,092	25,056	60.5	6,046	2,879	0.066
3.00		12,930	26,792	64.7	4,310	3,079	0.071
4.00		13,394	27,753	67.0	3,349	3,189	0.073
5.00		13,689	28,364	68.4	2,738	3,259	0.075
6.00	*NOTE: Be aware when BCA becomes too small to be feasible.*	13,893	28,786	69.5	2,316	3,308	0.076
7.00		14,042	29,096	70.2	2,006	3,343	0.077
8.00		14,157	29,332	70.8	1,770	3,371	0.077
9.00		14,247	29,519	71.2	1,583	3,392	0.078
10.00		14,319	29,670	71.6	1,432	3,409	0.078
11.00		14,380	29,795	71.9	1,307	3,424	0.079
12.00		14,430	29,899	72.2	1,203	3,436	0.079
13.00		14,473	29,989	72.4	1,113	3,446	0.079
14.00		14,510	30,065	72.6	1,036	3,455	0.079
15.00		14,543	30,132	72.7	970	3,463	0.079

WARNING: These are preliminary forecasts that must not be used to make final decisions.

1) These forecasts are not a substitute for the "due diligence" research that must be conducted to support the final definition of "unbuildable areas" above and the final decision to purchase land. This research includes, but is not limited to, verification of adequate subsurface soil, zoning, environmental clearance, access, title, utilities and water pressure, clearance from deed restriction, easement and right-of-way encumbrances, clearance from existing above and below ground facility conflicts, etc.

2) The most promising forecast(s) made on the basis of data entered in the design specification from "due diligence" research must be verified at the drawing board before funds are committed and land purchase decisions are made. Actual land shape ratios, dimensions and irregularities encountered may require adjustments to the general forecasts above.

3) The software licensee shall take responsibility for the design specification values entered and any advice given that is based on the forecast produced.

potential expansion area. Table 8.2 displays the context measurements that create the record for this project. Table 8.5 displays the design specification values derived from this record, and Exhibit 8.3 displays the forecast that would have have been produced if these values had been entered in the design specification panel of forecast model CG1L before the project was constructed. The development capacity option chosen from these single-system alternatives is outlined in the planning forecast panel of Exhibit 8.3.

Exhibit 8.4 reduces the project open space *S* allocation in Exhibit 8.3 to 50% in order to compare the development capacity implications of this option with that actually constructed. Exhibit 8.3 shows that the gross building area GBA capacity of the site at 78.6% open space was 293,062 sq ft when the building footprint BCA was 73,266 sq ft. When project open space is reduced to 50%, Exhibit 8.4 shows that the same 4-story building height FLR can now produce 759,659 sq ft of gross building area.[6] This is an increase of 259%, or 466,597 sq ft from the previous option. It also implies a significant decrease in the parklike context provided. If a library of context records and social research were available, this option could be compared with an existing project with a similar specification to understand the implications of the reduction. This could easily enhance the value and depth of the single-system comparisons made.

Variances

Parking variances can also be evaluated with this comparative process. Parking systems dominate land consumption because every parking space justifies additional building area. Project open space often represents the land left at the end of the process. The problem can be compounded when building and parking "don't fit" on the site and parking variances are requested.[7] Parking variances increase development capacity, and their effect can be seen in Exhibit 8.5. The parking

EXHIBIT 8.3 Post Road Development Capacity Forecast (Fig. 8.2)

Forecast Model CG1L

Development capacity forecast for ***NONRESIDENTIAL BUILDINGS*** *based on the use of an adjacent* ***GRADE PARKING LOT*** *located on the same premises. When s and a equal zero in the design specification below, the forecast pertains to conditions when* ***NO PARKING*** *is required.*

Given: *Gross land area.* ***To Find:*** *Maximum development capacity of the land area (gross building area potential) based on the design specification values entered below.* ***Premise:*** *All building floors considered equal in area.*

DESIGN SPECIFICATION

Enter values in boxed areas where text is bold and blue. Express all fractions as decimals.

Given:	**Gross land area**	GLA=	**42.260**	acres	1,840,846	SF
Land Variables:	Public/ private right-of-way & paved easements	W=	**0.000**	fraction of GLA	0	SF
	Net land area	NLA=	42.260	acres	1,840,846	SF
	Unbuildable and/or future expansion areas	U=	**0.000**	fraction of GLA	0	SF
	Gross land area reduction	X=	0.000	fraction of GLA	0	SF
	Buildable land area remaining	BLA=	42.260	acres	1,840,846	SF
Parking Variables:	Est. gross pkg. lot area per space in SF	s =	**361**	ENTER ZERO IF NO PARKING REQUIRED		
	Building SF permitted per parking space	a =	**411**	ENTER ZERO IF NO PARKING REQUIRED		
	No. of loading spaces	l=	**4**			
	Gross area per loading space	b =	**2,010**	SF	8,040	SF
Site Variables:	**Project open space as fraction of BLA**	S=	**0.786**		1,446,905	SF
	Private driveways as fraction of BLA	R=	**0.020**		36,817	SF
	Misc. pavement as fraction of BLA	M=	**0.010**		18,408	SF
	Loading area as fraction of BLA	L=	0.004		8,040	SF
	Total site support areas as a fraction of BLA	Su=	0.820		1,510,170	SF
Core:	**Core development area as fraction of BLA**	C=	0.180	C+Su must = 1	330,676	SF

PLANNING FORECAST

no. of floors **FLR**	CORE minimum land area for BCG & PLA	gross building area **GBA**	parking lot area PLA	pkg. spaces NPS	footprint **BCA**	bldg SF / acre SFAC function of BLA	flr area ratio FAR function of BLA
1.00	330,676	176,046	154,629	428.3	176,046	4,166	0.096
2.00		239,908	210,722	583.7	119,954	5,677	0.130
3.00		272,907	239,707	664.0	90,969	6,458	0.148
4.00		**293,062**	**257,410**	**713.0**	**73,266**	**6,935**	**0.159**
5.00		306,651	269,345	746.1	61,330	7,256	0.167
6.00		316,432	277,937	769.9	52,739	7,488	0.172
7.00	*NOTE: Be aware when BCA becomes too small to be feasible.*	323,810	284,417	787.9	46,259	7,662	0.176
8.00		329,573	289,479	801.9	41,197	7,799	0.179
9.00		334,199	293,542	813.1	37,133	7,908	0.182
10.00		337,995	296,876	822.4	33,799	7,998	0.184
11.00		341,165	299,661	830.1	31,015	8,073	0.185
12.00		343,852	302,021	836.6	28,654	8,137	0.187
13.00		346,160	304,048	842.2	26,628	8,191	0.188
14.00		348,162	305,807	847.1	24,869	8,239	0.189
15.00		349,917	307,348	851.4	23,328	8,280	0.190

WARNING: These are preliminary forecasts that must not be used to make final decisions.

1) These forecasts are not a substitute for the "due diligence" research that must be conducted to support the final definition of "unbuildable areas" above and the final decision to purchase land. This research includes, but is not limited to, verification of adequate subsurface soil, zoning, environmental clearance, access, title, utilities and water pressure, clearance from deed restriction, easement and right-of-way encumbrances, clearance from existing above and below ground facility conflicts, etc.

2) The most promising forecast(s) made on the basis of data entered in the design specification from "due diligence" research must be verified at the drawing board before funds are committed and land purchase decisions are made. Actual land shape ratios, dimensions and irregularities encountered may require adjustments to the general forecasts above.

3) The software licensee shall take responsibility for the design specification values entered and any advice given that is based on the forecast produced.

Exhibit 8.4 Post Road Open Space Reduction Forecast

Forecast Model CG1L

Development capacity forecast for ***NONRESIDENTIAL BUILDINGS*** *based on the use of an adjacent* ***GRADE PARKING LOT*** *located on the same premises. When s and a equal zero in the design specification below, the forecast pertains to conditions when* ***NO PARKING*** *is required.*

Given: *Gross land area.* ***To Find:*** *Maximum development capacity of the land area (gross building area potential) based on the design specification values entered below.* ***Premise:*** *All building floors considered equal in area.*

DESIGN SPECIFICATION

Enter values in boxed areas where text is bold and blue. Express all fractions as decimals.

Given:	**Gross land area**	GLA=	**42.260**	acres	1,840,846	SF
Land Variables:	Public/ private right-of-way & paved easements	W=	**0.000**	fraction of GLA	0	SF
	Net land area	NLA=	42.260	acres	1,840,846	SF
	Unbuildable and/or future expansion areas	U=	**0.000**	fraction of GLA	0	SF
	Gross land area reduction	X=	0.000	fraction of GLA	0	SF
	Buildable land area remaining	BLA=	42.260	acres	1,840,846	SF
Parking Variables:	Est. gross pkg. lot area per space in SF	s =	**361**	ENTER ZERO IF NO PARKING REQUIRED		
	Building SF permitted per parking space	a =	**411**	ENTER ZERO IF NO PARKING REQUIRED		
	No. of loading spaces	l=	**4**			
	Gross area per loading space	b =	**2,010**	SF	8,040	SF
Site Variables:	**Project open space as fraction of BLA**	S=	**0.500**		920,423	SF
	Private driveways as fraction of BLA	R=	**0.020**		36,817	SF
	Misc. pavement as fraction of BLA	M=	**0.010**		18,408	SF
	Loading area as fraction of BLA	L=	0.004		8,040	SF
	Total site support areas as a fraction of BLA	Su=	0.534		983,688	SF
Core:	**Core development area as fraction of BLA**	C=	0.466	C+Su must = 1	857,157	SF

PLANNING FORECAST

no. of floors **FLR**	CORE minimum land area for BCG & PLA	gross building area **GBA**	parking lot area PLA	pkg. spaces NPS	footprint **BCA**	bldg SF / acre SFAC function of BLA	flr area ratio FAR function of BLA
1.00	857,157	456,336	400,821	1110.3	456,336	10,798	0.248
2.00		621,874	546,220	1513.1	310,937	14,715	0.338
3.00		707,413	621,353	1721.2	235,804	16,740	0.384
4.00		**759,659**	**667,243**	**1848.3**	**189,915**	**17,976**	**0.413**
5.00		794,882	698,181	1934.0	158,976	18,809	0.432
6.00		820,237	720,451	1995.7	136,706	19,409	0.446
7.00	*NOTE: Be aware when BCA becomes too small to be feasible.*	839,361	737,249	2042.2	119,909	19,862	0.456
8.00		854,299	750,370	2078.6	106,787	20,215	0.464
9.00		866,291	760,903	2107.8	96,255	20,499	0.471
10.00		876,130	769,544	2131.7	87,613	20,732	0.476
11.00		884,347	776,762	2151.7	80,395	20,926	0.480
12.00		891,314	782,881	2168.6	74,276	21,091	0.484
13.00		897,295	788,135	2183.2	69,023	21,233	0.487
14.00		902,486	792,694	2195.8	64,463	21,356	0.490
15.00		907,033	796,689	2206.9	60,469	21,463	0.493

WARNING: These are preliminary forecasts that must not be used to make final decisions.

1) These forecasts are not a substitute for the "due diligence" research that must be conducted to support the final definition of "unbuildable areas" above and the final decision to purchase land. This research includes, but is not limited to, verification of adequate subsurface soil, zoning, environmental clearance, access, title, utilities and water pressure, clearance from deed restriction, easement and right-of-way encumbrances, clearance from existing above and below ground facility conflicts, etc.

2) The most promising forecast(s) made on the basis of data entered in the design specification from "due diligence" research must be verified at the drawing board before funds are committed and land purchase decisions are made. Actual land shape ratios, dimensions and irregularities encountered may require adjustments to the general forecasts above.

3) The software licensee shall take responsibility for the design specification values entered and any advice given that is based on the forecast produced.

Exhibit 8.5 Cramer Creek Parking Variance Alternative

Forecast Model CG1L

*Development capacity forecast for **NONRESIDENTIAL BUILDINGS** based on the use of an adjacent **GRADE PARKING LOT** located on the same premises. When s and a equal zero in the design specification below, the forecast pertains to conditions when **NO PARKING** is required.*

***Given:** Gross land area. **To Find:** Maximum development capacity of the land area (gross building area potential) based on the design specification values entered below. **Premise:** All building floors considered equal in area.*

DESIGN SPECIFICATION

Enter values in boxed areas where text is bold and blue. Express all fractions as decimals.

Given:	**Gross land area**	**GLA=**	**4.200**	acres	182,952	SF
Land Variables:	Public/ private right-of-way & paved easements	W=	**0.000**	fraction of GLA	0	SF
	Net land area	NLA=	4.200	acres	182,952	SF
	Unbuildable and/or future expansion areas	U=	**0.000**	fraction of GLA	0	SF
	Gross land area reduction	X=	0.000	fraction of GLA	0	SF
	Buildable land area remaining	**BLA=**	4.200	acres	182,952	SF
Parking Variables:	Est. gross pkg. lot area per space in SF	s =	**414.4**	ENTER ZERO IF NO PARKING REQUIRED		
	Building SF permitted per parking space	a =	**250**			
	No. of loading spaces	l=	**0**			
	Gross area per loading space	b =	**0**	SF	0	SF
Site Variables:	**Project open space as fraction of BLA**	**S=**	**0.315**		57,630	SF
	Private driveways as fraction of BLA	R=	**0.010**		1,830	SF
	Misc. pavement as fraction of BLA	M=	**0.020**		3,659	SF
	Loading area as fraction of BLA	L=	0.000		0	SF
	Total site support areas as a fraction of BLA	Su=	0.345		63,118	SF
Core:	**Core development area as fraction of BLA**	**C=**	0.655	C+Su must = 1	119,834	SF

PLANNING FORECAST

no. of floors **FLR**	CORE minimum land area for BCG & PLA	gross building area **GBA**	parking lot area PLA	pkg. spaces NPS	footprint **BCA**	bldg SF / acre SFAC function of BLA	flr area ratio FAR function of BLA
1.00	**119,834**	**45,091**	**74,743**	**180.4**	**45,091**	**10,736**	**0.246**
2.00		55,540	92,063	222.2	27,770	13,224	0.304
3.00		60,190	99,770	240.8	20,063	14,331	0.329
4.00		62,819	104,129	251.3	15,705	14,957	0.343
5.00		64,510	106,932	258.0	12,902	15,359	0.353
6.00	*NOTE: Be aware when BCA becomes too small to be feasible.*	65,689	108,885	262.8	10,948	15,640	0.359
7.00		66,557	110,325	266.2	9,508	15,847	0.364
8.00		67,224	111,431	268.9	8,403	16,006	0.367
9.00		67,752	112,306	271.0	7,528	16,131	0.370
10.00		68,180	113,016	272.7	6,818	16,233	0.373
11.00		68,535	113,603	274.1	6,230	16,318	0.375
12.00		68,833	114,097	275.3	5,736	16,389	0.376
13.00		69,087	114,519	276.3	5,314	16,449	0.378
14.00		69,307	114,883	277.2	4,950	16,502	0.379
15.00		69,498	115,200	278.0	4,633	16,547	0.380

WARNING: These are preliminary forecasts that must not be used to make final decisions.
1) These forecasts are not a substitute for the "due diligence" research that must be conducted to support the final definition of "unbuildable areas" above and the final decision to purchase land. This research includes, but is not limited to, verification of adequate subsurface soil, zoning, environmental clearance, access, title, utilities and water pressure, clearance from deed restriction, easement and right-of-way encumbrances, clearance from existing above and below ground facility conflicts, etc.
2) The most promising forecast(s) made on the basis of data entered in the design specification from "due diligence" research must be verified at the drawing board before funds are committed and land purchase decisions are made. Actual land shape ratios, dimensions and irregularities encountered may require adjustments to the general forecasts above.
3) The software licensee shall take responsibility for the design specification values entered and any advice given that is based on the forecast produced.

requirement *a* shown in Exhibit 8.1 was 200.[8] If this requirement were reduced to 250 in Exhibit 8.5, the 1-story building area could be increased 16% from the 39,008 sq ft constructed to the 45,091 sq ft shown in Exhibit 8.5. The context would not visibly change unless the parking reduction produced congestion, but the economics would immediately change. As a word of caution, however, it is possible to produce congestion when such variances are too liberal, which produces an unacceptable level of intensity.

Open Space

Increasing development capacity with parking variances is common, and not always desirable, but this comparative example has been based on one significant departure from the typical process. Project open space *S* was entered as part of the design specification and was subtracted from the buildable land available BLA before capacity was evaluated. This is quite different from the more common *left-over approach*. If the left-over approach were used with the forecast system being discussed, project open space *S* would be entered as close to zero as permitted by the zoning ordinance having jurisdiction.[9] All other development values would be entered and the gross building area GBA forecast would be located in the planning forecast panel along the row corresponding to the number of building floors FLR being contemplated. If the GBA value were greater than that needed, or desired, the project open space *S* value could be increased until the GBA value declined to that required. In this scenario, project open space is the area left over after all development cover objectives are satisfied, and is not specified to produce a predetermined environmental objective.

The implication of the preceding example is that predetermined project open space objectives can be evaluated with the single-system comparative process to assess their development capacity implications. They can then be specified, since they are too important to remain an afterthought in the development process. The development capacity impact of project open space objectives can be quickly and easily evaluated with the collection of forecast models included on the attached CD-ROM, and the debate that ensues can be based on a quantitative evaluation of the options available and the implications revealed through the comparative process.

Indexes

The two context records in Tables 8.1 and 8.2 define the widely divergent environments photographed in Figs. 8.1 and 8.2. These records are indexed with three values noted at the bottom of each: (1) a *balance index* BNX that records the total development cover percentage introduced; (2) an *intensity index* INX that combines the building height present, expressed as the number of floors present, with the development cover introduced; and (3) a *capacity index* SFAC that reports the gross square feet of building area that has been constructed per acre of buildable land area. A quick glance at the indexes immediately tells the reader that one project is low-rise with a modest amount of open space and that the other is a mid-rise project in a parklike setting. This indexing system is not a substitute for the context record and design specification it represents, but it can be a helpful organizing tool and a simple regulatory system when supplemented with more specific design guidelines.

The Residential Family

The concept of single-system development capacity evaluation and comparison can also be used with the residential collection of forecast models, but the context record measurements and design specification values that result include several unique residential characteristics that expand the list of components measured and specification values defined.[10] The additional measurements are shown in Tables 8.3 and 8.4. These measurements represent the projects shown in Figs. 8.3 and 8.4, and Table 8.6 lists the design specification values that were either entered or calculated within form CR, based on the measurements entered in the open boxes. A quick comparison of Tables 8.5 and 8.6 illustrates the expanded list of design specification values produced from these measurements.

Jefferson Apartments

The specification values defined in Tables 8.5 and 8.6 were entered in Exhibits 8.6 and 8.8 to create representative forecast models for the Jefferson and Asherton projects measured. The Jefferson forecast model (Exhibit 8.6) shows the results that would have been predicted if these specification values had been entered during the planning and design phase. The forecast results are not identical to the project measurements in Table 8.3 due to the precision of mathematical calculations and rounding, but are within small tolerance ranges. For instance, the density per buildable acre dBA forecast is 40.96; the density achieved was 41.0. The number of dwelling units NDU forecast is 71.9; the number achieved was 72. The gross building area GBA forecast was 62,359 sq ft; the area measured was 62,400 sq ft. Exact comparisons will be unusual due to the variables involved in design development, but the results forecast can be very useful when establishing the planning, design, and context parameters for land use areas and building construction programs.

A review of the planning forecast panel in the Jefferson model, Exhibit 8.6, again shows that a wide range of options was available based on the design specification used. The density per buildable acre dBA could have easily increased to 48.66 if a 5-story building had been constructed with a reduced footprint area. This would have been unfortunate since the open space provided is only 30.4%, and the landscape treatment of this open space is minimal. Additional height may have been limited by zoning regulations, but if it had been permitted by variance, the intensity INX would have increased from 3.696 to 5.696. This would have been

Exhibit 8.6 Jefferson Development Capacity Forecast (Fig. 8.3)

Forecast Model RG1L

*Development capacity forecast for **APARTMENTS** based on the use of an adjacent **GRADE PARKING LOT** located on the same premises. When s and u equal zero in the design specification below, the forecast pertains to conditions when **NO PARKING** is required.*

***Given:** Gross land area. **To Find:** Maximum dwelling unit capacity of the land area given based on the design specification values entered below. **Premise:** All building floors considered equal in area.*

DESIGN SPECIFICATION *Enter values in boxed areas where text is bold and blue. Express all fractions as decimals.*

Given:	**Gross land area**	GLA=	**1.756**	acres	76,491	SF
Land Variables:	Public/ private right-of-way & paved easements	W=	**0.000**	fraction of GLA	0	SF
	Net land area	NLA=	1.756	acres	76,491	SF
	Unbuildable and/or future expansion areas	U=	**0.000**	fraction of GLA	0	SF
	Gross Land Area Reduction	X=	0.000	fraction of GLA	0	SF
	Buildable land area remaining	BLA=	**1.756**	acres	76,491	SF
Parking Variables:	Est. gross pkg. lot area per pkg. space in SF	s =	**311**	ENTER ZERO IF NO PARKING REQUIRED		
	Parking lot spaces planned or required per dwelling unit	u=	**1.42**	ENTER ZERO IF NO PARKING REQUIRED		
	Garage parking spaces planned or required per dwelling unit	Gn=	**0**	ENTER ZERO IF NO PARKING REQUIRED		
	Gross building area per garage space	Ga=	**0**	ENTER ZERO IF NO PARKING REQUIRED		
	No. of loading spaces	l =	**0**			
	Gross area per loading space	b =	**0**	SF	0	SF
Site Variables:	**Project open space as fraction of BLA**	S=	**0.304**		23,253	SF
	Private driveways as fraction of BLA	R=	**0.008**		612	SF
	Misc. pavement as fraction of BLA	M=	**0.001**		76	SF
	Loading area as fraction of BLA	L=	0.000		0	SF
	Total site support areas as a fraction of BLA	Su=	0.313		23,942	SF
Core:	**Core development area as fraction of BLA**	C=	0.687	C+Su must = 1	52,550	SF
Building Variables:	Building efficiency as percentage of GBA	Be=	**0.900**			
	Building support as fraction of GBA	Bu=	0.100	Be + Bu must = 1		

Dwelling Unit Mix Table:

DU dwelling unit type	GDA gross du area	CDA=GDA/Be comprehensive du area	MIX du mix	PDA = (CDA)MIX Prorated du area

Dwelling unit mix not svailable

Aggregate average dwelling unit area (AGG) found by dividing gross building area by number of dwelling units present

Aggregate avg. dwelling unit area (AGG) = **867**

GBA sf per parking space a= 611

PLANNING FORECAST

no. of floors FLR	CORE minimum lot area for BCG & PLA	density per net acre dNA	dwelling units NDU	pkg. lot spaces NPS	parking lot area PLA	garage spaces GPS	garage area GAR	gross bldg area GBA no garages	footprint BCA	density per dBA bldable acre
1.00	52,550	22.87	40.2	57.0	17,734	n/a	n/a	34,816	34,816	22.87
2.00		34.20	60.0	85.3	26,519	n/a	n/a	52,062	26,031	34.20
3.00		**40.96**	**71.9**	**102.1**	**31,763**	**n/a**	**n/a**	**62,359**	**20,786**	**40.96**
4.00		45.45	79.8	113.3	35,249	n/a	n/a	69,202	17,300	45.45
5.00		48.66	85.4	121.3	37,734	n/a	n/a	74,080	14,816	48.66
6.00		51.06	89.7	127.3	39,594	n/a	n/a	77,732	12,955	51.06
7.00	*NOTE: Be aware when BCA becomes too small to be feasible.*	52.92	92.9	132.0	41,040	n/a	n/a	80,570	11,510	52.92
8.00		54.41	95.5	135.7	42,195	n/a	n/a	82,838	10,355	54.41
9.00		55.63	97.7	138.7	43,139	n/a	n/a	84,692	9,410	55.63
10.00		56.64	99.5	141.2	43,926	n/a	n/a	86,237	8,624	56.64
11.00		57.50	101.0	143.4	44,591	n/a	n/a	87,543	7,958	57.50
12.00		58.24	102.3	145.2	45,161	n/a	n/a	88,661	7,388	58.24
13.00		58.87	103.4	146.8	45,655	n/a	n/a	89,631	6,895	58.87
14.00		59.43	104.4	148.2	46,087	n/a	n/a	90,479	6,463	59.43

WARNING: These are preliminary forecasts that must not be used to make final decisions.

1) These forecasts are not a substitute for the "due diligence" research that must be conducted to support the final definition of "unbuildable areas" above and the final decision to purchase land. This research includes, but is not limited to, verification of adequate subsurface soil, zoning, environmental clearance, access, title, utilities and water pressure, clearance from deed restriction, easement and right-of-way encumbrances, clearance from existing above and below ground facility conflicts, etc.

2) The most promising forecast(s) made on the basis of data entered in the design specification from "due diligence" research must be verified at the drawing board before funds are committed and land purchase decisions are made. Actual land shape ratios, dimensions and irregularities encountered may require adjustments to the general forecasts above.

3) The software licensee shall take responsibility for the design specification values entered and any advice given that is based on the forecast produced.

a significant increase to what is already, in the author's opinion, a marginal residential parking lot context. The Jefferson lot surrounds dwelling units with pavement and cars, provides no exterior places for people, and omits grading and landscape relief in the open space that remains. Exhibit 8.7 predicts that the same number of dwelling units and density could have been achieved with a 50% open space provision if the building height had been increased to 11 floors and the total building footprint had been reduced to 5688 sq ft. This is a significant increase in height that would trigger more restrictive building code provisions, increase construction cost, and might violate zoning ordinance regulations. Building code requirements are a cost of doing business, but building height limits can often be a contentious political issue. The open space quantity that is a tradeoff for increased building height is a factor in the analysis, however, and one that can be forecast with single-system comparative analysis. This makes the desirability and implications of the context implied capable of quantitative analysis and debate.

The planning forecast panel in Exhibit 8.7 shows that the Jefferson land area could support 51.4 dwelling units when 50% open space is provided and building height is limited to 3 floors, assuming all other design specification values from Exhibit 8.6 are held constant. It also shows that a 5-story building could contain 61.1 dwelling units under these circumstances. The 3-story complex actually constructed provided 72 dwelling units with a 30% open space provision. The traditional planning debate over acceptable levels of density is therefore engaged with a series of quantitatively comparable forecasts that include, but are not limited to, the following:

- 3-story building height, 72 dwelling units, and 30% open space
- 3-story building height, 51.4 dwelling units, and 50% open space
- 5-story building height, 61.1 dwelling units, and 50% open space

Exhibit 8.7 Jefferson Development Capacity with 50% Open Space

Forecast Model RG1L

Development capacity forecast for ***APARTMENTS*** *based on the use of an adjacent* ***GRADE PARKING LOT*** *located on the same premises. When s and u equal zero in the design specification below, the forecast pertains to conditions when* ***NO PARKING*** *is required.*

Given: *Gross land area.* ***To Find:*** *Maximum dwelling unit capacity of the land area given based on the design specification values entered below.* ***Premise:*** *All building floors considered equal in area.*

DESIGN SPECIFICATION *Enter values in boxed areas where text is bold and blue. Express all fractions as decimals.*

Given:	**Gross land area**	GLA=	**1.756**	acres	76,491	SF
Land Variables:	Public/ private right-of-way & paved easements	W=	**0.000**	fraction of GLA	0	SF
	Net land area	NLA=	1.756	acres	76,491	SF
	Unbuildable and/or future expansion areas	U=	**0.000**	fraction of GLA	0	SF
	Gross land area reduction	X=	0.000	fraction of GLA	0	SF
	Buildable land area remaining	BLA=	**1.756**	acres	76,491	SF
Parking Variables:	Est. gross pkg. lot area per pkg. space in SF	s =	**311**	ENTER ZERO IF NO PARKING REQUIRED		
	Parking lot spaces planned or required per dwelling unit	u=	**1.42**	ENTER ZERO IF NO PARKING REQUIRED		
	Garage parking spaces planned or required per dwelling unit	Gn=	**0**	ENTER ZERO IF NO PARKING REQUIRED		
	Gross building area per garage space	Ga=	**0**	ENTER ZERO IF NO PARKING REQUIRED		
	No. of loading spaces	l =	**0**			
	Gross area per loading space	b =	**0**	SF	0	SF
Site Variables:	**Project open space as fraction of BLA**	S=	**0.500**		38,246	SF
	Private driveways as fraction of BLA	R=	**0.008**		612	SF
	Misc. pavement as fraction of BLA	M=	**0.001**		76	SF
	Loading area as fraction of BLA	L=	0.000		0	SF
	Total site support areas as a fraction of BLA	Su=	0.509		38,934	SF
Core:	**Core development area as fraction of BLA**	C=	0.491	C+Su must = 1	37,557	SF
Building Variables:	Building efficiency as percentage of GBA	Be=	**0.900**			
	Building support as fraction of GBA	Bu=	0.100	Be + Bu must = 1		

Dwelling Unit Mix Table:

DU dwelling unit type	GDA gross du area	CDA=GDA/Be comprehensive du area	MIX du mix	PDA = (CDA)MIX Pro-rated du area

Dwelling unit mix not available

Aggregate average dwelling unit area (AGG) found by dividing gross building area by number of dwelling units present

Aggregate avg. dwelling unit area (AGG) = **867**

GBA sf per parking space a= 611

PLANNING FORECAST

no. of floors FLR	CORE minimum lot area for BCG & PLA	density per net acre dNA	dwelling units NDU	pkg. lot spaces NPS	parking lot area PLA	garage spaces GPS	garage area GAR	gross bldg area GBA no garages	footprint BCA	density per dBA bldable acre
1.00	37,557	16.34	28.7	40.8	12,674	n/a	n/a	24,883	24,883	16.34
2.00		24.44	42.9	60.9	18,953	n/a	n/a	37,209	18,604	24.44
3.00		29.27	51.4	73.0	22,701	n/a	n/a	44,568	14,856	29.27
4.00		32.49	57.0	81.0	25,193	n/a	n/a	49,459	12,365	32.49
5.00	*NOTE: Be aware when BCA becomes too small to be feasible.*	34.78	61.1	86.7	26,968	n/a	n/a	52,945	10,589	34.78
6.00		36.49	64.1	91.0	28,298	n/a	n/a	55,555	9,259	36.49
7.00		37.82	66.4	94.3	29,331	n/a	n/a	57,583	8,226	37.82
8.00		38.89	68.3	97.0	30,157	n/a	n/a	59,204	7,401	38.89
9.00		39.76	69.8	99.1	30,832	n/a	n/a	60,530	6,726	39.76
10.00		40.48	71.1	100.9	31,394	n/a	n/a	61,633	6,163	40.48
11.00		**41.10**	**72.2**	**102.5**	**31,869**	**n/a**	**n/a**	**62,567**	**5,688**	**41.10**
12.00		41.62	73.1	103.8	32,277	n/a	n/a	63,366	5,281	41.62
13.00		42.08	73.9	104.9	32,630	n/a	n/a	64,059	4,928	42.08
14.00		42.47	74.6	105.9	32,938	n/a	n/a	64,665	4,619	42.47

WARNING: These are preliminary forecasts that must not be used to make final decisions.

1) These forecasts are not a substitute for the "due diligence" research that must be conducted to support the final definition of "unbuildable areas" above and the final decision to purchase land. This research includes, but is not limited to, verification of adequate subsurface soil, zoning, environmental clearance, access, title, utilities and water pressure, clearance from deed restriction, easement and right-of-way encumbrances, clearance from existing above and below ground facility conflicts, etc.

2) The most promising forecast(s) made on the basis of data entered in the design specification from "due diligence" research must be verified at the drawing board before funds are committed and land purchase decisions are made. Actual land shape ratios, dimensions and irregularities encountered may require adjustments to the general forecasts above.

3) The software licensee shall take responsibility for the design specification values entered and any advice given that is based on the forecast produced.

- 11-story building height, 72 dwelling units, and 50% open space

The dwelling unit capacity of the Jefferson land area is a function of the design specification values contemplated.[11] The context created is also a function of these values. The debate over density limits and reasonable development capacity has often been more emotional that factual, but it is possible to improve the factual level of the debate by more precisely forecasting the development capacity implications of the decisions under consideration. Context record comparisons can also improve evaluation of the lifestyles and quality of life implied by the specification values under consideration.

Asherton

The Asherton project, photographed in Fig. 8.4 and recorded in Table 8.4, has been included as a contrast to the Jefferson project, photographed in Fig. 8.3. The *S* allocation in the Asherton project is 79.4%, the dBA is 5.02, and the INX is 2.206. It represents a low-density clustered approach to multifamily housing that consumes a great deal of land in relation to the housing provided, and has a dBA and INX that are no greater than many single-family detached housing neighborhoods. In a popularity contest, the Asherton project would probably win by a large majority, but the 79.4% *S* provision represents a recipe for suburban sprawl that can threaten our sustainable future. This is not an argument for increased building height, however. It is simply meant to point out the broad spectrum of housing opportunities that can be quantitatively planned to offer a wider variety of lifestyles that represent an equal but different quality of life. Housing does not need to be low density to represent quality, but it represents the safest marketable commodity when knowledge about equivalent lifestyles is limited.

The Asherton context record has been entered in Exhibit 8.8 as a design specification to produce a low-density forecast example, but no additional single-system forecasts have been produced by changing these design specification values. If the reader is interested in this exploration, he or she simply needs to select forecast model RG1L from the development forecast collection and enter the design specification values presented in Exhibit 8.8. The resulting forecast can be revised by changing any one or more of the bold, blue, and boxed values in the design specification panel of model RG1L. For instance, more dwelling units could be produced if the project open space *S* were reduced from 79.4% to 60%, but the impact on the water-oriented context and marketability of the project would have to be carefully evaluated in relation to the increased housing produced. These are not easy assessments, but the knowledge derived is an essential foundation for the sustainable future that must be planned.

Notes

1. The amount of building area that can be constructed on a given land area.
2. The Decision Guides included as Figs. 1.1 and 1.2 are designed to lead the user to an applicable forecast model based on the information given and the parking system contemplated.
3. Generally, the gross land area GLA, rights-of-way *W*, and unbuildable areas *U* of a site do not change with each comparison. Miscellaneous pavement *M*, driveway *R*, and loading areas *l* and *b* often represent minor amounts of land consumption, and garages Gn and Ga may often be omitted.
4. In the case of residential land use, this value is designated as *u*.
5. The glossary contains explanations of these values.
6. When the building footprint BCA is 189,915 sq ft.
7. A parking variance is one of the most common cases to appear before a board of zoning adjustment, and is frequently granted because the quantities specified in a zoning ordinance are approximations of need that have not been precisely confirmed with established research.
8. This means that 200 sq ft of building area is permitted for every parking space provided.

9. Most zoning ordinances specify yard areas, but then permit construction, especially parking, to be placed in some, if not all, of these yard areas.
10. The same context record form has been used for both, even though form GT could have been used for Table 8.4.
11. Don't overlook the aggregate average dwelling unit area AGG planned. This is an important value in forecasting residential development capacity.

Multisystem Comparisons

A forecast model represents a land use family and a parking system. Single-system comparisons address one parking method and focus on the development capacity[1] results produced when design specification values are changed within the forecast model that pertains. Multisystem comparisons expand on this format by choosing two or more forecast models for evaluation. Design specification values can be changed within each model, or they can be held constant to compare the advantages and disadvantages of the parking systems represented. In other words, multisystem comparisons contrast the development capacity results that can be produced by two or more parking system alternatives.

The choice of a parking system dramatically affects the capacity of a given land area, its development cost, and its return on investment. Comparing parking system alternatives has traditionally been a time-consuming effort requiring at least two separate design solutions drawn in architectural detail in order to compensate for the lack of mathematical design specifications and forecasting tools. This has limited the alternatives that can be evaluated within a reasonable time frame. Since forecast models are indexed by parking system, several systems can be chosen for comparison by choosing the models of interest. These multisystem mathematical comparisons can reduce time, improve the scope of evaluation conducted, and build confidence in the alternatives chosen for further drawing board investigation.

Parking System Comparisons Involving Nonresidential Land Uses

This case study will compare the capacity limits of a common G1 parking system (surface parking lot) with the much less common S1 parking system (adjacent parking structure). Common sense tells us that the S1 system will have greater capacity, but common sense cannot forecast the extent of these differences. Figures 9.1 and 9.2 illustrate the two parking systems under consideration. Figure 9.1 is called the Office, and Fig. 9.2 is called the Federal Building. Tables 9.1 and 9.2 present the context records for these two projects. Exhibits 9.1 and 9.2 are forecast models that reflect the choices that were available prior to the design and construction of these projects. The choices made are identified by the outlined rows in each model. These two projects can be compared immediately by looking at the summary panel in each context record and locating the SFAC column. The G1 system used by the Office supports 17,718 sq ft of gross building area per acre when three building floors are constructed and 23.6% project open space is provided. The S1 system used by the Federal Building supports 178,190 sq ft of building area per acre when seven building floors and eight parking levels are constructed with no project open space provided. This tenfold increase in development capacity portrays the differences in system potential represented. These statistics may not be as meaningful to some, however, as pointing out that 10 times the building area has been constructed on 30% of the land area when an S1 parking system is used with no project open space provi-

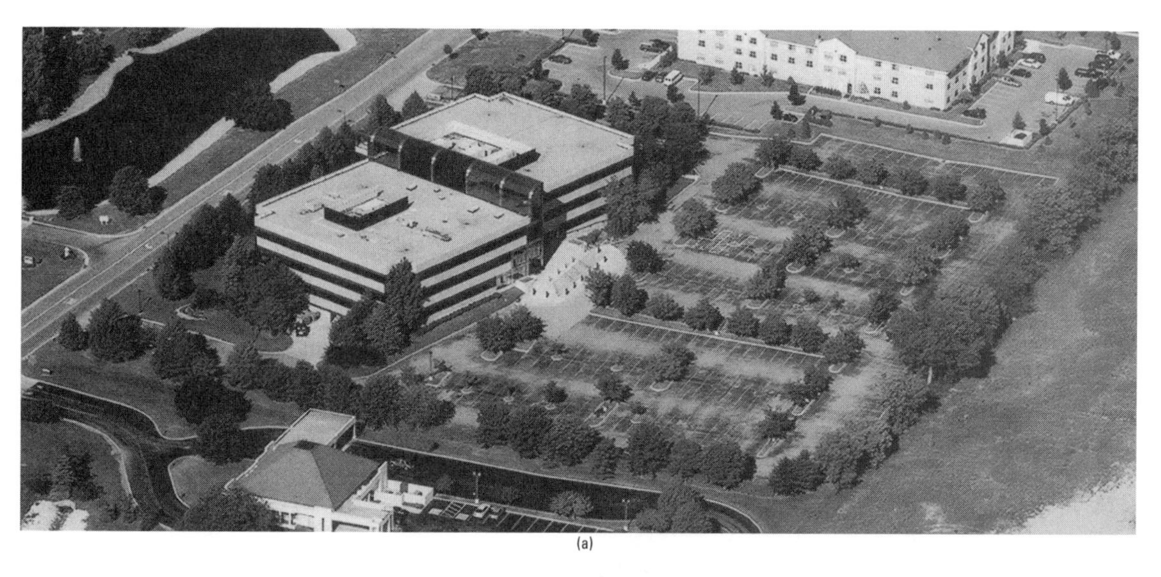

(a)

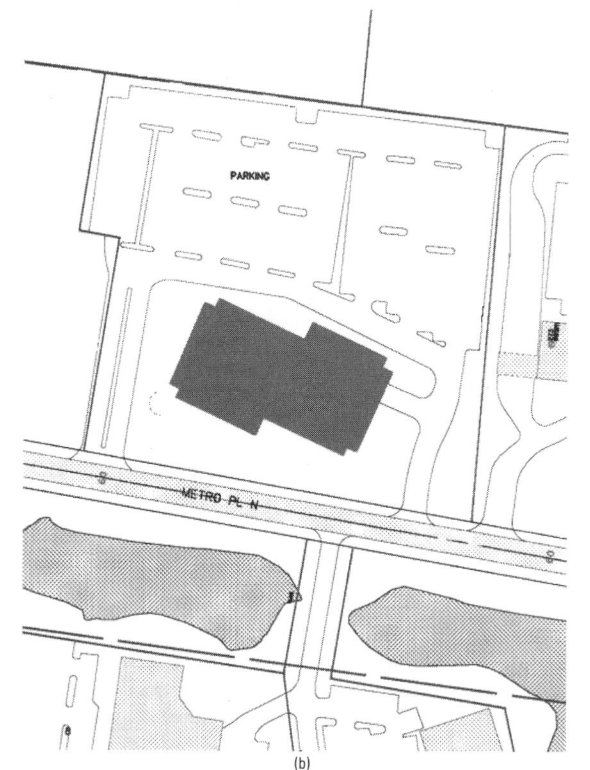

(b)

(c)

(d)

Figure 9.1 The Office—nonresidential G1 parking system. This project provides only 24% project open space, but benefits from a water feature within the common open space of its executive office park. The entry side of the building is quite different from the curb side. (*a*) Overview; (*b*) site plan; (*c*) parking area; (*d*) curbside view.

(Photos by W. M. Hosack)

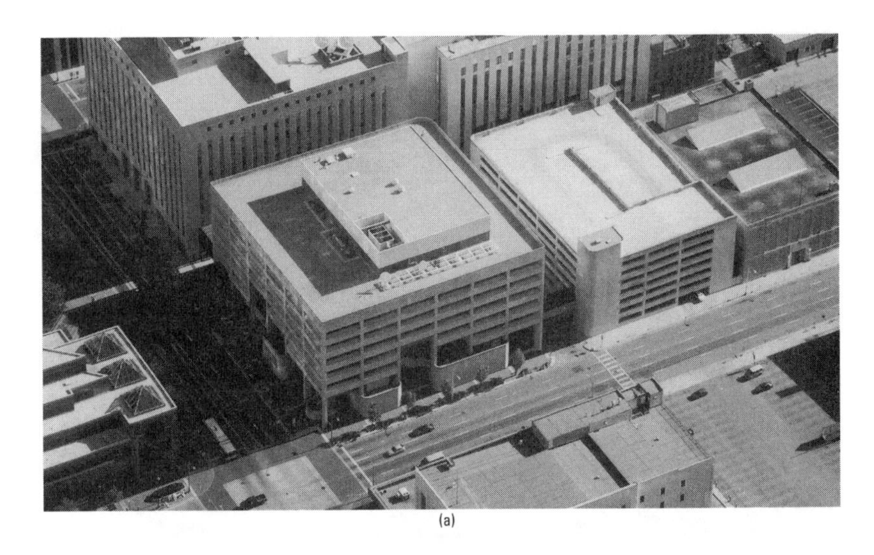

(a)

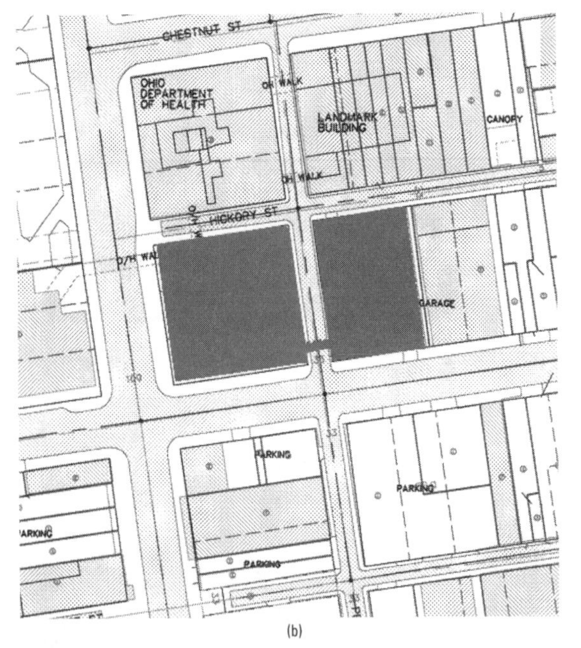

(b)

(c)

Figure 9.2 Federal Office Building—S1 adjacent parking structure system. This is a classic example of a building and associated parking structure occupying the entire area of an urban site. For the sake of this example, the entire site includes the public alley shown. This alley has been subtracted from the gross land area to find the buildable land area available. (*a*) Overview; (*b*) site plan; (*c*) streetfront view.

(Photos by W. M. Hosack)

Exhibit 9.1 The Office Development Capacity Forecast

Forecast Model CG1L

Development capacity forecast for ***NONRESIDENTIAL BUILDINGS*** *based on the use of an adjacent* ***GRADE PARKING LOT*** *located on the same premises. When s and a equal zero in the design specification below, the forecast pertains to conditions when* ***NO PARKING*** *is required.*

Given: *Gross land area.* ***To Find:*** *Maximum development capacity of the land area (gross building area potential) based on the design specification values entered below.* ***Premise:*** *All building floors considered equal in area.*

DESIGN SPECIFICATION

Enter values in boxed areas where text is bold and blue. Express all fractions as decimals.

Given:	**Gross land area**	**GLA=**	**5.288**	acres	230,345	SF
Land Variables:	Public/ private right-of-way & paved easements	W=	**0.000**	fraction of GLA	0	SF
	Net land area	NLA=	5.288	acres	230,345	SF
	Unbuildable and/or future expansion areas	U=	**0.000**	fraction of GLA	0	SF
	Gross land area reduction	X=	0.000	fraction of GLA	0	SF
	Buildable land area remaining	**BLA=**	5.288	acres	230,345	SF
Parking Variables:	Est. gross pkg. lot area per space in SF	s =	**341.4**	ENTER ZERO IF NO PARKING REQUIRED		
	Building SF permitted per parking space	a =	**285.4**	ENTER ZERO IF NO PARKING REQUIRED		
	No. of loading spaces	l=	**1**			
	Gross area per loading space	b =	**1,800**	SF	1,800	SF
Site Variables:	**Project open space as fraction of BLA**	**S=**	**0.237**		54,592	SF
	Private driveways as fraction of BLA	R=	**0.115**		26,490	SF
	Misc. pavement as fraction of BLA	M=	**0.018**		4,146	SF
	Loading area as fraction of BLA	L=	0.008		1,800	SF
	Total site support areas as a fraction of BLA	Su=	0.378		87,028	SF
Core:	**Core development area as fraction of BLA**	**C=**	0.622	C+Su must = 1	143,318	SF

PLANNING FORECAST

no. of floors **FLR**	CORE (minimum land area for BCG & PLA)	gross building area **GBA**	parking lot area PLA	pkg. spaces NPS	footprint **BCA**	bldg SF / acre SFAC (function of BLA)	flr area ratio FAR (function of BLA)
1.00	143,318	65,254	78,064	228.7	65,254	12,340	0.283
2.00		84,488	101,074	296.1	42,244	15,977	0.367
3.00		**93,693**	**112,086**	**328.3**	**31,231**	**17,718**	**0.407**
4.00		99,092	118,545	347.2	24,773	18,739	0.430
5.00		102,640	122,790	359.7	20,528	19,410	0.446
6.00	*NOTE: Be aware when BCA becomes too small to be feasible.*	105,150	125,793	368.5	17,525	19,885	0.456
7.00		107,020	128,029	375.0	15,289	20,238	0.465
8.00		108,466	129,759	380.1	13,558	20,512	0.471
9.00		109,618	131,138	384.1	12,180	20,730	0.476
10.00		110,558	132,262	387.4	11,056	20,907	0.480
11.00		111,339	133,196	390.2	10,122	21,055	0.483
12.00		111,998	133,984	392.5	9,333	21,180	0.486
13.00		112,562	134,659	394.4	8,659	21,286	0.489
14.00		113,049	135,243	396.2	8,075	21,378	0.491
15.00		113,476	135,752	397.6	7,565	21,459	0.493

WARNING: These are preliminary forecasts that must not be used to make final decisions.
1) These forecasts are not a substitute for the "due diligence" research that must be conducted to support the final definition of "unbuildable areas" above and the final decision to purchase land. This research includes, but is not limited to, verification of adequate subsurface soil, zoning, environmental clearance, access, title, utilities and water pressure, clearance from deed restriction, easement and right-of-way encumbrances, clearance from existing above and below ground facility conflicts, etc.
2) The most promising forecast(s) made on the basis of data entered in the design specification from "due diligence" research must be verified at the drawing board before funds are committed and land purchase decisions are made. Actual land shape ratios, dimensions and irregularities encountered may require adjustments to the general forecasts above.
3) The software licensee shall take responsibility for the design specification values entered and any advice given that is based on the forecast produced.

Exhibit 9.2 Federal Office Building Development Capacity Forecast

Forecast Model CS1L

Development capacity forecast for ***NONRESIDENTIAL BUILDINGS*** *using an* ***ADJACENT PARKING STRUCTURE*** *located on the same premise.*
Given: *Gross land area available.* ***To Find:*** *Maximum development capacity of the land area given (gross building area potential) based on the design specification values entered below.*
Design Premise: *Building footprint adjacent to parking garage footprint within the core development area. All similar floors considered equal in area.*

Enter values in boxed areas where text is bold and blue. Express all fractions as decimals.

DESIGN SPECIFICATION

Given:	**Gross land area**	GLA=	**1.549**	acres	67,474	SF
Land Variables:	Public/ private right-of-way & paved easements	W=	**0.084**	fraction of GLA	5,700	SF
	Net land area	NLA=	1.418	acres	61,775	SF
	Unbuildable and/or future expansion areas	U=	**0.000**	fraction of GLA	0	SF
	Gross land area reduction	X=	0.084	fraction of GLA	5,700	SF
	Buildable land area remaining	BLA=	1.418	acres	61,775	SF
Parking Variables:	Est. net pkg. structure area per pkg. space	s =	**340.0**	SF		
	Building SF permitted per parking space	a =	**492.1**	SF		
	Number of parking levels to be evaluated	p=	**8.00**			
	Parking support as fraction of gross pkg. structure area (GPA)	Pu=	**0.150**			
	Net area for parking & circulation as fraction of GPA	Pe=	0.850	Pu+Pe must = 1		
	No. of loading spaces	l=	**0**			
	Gross area per loading space	b =	**1,000**	SF	0	SF
Site Variables:	**Project open space as fraction of BLA**	S=	**0.000**		0	SF
	Private driveways as fraction of BLA	R=	**0.000**		0	SF
	Misc. pavement as fraction of BLA	M=	**0.000**		0	SF
	Loading area as fraction of BLA	L=	0.000		0	SF
	Total site support areas as a fraction of BLA	Su=	0.000		0	SF
Core:	**Core development area as fraction of BLA**	C=	1.000	C+Su must = 1	61,775	SF

NOTE: p=1 is a grade parking lot based on design premise. Increase the number of parking levels to increase the capacity forecast forecast below. Other variables in blue, box and bold may also be changed.

PLANNING FORECAST

no. of **FLR** bldg floors	min land area CORE for BCG & PSA	gross building area **GBA**	gross pkg GPA struc area	pkg struct cover **GPL** per level	net parking NPA area	pkg spaces NPS	bldg cover **BCA** footprint	bldg SF per SFAC bldable acre	flr area ratio FAR function of BLA
1.00	61,775	56,077	45,580	5,698	38,743	113.95	56,077	39,542	0.908
2.00		102,684	83,463	10,433	70,944	208.66	51,342	72,407	1.662
3.00		142,032	115,446	14,431	98,129	288.62	47,344	100,153	2.299
4.00		175,696	142,808	17,851	121,387	357.02	43,924	123,890	2.844
5.00		204,823	166,483	20,810	141,510	416.21	40,965	144,429	3.316
6.00		230,273	187,169	23,396	159,094	467.92	38,379	162,375	3.728
7.00		**252,700**	**205,399**	**25,675**	**174,589**	**513.50**	**36,100**	**178,189**	**4.091**
8.00		272,614	221,585	27,698	188,347	553.96	34,077	192,231	4.413
9.00		290,414	236,053	29,507	200,645	590.13	32,268	204,783	4.701
10.00		306,420	249,063	31,133	211,703	622.66	30,642	216,069	4.960
11.00		320,890	260,824	32,603	221,701	652.06	29,172	226,273	5.195
12.00		334,035	271,509	33,939	230,782	678.77	27,836	235,542	5.407
13.00		346,029	281,258	35,157	239,069	703.14	26,618	244,000	5.601
14.00		357,017	290,189	36,274	246,661	725.47	25,501	251,748	5.779
15.00		367,121	298,401	37,300	253,641	746.00	24,475	258,872	5.943

WARNING: These are preliminary forecasts that must not be used to make final decisions.
1) These forecasts are not a substitute for the "due diligence" research that must be conducted to support the final definition of "unbuildable areas" above and the final decision to purchase land. This research includes, but is not limited to, verification of adequate subsurface soil, zoning, environmental clearance, access, title, utilities and water pressure, clearance from deed restriction, easement and right-of-way encumbrances, clearance from existing above and below ground facility conflicts, etc.
2) The most promising forecast(s) made on the basis of data entered in the design specification from "due diligence" research must be verified at the drawing board before funds are committed and land purchase decisions are made. Actual land shape ratios, dimensions and irregularities encountered may require adjustments to the general forecasts above.
3) The software licensee shall take responsibility for the design specification values entered and any advice given that is based on the forecast produced.

Table 9.1 The Office Context Record with G1 Parking System

Form CR

Name	**The Office**
Address or Location	**400 Metro Place North**
Land Use	**Office CG1**

Parking System	*check box*
Surface parking around but not under building	**x**
Surface parking around and under building	
No parking	
Notes	

		Description	Project Measurements		Context Summary SF	%	
Site							
		Gross land area in acres	**5.288**	GLA	230,345		
		Public/ private ROW and paved easements in SF		W		0.0%	of GLA
		Net lot area		NLA	230,345	100.0%	of GLA
		Total unbuildable area in SF	**0**	U		0.0%	of GLA
		Water area within unbuildable area in SF	**0**	WAT			of U
		Buildable land area		BLA	230,345	100.0%	of GLA
Development Cover							
		Gross building area in SF	**93,693**	GBA			
		Total building footprint in SF	**31,231**	B		13.6%	of BLA
		Total bldg. support in SF (stairs, corridors, elevators, etc.)		BSU			
		Building efficiency		Be			of GBA
	Residential only	Number of dwelling units		DU			
		Aggregate average dwelling unit area in SF		AGG			
		Number of building floors	**3.00**	FLR			
	G2 only	Unenclosed surface parking area under building in SF		PUB			
		Total surface parking area in SF**	**112,086**	P or G		48.7%	of BLA
		Number of surface parking spaces on premise	**328**	NPSs			
		Building cover over surface parking area		AIR			of GPA
		Gross surface parking area per parking space in SF		ss	342		
		Gross building area per surface parking space in SF		as	286		
		Surface parking spaces per dwelling unit		us	0.00		
	Pkg. struc. only	Total parking structure footprint in SF		P		0.0%	of BLA
	Pkg. struc. only	Parking structure area below grade in SF		PSC			
	Pkg. struc. only	Parking structure area above grade under building in SF		UNG			
	Pkg. struc. only	Number of parking structure floors		p			
	Pkg. struc. only	Total parking structure area in SF		GPA			
	Pkg. struc. only	Total parking structure spaces		NPSg			
	Pkg. struc. only	Total pkg. struc. support in SF (stairs, elevators, etc.)		PSS			
	Pkg. struc. only	Parking structure support percentage		Pu			of GPA
	Pkg. struc. only	Gross parking structure area per space in SF		sg	0		
	Pkg. struc. only	Gross building area per structure parking space in SF		ag	0		
	Pkg. struc. only	Parking structure spaces per dwelling unit		ug	0.00		
	Residential only	Total dwelling unit garage area in SF		GPA			
	Residential only	Dwelling unit garage above grade under builiding in SF		GUB			
	Residential only	Number of dwelling unit garage parking spaces		GPS			
	Residential only	Garage parking spaces per dwelling		Gn	0.0		
	Residential only	Total dwelling unit garage area per space in SF		Ga			
	Residential only	Total dwelling unit garage cover in addition to building cover		Gc		0.0%	of BLA
		Total parking spaces		NPS	328.0		
	Residential only	Total parking spaces per dwelling unit		u			
		Gross building area per parking space total		a	285.6		
		Number of loading spaces	**1**	l			
		Total loading area in SF	**1,800**	LDA			
		Gross area per loading space in SF		b	1,800		
		Loading area percentage		L		0.8%	of BLA
		Driveway areas in SF	**26,600**	R		11.5%	of BLA
		Misc. pavement and social center pvmt. in SF	**4,200**	M		1.8%	of BLA
		TOTAL DEVELOPMENT COVER		D	175,917	**76.4%**	of BLA
Project Open Space							
		PROJECT OPEN SPACE		S	54,428	**23.6%**	of BLA
		Water area within project open space in SF	0.0%	wat			of S
Summary							
		Development balance index		BNX		0.764	
		Development intensity index		INX		3.764	
		Floor area ratio per buildable acre		BFAR		0.407	
		Capacity index (gross building sq. ft. per buildable acre)		SFAC		17,718	
		Density per buildable acre if applicable		dBA			
		Density per net acre if applicable		dNA			
		Density per gross acre if applicable		dGA			

** Include internal parking lot landscaping, circulation aisles and private roadway parking space areas. Express private roads separately as driveway areas..

Table 9.2 The Federal Office Building Context Record with S1 Parking System

Name	Federal Office Building
Address or Location	200 N. High Street
Land Use	Office CS1

Parking System	check box
Structure parking adjacent to building	x
Structure parking underground	
Structure parking above grade under building	
Notes	

	Description	Project Measurements		Context Summary SF	%	
Site						
	Gross land area in acres	1.549	GLA	67,474		
	Public/ private ROW and paved easements in SF	5,700	W		8.4%	of GLA
	Net lot area		NLA	61,774	91.6%	of GLA
	Total unbuildable area in SF	0	U		0.0%	of GLA
	Water area within unbuildable area in SF	0	WAT			of U
	Buildable land area		BLA	61,774	91.6%	of GLA
Development Cover						
	Gross building area in SF	252,700	GBA			
	Total building footprint in SF	36,100	B		58.4%	of BLA
	Total bldg. support in SF (stairs, corridors, elevators, etc.)		BSU			
	Building efficiency		Be			of GBA
Residential only	Number of dwelling units		DU			
	Aggregate average dwelling unit area in SF		AGG			
	Number of building floors	7.00	FLR			
G2 only	Unenclosed surface parking area under building in SF		PUB			
	Total surface parking area in SF**		P or G		0.0%	of BLA
	Number of surface parking spaces on premise		NPSs			
	Building cover over surface parking area		AIR			of GPA
	Gross surface parking area per parking space in SF		ss			
	Gross building area per surface parking space in SF		as			
	Surface parking spaces per dwelling unit		us	0.00		
Pkg. struc. only	Total parking structure footprint in SF	25,650	P		41.5%	of BLA
Pkg. struc. only	Parking structure area below grade in SF		UNG			
Pkg. struc. only	Parking structure area above grade under building in SF		PSC			
Pkg. struc. only	Number of parking structure floors	8	p			
Pkg. struc. only	Total parking structure area in SF	205,200	GPA			
Pkg. struc. only	Total parking structure spaces	513	NPSg			
Pkg. struc. only	Total pkg. struc. support in SF (stairs, elevators, etc.)		PSS			
Pkg. struc. only	Parking structure support percentage		Pu			of GPA
Pkg. struc. only	Gross parking structure area per space in SF		sg	400		
Pkg. struc. only	Gross building area per structure parking space in SF		ag	492.6		
Pkg. struc. only	Parking structure spaces per dwelling unit		ug	0.00		
Residential only	Total dwelling unit garage area in SF		GPA			
Residential only	Dwelling unit garage above grade under builiding in SF		GUB			
Residential only	Number of dwelling unit garage parking spaces		GPS			
Residential only	Garage parking spaces per dwelling		Gn	0.0		
Residential only	Total dwelling unit garage area per space in SF		Ga			
Residential only	Total dwelling unit garage cover in addition to building cover		Gc		0.0%	of BLA
	Total parking spaces		NPS	513.0		
Residential only	Total parking spaces per dwelling unit		u			
	Gross building area per parking space total		a	492.6		
	Number of loading spaces		l			
	Total loading area in SF		LDA			
	Gross area per loading space in SF		b			
	Loading area percentage		L		0.0%	of BLA
	Driveway areas in SF		R		0.0%	of BLA
	Misc. pavement and social center pvmt. in SF		M		0.0%	of BLA
	TOTAL DEVELOPMENT COVER		D	61,750	**100.0%**	of BLA
Project Open Space						
	PROJECT OPEN SPACE		S	24	**0.0%**	of BLA
	Water area within project open space in SF	0.0%	wat			of S
Summary						
	Development balance index		BNX		1.000	
	Development intensity index		INX		8.000	
	Floor area ratio per buildable acre		BFAR		4.091	
	Capacity index (gross building sq. ft. per buildable acre)		SFAC		178,190	
	Density per buildable acre if applicable		dBA			
	Density per net acre if applicable		dNA			
	Density per gross acre if applicable		dGA			

** Include internal parking lot landscaping, circulation aisles and private roadway parking space areas. Express private roads separately as driveway areas..

sion. This capacity does not expand without cost, however, since each parking garage space can be 10, 20, or more times the cost of a surface parking space.

In the case of the Office G1 system, the development intensity index is 3.764. In the case of the Federal Building S1 system, the index is 8.000.[2] A simple doubling of the index, however, has produced a dramatic change in the context and lifestyles illustrated by Figs. 9.1 and 9.2.

This comparison might be clearer if the same land area were involved. This can be accomplished by changing the gross land area value in the design specification of Exhibit 9.1 from 5.288 acres to equal the 1.549 acres used in Exhibit 9.2. This change is shown in Exhibit 9.3 and reduces the gross building area potential of the Office from 93,693 to 26,613 sq ft. The S1 system provides no project open space, however, and the Office G1 system provides 23.7%. If The Federal Building S1 system were to provide the same amount of project open space, Exhibit 9.4 shows that its gross building area capacity would decrease to 192,810 sq ft from the 252,700 sq ft shown in Exhibit 9.1. Exhibit 9.4 also shows that the original 254,869 gross sq ft of federal building area could be provided when 23.7% project open space is introduced if the building height were increased from 7 to 12 floors. If the building height is increased, however, this exhibit also shows that the footprint would have to be reduced from 27,544 to 21,239 sq ft. This reduced footprint, when combined with increased height, is able to produce a building area that is roughly equal to the original forecast, but includes much more open space.

If The Office had been built on the same land area as the Federal Building with the same lack of open space, the preceding comparison could be made in reverse. If the Office adopted the Federal Building design specification, the result would be a "sea of asphalt" that few find desirable in a suburban environment. This condition is forecast in Exhibit 9.5, and shows that the gross building area increases from

Exhibit 9.3 The Office Capacity Forecast Using 1.5 Acres and 23.7% Open Space

Forecast Model CG1L

Development capacity forecast for ***NONRESIDENTIAL BUILDINGS*** *based on the use of an adjacent* ***GRADE PARKING LOT*** *located on the same premises. When s and a equal zero in the design specification below, the forecast pertains to conditions when* ***NO PARKING*** *is required.*

Given: *Gross land area.* ***To Find:*** *Maximum development capacity of the land area (gross building area potential) based on the design specification values entered below.* ***Premise:*** *All building floors considered equal in area.*

DESIGN SPECIFICATION *Enter values in boxed areas where text is bold and blue. Express all fractions as decimals.*

Given:	**Gross land area**	GLA=	**1.549**	acres	67,474	SF
Land Variables:	Public/ private right-of-way & paved easements	W=	**0.000**	fraction of GLA	0	SF
	Net land area	NLA=	1.549	acres	67,474	SF
	Unbuildable and/or future expansion areas	U=	**0.000**	fraction of GLA	0	SF
	Gross land area reduction	X=	0.000	fraction of GLA	0	SF
	Buildable land area remaining	BLA=	1.549	acres	67,474	SF
Parking Variables:	Est. gross pkg. lot area per space in SF	s =	**341.387**	ENTER ZERO IF NO PARKING REQUIRED		
	Building SF permitted per parking space	a =	**285.3658**	ENTER ZERO IF NO PARKING REQUIRED		
	No. of loading spaces	l=	**1**			
	Gross area per loading space	b =	**1,800**	SF	1,800	SF
Site Variables:	**Project open apace as fraction of BLA**	S=	**0.237**		15,991	SF
	Private sriveways as fraction of BLA	R=	**0.115**		7,760	SF
	Misc. pavement as fraction of BLA	M=	**0.018**		1,215	SF
	Loading area as fraction of BLA	L=	0.027		1,800	SF
	Total site support areas as a fraction of BLA	Su=	0.397		26,766	SF
Core:	**Core development area as fraction of BLA**	C=	0.603	C+Su must = 1	40,709	SF

PLANNING FORECAST

no. of floors **FLR**	CORE (minimum land area for BCG & PLA)	gross building area **GBA**	parking lot area PLA	pkg spaces NPS	footprint **BCA**	bldg SF / acre SFAC (function of BLA)	flr area ratio FAR (function of BLA)
1.00	40,709	18,535	22,174	65.0	18,535	11,966	0.275
2.00		23,998	28,710	84.1	11,999	15,493	0.356
3.00		**26,613**	**31,838**	**93.3**	**8,871**	**17,181**	**0.394**
4.00		28,147	33,672	98.6	7,037	18,171	0.417
5.00		29,155	34,878	102.2	5,831	18,822	0.432
6.00	*NOTE: Be aware when BCA becomes too small to be feasible.*	29,868	35,731	104.7	4,978	19,282	0.443
7.00		30,399	36,366	106.5	4,343	19,625	0.451
8.00		30,809	36,858	108.0	3,851	19,890	0.457
9.00		31,137	37,249	109.1	3,460	20,101	0.461
10.00		31,404	37,569	110.0	3,140	20,273	0.465
11.00		31,625	37,834	110.8	2,875	20,417	0.469
12.00		31,813	38,058	111.5	2,651	20,538	0.471
13.00		31,973	38,249	112.0	2,459	20,641	0.474
14.00		32,111	38,415	112.5	2,294	20,730	0.476
15.00		32,232	38,560	113.0	2,149	20,809	0.478

WARNING: These are preliminary forecasts that must not be used to make final decisions.
1) These forecasts are not a substitute for the "due diligence" research that must be conducted to support the final definition of "unbuildable areas" above and the final decision to purchase land. This research includes, but is not limited to, verification of adequate subsurface soil, zoning, environmental clearance, access, title, utilities and water pressure, clearance from deed restriction, easement and right-of-way encumbrances, clearance from existing above and below ground facility conflicts, etc.
2) The most promising forecast(s) made on the basis of data entered in the design specification from "due diligence" research must be verified at the drawing board before funds are committed and land purchase decisions are made. Actual land shape ratios, dimensions and irregularities encountered may require adjustments to the general forecasts above.
3) The software licensee shall take responsibility for the design specification values entered and any advice given that is based on the forecast produced.

Exhibit 9.4 Federal Office Capacity Forecast Using 1.5 Acres and 23.7% Open Space

Forecast Model CS1L

Development capacity forecast for ***NONRESIDENTIAL BUILDINGS*** *using an* ***ADJACENT PARKING STRUCTURE*** *locatedon the same premise.*
Given: *Gross land area available.* ***To Find:*** *Maximum development capacity of the land area given (gross building area potential) based on the design specification values entered below.*
Design Premise: *Building footprint adjacent to parking garage footprint within the core development area. All similar floors considered equal in area.*

Enter values in boxed areas where text is bold and blue. Express all fractions as decimals.

DESIGN SPECIFICATION

Given:	**Gross land area**	**GLA=**	**1.549**	acres	67,474	SF
Land Variables:	Public/ private right-of-way & paved easements	W=	**0.084**	fraction of GLA	5,700	SF
	Net land area	NLA=	1.418	acres	61,775	SF
	Unbuildable and/or future expansion areas	U=	**0.000**	fraction of GLA	0	SF
	Gross land area reduction	X=	0.084	fraction of GLA	5,700	SF
	Buildable land area remaining	**BLA=**	1.418	acres	61,775	SF
Parking Variables:	Est. net pkg. structure area per pkg. space	s =	**340.0**	SF		
	Building SF permitted per parking space	a =	**492.1**	SF		
	Number of parking levels to be evaluated	**p=**	**8.00**			
	Pkg. support as fraction of gross pkg. structure area (GPA)	Pu=	**0.150**			
	Net area for parking & circulation as fraction of GPA	Pe=	0.850	Pu+Pe must = 1		
	No. of loading spaces	l=	**0**			
	Gross area per loading space	b =	**1,000**	SF	0	SF
Site Variables:	**Project open space as fraction of BLA**	**S=**	**0.237**		14,641	SF
	Private driveways as fraction of BLA	R=	**0.000**		0	SF
	Misc. pavement as fraction of BLA	M=	**0.000**		0	SF
	Loading area as fraction of BLA	L=	0.000		0	SF
	Total site support areas as a fraction of BLA	Su=	0.237		14,641	SF
Core:	**Core development area as fraction of BLA**	**C=**	0.763	C+Su must = 1	47,134	SF

NOTE: p=1 is a grade parking lot based on design premise. Increase the number of parking levels to increase the capacity forecast forecast below. Other variables in blue, box and bold may also be changed.

PLANNING FORECAST

no. of **FLR** bldg floors	min land area CORE for BCG & PSA	gross building area **GBA**	gross pkg GPA struc area	pkg struct cover **GPL** per level	net parking NPA area	pkg spaces NPS	bldg cover **BCA** footprint	bldg SF per SFAC bldable acre	flr area ratio FAR function of BLA
1.00	47,134	42,787	34,778	4,347	29,561	86.94	42,787	30,171	0.693
2.00		78,348	63,682	7,960	54,130	159.21	39,174	55,246	1.268
3.00		108,371	88,085	11,011	74,873	220.21	36,124	76,417	1.754
4.00		134,056	108,962	13,620	92,618	272.41	33,514	94,528	2.170
5.00		156,280	127,026	15,878	107,972	317.57	31,256	110,199	2.530
6.00		175,698	142,810	17,851	121,388	357.02	29,283	123,892	2.844
7.00		192,810	156,719	19,590	133,211	391.80	27,544	135,959	3.121
8.00		208,005	169,069	21,134	143,709	422.67	26,001	146,673	3.367
9.00		221,586	180,108	22,514	153,092	450.27	24,621	156,249	3.587
10.00		233,799	190,035	23,754	161,530	475.09	23,380	164,861	3.785
11.00		244,839	199,009	24,876	169,158	497.52	22,258	172,646	3.963
12.00		**254,869**	**207,161**	**25,895**	**176,087**	**517.90**	**21,239**	**179,719**	**4.126**
13.00		264,020	214,600	26,825	182,410	536.50	20,309	186,172	4.274
14.00		272,404	221,414	27,677	188,202	553.54	19,457	192,083	4.410
15.00		280,113	227,680	28,460	193,528	569.20	18,674	197,519	4.534

WARNING: These are preliminary forecasts that must not be used to make final decisions.
1) These forecasts are not a substitute for the "due diligence" research that must be conducted to support the final definition of "unbuildable areas" above and the final decision to purchase land. This research includes, but is not limited to, verification of adequate subsurface soil, zoning, environmental clearance, access, title, utilities and water pressure, clearance from deed restriction, easement and right-of-way encumbrances, clearance from existing above and below ground facility conflicts, etc.
2) The most promising forecast(s) made on the basis of data entered in the design specification from "due diligence" research must be verified at the drawing board before funds are committed and land purchase decisions are made. Actual land shape ratios, dimensions and irregularities encountered may require adjustments to the general forecasts above.
3) The software licensee shall take responsibility for the design specification values entered and any advice given that is based on the forecast produced.

Exhibit 9.5 The Office Capacity Forecast Using 1.5 Acres and 0.0% Open Space

Forecast Model CG1L

Development capacity forecast for ***NONRESIDENTIAL BUILDINGS*** *based on the use of an adjacent* ***GRADE PARKING LOT*** *located on the same premises. When s and a equal zero in the design specification below, the forecast pertains to conditions when* ***NO PARKING*** *is required.*

Given: *Gross land area.* ***To Find:*** *Maximum development capacity of the land area (gross building area potential) based on the design specification values entered below.* ***Premise:*** *All building floors considered equal in area.*

DESIGN SPECIFICATION

Enter values in boxed areas where text is bold and blue. Express all fractions as decimals.

Given:	**Gross land area**	GLA=	**1.549**	acres	67,474	SF
Land Variables:	Public/ private right-of-way & paved easements	W=	**0.000**	fraction of GLA	0	SF
	Net land area	NLA=	1.549	acres	67,474	SF
	Unbuildable and/or future expansion areas	U=	**0.000**	fraction of GLA	0	SF
	Gross land area reduction	X=	0.000	fraction of GLA	0	SF
	Buildable land area remaining	BLA=	1.549	acres	67,474	SF
Parking Variables:	Est. gross pkg. lot area per space in SF	s =	**341.387**	ENTER ZERO IF NO PARKING REQUIRED		
	Building SF permitted per parking space	a =	**285.3658**	ENTER ZERO IF NO PARKING REQUIRED		
	No. of loading spaces	l=	**1**			
	Gross area per loading space	b =	**1,800**	SF	1,800	SF
Site Variables:	**Project open space as fraction of BLA**	S=	**0.000**		0	SF
	Private driveways as fraction of BLA	R=	**0.115**		7,760	SF
	Misc. pavement as fraction of BLA	M=	**0.018**		1,215	SF
	Loading area as fraction of BLA	L=	0.027		1,800	SF
	Total site support areas as a fraction of BLA	Su=	0.160		10,774	SF
Core:	**Core development area as fraction of BLA**	C=	0.840	C+Su must = 1	56,700	SF

PLANNING FORECAST

no. of floors **FLR**	CORE minimum land area for BCG & PLA	gross building area **GBA**	parking lot area PLA	pkg. spaces NPS	footprint **BCA**	bldg SF / acre SFAC function of BLA	flr area ratio FAR function of BLA
1.00	56,700	25,816	30,884	90.5	25,816	16,666	0.383
2.00		33,426	39,988	117.1	16,713	21,579	0.495
3.00		**37,068**	**44,344**	**129.9**	**12,356**	**23,930**	**0.549**
4.00		39,203	46,900	137.4	9,801	25,309	0.581
5.00		40,607	48,579	142.3	8,121	26,215	0.602
6.00	*NOTE: Be aware when BCA becomes too small to be feasible.*	41,600	49,767	145.8	6,933	26,856	0.617
7.00		42,340	50,652	148.4	6,049	27,334	0.627
8.00		42,912	51,336	150.4	5,364	27,703	0.636
9.00		43,368	51,882	152.0	4,819	27,997	0.643
10.00		43,740	52,326	153.3	4,374	28,237	0.648
11.00		44,049	52,696	154.4	4,004	28,437	0.653
12.00		44,309	53,008	155.3	3,692	28,605	0.657
13.00		44,532	53,275	156.1	3,426	28,749	0.660
14.00		44,725	53,506	156.7	3,195	28,874	0.663
15.00		44,894	53,707	157.3	2,993	28,983	0.665

WARNING: These are preliminary forecasts that must not be used to make final decisions.

1) These forecasts are not a substitute for the "due diligence" research that must be conducted to support the final definition of "unbuildable areas" above and the final decision to purchase land. This research includes, but is not limited to, verification of adequate subsurface soil, zoning, environmental clearance, access, title, utilities and water pressure, clearance from deed restriction, easement and right-of-way encumbrances, clearance from existing above and below ground facility conflicts, etc.

2) The most promising forecast(s) made on the basis of data entered in the design specification from "due diligence" research must be verified at the drawing board before funds are committed and land purchase decisions are made. Actual land shape ratios, dimensions and irregularities encountered may require adjustments to the general forecasts above.

3) The software licensee shall take responsibility for the design specification values entered and any advice given that is based on the forecast produced.

26,613 to 37,068 sq ft when the Office open space allocation of 23.7% is eliminated. This 42% increase in gross building area implies a heavy price to the community in terms of overbuilt context, however, and is one of the excesses that front yard zoning setback requirements originally attempted to mitigate.

Exhibit 9.5 still uses an urban parking provision, *a*, that is much less than the suburban provision. The urban provision shown in Exhibit 9.2 for the Federal Building is 492, and the suburban provision in Exhibit 9.1 for the Office is 285.[3] These urban provisions are often unrealistic and depend on supplemental remote parking lots and garages in the surrounding business district. It is also one reason for the popularity of suburban office environments where both parking and open space are more plentiful.[4] If the Office parking requirement were reduced to 492 to equal the urban office requirement, and if no open space were provided, Exhibit 9.6 shows that the gross building area on a 1.5-acre lot would increase from 37,068 to 60,343 sq ft. This, however, represents a classic case of suburban overdevelopment that not only produces a sea of asphalt, but inadequate parking accommodation, street congestion, and buildings that are much too large for the context planned. These comparisons illustrate how the forecast model collection can be used to predict and compare development capacity alternatives, the context that will be created, and the quality of life that will result.

Parking system comparisons and choices imply development capacity alternatives that have physical, social, and economic consequences. The physical implications of parking system choices involve the context that will be created, the building area that will be constructed, and the environment that will be preserved. The social implications involve the shelter that can be produced, the populations that can be protected, the lifestyles that evolve, and the quality of life that results. The economic implications involve the

Exhibit 9.6 The Office Specification Equals the Federal Building Specification

Forecast Model CG1L

*Development capacity forecast for **NONRESIDENTIAL BUILDINGS** based on the use of an adjacent **GRADE PARKING LOT** located on the same premises. When s and a equal zero in the design specification below, the forecast pertains to conditions when **NO PARKING** is required.*

Given: *Gross land area.* ***To Find:*** *Maximum development capacity of the land area (gross building area potential) based on the design specification values entered below.* ***Premise:*** *All building floors considered equal in area.*

DESIGN SPECIFICATION *Enter values in boxed areas where text is bold and blue. Express all fractions as decimals.*

Given:	**Gross land area**	GLA=	**1.549**	acres	67,474	SF
Land Variables:	Public/ private right-of-way & paved easements	W=	**0.084**	fraction of GLA	5,668	SF
	Net land area	NLA=	1.419	acres	61,807	SF
	Unbuildable and/or future expansion areas	U=	**0.000**	fraction of GLA	0	SF
	Gross land area reduction	X=	0.084	fraction of GLA	5,668	SF
	Buildable land area remaining	**BLA=**	1.419	acres	61,807	SF
Parking Variables:	Est. gross pkg. lot area per space in SF	s =	**340**	ENTER ZERO IF NO PARKING REQUIRED		
	Building SF permitted per parking space	a =	**492.1**	ENTER ZERO IF NO PARKING REQUIRED		
	No. of loading spaces	l=	**0**			
	Gross area per loading space	b =	**1,800**	SF	0	SF
Site Variables:	**Project open space as fraction of BLA**	**S=**	**0.000**		0	SF
	Private driveways as fraction of BLA	R=	**0.000**		0	SF
	Misc. pavement as fraction of BLA	M=	**0.000**		0	SF
	Loading area as fraction of BLA	L=	0.000		0	SF
	Total site support areas as a fraction of BLA	Su=	0.000		0	SF
Core:	**Core development area as fraction of BLA**	**C=**	1.000	C+Su must = 1	61,807	SF

PLANNING FORECAST

no. of floors **FLR**	CORE minimum land area for BCG & PLA	gross building area **GBA**	parking lot area PLA	pkg. spaces NPS	footprint **BCA**	bldg SF / acre SFAC function of BLA	flr area ratio FAR function of BLA
1.00	61,807	36,552	25,254	74.3	36,552	25,761	0.591
2.00		51,898	35,857	105.5	25,949	36,577	0.840
3.00		**60,343**	**41,692**	**122.6**	**20,114**	**42,529**	**0.976**
4.00		65,688	45,385	133.5	16,422	46,295	1.063
5.00		69,374	47,932	141.0	13,875	48,893	1.122
6.00	*NOTE: Be aware when BCA becomes too small to be feasible.*	72,071	49,795	146.5	12,012	50,794	1.166
7.00		74,129	51,217	150.6	10,590	52,244	1.199
8.00		75,751	52,338	153.9	9,469	53,388	1.226
9.00		77,063	53,244	156.6	8,563	54,312	1.247
10.00		78,146	53,992	158.8	7,815	55,075	1.264
11.00		79,054	54,620	160.6	7,187	55,716	1.279
12.00		79,828	55,154	162.2	6,652	56,261	1.292
13.00		80,494	55,615	163.6	6,192	56,731	1.302
14.00		81,074	56,016	164.8	5,791	57,139	1.312
15.00		81,584	56,368	165.8	5,439	57,499	1.320

WARNING: These are preliminary forecasts that must not be used to make final decisions.
1) These forecasts are not a substitute for the "due diligence" research that must be conducted to support the final definition of "unbuildable areas" above and the final decision to purchase land. This research includes, but is not limited to, verification of adequate subsurface soil, zoning, environmental clearance, access, title, utilities and water pressure, clearance from deed restriction, easement and right-of-way encumbrances, clearance from existing above and below ground facility conflicts, etc.
2) The most promising forecast(s) made on the basis of data entered in the design specification from "due diligence" research must be verified at the drawing board before funds are committed and land purchase decisions are made. Actual land shape ratios, dimensions and irregularities encountered may require adjustments to the general forecasts above.
3) The software licensee shall take responsibility for the design specification values entered and any advice given that is based on the forecast produced.

financial feasibility of the investment and the yield per acre that can be reinvested by government to support a community's way of life. Therefore, these development capacity comparisons go far beyond the obvious and directly affect the public health, safety, and welfare.

Parking System Comparisons Involving Residential Land Uses

Multisystem development comparison may also be applied to the residential land use family by following the same process of forecast model selection and evaluation described previously. Figures 9.3 and 9.4 illustrate two forms of residential apartment housing that are worlds apart, but that also contain similarities that help to illustrate the meaning of context and intensity. Figure 9.3 is called Lane Crest; Fig. 9.4 is called Waterford Tower. The context records for these two projects are included as Tables 9.3 and 9.4. Both are built on roughly the same land area; both provide approximately the same number of parking spaces per dwelling unit; and both provide similar amounts of project open space. One is over six times as tall, however, and has twice as many dwelling units. It is located downtown, uses an adjacent parking structure, has an aggregate average dwelling unit area that is three times the other, and is an exclusive urban condominium on the edge of a downtown business district overlooking a river. The other is a suburban apartment building surrounded by a parking lot and hidden behind a strip commercial shopping district. The intensity index for the condominium is 19.971 and the intensity for the apartment is 3.772. A higher index does not always indicate an inferior context, however. In this case, the Lane Crest density of 68.57 dwelling units per buildable acre, the aggregate average dwelling unit area of 663 sq ft, and the strip commercial location combine to produce a lifestyle and quality of life that is less desirable to most. This is the case,

(a)

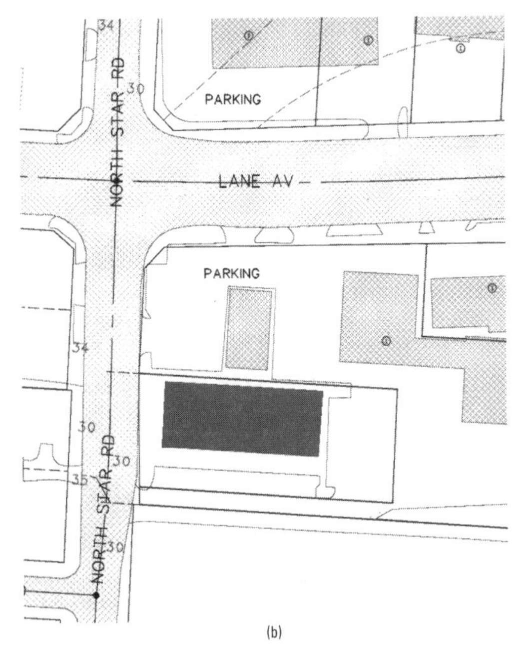

(b)

(c)

Figure 9.3 Lane Crest Apartments—residential apartment G1 parking system. A shared parking lot and circulation aisle combined with a low aggregate average dwelling unit area, reduced parking allocation, 2.5-story building height, and 23% project open space make it possible to maximize the achievable density on this site within zoning ordinance building height limits. (*a*) Street view; (*b*) site plan; (*c*) overview.

(Photos by W. M. Hosack)

(a)

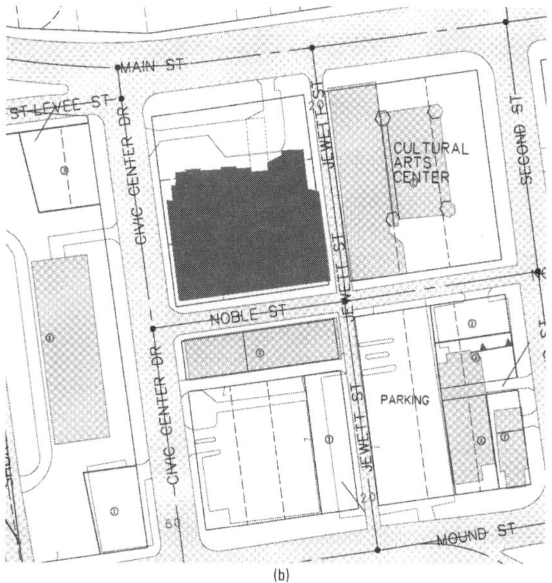

(b)

(c)

Figure 9.4 Waterford Tower—residential S1 parking system. This adjacent parking garage system attempts to conserve project open space while also providing parking in an urban area. To accomplish this, the building is increased in height and a service side of the building is created. (*a*) Overview; (*b*) site plan; (*c*) streetfront view.

(Photos by W. M. Hosack and B. Higgins)

Table 9.3 Lane Apartment Context Record with G1 Parking System

Name	**LaneCrest**
Address or Location	**2376 N. Star**
Land Use	**Residential RG1**
Surveyor	
Parking System	*check box*
Surface parking around but not under building	
Surface parking around and under building	
No parking	
Notes	

	Description	Project Measurements		Context Summary SF	%	
Site				SF	%	
	Gross land area in acres	**0.525**	GLA	22,869		
	Public/ private ROW and paved easements in SF	**0**	W		0.0%	of GLA
	Net lot area		NLA	22,869	100.0%	of GLA
	Total unbuildable area in SF	**0**	U		0.0%	of GLA
	Water area within unbuildable area in SF	**0**	WAT			of U
	Buildable land area		BLA	22,869	100.0%	of GLA
Development Cover						
	Gross building area in SF	**23,856**	GBA			
	Total building footprint in SF	**7,952**	B		34.8%	of BLA
	Total bldg. support in SF (stairs, corridors, elevators, etc.)	**450**	BSU			
	Building efficiency		Be		98.1%	of GBA
Residential only	Number of dwelling units	**36**	DU			
	Aggregate average dwelling unit area in SF		AGG	663		
	Number of building floors	**3.00**	FLR			
G2 only	Unenclosed surface parking area under building in SF		PUB			
	Total surface parking area in SF**	**9,000**	P or G		39.4%	of BLA
	Number of surface parking spaces on premise	**24**	NPSs			
	Building cover over surface parking area		AIR			of GPA
	Gross surface parking area per parking space in SF		ss	375		
	Gross building area per surface parking space in SF		as	994		
	Surface parking spaces per dwelling unit		us	0.67		
Pkg. struc. only	Total parking structure footprint in SF		P		0.0%	of BLA
Pkg. struc. only	Parking structure area below grade in SF		UNG			
Pkg. struc. only	Parking structure area above grade under building in SF		PSC			
Pkg. struc. only	Number of parking structure floors		p			
Pkg. struc. only	Total parking structure area in SF		GPA			
Pkg. struc. only	Total parking structure spaces		NPSg			
Pkg. struc. only	Total pkg. struc. support in SF (stairs, elevators, etc.)		PSS			
Pkg. struc. only	Parking structure support percentage		Pu			of GPA
Pkg. struc. only	Gross parking structure area per space in SF		sg	0		
Pkg. struc. only	Gross building area per structure parking space in SF		ag	0		
Pkg. struc. only	Parking structure spaces per dwelling unit		ug	0.00		
Residential only	Total dwelling unit garage area in SF		GPA			
Residential only	Dwelling unit garage above grade under builiding in SF		GUB			
Residential only	Number of dwelling unit garage parking spaces		GPS			
Residential only	Garage parking spaces per dwelling		Gn	0.0		
Residential only	Total dwelling unit garage area per space in SF		Ga			
Residential only	Total dwelling unit garage cover in addition to building cover		Gc		0.0%	of BLA
	Total parking spaces		NPS	24.0		
Residential only	Total parking spaces per dwelling unit		u	0.7		
	Gross building area per parking space total		a	994.0		
	Number of loading spaces		l			
	Total loading area in SF		LDA			
	Gross area per loading space in SF		b			
	Loading area percentage		L		0.0%	of BLA
	Driveway areas in SF		R		0.0%	of BLA
	Misc. pavement and social center pvmt. in SF	**700**	M		3.1%	of BLA
	TOTAL DEVELOPMENT COVER		D	17,652	**77.2%**	of BLA
Project Open Space						
	PROJECT OPEN SPACE		S	5,217	**22.8%**	of BLA
	Water area within project open space in SF		wat			of S
Summary		0.0%				
	Development balance index		BNX		0.772	
	Development intensity index		INX		3.772	
	Floor area ratio per buildable acre		BFAR		1.043	
	Capacity index (gross building sq. ft. per buildable acre)		SFAC		45,440	
	Density per buildable acre if applicable		dBA		68.57	
	Density per net acre if applicable		dNA		68.57	
	Density per gross acre if applicable		dGA		68.57	

** Include internal parking lot landscaping, circulation aisles and private roadway parking space areas. Express private roads separately as driveway areas..

Table 9.4 Waterford Context Record with S1 Parking System

Name	**Waterford Tower**	
Address or Location	**155 W. Main Street**	
Land Use	**Residential RS1**	
Surveyor	**Higgins**	
Parking System		*check box*
Structure parking adjacent to building		x
Structure parking underground		
Structure parking above grade under building		
Notes		

	Description	Project Measurements		Context Summary SF	Context Summary %	
Site				SF	%	
	Gross land area in acres	0.658	GLA	28,662		
	Public/ private ROW and paved easements in SF	0	W		0.0%	of GLA
	Net lot area		NLA	28,662	100.0%	of GLA
	Total unbuildable area in SF	0	U		0.0%	of GLA
	Water area within unbuildable area in SF	0	WAT			of U
	Buildable land area		BLA	28,662	100.0%	of GLA
Development Cover						
	Gross building area in SF	171,000	GBA			
	Total building footprint in SF	9,000	B		31.4%	of BLA
	Total bldg. support in SF (stairs, corridors, elevators, etc.)		BSU			
	Building efficiency		Be			of GBA
Residential only	Number of dwelling units	96	DU			
	Aggregate average dwelling unit area in SF		AGG	1,781		
	Number of building floors	19.00	FLR			
G2 only	Unenclosed surface parking area under building in SF		PUB			
	Total surface parking area in SF**		P or G		0.0%	of BLA
	Number of surface parking spaces on premise		NPSs			
	Building cover over surface parking area		AIR			of GPA
	Gross surface parking area per parking space in SF		ss			
	Gross building area per surface parking space in SF		as			
	Surface parking spaces per dwelling unit		us	0.00		
Pkg. struc. only	Total parking structure footprint in SF	9,000	P		31.4%	of BLA
Pkg. struc. only	Parking structure area below grade in SF	0	UNG			
Pkg. struc. only	Parking structure area above grade under building in SF	0	PSC			
Pkg. struc. only	Number of parking structure floors	3	p			
Pkg. struc. only	Total parking structure area in SF	27,000	GPA			
Pkg. struc. only	Total parking structure spaces	75	NPSg			
Pkg. struc. only	Total pkg. struc. support in SF (stairs, elevators, etc.)	1,350	PSS			
Pkg. struc. only	Parking structure support percentage		Pu		5.0%	of GPA
Pkg. struc. only	Gross parking structure area per space in SF		sg	360		
Pkg. struc. only	Gross building area per structure parking space in SF		ag	2,280		
Pkg. struc. only	Parking structure spaces per dwelling unit		ug	0.78		
Residential only	Total dwelling unit garage area in SF		GPA			
Residential only	Dwelling unit garage above grade under builiding in SF		GUB			
Residential only	Number of dwelling unit garage parking spaces		GPS			
Residential only	Garage parking spaces per dwelling		Gn	0.0		
Residential only	Total dwelling unit garage area per space in SF		Ga			
Residential only	Total dwelling unit garage cover in addition to building cover		Gc		0.0%	of BLA
	Total parking spaces		NPS	75.0		
Residential only	Total parking spaces per dwelling unit		u	0.8		
	Gross building area per parking space total		a	2,280.0		
	Number of loading spaces		l			
	Total loading area in SF		LDA			
	Gross area per loading space in SF		b			
	Loading area percentage		L		0.0%	of BLA
	Driveway areas in SF	4,430	R		15.5%	of BLA
	Misc. pavement and social center pvmt. in SF	250	M		0.9%	of BLA
	TOTAL DEVELOPMENT COVER		D	22,680	**79.1%**	of BLA
Project Open Space						
	PROJECT OPEN SPACE		S	5,982	**20.9%**	of BLA
	Water area within project open space in SF		wat			of S
Summary		0.0%				
	Development balance index		BNX		0.791	
	Development intensity index		INX		19.791	
	Floor area ratio per buildable acre		BFAR		5.966	
	Capacity index (gross building sq. ft. per buildable acre)		SFAC		259,878	
	Density per buildable acre if applicable		dBA		145.90	
	Density per net acre if applicable		dNA		145.90	
	Density per gross acre if applicable		dGA		145.90	

** Include internal parking lot landscaping, circulation aisles and private roadway parking space areas. Express private roads separately as driveway areas..

even though the Waterford Tower density of 145.9 dwelling units per buildable acre and its intensity level of 19.791 are much greater. The Waterford Tower parking structure system makes the double density and triple intensity possible, but the quality of life that makes this condominium lifestyle feasible is established by the context in which it is located, and which it complements. If Waterford Tower were located in the strip commercial context of Lane Crest, it could easily become just as tired just as quickly with few interested in the units offered. From the standpoint of context compatibility, neither the apartment nor the condominium is comfortable in a suburban setting. Both have small amounts of open space, high residential densities, and less than one parking space per dwelling unit. Lane Crest however, combines high-density housing with a low-aggregate average dwelling unit area and an exposed parking lot to produce a context contradiction that is an extension of its strip commercial neighbor. It does not function as a transition to lower-density housing since its intensity index is as high as the strip commercial that abuts. Its appearance also suffers and cannot be supplemented with grading, trees, and landscape since little open space is available. These conditions could be avoided with more compatible design specifications that are based on development capacity comparison and evaluation.

Exhibit 9.7 has been created to explore this question. It is based on the concept that an apartment transition from strip commercial could exceed a suburb's building height limitations if it provided more residential amounts of open space in return.[5] This is a design hypothesis that can be evaluated with the help of forecast model RG1L.

Exhibit 9.7 uses model RG1L to evaluate the impact of providing 50% project open space on the Lane Crest site instead of the 22.8% present. This evaluation is based on the hypothesis that a greater amount of open space will form a better transition to the adjacent 70% residential neighbor-

Exhibit 9.7 Lane Apartments with 50% Open Space

Forecast Model RG1L

*Development capacity forecast for **APARTMENTS** based on the use of an adjacent **GRADE PARKING LOT** located on the same premises. When s and u equal zero in the design specification below, the forecast pertains to conditions when **NO PARKING** is required.*

***Given:** Gross land area. **To Find:** Maximum dwelling unit capacity of the land area given based on the design specification values entered below. **Premise:** All building floors considered equal in area.*

DESIGN SPECIFICATION *Enter values in boxed areas where text is bold and blue. Express all fractions as decimals.*

Given:	**Gross land area**	GLA=	**0.525**	acres	22,869	SF
Land Variables:	Public/ private right-of-way & paved easements	W=	**0.000**	fraction of GLA	0	SF
	Net land area	NLA=	0.525	acres	22,869	SF
	Unbuildable and/or future expansion areas	U=	**0.000**	fraction of GLA	0	SF
	Gross land area reduction	X=	0.000	fraction of GLA	0	SF
	Buildable land area remaining	BLA=	**0.525**	acres	22,869	SF
Parking Variables:	Est. gross pkg. lot area per pkg. space in SF	s =	**375**	ENTER ZERO IF NO PARKING REQUIRED		
	Parking lot spaces planned or required per dwelling unit	u=	**0.67**	ENTER ZERO IF NO PARKING REQUIRED		
	Garage parking spaces planned or required per dwelling unit	Gn=	**0**	ENTER ZERO IF NO PARKING REQUIRED		
	Gross building area per garage space	Ga=	**0**	ENTER ZERO IF NO PARKING REQUIRED		
	No. of loading spaces	l =	**0**			
	Gross area per loading space	b =	**0**	SF	0	SF
Site Variables:	**Project open space as fraction of BLA**	S=	**0.500**		11,435	SF
	Private driveways as fraction of BLA	R=	**0.000**		0	SF
	Misc. pavement as fraction of BLA	M=	**0.031**		709	SF
	Loading area as fraction of BLA	L=	0.000		0	SF
	Total site support areas as a fraction of BLA	Su=	0.531		12,143	SF
Core:	**Core development area as fraction of BLA**	C=	0.469	C+Su must = 1	10,726	SF
Building Variables:	Building efficiency as percentage of GBA	Be=	**0.981**			
	Building support as fraction of GBA	Bu=	0.019	Be + Bu must = 1		

Dwelling Unit Mix Table:

DU dwelling unit type	GDA gross du area	CDA=GDA/Be comprehensive du area	MIX du mix	PDA = (CDA)MIX Pro-rated du area

Dwelling unit mix not available

Aggregate average dwelling unit area (AGG) found by dividing gross building area by number of dwelling units present

Aggregate avg. dwelling unit area (AGG) = **663**

GBA sf per parking space a= 990

PLANNING FORECAST

no. of floors FLR	CORE minimum lot area for BCG & PLA	density per net acre dNA	dwelling units **NDU**	pkg. lot spaces NPS	parking lot area **PLA**	garage spaces GPS	garage area **GAR**	gross bldg area GBA no garages	footprint **BCA**	density per dBA bldable acre
1.00	10,726	22.35	11.7	7.9	2,948	n/a	n/a	7,778	7,778	22.35
2.00		35.06	18.4	12.3	4,624	n/a	n/a	12,203	6,101	35.06
3.00		43.26	22.7	15.2	5,706	n/a	n/a	15,058	5,019	43.26
4.00		48.99	25.7	17.2	6,462	n/a	n/a	17,053	4,263	48.99
5.00		53.22	27.9	18.7	7,020	n/a	n/a	18,526	3,705	53.22
6.00		**56.47**	**29.6**	**19.9**	**7,449**	**n/a**	**n/a**	**19,657**	**3,276**	**56.47**
7.00	*NOTE: Be aware when BCA becomes too small to be feasible.*	59.05	31.0	20.8	7,789	n/a	n/a	20,554	2,936	59.05
8.00		61.14	32.1	21.5	8,065	n/a	n/a	21,283	2,660	61.14
9.00		62.88	33.0	22.1	8,294	n/a	n/a	21,886	2,432	62.88
10.00		64.34	33.8	22.6	8,486	n/a	n/a	22,393	2,239	64.34
11.00		65.58	34.4	23.1	8,650	n/a	n/a	22,827	2,075	65.58
12.00		66.65	35.0	23.4	8,792	n/a	n/a	23,201	1,933	66.65
13.00		67.59	35.5	23.8	8,916	n/a	n/a	23,527	1,810	67.59
14.00		**68.42**	**35.9**	**24.1**	**9,025**	**n/a**	**n/a**	**23,814**	**1,701**	**68.42**

WARNING: These are preliminary forecasts that must not be used to make final decisions.
1) These forecasts are not a substitute for the "due diligence" research that must be conducted to support the final definition of "unbuildable areas" above and the final decision to purchase land. This research includes, but is not limited to, verification of adequate subsurface soil, zoning, environmental clearance, access, title, utilities and water pressure, clearance from deed restriction, easement and right-of-way encumbrances, clearance from existing above and below ground facility conflicts, etc.
2) The most promising forecast(s) made on the basis of data entered in the design specification from "due diligence" research must be verified at the drawing board before funds are committed and land purchase decisions are made. Actual land shape ratios, dimensions and irregularities encountered may require adjustments to the general forecasts above.
3) The software licensee shall take responsibility for the design specification values entered and any advice given that is based on the forecast produced.

hoods. It also suggests that the effect of increased building height will be mitigated by the increased open space, and that the combination of increased height and open space will more effectively relate the high-intensity strip commercial activity to the low-density residential environment beyond.

Design specification values can be used to plan this neighborhood context, and to guide its construction one lot at a time. The planning forecast panel in Exhibit 9.7 shows, however, that it would take a 14-story building to produce the same number of dwelling units on the site when 50% project open space is provided. It also shows that a 14-story building would be limited to a 1701-sq-ft footprint. Since the aggregate average dwelling unit area of the current project is 663 sq ft, this means that the height is relatively unrealistic since only two dwelling units could fit on a floor.

The planning forecast panel shows that a 6-story building could shelter approximately 30 dwelling units when 50% open space is provided and the footprint is limited to 3276 sq ft. This is six units less than the project constructed, and it raises the inevitable debate over adequate density levels and acceptable building height limits.

Conclusion

The objective of this book is not to answer the question of acceptable density levels, since communities will reach separate conclusions. The objective is to produce quantitative tools in the form of a Context Record System and Development Forecast Collection that can help others answer this question for themselves. With these tools, research can be conducted and conclusions can be expressed in common terms that have a reasonable chance of achieving the objectives defined. The comparative system that evolves can be used to improve development results and regulations. It may also be able to reduce the level of emotional conflict in some circumstances.

Notes

1. *Development capacity* means the amount of gross building area that can be placed on a given land area or the land area required to accommodate a given building area.
2. An index of 8.000 is more intense than 8.8 for instance, since 8.8 indicates that 20% open space and 80% development cover has been provided. An index of 8.0 indicates that no open space has been provided and that development cover is 100%.
3. This equals the amount of building square feet permitted per parking space. Residential buildings generally provide one parking space for every 800 to 1500 gross sq ft constructed, depending on the aggregate average dwelling unit area planned. Nonresidential buildings generally provide one parking space for every 50 to 400 gross sq ft constructed depending on the land use activity involved.
4. This does not mean that it is impossible to provide more parking and open space in an urban environment.
5. In fact, additional height could serve as a more effective buffer if combined with adequate amounts of open space and attention to parking lot screening techniques.

Parking Structure Comparisons

Chapter 8 focuses on single-system development capacity comparisons based on surface parking lot specification options. Chapter 9 compares S2 adjacent parking structure options to G1 surface parking lot options in multisystem development capacity evaluations. This chapter compares different parking structure systems and the development capacity implications of these options.

The S1 adjacent parking structure system is popular because it is a separate structure above grade that does not require extensive excavation. It also does not increase building height or impact building appearance by being placed above grade under the building in an S3 configuration. The disadvantages of an S1 system are that it reduces the land area that can be occupied by the building, reduces the development capacity of the land as a result, and discourages the introduction of places for people at street level.[1] If open space is provided, S1 parking levels can be increased to offset the loss, but this simply increases intensity and can never compete with the potential of the other two systems, given the same design specification. The following context records show that the S1, S2, and S3 parking structure systems are quite different for very specific reasons, and are not always chosen for their development capacity characteristics.

Nonresidential Parking Structure Comparisons

Figure 10.1 should look familiar. It is the same S1 adjacent parking structure system used in Chap. 9. Figure 10.2 is an underground parking structure system (S2) designed to preserve open space by using the entire underground area as a two-level parking garage. The S2 approach is often used to conserve surface land area for people and, in this case, as a parklike setting for state government as well. Figure 10.3 illustrates a parking structure above grade under a building. This design approach has been referred to as an S3 parking system. It is an unusual but efficient system that allows the same land area to be used for both parking cover and building cover since the facilities are stacked. In this case, the land conserved by this approach has been used to introduce planting boxes and gathering places for people at street level.

The context records for these three systems are included as Tables 10.1, 10.2, and 10.3. At first glance, the systems are not comparable since they appear quite different and are located on land areas of dissimilar size. The indexes at the bottom of each record, however, are designed to calculate comparable data from the measurements entered. In this case, the Federal Office in Table 10.1 has an intensity index INX of[2] 8.000 and a capacity index SFAC of 178,190 gross sq ft of building area per buildable acre. The Statehouse in Table 10.2 has an INX of[3] 2.338 and an SFAC of 19,981 gross sq ft of building area per buildable acre. The IBM building in Table 10.3 has an INX of[4] 15.678 and an SFAC of 132,425 gross sq ft of building area per buildable acre.

This is not the capacity hierarchy that would be expected from these systems. An intuitive expectation would assume

(a)

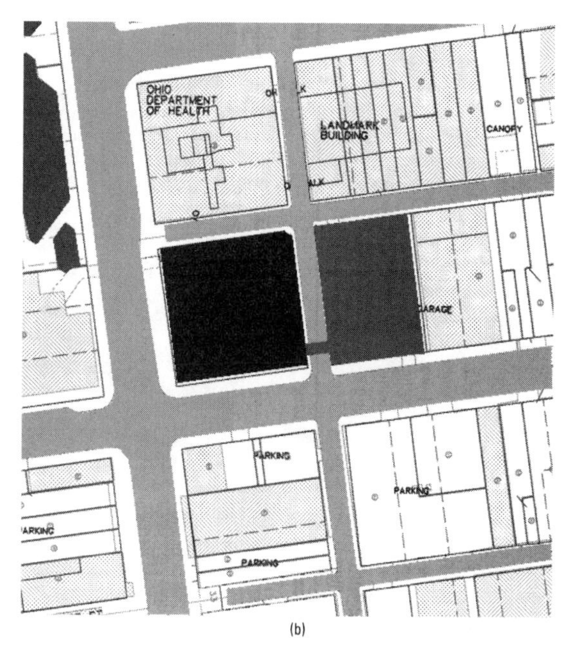

(b)

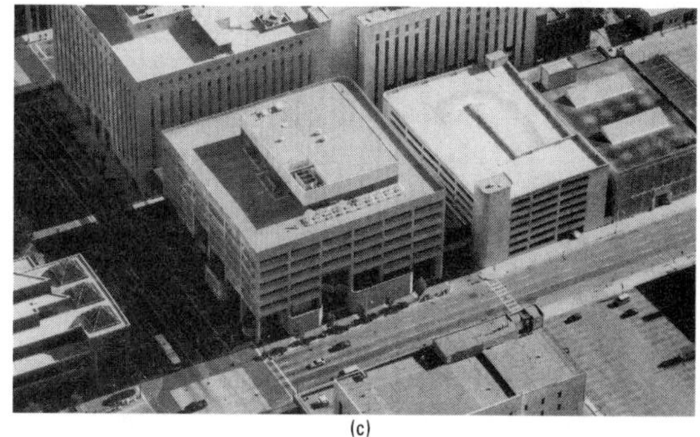

(c)

Figure 10.1 Federal Office building—S1 adjacent parking structure system. (*a*) Street view. (*b*) The site plan illustrates a building and associated S1 parking structure occupying the entire area of an urban site. In the context record for this example, the alley shown has been subtracted from the gross land area to find the buildable land area available. (*c*) Overview.

(Photos by W. M. Hosack)

that the Statehouse and the IBM parking systems would produce comparable development capacity results, and that the Federal Office system would produce lower results. This assumption would be based on the fact that the Federal Office site must be divided between building cover and parking cover while the other two systems are capable of using the entire site for building and parking by stacking one above the other. The design specifications for the Statehouse and IBM systems include project open space and the

(a)

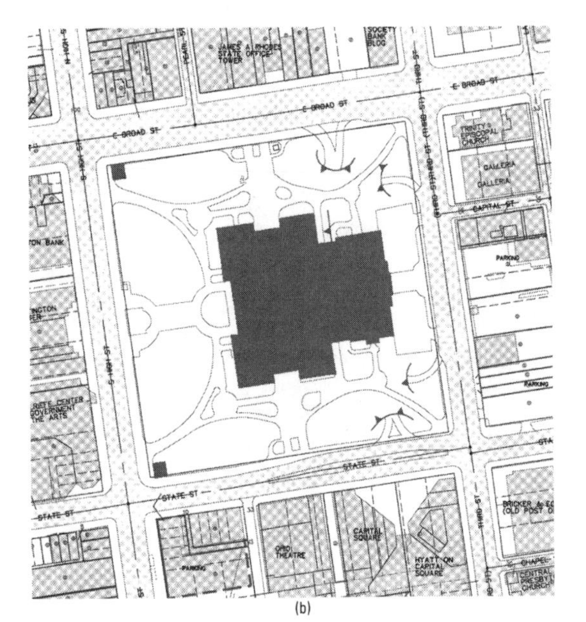

(b)

(c)

Figure 10.2 Statehouse—S2 underground parking structure system. (*a*) Underground entries are the only visible hint of an automobile presence. The open space preserved is a characteristic advantage of this parking system, and its parking capacity makes it possible to serve both the building and the surrounding area. This often extends the context benefit of this system beyond its boundaries. (*b*) Site plan. (*c*) Overview.

(Photos by W. M. Hosack)

Federal Office does not. The Federal Office also includes a greater number of parking levels. As a result, the Statehouse and IBM intensity levels and capacity indexes are lower than the Federal Office system, even though their systems have the ability to outperform this system.

The Statehouse underground parking garage design specification implicitly identifies two development objectives

(a)

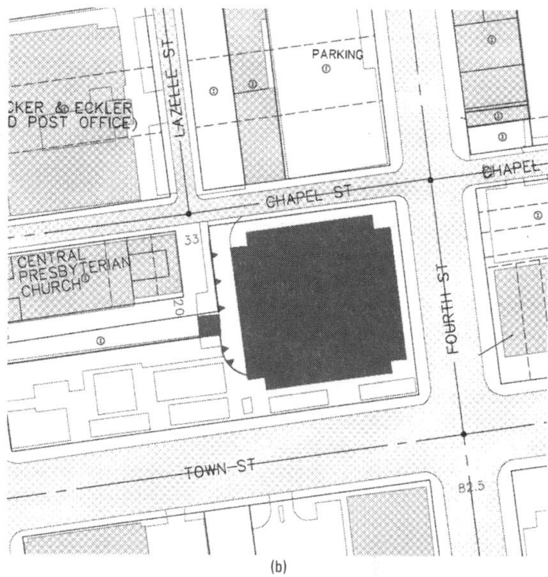

(b)

(c)

Figure 10.3 IBM—S3 above-grade under–building parking structure system. (*a*) The photograph illustrates an S3 parking system that adds to the corporate image projected. This image can be compromised when context integrity is not extended beyond property lines. (*b*) The IBM site plan illustrates a building area stacked above an S3 parking structure to conserve open space. (*c*) Increased building height and a Town St. garden result from this approach.

(Photos by W. M. Hosack)

that have produced the lower capacity photographed, even though the ultimate capacity of the Statehouse system is much greater. First, the Statehouse parking allocation is 167 sq ft of building area per parking space, which provides three times the spaces provided by the Federal Office allocation of 492 sq ft of building area per parking space. The Federal Office allocation is similar to other central business

Table 10.1 Federal Office Context Record

Name	**Federal Office Building**
Address or Location	**200 N. High Street**
Land Use	**Office CS1**

Parking System	check box
Structure parking adjacent to building	x
Structure parking underground	
Structure parking above grade under building	
Notes	

	Description	Project Measurements		Context Summary SF	%	
Site						
	Gross land area in acres	**1.549**	GLA	67,474		
	Public/ private ROW and paved easements in SF	**5,700**	W		8.4%	of GLA
	Net lot area		NLA	61,774	91.6%	of GLA
	Total unbuildable area in SF	**0**	U		0.0%	of GLA
	Water area within unbuildable area in SF	**0**	WAT			of U
	Buildable land area		BLA	61,774	91.6%	of GLA
Development Cover						
	Gross building area in SF	**252,700**	GBA			
	Total building footprint in SF	**36,100**	B		58.4%	of BLA
	Total bldg. support in SF (stairs, corridors, elevators, etc.)		BSU			
	Building efficiency		Be			of GBA
Residential only	Number of dwelling units		DU			
	Aggregate average dwelling unit area in SF		AGG			
	Number of building floors	**7.00**	FLR			
G2 only	Unenclosed surface parking area under building in SF		PUB			
	Total surface parking area in SF**		P or G		0.0%	of BLA
	Number of surface parking spaces on premise		NPSs			
	Building cover over surface parking area		AIR			of GPA
	Gross surface parking area per parking space in SF		ss			
	Gross building area per surface parking space in SF		as			
	Surface parking spaces per dwelling unit		us	0.00		
Pkg. struc. only	Total parking structure footprint in SF	25,650	P		41.5%	of BLA
Pkg. struc. only	Parking structure area below grade in SF					
Pkg. struc. only	Parking structure area above grade under building in SF					
Pkg. struc. only	Number of parking structure floors	8	p			
Pkg. struc. only	Total parking structure area in SF	205,200	GPA			
Pkg. struc. only	Total parking structure spaces	513	NPSg			
Pkg. struc. only	Total pkg. struc. support in SF (stairs, elevators, etc.)		PSS			
Pkg. struc. only	Parking structure support percentage		Pu			of GPA
Pkg. struc. only	Gross parking structure area per space in SF		sg	400		
Pkg. struc. only	Gross building area per structure parking space in SF		ag	492.6		
Pkg. struc. only	Parking structure spaces per dwelling unit		ug	0.00		
Residential only	Total dwelling unit garage area in SF		GPA			
Residential only	Dwelling unit garage above grade under builiding in SF		GUB			
Residential only	Number of dwelling unit garage parking spaces		GPS			
Residential only	Garage parking spaces per dwelling		Gn	0.0		
Residential only	Total dwelling unit garage area per space in SF		Ga			
Residential only	Total dwelling unit garage cover in addition to building cover		Gc		0.0%	of BLA
	Total parking spaces			513.0		
Residential only	Total parking spaces per dwelling unit		u			
	Gross building area per parking space total		a	492.6		
	Number of loading spaces		l			
	Total loading area in SF		LDA			
	Gross area per loading space in SF		b			
	Loading area percentage		L		0.0%	of BLA
	Driveway areas in SF		R		0.0%	of BLA
	Misc. pavement and social center pvmt. in SF		M		0.0%	of BLA
	TOTAL DEVELOPMENT COVER		D	61,750	**100.0%**	of BLA
Project Open Space						
	PROJECT OPEN SPACE		S	24	**0.0%**	of BLA
	Water area within project open space in SF	0.0%	wat			of S
Summary						
	Development balance index		BNX		1.000	
	Development intensity index		INX		8.000	
	Floor area ratio per buildable acre		BFAR		4.091	
	Capacity index (gross building sq. ft. per buildable acre)		SFAC		178,190	
	Density per buildable acre if applicable		dBA			
	Density per net acre if applicable		dNA			
	Density per gross acre if applicable		dGA			

** Include internal parking lot landscaping, circulation aisles and private roadway parking space areas. Express private roads separately as driveway areas..

Table 10.2 Statehouse Context Record

Name	**State House**
Address or Location	**Broad and High St.**
Land Use	**Institutional CS2**

Parking System	*check box*
Structure parking adjacent to building	
Structure parking underground	x
Structure parking above grade under building	
Notes	

		Description	Project Measurements		Context Summary SF	%	
Site							
		Gross land area in acres	10.000	GLA	435,600		
		Public/ private ROW and paved easements in SF	0	W		0.0%	of GLA
		Net lot area		NLA	435,600	100.0%	of GLA
		Total unbuildable area in SF	0	U		0.0%	of GLA
		Water area within unbuildable area in SF	0	WAT			of U
		Buildable land area		BLA	435,600	100.0%	of GLA
Development Cover							
		Gross building area in SF	199,806	GBA			
		Total building footprint in SF	99,903	B		22.9%	of BLA
		Total bldg. support in SF (stairs, corridors, elevators, etc.)		BSU			
		Building efficiency		Be			of GBA
	Residential only	Number of dwelling units		DU			
		Aggregate average dwelling unit area in SF		AGG			
		Number of building floors	2.00	FLR			
	G2 only	Unenclosed surface parking area under building in SF		PUB			
		Total surface parking area in SF (include internal landscape)		P or G		0.0%	of BLA
		Number of surface parking spaces		NPSs			
		Building cover over surface parking area		AIR			of GPA
		Gross surface parking area per parking space in SF		ss			
		Gross building area per surface parking space in SF		as			
		Surface parking spaces per dwelling unit		us	0.00		
	Pkg. struc. only	Total parking structure footprint in SF	209,400	P		0.0%	of BLA
	Pkg. struc. only	Parking structure area below grade in SF	209,400				
	Pkg. struc. only	Parking structure area above grade under building in SF	0				
	Pkg. struc. only	Number of parking structure floors	2	p			
	Pkg. struc. only	Total parking structure area in SF	408,800	GPA			
	Pkg. struc. only	Total parking structure spaces	1,200	NPSg			
	Pkg. struc. only	Total pkg. struc. support in SF (stairs, elevators, etc.)	10,000	PSS			
	Pkg. struc. only	Parking structure support percentage		Pu		2.4%	of GPA
	Pkg. struc. only	Gross parking structure area per space in SF		sg	341		
	Pkg. struc. only	Gross building area per structure parking space in SF		ag	167		
	Pkg. struc. only	Parking structure spaces per dwelling unit		ug	0.00		
	Residential only	Total dwelling unit garage area in SF		GPA			
	Residential only	Dwelling unit garage under builiding in SF		GUB			
	Residential only	Number of dwelling unit garage parking spaces		GPS			
	Residential only	Garage parking spaces per dwelling		Gn	0.0		
	Residential only	Total dwelling unit garage area per space in SF		Ga			
	Residential only	Total dwelling unit garage cover in addition to building cover		Gc		0.0%	of BLA
		Total parking spaces			1,200.0		
	Residential only	Total parking spaces per dwelling unit		u			
		Gross building area per parking space total		a	166.5		
		Number of loading spaces		l			
		Total loading area in SF		LDA			
		Gross area per loading space in SF		b			
		Loading area percentage		L		0.0%	of BLA
		Driveway areas in SF	4,500	R		1.0%	of BLA
		Misc. pavement and social center pvmt. in SF	43,000	M		9.9%	of BLA
		TOTAL DEVELOPMENT COVER		D	147,403	**33.8%**	of BLA
Project Open Space							
		PROJECT OPEN SPACE		S	288,197	**66.2%**	of BLA
		Water area within project open space in SF		wat			of S
Summary			0.0%				
		Development balance index		BNX		0.338	
		Development intensity index		INX		2.338	
		Floor area ratio per buildable acre		BFAR		0.459	
		Capacity index (gross building sq. ft. per buildable acre)		SFAC		19,981	
		Density per buildable acre if applicable		dBA			
		Density per net acre if applicable		dNA			
		Density per gross acre if applicable		dGA			

Table 10.3 IBM Building Context Record

Name	**IBM**
Address or Location	**140 E. Town Street**
Land Use	**Office**

Parking System	*check box*
Structure parking adjacent to building	**x**
Structure parking underground	
Structure parking above grade under building	
Notes	

		Description	Project Measurements		Context Summary SF	%	
Site							
		Gross land area in acres	**1.241**	GLA	54,058		
		Public/ private ROW and paved easements in SF	**0**	W		0.0%	of GLA
		Net lot area		NLA	54,058	100.0%	of GLA
		Total unbuildable area in SF	**0**	U		0.0%	of GLA
		Water area within unbuildable area in SF	**0**	WAT			of U
		Buildable land area		BLA	54,058	100.0%	of GLA
Development Cover							
		Gross building area in SF	**164,340**	GBA			
		Total building footprint in SF	**16,434**	B		0.0%	of BLA
		Total bldg. support in SF (stairs, corridors, elevators, etc.)		BSU			
		Building efficiency		Be			of GBA
	Residential only	Number of dwelling units		DU			
		Aggregate average dwelling unit area in SF		AGG			
		Number of building floors	**10.00**	FLR			
	G2 only	Unenclosed surface parking area under building in SF		PUB			
		Total surface parking area in SF**		P or G		0.0%	of BLA
		Number of surface parking spaces on premise		NPSs			
		Building cover over surface parking area		AIR			of GPA
		Gross surface parking area per parking space in SF		ss			
		Gross building area per surface parking space in SF		as			
		Surface parking spaces per dwelling unit		us	0.00		
	Pkg. struc. only	Total parking structure footprint in SF	**23,596**	P		43.6%	of BLA
	Pkg. struc. only	Parking structure area below grade in SF	**0**	UNG			
	Pkg. struc. only	Parking structure area above grade under building in SF	**16,434**	PSC			
	Pkg. struc. only	Number of parking structure floors	**5**	p			
	Pkg. struc. only	Number of parking structure floors beneath building	**5**				
	Pkg. struc. only	Total parking structure area in SF	**117,980**	GPA			
	Pkg. struc. only	Total parking structure spaces	**288**	NPSg			
	Pkg. struc. only	Total pkg. struc. support in SF (stairs, elevators, etc.)		PSS			
	Pkg. struc. only	Parking structure support percentage		Pu			of GPA
	Pkg. struc. only	Gross parking structure area per space in SF		sg	410		
	Pkg. struc. only	Gross building area per structure parking space in SF		ag	571		
	Residential only	Parking structure spaces per dwelling unit		ug	0.00		
	Residential only	Total dwelling unit garage area in SF		GPA			
	Residential only	Dwelling unit garage above grade under builiding in SF		GUB			
	Residential only	Number of dwelling unit garage parking spaces		GPS			
	Residential only	Garage parking spaces per dwelling		Gn	0.0		
	Residential only	Total dwelling unit garage area per space in SF		Ga			
	Residential only	Total dwelling unit garage cover in addition to building cover		Gc		0.0%	of BLA
		Total parking spaces		NPS	288.0		
	Residential only	Total parking spaces per dwelling unit		u			
		Gross building area per parking space total		a	570.6		
		Number of loading spaces	**2**	l			
		Total loading area in SF	**484**	LDA			
		Gross area per loading space in SF		b	242		
		Loading area percentage		L		0.9%	of BLA
		Driveway areas in SF	**3,100**	R		5.7%	of BLA
		Misc. pavement and social center pvmt. in SF	**9,488**	M		17.6%	of BLA
		TOTAL DEVELOPMENT COVER		D	36,668	**67.8%**	of BLA
Project Open Space							
		PROJECT OPEN SPACE		S	17,390	**32.2%**	of BLA
		Water area within project open space in SF	**0** 0.0%	wat			of S
Summary							
		Development balance index		BNX		67.8%	
		Development intensity index		INX		15.678	
		Floor area ratio per buildable acre		BFAR		3.040	
		Capacity index (gross building sq. ft. per buildable acre)		SFAC		132,425	
		Density per buildable acre if applicable		dBA			
		Density per net acre if applicable		dNA			
		Density per gross acre if applicable		dGA			

** Include internal parking lot landscaping, circulation aisles and private roadway parking space areas. Express private roads separately as driveway areas..

district (CBD) offices, when they provide parking at all, and the Statehouse provision is greater than typical suburban standards of 250 sq ft of building area per parking space. This implies that the Statehouse objective was to serve both of the aforementioned offices and the parking needs of the surrounding business district. Project open space appears to be the second objective, since the Statehouse allocation is 66.2%. This is far above typical CBD standards, which are often 0.0%, and implies that the objective was to provide gathering places for people and a suitable setting for state government[5] in the heart of the urban area. The site plan in Fig. 10.2 illustrates this open space arrangement. The height of the Statehouse is well below CBD standards, which makes the gross building area low in relation to the number of parking spaces provided, and also implies that the parking and open space were not provided to increase the development capacity of the site. In this case, it appears safe to assume that the Statehouse is a historic symbol of the state that has been preserved and that development capacity has not been an issue. It also appears that the 66.2% open space provided and the dual use of the underground parking garage have been introduced to serve the public need for gathering, parking, open space, and symbolism. The intensity index of 2.338 and the development capacity index of 19,981 gross sq ft of building area per buildable acre are much lower than those of the S1 option, but the underground parking structure system was an ideal choice capable of meeting the mentioned objectives, since its capacity potential serves the surrounding district and not the property itself.

The IBM building also has a lower development capacity than it could have produced. The Table 10.3 context record for IBM shows that this index is 132,425 and the public open space provided is 32.2% of its buildable area. The intensity index of 15.678 illustrates the greater building height typically involved with the S3 parking system, and the fact that 67.8% development cover was constructed. The Fig. 10.3 site plan illustrates this arrangement and the open space that remains. The implication from this site plan and context record is either that the owner consciously intended to provide this open space, or that it was not needed to produce the gross building area required. If the latter were the case, however, the land would probably have been sold. The fact that the open space was provided and the S3 system was chosen implies that the intent was to preserve street-level open space. The S3 system accomplishes this by stacking building and parking garage above grade to avoid the expensive excavation associated with underground parking garages. The characteristics of the system chosen therefore, and the open space provided, indicate that urban open space preservation was an objective from the beginning.

The Federal Office in Fig. 10.1 is at the other end of the spectrum. It uses every available square foot of land area for development cover and divides this land use among building cover and parking cover. This arrangement can be seen in Fig. 10.1. The development capacity index (SFAC) produced by this arrangement is recorded as 178,190 gross sq ft of building area per buildable acre in the summary section of Table 10.1. No open space has been introduced, eight parking levels have been provided, and seven building floors have been constructed to achieve this capacity. This is not close to the ultimate capacity of the site, but it is near the reasonable limit when eight adjacent parking levels are considered.[6]

The planning forecast panel in Exhibit 10.1 illustrates the options surrounding the 7-story building and 8-level parking garage configuration chosen. The actual choice constructed is shown in row 7 of the planning forecast panel and produced 252,700 gross sq ft of building area. The panel also shows that a 15-story building and an 8-level garage could have produced 367,121 sq ft of gross building

Exhibit 10.1 Federal Office Building Forecast Record

Forecast Model CS1L

Development capacity forecast for ***NONRESIDENTIAL BUILDINGS*** *using an* ***ADJACENT PARKING STRUCTURE*** *located on the same premise.*
Given: *Gross land area available.* ***To Find:*** *Max. development capacity of the land area given (gross building area potential) based on the design specification values entered below.*
Design Premise: *Building footprint adjacent to parking garage footprint within the core development area. All similar floors considered equal in area.*

Enter values in boxed areas where text is bold and blue. Express all fractions as decimals.

DESIGN SPECIFICATION

Given:	**Gross land area**	GLA=	**1.549**	acres	67,474	SF
Land Variables:	Public/ private right-of-way & paved easements	W=	**0.084**	fraction of GLA	5,700	SF
	Net land area	NLA=	1.418	acres	61,775	SF
	Unbuildable and/or future expansion areas	U=	**0.000**	fraction of GLA	0	SF
	Gross land area reduction	X=	0.084	fraction of GLA	5,700	SF
	Buildable land area remaining	BLA=	1.418	acres	61,775	SF
Parking Variables:	Est. net pkg. structure area per pkg. space	s =	**340.0**	SF		
	Building SF permitted per parking space	a =	**492.1**	SF		
	Number of parking levels to be evaluated	p=	**8.00**			
	Pkg. support as fraction of gross pkg. structure area (GPA)	Pu=	**0.150**			
	Net area for parking & circulation as fraction of GPA	Pe=	0.850	Pu+Pe must = 1		
	No. of loading spaces	l=	**0**			
	Gross area per loading space	b =	**1,000**	SF	0	SF
Site Variables:	**Project open space as fraction of BLA**	S=	**0.000**		0	SF
	Private driveways as fraction of BLA	R=	**0.000**		0	SF
	Misc. pavement as fraction of BLA	M=	**0.000**		0	SF
	Loading area as fraction of BLA	L=	0.000		0	SF
	Total site support areas as a fraction of BLA	Su=	0.000		0	SF
Core:	**Core development area as fraction of BLA**	C=	1.000	C+Su must = 1	61,775	SF

NOTE: p=1 is a grade parking lot based on design premise. Increase the number of parking levels to increase the capacity forecast forecast below. Other variables in blue, box and bold may also be changed.

PLANNING FORECAST

no. of **FLR** bldg floors	min land area CORE for BCG & PSA	gross building area **GBA**	gross pkg GPA struc area	pkg struct cover **GPL** per level	net parking NPA area	pkg spaces NPS	bldg cover **BCA** footprint	bldg SF per SFAC bldable acre	flr area ratio FAR function of BLA
1.00	61,775	56,077	45,580	5,698	38,743	113.95	56,077	39,542	0.908
2.00		102,684	83,463	10,433	70,944	208.66	51,342	72,407	1.662
3.00		142,032	115,446	14,431	98,129	288.62	47,344	100,153	2.299
4.00		175,696	142,808	17,851	121,387	357.02	43,924	123,890	2.844
5.00		204,823	166,483	20,810	141,510	416.21	40,965	144,429	3.316
6.00		230,273	187,169	23,396	159,094	467.92	38,379	162,375	3.728
7.00		**252,700**	**205,399**	**25,675**	**174,589**	**513.50**	**36,100**	**178,189**	**4.091**
8.00		272,614	221,585	27,698	188,347	553.96	34,077	192,231	4.413
9.00		290,414	236,053	29,507	200,645	590.13	32,268	204,783	4.701
10.00		306,420	249,063	31,133	211,703	622.66	30,642	216,069	4.960
11.00		320,890	260,824	32,603	221,701	652.06	29,172	226,273	5.195
12.00		334,035	271,509	33,939	230,782	678.77	27,836	235,542	5.407
13.00		346,029	281,258	35,157	239,069	703.14	26,618	244,000	5.601
14.00		357,017	290,189	36,274	246,661	725.47	25,501	251,748	5.779
15.00		367,121	298,401	37,300	253,641	746.00	24,475	258,872	5.943

WARNING: These are preliminary forecasts that must not be used to make final decisions.
1) These forecasts are not a substitute for the "due diligence" research that must be conducted to support the final definition of "unbuildable areas" above and the final decision to purchase land. This research includes, but is not limited to, verification of adequate subsurface soil, zoning, environmental clearance, access, title, utilities and water pressure, clearance from deed restriction, easement and right-of-way encumbrances, clearance from existing above and below ground facility conflicts, etc.
2) The most promising forecast(s) made on the basis of data entered in the design specification from "due diligence" research must be verified at the drawing board before funds are committed and land purchase decisions are made. Actual land shape ratios, dimensions and irregularities encountered may require adjustments to the general forecasts above.
3) The software licensee shall take responsibility for the design specification values entered and any advice given that is based on the forecast produced.

area. This could have been accomplished if the building footprint had been shrunk from 36,100 to 24,475 sq ft while increasing the parking garage footprint from 25,675 to 37,300 sq ft. These revised site plan relationships could have produced an additional 114,421 gross sq ft of building area. This 45.2% increase indicates that more could have been done with the land available, but two very practical constraints may have produced the results photographed. The first and foremost is generally an owner who has a specific budget and no greater need. This good-for-the-moment strategy can produce underdevelopment, demolition, reconstruction, and conflict over historic preservation as development capacity priorities, housing needs, and profit potential change over time. The second constraint is much easier to address and is shown as a public alley that bisects the property in Fig. 10.1. This alley could make reallocation of land areas for expanded parking structure cover difficult, but it is often possible to vacate or relocate alleys. Such extraordinary measures are not normally pursued, however, if the owner's need can be met with fewer complications.

Exhibits 10.2 and 10.3 have been prepared to demonstrate the concept of ultimate development capacity by using the Federal Office building as an example. Table 10.1 defined the development capacity index (SFAC) of the project constructed as 178,190 sq ft of building area per buildable acre. Exhibit 10.1 forecasts that this SFAC capacity index could have increased to 258,872 gross sq ft of building area per buildable acre if a 15-story building had been constructed with an S1 adjacent parking garage system and the allocation between building footprint area and parking footprint area had been adjusted. Exhibit 10.2 forecasts that a 7-story building covering the entire site would require a 5-level S2 underground parking garage system. This underground parking approach would increase the SFAC development capacity index to 304,920 gross sq ft of build-

Exhibit 10.2 Federal Office Building S2 Parking System Comparison Forecast

Forecast Model CS2L

*Development capacity forecast for **NONRESIDENTIAL BUILDINGS** based on the use of an **UNDERGROUND PARKING STRUCTURE.***

***Given:** Gross land area. **To Find:** Maximum development capacity of the land area given (gross building area potential) based on the design specification values entered below.*
***Design Premise:** Underground parking footprint may be larger, smaller, or equal to the building footprint above. All similar floors considered equal in area.*

DESIGN SPECIFICATION

Enter values in boxed areas where text is bold and blue. Express all fractions as decimals.

Given:	**Gross land area**	GLA=	1.549	acres	67,474	SF
Land Variables:	Public/ private right-of-way & paved easements	W=	0.084	fraction of GLA	5,668	SF
	Net land area	NLA=	1.419	acres	61,807	SF
	Future expansion and/or unbuildalbe areas	U=	0.000	fraction of GLA	0	SF
	Gross lot area reduction	X=	0.084	fraction of GLA	5,668	SF
	Buildable lot area remaining	BLA=	1.419	acres	61,807	SF
Parking Variables:	Estimated net pkg. structure area per parking space	s =	340.0	SF		
	Building SF permitted per parking space	a =	492.1	SF		
	No. of loading spaces	l =	0			
	Gross area per loading space	b =	0	SF	0	SF
Site Variables at Grade:	**Project open space as fraction of BLA**	S=	0.000		0	SF
	Private driveways as fraction of BLA	R=	0.000		0	SF
	Misc. pavement as fraction of BLA	M=	0.000		0	SF
	Loading area as fraction of BLA	L=	0.000		0	SF
	Total site support areas at grade as a fraction of BLA	Su=	0.000		0	SF
Core:	**Core development area as fraction of BLA**	C=	1.000	C+Su must = 1	61,807	SF
Below Grade:	Gross underground parking (UNG) as fraction of BLA	G=	1.000			
	Parking support within parking structure as fraction of UNG	Pu=	0.050			
	Net pkg. structure area for parking & circulation as fraction of UNG	Pe=	0.950	Pe+Pu must = 1		

PLANNING FORECAST

NOTE: p=1 is one level below grade and is not a surface parking lot based on design premise

no. of bldg flrs. FLR	no. of pkg levels p	development area CORE at grade	gross building area GBA	net pkg area NPA	pkg spaces NPS	gross pkg area GPA	bldg SF / acre SFAC	flr area ratio FAR function of BLA
1.00	0.73	61,807	61,807	42,703	126	44,951	43,560	1.000
2.00	1.45		123,613	85,406	251	89,901	87,120	2.000
3.00	2.18		185,420	128,110	377	134,852	130,680	3.000
4.00	2.91		247,226	170,813	502	179,803	174,240	4.000
5.00	3.64		309,033	213,516	628	224,754	217,800	5.000
6.00	4.36		370,840	256,219	754	269,704	261,360	6.000
7.00	**5.09**		**432,646**	**298,922**	**879**	**314,655**	**304,920**	**7.000**
8.00	5.82	footprint at grade	494,453	341,626	1,005	359,606	348,480	8.000
9.00	6.55	BCA = CORE	556,259	384,329	1,130	404,557	392,040	9.000
10.00	7.27	61,807	618,066	427,032	1,256	449,507	435,600	10.000
11.00	8.00		679,872	469,735	1,382	494,458	479,160	11.000
12.00	8.73	net pkg. area / underground level	741,679	512,438	1,507	539,409	522,720	12.000
13.00	9.45	NPL	803,486	555,141	1,633	584,359	566,280	13.000
14.00	10.18	58,716	865,292	597,845	1,758	629,310	609,840	14.000
15.00	10.91		927,099	640,548	1,884	674,261	653,400	15.000

WARNING: These are preliminary forecasts that must not be used to make final decisions.
1) These forecasts are not a substitute for the "due diligence" research that must be conducted to support the final definition of "unbuildable areas" above and the final decision to purchase land. This research includes, but is not limited to, verification of adequate subsurface soil, zoning, environmental clearance, access, title, utilities and water pressure, clearance from deed restriction, easement and right-of-way encumbrances, clearance from existing above and below ground facility conflicts, etc.
2) The most promising forecast(s) made on the basis of data entered in the design specification from "due diligence" research must be verified at the drawing board before funds are committed and land purchase decisions are made. Actual land shape ratios, dimensions and irregularities encountered may require adjustments to the general forecasts above.
3) The software licensee shall take responsibility for the design specification values entered and any advice given that is based on the forecast produced.

Exhibit 10.3 Federal Office Building S2 Parking System Comparison Forecast

Forecast Model CS2L

Development capacity forecast for ***NONRESIDENTIAL BUILDINGS*** *based on the use of an* ***UNDERGROUND PARKING STRUCTURE.***

Given: *Gross land area.* ***To Find:*** *Maximum development capacity of the land area given (gross building area potential) based on the design specification values entered below.*
Design Premise: *Underground parking footprint may be larger, smaller or equal to the building footprint above. All similar floors considered equal in area.*

DESIGN SPECIFICATION

Enter values in boxed areas where text is bold and blue. Express all fractions as decimals.

Given:	**Gross land area**	GLA=	**1.549**	acres	67,474	SF
Land Variables:	Public/ private right-of-way & paved easements	W=	**0.084**	fraction of GLA	5,668	SF
	Net land area	NLA=	1.419	acres	61,807	SF
	Future expansion and/or unbuildalbe areas	U=	**0.000**	fraction of GLA	0	SF
	Gross lot area reduction	X=	0.084	fraction of GLA	5,668	SF
	Buildable lot area remaining	BLA=	1.419	acres	61,807	SF
Parking Variables:	Estimated net pkg. structure area per parking space	s =	**340.0**	SF		
	Building SF permitted per parking space	a =	**492.1**	SF		
	No. of loading spaces	l =	**0**			
	Gross area per loading space	b =	**0**	SF	0	SF
Site Variables at Grade:	**Project open space as fraction of BLA**	S=	**0.000**		0	SF
	Private driveways as fraction of BLA	R=	**0.000**		0	SF
	Misc. pavement as fraction of BLA	M=	**0.000**		0	SF
	Loading area as fraction of BLA	L=	0.000		0	SF
	Total site support areas at grade as a fraction of BLA	Su=	0.000		0	SF
Core:	**Core development area as fraction of BLA**	C=	1.000	C+Su must = 1	61,807	SF
Below Grade:	Gross underground parking (UNG) as fraction of BLA	G=	**1.000**			
	Parking support within parking structure as fraction of UNG	Pu=	**0.050**			
	Net pkg. structure area for parking & circulation as fraction of UNG	Pe=	0.950	Pe+Pu must = 1		

PLANNING FORECAST

NOTE: p=1 is one level below grade and is not a surface parking lot based on design premise

no. of bldg flrs. **FLR**	no. of pkg levels **p**	development area CORE at grade	gross building area **GBA**	net pkg area NPA	pkg spaces NPS	gross pkg area GPA	bldg SF / acre SFAC	flr area ratio FAR function of BLA
1.00	0.73	61,807	61,807	42,703	126	44,951	43,560	1.000
2.00	1.45		123,613	85,406	251	89,901	87,120	2.000
3.00	2.18		185,420	128,110	377	134,852	130,680	3.000
4.00	2.91		247,226	170,813	502	179,803	174,240	4.000
5.00	3.64		309,033	213,516	628	224,754	217,800	5.000
6.00	4.36		370,840	256,219	754	269,704	261,360	6.000
7.00	5.09		432,646	298,922	879	314,655	304,920	7.000
8.00	5.82	footprint at grade	494,453	341,626	1,005	359,606	348,480	8.000
9.00	6.55	BCA = CORE	556,259	384,329	1,130	404,557	392,040	9.000
10.00	7.27	61,807	618,066	427,032	1,256	449,507	435,600	10.000
11.00	8.00		679,872	469,735	1,382	494,458	479,160	11.000
12.00	8.73	net pkg. area / underground level	741,679	512,438	1,507	539,409	522,720	12.000
13.00	9.45	NPL	803,486	555,141	1,633	584,359	566,280	13.000
14.00	10.18	58,716	865,292	597,845	1,758	629,310	609,840	14.000
15.00	**10.91**		**927,099**	**640,548**	**1,884**	**674,261**	**653,400**	**15.000**

WARNING: These are preliminary forecasts that must not be used to make final decisions.
1) These forecasts are not a substitute for the "due diligence" research that must be conducted to support the final definition of "unbuildable areas" above and the final decision to purchase land. This research includes, but is not limited to, verification of adequate subsurface soil, zoning, environmental clearance, access, title, utilities and water pressure, clearance from deed restriction, easement and right-of-way encumbrances, clearance from existing above and below ground facility conflicts, etc.
2) The most promising forecast(s) made on the basis of data entered in the design specification from "due diligence" research must be verified at the drawing board before funds are committed and land purchase decisions are made. Actual land shape ratios, dimensions and irregularities encountered may require adjustments to the general forecasts above.
3) The software licensee shall take responsibility for the design specification values entered and any advice given that is based on the forecast produced.

ing area per buildable acre. Exhibit 10.3 shows that a 15-story building would require a 10.9-level garage and would produce an SFAC capacity index of 653,400 gross sq ft of building area per buildable acre. Exhibit 10.4 forecasts that an S3 parking garage system with 8 parking levels above grade over the entire Federal Office site could serve 11 building floors above the garage and produce an SFAC development capacity index of 479,155 gross sq ft of building area per buildable acre. Exhibit 10.5 shows that if the parking levels were increased to 15, the building floors above could increase to 20.6 and the SFAC capacity index would increase to 898,415 gross sq ft of building area per buildable acre.

All of these forecasts are based on the same design specification, but different parking structure systems have been chosen to illustrate the range of development capacity potential available on the same site. All of these forecasts are also based on the assumption that no open space is provided in order to compare ultimate development capacity. This is not meant to imply that this is desirable. It is simply meant to illustrate the range of possibilities that are available and the implications this has for sheltering an expanding population. Design specifications may be written in response to these forecasts in order to harness this capacity to reduce land consumption, increase housing capacity, improve conservation, and produce the quality of life we desire within the communities we build. The three examples in this chapter illustrate how design priorities can influence potential capacity. Judging from the project context records measured, the priorities that limited capacity in these three cases were those suggested in Table 10.4.

The Statehouse development capacity in Table 10.2 is less than twice that of a typical suburban office project, even though it uses an underground parking garage and is located in the heart of a major city. This results from the

Exhibit 10.4 Federal Office Buildng S3 Parking System Comparison Forecast

Forecast Model CS3L

Development capacity forecast for ***NONRESIDENTIAL BUILDINGS*** *based on the use of a* ***PARKING STRUCTURE*** *either partially or completely* ***ABOVE*** *grade* ***UNDER*** *the building or buildings.*

Given: *Gross land area.* ***To Find:*** *Maximum development capacity (given building area potential) of the land area given based on the design specification values entered below.*
Design Premise: *Parking footprint is beneath building footprint. The footprints may equal or exceed each other within the core development area. All similar floors considered equal in area.*

DESIGN SPECIFICATION

Enter values in boxed areas where text is bold and blue. Express all fractions as decimals.

Given:	**Gross land area**	GLA=	**1.549**	acres	67,474	SF
Land Variables:	Public/ private right-of-way & paved easements	W=	**0.084**	fraction of GLA	5,668	SF
	Net land area	NLA=	1.419	acres	61,807	SF
	Future expansion and/or unbuildalbe areas	U=	**0.000**	fraction of GLA	0	SF
	Gross land area reduction	X=	0.084	fraction of GLA	5,668	SF
	Buildable land area remaining	**BLA=**	1.419	acres	61,807	SF
Parking Variables:	Est. net pkg. structure area per pkg. space	s =	**340**	SF		
	Building SF permitted per Parking Space	a =	**492.1**	SF		
	No. of loading spaces	l=	**0**			
	Gross area per loading space	b =	**0**	SF	0	SF
Site Variables:	**Project open space as fraction of BLA**	**S=**	**0.000**		0	SF
	Private driveways as fraction of BLA	R=	**0.000**		0	SF
	Misc. pavement as fraction of BLA	M=	**0.000**		0	SF
	Loading area as fraction of BLA	L=	0.000		0	SF
	Total support areas as a fraction of BLA	Su=	0.000		0	SF
Core:	**Core development area as fraction of BLA**	**C=**	1.000	C+Su must = 1	61,807	SF
Pkg. Structure Variables:	Gross pkg. struc. cover as fraction of core area (C)	P=	**1.000**	DO NOT EXCEED 1.0		
	Parking support as fraction of GPA	Pu=	**0.050**			
	Net area for parking & circulation as fraction of GPA	Pe=	0.950	Pe+Pu must = 1	58,716	SF
Building Variable:	Bldg. footprint over pkg. structure as fraction of C	B=	**1.000**	DO NOT EXCEED 1.0		

PLANNING FORECAST

no. of pkg. levels **p**	development area CORE	total pkg. struc GPA area all levels	net pkg. area NPA all levels	pkg. spaces NPS	gross bldg. **GBA** area	no. of bldg. **FLR** floors	total floors **F** pkg + bldg	bldg SF / acre SFAC	flr area ratio FAR function of BLA
NOTE: p <= 1 = grade parking lot beneath building									
1.00	61,807	61,807	58,716	172.7	84,983	1.4	2.4	59,894	1.375
2.00		123,613	117,433	345.4	169,966	2.7	4.7	119,789	2.750
3.00	footprint	185,420	176,149	518.1	254,949	4.1	7.1	179,683	4.125
4.00	BCA	247,226	234,865	690.8	339,933	5.5	9.5	239,577	5.500
5.00	61,807	309,033	293,581	863.5	424,916	6.9	11.9	299,472	6.875
6.00		370,840	352,298	1036.2	509,899	8.2	14.2	359,366	8.250
7.00		432,646	411,014	1208.9	594,882	9.6	16.6	419,261	9.625
8.00		**494,453**	**469,730**	**1381.6**	**679,865**	**11.0**	**19.0**	**479,155**	**11.000**
9.00		556,259	528,446	1554.3	764,848	12.4	21.4	539,049	12.375
10.00	gross pkg. area/ level	618,066	587,163	1726.9	849,831	13.7	23.7	598,944	13.750
11.00	GPL	679,872	645,879	1899.6	934,815	15.1	26.1	658,838	15.125
12.00	61,807	741,679	704,595	2072.3	1,019,798	16.5	28.5	718,732	16.500
13.00	net pkg area/ level	803,486	763,311	2245.0	1,104,781	17.9	30.9	778,627	17.875
14.00	NPL	865,292	822,028	2417.7	1,189,764	19.2	33.2	838,521	19.250
15.00	58,716	927,099	880,744	2590.4	1,274,747	20.6	35.6	898,415	20.625

WARNING: These are preliminary forecasts that must not be used to make final decisions.
1) These forecasts are not a substitute for the "due diligence" research that must be conducted to support the final definition of "unbuildable areas" above and the final decision to purchase land. This research includes, but is not limited to, verification of adequate subsurface soil, zoning, environmental clearance, access, title, utilities and water pressure, clearance from deed restriction, easement and right-of-way encumbrances, clearance from existing above and below ground facility conflicts, etc.
2) The most promising forecast(s) made on the basis of data entered in the design specification from "due diligence" research must be verified at the drawing board before funds are committed and land purchase decisions are made. Actual land shape ratios, dimensions and irregularities encountered may require adjustments to the general forecasts above.
3) The software licensee shall take responsibility for the design specification values entered and any advice given that is based on the forecast produced.

Exhibit 10.5 Federal Office Buildng S3 Parking System Comparison Forecast

Forecast Model CS3L

Development capacity forecast for ***NONRESIDENTIAL BUILDINGS*** *based on the use of a* ***PARKING STRUCTURE*** *either partially or completely* ***ABOVE*** *grade* ***UNDER*** *the building or buildings.*

Given: *Gross land area.* ***To Find:*** *Maximum development capacity (given building area potential) of the land area given based on the design specification values entered below.*
Design Premise: *Parking footprint is beneath building footprint. The footprints may equal or exceed each other within the core development area. All similar floors considered equal in area.*

DESIGN SPECIFICATION

Enter values in boxed areas where text is bold and blue. Express all fractions as decimals.

Given:	**Gross land area**	GLA=	1.549	acres	67,474	SF
Land Variables:	Public/ private right-of-way & paved easements	W=	0.084	fraction of GLA	5,668	SF
	Net land area	NLA=	1.419	acres	61,807	SF
	Future expansion and/or unbuildalbe areas	U=	0.000	fraction of GLA	0	SF
	Gross land area reduction	X=	0.084	fraction of GLA	5,668	SF
	Buildable land area remaining	BLA=	1.419	acres	61,807	SF
Parking Variables:	Est. net pkg. structure area per pkg. space	s =	340	SF		
	Building SF permitted per Parking Space	a =	492.1	SF		
	No. of loading spaces	l=	0			
	Gross area per loading space	b =	0	SF	0	SF
Site Variables:	**Project open space as fraction of BLA**	S=	0.000		0	SF
	Private driveways as fraction of BLA	R=	0.000		0	SF
	Misc. pavement as fraction of BLA	M=	0.000		0	SF
	Loading area as fraction of BLA	L=	0.000		0	SF
	Total support areas as a fraction of BLA	Su=	0.000		0	SF
Core:	**Core development area as fraction of BLA**	C=	1.000	C+Su must = 1	61,807	SF
Pkg. Structure Variables:	Gross pkg. struc. cover as fraction of core area (C)	P=	1.000	DO NOT EXCEED 1.0		
	Parking support as fraction of GPA	Pu=	0.050			
	Net area for parking & circulation as fraction of GPA	Pe=	0.950	Pe+Pu must = 1	58,716	SF
Building Variable:	Bldg. footprint over pkg. structure as fraction of C	B=	1.000	DO NOT EXCEED 1.0		

PLANNING FORECAST

no. of pkg. levels **p**	development area CORE	total pkg. struc GPA area all levels	net pkg. area NPA all levels	pkg. spaces NPS	gross bldg. **GBA** area	no. of bldg. **FLR** floors	total floors **F** pkg + bldg	bldg SF / acre SFAC	flr area ratio FAR function of BLA
NOTE: p <= 1 = grade parking lot beneath building									
1.00	61,807	61,807	58,716	172.7	84,983	1.4	2.4	59,894	1.375
2.00		123,613	117,433	345.4	169,966	2.7	4.7	119,789	2.750
3.00	footprint	185,420	176,149	518.1	254,949	4.1	7.1	179,683	4.125
4.00	BCA	247,226	234,865	690.8	339,933	5.5	9.5	239,577	5.500
5.00	61,807	309,033	293,581	863.5	424,916	6.9	11.9	299,472	6.875
6.00		370,840	352,298	1036.2	509,899	8.2	14.2	359,366	8.250
7.00	gross pkg. area/ level	432,646	411,014	1208.9	594,882	9.6	16.6	419,261	9.625
8.00	GPL	494,453	469,730	1381.6	679,865	11.0	19.0	479,155	11.000
9.00	61,807	556,259	528,446	1554.3	764,848	12.4	21.4	539,049	12.375
10.00	net pkg area/ level	618,066	587,163	1726.9	849,831	13.7	23.7	598,944	13.750
11.00	NPL	679,872	645,879	1899.6	934,815	15.1	26.1	658,838	15.125
12.00	58,716	741,679	704,595	2072.3	1,019,798	16.5	28.5	718,732	16.500
13.00		803,486	763,311	2245.0	1,104,781	17.9	30.9	778,627	17.875
14.00		865,292	822,028	2417.7	1,189,764	19.2	33.2	838,521	19.250
15.00		**927,099**	**880,744**	**2590.4**	**1,274,747**	**20.6**	**35.6**	**898,415**	**20.625**

WARNING: These are preliminary forecasts that must not be used to make final decisions.
1) These forecasts are not a substitute for the "due diligence" research that must be conducted to support the final definition of "unbuildable areas" above and the final decision to purchase land. This research includes, but is not limited to, verification of adequate subsurface soil, zoning, environmental clearance, access, title, utilities and water pressure, clearance from deed restriction, easement and right-of-way encumbrances, clearance from existing above and below ground facility conflicts, etc.
2) The most promising forecast(s) made on the basis of data entered in the design specification from "due diligence" research must be verified at the drawing board before funds are committed and land purchase decisions are made. Actual land shape ratios, dimensions and irregularities encountered may require adjustments to the general forecasts above.
3) The software licensee shall take responsibility for the design specification values entered and any advice given that is based on the forecast produced.

Table 10.4 Urban Design Priorities

Priorities	Capacity Index	Intensity Index	Project Open Space Provided
Statehouse (S2) Public image, public open space, public parking	19,981	2.338	66.2%
IBM (S3) Corporate image, public open space, office capacity, private parking	132,145	5.678	32.2%
Federal Office (S1) Office capacity, private parking, cost containment	178,190	8.000	0.0%

project priorities suggested in Table 10.4 that produce 66.2% open space. The IBM building in Table 10.3 produces much greater development capacity with a blend of image, office area, and open space (32.2%). The Federal Office building in Table 10.1 produces the greatest development capacity, focuses on office space, provides no open space, and compensates with an Indiana limestone appearance. Even though the Federal Office building consumes all available land for development cover, it has much greater potential development capacity than that constructed because of the parking structure involved. In fact, the development capacity of all three is much greater than that constructed and has been limited by a variety of public and private priorities, some of which are suggested in Table 10.4. The context and intensity that has resulted from these priorities may be inadvertent, and unacceptable to some, but it is possible to translate priorities into environmental results using context record comparisons, development capacity forecasting, and design specification limits.

It is actually a simple set of relationships that may sound overly complex because of the detail involved. A given land area has many development options that are a function of the parking system chosen and the design specification written. Some options will produce desirable lifestyles and a pleasant quality of life while others will produce overdevelopment. The entire spectrum of development capacity options available represents context potential. Design specifications can define varying levels of intensity that limit context potential and produce context variety and capacity. Context variety shapes lifestyle alternatives. These alternatives combine to form cities and establish population capacity. Context intensity can enhance or detract from the quality of life associated with the lifestyles within, and design specifications can be used to define acceptable context environments at all reasonable levels of intensity. Context record comparison and development capacity forecasting can be used to evaluate the results produced by design specification values. The entire process can be used to plan and build a sustainable habitat through the quantitative methods of development capacity evaluation.

Residential Parking Structure Comparisons

Two residential apartment projects have been selected to illustrate how forecast models RS1L and RS2L can predict the advantages of residential parking structure systems. These advantages have a price, but they can improve our ability to offer housing options and produce building areas that shelter diverse activities and populations in more compact, sustainable patterns. Figure 10.4 illustrates the 24-story Summit Chase residential condominium project. It combines an underground parking structure with a surface parking lot to produce 16.2% development cover and 83.8% open space. Figure 10.5 illustrates the exclusive 19-story Waterford residential housing project. It uses an adjacent parking structure to produce 79.1% development cover and 20.9% open space. The Summit Chase project emphasizes open space. The Waterford project emphasizes high-density exclusivity and minimal open space. Tables 10.5 and 10.6 present the context records for each. The sum-

(a)

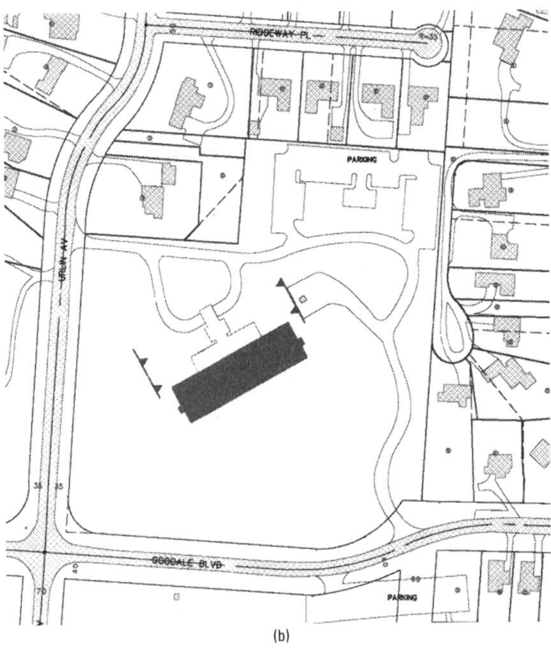

(b)

(c)

Figure 10.4 Summit Chase—S2 underground parking system. (*a*) This high-rise residential apartment supplements surface parking with an underground parking structure to conserve project open space for residential enjoyment. (*b*) Site plan. (*c*) Overview.

(Photos by W. M. Hosack)

mary at the bottom of each record shows that Summit Chase produced 23.05 dwelling units and 50,038 gross sq ft of building area per buildable acre, while Waterford produced 145.90 dwelling units and 259,878 gross sq ft of building area per buildable acre. The aggregate average dwelling unit at Summit Chase is 2171 sq ft in area, and Waterford pro-

(a)

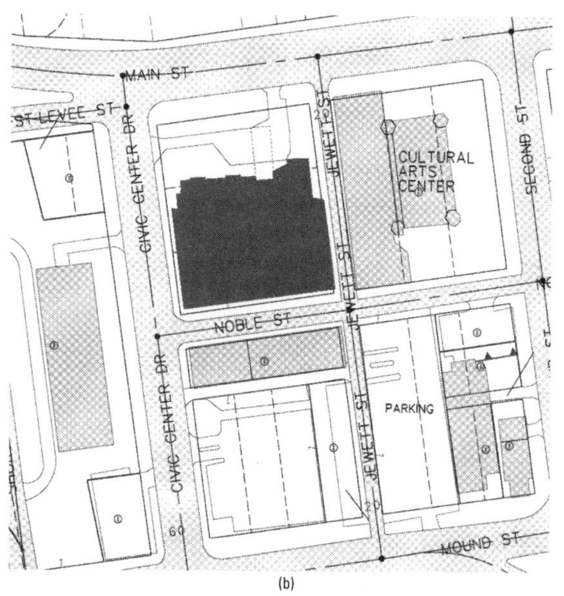

(b)

(c)

Figure 10.5 Waterford—S1 adjacent parking structure system. (*a*) An adjacent parking garage and low parking ratio is combined with increased building height and modest on-site open space to produce greater residential housing capacity. (*b*) Site plan. (*c*) Streetview.

(Photos by B. Higgins)

Table 10.5 Summit Chase Context Record

Name	**Summit Chase**
Address or Location	**1000 Urlin Avenue**
Land Use	**Apartmenthouse**

Parking System	*check box*
Structure parking adjacent to building	
Structure parking underground	x
Structure parking above grade under building	
Notes	Also valet surface parking

	Description	Project Measurements		Context Summary SF	%	
Site						
	Gross land area in acres	7.814	GLA	340,378		
	Public/ private ROW and paved easements in SF	0	W		0.0%	of GLA
	Net lot area		NLA	340,378	100.0%	of GLA
	Total unbuildable area in SF	4,000	U		1.2%	of GLA
	Water area within unbuildable area in SF	0	WAT			of U
	Buildable land area		BLA	336,378	98.8%	of GLA
Development Cover						
	Gross building area in SF	386,400	GBA			
	Total building footprint in SF	16,100	B		4.8%	of BLA
	Total bldg. support in SF (stairs, corridors, elevators, etc.)		BSU			
	Building efficiency		Be			of GBA
Residential only	Number of dwelling units	178	DU			
	Aggregate average dwelling unit area in SF		AGG	2,171		
	Number of building floors	24.00	FLR			
G2 only	Unenclosed surface parking area under building in SF	0	PUB			
	Total surface parking area in SF**	27,600	P or G		8.2%	of BLA
	Number of surface parking spaces on premise	78	NPSs			
	Building cover over surface parking area		AIR			of GPA
	Gross surface parking area per parking space in SF		ss	354		
	Gross building area per surface parking space in SF		as	4,954		
	Surface parking spaces per dwelling unit		us	0.44		
Pkg. struc. only	Total parking structure footprint in SF	33,600	P		0.0%	of BLA
Pkg. struc. only	Parking structure area below grade in SF	33,600	UNG			
Pkg. struc. only	Parking structure area above grade under building in SF	0	PSC			
Pkg. struc. only	Number of parking structure floors	2	p			
Pkg. struc. only	Total parking structure area in SF	67,200	GPA			
Pkg. struc. only	Total parking structure spaces	188	NPSg			
Pkg. struc. only	Total pkg. struc. support in SF (stairs, elevators, etc.)		PSS			
Pkg. struc. only	Parking structure support percentage		Pu			of GPA
Pkg. struc. only	Gross parking structure area per space in SF		sg	357		
Pkg. struc. only	Gross building area per structure parking space in SF		ag	2,055		
Pkg. struc. only	Parking structure spaces per dwelling unit		ug	1.06		
Residential only	Total dwelling unit garage area in SF		GPA			
Residential only	Dwelling unit garage above grade under builiding in SF		GUB			
Residential only	Number of dwelling unit garage parking spaces		GPS			
Residential only	Garage parking spaces per dwelling		Gn	0.0		
Residential only	Total dwelling unit garage area per space in SF		Ga			
Residential only	Total dwelling unit garage cover in addition to building cover		Gc		0.0%	of BLA
	Total parking spaces		NPS	266.0		
Residential only	Total parking spaces per dwelling unit		u	1.5		
	Gross building area per parking space total		a	1,452.6		
	Number of loading spaces	1	l			
	Total loading area in SF	1,700	LDA			
	Gross area per loading space in SF		b	1,700		
	Loading area percentage		L		0.5%	of BLA
	Driveway areas in SF	7,900	R		2.3%	of BLA
	Misc. pavement and social center pvmt. in SF	1,185	M		0.4%	of BLA
	TOTAL DEVELOPMENT COVER		D	54,485	**16.2%**	of BLA
Project Open Space						
	PROJECT OPEN SPACE		S	281,893	**83.8%**	of BLA
	Water area within project open space in SF		wat			of S
Summary		0.0%				
	Development balance index		BNX		0.162	
	Development intensity index		INX		24.162	
	Floor area ratio per buildable acre		BFAR		1.149	
	Capacity index (gross building sq. ft. per buildable acre)		SFAC		50,038	
	Density per buildable acre if applicable		dBA		23.05	
	Density per net acre if applicable		dNA		22.78	
	Density per gross acre if applicable		dGA		22.78	

** Include internal parking lot landscaping, circulation aisles, and private roadway parking space areas. Express private roads separately as driveway areas..

Table 10.6 Waterford Context Record

Name	**Waterford Tower**
Address or Location	**155 W. Main Street**
Land Use	**Apartmenthouse**

Parking System	check box
Structure parking adjacent to building	x
Structure parking underground	
Structure parking above grade under building	
Notes	

		Description	Project Measurements		Context Summary SF	Context Summary %	
Site							
		Gross land area in acres	**0.658**	GLA	28,662		
		Public/ private ROW and paved easements in SF	**0**	W		0.0%	of GLA
		Net lot area		NLA	28,662	100.0%	of GLA
		Total unbuildable area in SF	**0**	U		0.0%	of GLA
		Water area within unbuildable area in SF	**0**	WAT			of U
		Buildable land area		BLA	28,662	100.0%	of GLA
Development Cover							
		Gross building area in SF	**171,000**	GBA			
		Total building footprint in SF	**9,000**	B		31.4%	of BLA
		Total bldg. support in SF (stairs, corridors, elevators, etc.)		BSU			
		Building efficiency		Be			of GBA
	Residential only	Number of dwelling units	**96**	DU			
		Aggregate average dwelling unit area in SF		AGG	1,781		
		Number of building floors	**19.00**	FLR			
	G2 only	Unenclosed surface parking area under building in SF		PUB			
		Total surface parking area in SF**		P or G		0.0%	of BLA
		Number of surface parking spaces on premise		NPSs			
		Building cover over surface parking area		AIR			of GPA
		Gross surface parking area per parking space in SF		ss			
		Gross building area per surface parking space in SF		as			
		Surface parking spaces per dwelling unit		us	0.00		
	Pkg. struc. only	Total parking structure footprint in SF	9,000	P		31.4%	of BLA
	Pkg. struc. only	Parking structure area below grade in SF	0	UNG			
	Pkg. struc. only	Parking structure area above grade under building in SF	0	PSC			
	Pkg. struc. only	Number of parking structure floors	3	p			
	Pkg. struc. only	Total parking structure area in SF	27,000	GPA			
	Pkg. struc. only	Total parking structure spaces	75	NPSg			
	Pkg. struc. only	Total pkg. struc. support in SF (stairs, elevators, etc.)	1,350	PSS			
	Pkg. struc. only	Parking structure support percentage		Pu		5.0%	of GPA
	Pkg. struc. only	Gross parking structure area per space in SF		sg	360		
	Pkg. struc. only	Gross building area per structure parking space in SF		ag	2,280		
	Pkg. struc. only	Parking structure spaces per dwelling unit		ug	0.78		
	Residential only	Total dwelling unit garage area in SF		GPA			
	Residential only	Dwelling unit garage above grade under builiding in SF		GUB			
	Residential only	Number of dwelling unit garage parking spaces		GPS			
	Residential only	Garage parking spaces per dwelling		Gn	0.0		
	Residential only	Total dwelling unit garage area per space in SF		Ga			
	Residential only	Total dwelling unit garage cover in addition to building cover		Gc		0.0%	of BLA
		Total parking spaces		NPS	75.0		
	Residential only	Total parking spaces per dwelling unit		u	0.8		
		Gross building area per parking space total		a	2,280.0		
		Number of loading spaces		l			
		Total loading area in SF		LDA			
		Gross area per loading space in SF		b			
		Loading area percentage		L		0.0%	of BLA
		Driveway areas in SF	**4,430**	R		15.5%	of BLA
		Misc. pavement and social center pvmt. in SF	**250**	M		0.9%	of BLA
		TOTAL DEVELOPMENT COVER		D	22,680	**79.1%**	of BLA
Project Open Space							
		PROJECT OPEN SPACE		S	5,982	**20.9%**	of BLA
		Water area within project open space in SF	0.0%	wat			of S
Summary							
		Development balance index		BNX		0.791	
		Development intensity index		INX		19.791	
		Floor area ratio per buildable acre		BFAR		5.966	
		Capacity index (gross building sq. ft. per buildable acre)		SFAC		259,878	
		Density per buildable acre if applicable		dBA		145.90	
		Density per net acre if applicable		dNA		145.90	
		Density per gross acre if applicable		dGA		145.90	

** Include internal parking lot landscaping, circulation aisles, and private roadway parking space areas. Express private roads separately as driveway areas..

vides 1781 sq ft.[7] The average Waterford dwelling unit is not only smaller but also located in a CBD, while the Summit Chase dwelling unit is larger and located in a first-ring suburb.

Exhibits 10.6 and 10.7 are forecast models that contain the design specifications used by each project. These specifications were derived from the measurements entered in context records, Tables 10.5 and 10.6. A change to any value in either specification will produce a different density dBA and development capacity index SFAC forecast. This serves to illustrate that these indexes do not produce context results. They record the results produced by design specification values and cannot be expected to be an adequate substitute that will produce consistent results.

By now, you are familiar with the process of changing design specification values to produce alternate development capacity forecasts. If the Summit Chase emphasis on open space had been reduced from 83.8% to 70%, Exhibit 10.8 forecasts a density increase to 89.5 dwelling units per buildable acre dBA. It also forecasts a parking level increase to 7.76 levels, assuming all other design specification values remain constant. This 14% open space adjustment may or may not affect the context that results, but it quadruples the density produced. The project could also be made more feasible by increasing the underground parking structure area and reducing the number of parking levels required.

If the Waterford open space provision had been increased from 20.9% to 30%, Exhibit 10.9 forecasts a reduction in density from 145.95 to 124.77 dBA and a reduction in the number of dwelling units from 96 to 82. This forecast assumes all other design specification values remain constant as shown, including building height. The open space increase reduces development capacity and slightly changes the potential context involved. The degree to which this reduction is desirable is a function of the neighborhood context plan and the investment forecast of the developer. The

Exhibit 10.6 Summit Chase Forecast Record

Forecast Model RS2L

Development capacity forecast for ***APARTMENTS*** *based on the use of an* ***UNDERGROUND PARKING STRUCTURE.***
Given: *Gross land area.* ***To Find:*** *Dwelling unit capacity of the land area given based on the design specification values entered below.*
Design Premise: *Underground parking footprint may be larger, smaller, or equal to the building footprint above. All similar floors considered equal in area.*

DESIGN SPECIFICATION

Enter values in boxed areas where text is bold and blue. Express all fractions as decimals.

Given:	**Gross land area**	GLA=	**7.814**	acres	340,378	SF
Land Variables:	Public/ private right-of-way & paved easements	W=	**0.000**	fraction of GLA	0	SF
	Net land area	NLA=	7.814	acres	340,378	SF
	Future expansion and/or unbuildalbe areas	U=	**0.012**	fraction of GLA	3,982	SF
	Gross land area reduction	X=	0.012	fraction of GLA	3,982	SF
	Buildable land area remaining	BLA=	**7.723**	acres	336,395	SF
Parking Variables:	Est. net parking structure area per parking space	s =	**321.7**	SF		
*	Structure parking spaces provided per dwelling unit	u =	**1.056**			
	No. of loading spaces	l =	**1**			
	Gross area per loading space	b =	**1,700**		1,700	SF
Site Variables at Grade:	**Project open space as fraction of BLA**	S=	**0.838**		281,899	SF
	Private driveways as fraction of BLA	R=	**0.023**		7,737	SF
	Misc. pavement as fraction of BLA (includes surface parking area)	M=	**0.086**		28,930	SF
	Loading area as fraction of BLA	L=	0.005		1,700	SF
	Total site support areas at grade as a fraction of BLA	Su=	0.952		320,266	SF
Core:	**Core development area at grade as fraction of BLA**	C=	0.048	C+Su must = 1	16,129	SF
Below Grade:	Gross underground parking footprint (UNG) as fraction of BLA	G=	**0.100**		33,640	SF
*	Parking support within parking structure as fraction of UNG	Pu=	**0.100**			
	Net parking structure for pkg. & circulation as fraction of UNG	Pe=	0.900			
Building Variables:	Building efficiency as fraction of GBA	Be=				
	Building support as fraction of GBA	Bu=		Be+Bu must = 1		

Dwelling Unit Mix Table

DU dwelling unit type	GDA gross du area	CDA=GDA/Be comprehensive du area	MIX du mix	PDA = (CDA)MIX Pro-rated du area

Dwelling unit mix not available

Aggregate average dwelling unit area (AGG) found by dividing gross building area by number of dwelling units present

Aggregate avg. dwelling unit area (AGG) = **2,171**

PLANNING FORECAST

no. of bldg floors **FLR**	no. of pkg levels **p**	development area CORE at grade	density dBA per bldable acre	dwelling units **NDU**	pkg spaces NPS	net pkg area NPA	underground GPA incl BSU	gross bldg area **GBA**	gross bldg SF **SFAC** per buildable AC
1.00	0.08	16,129	1.0	7.4	7.8	2,524	2,804	16,129	2,089
2.00	0.17		1.9	14.9	15.7	5,048	5,609	32,258	4,177
3.00	0.25	footprint	2.9	22.3	23.5	7,572	8,413	48,387	6,266
4.00	0.33	BCA = CORE	3.8	29.7	31.4	10,095	11,217	64,516	8,354
5.00	0.42	16,129	4.8	37.1	39.2	12,619	14,021	80,645	10,443
6.00	0.50		5.8	44.6	47.1	15,143	16,826	96,774	12,531
7.00	0.58		6.7	52.0	54.9	17,667	19,630	112,903	14,620
8.00	0.67	pkg struc area / level	7.7	59.4	62.8	20,191	22,434	129,032	16,708
9.00	0.75	NPL	8.7	66.9	70.6	22,715	25,238	145,161	18,797
10.00	0.83	30,276	9.6	74.3	78.5	25,238	28,043	161,290	20,885
15.00	1.25		14.4	111.4	117.7	37,858	42,064	241,934	31,328
20.00	1.67		19.2	148.6	156.9	50,477	56,085	322,579	41,771
24.00	**2.00**		**23.1**	**178.3**	**188.3**	**60,572**	**67,302**	**387,095**	**50,125**
25.00	2.08		24.1	185.7	196.1	63,096	70,107	403,224	52,214
30.00	2.50		28.9	222.9	235.4	75,715	84,128	483,869	62,656

WARNING: These are preliminary forecasts that must not be used to make final decisions.
1) These forecasts are not a substitute for the "due diligence" research that must be conducted to support the final definition of "unbuildable areas" above and the final decision to purchase land. This research includes, but is not limited to, verification of adequate subsurface soil, zoning, environmental clearance, access, title, utilities and water pressure, clearance from deed restriction, easement and right-of-way encumbrances, clearance from existing above and below ground facility conflicts, etc.
2) The most promising forecast(s) made on the basis of data entered in the design specification from "due diligence" research must be verified at the drawing board before funds are committed and land purchase decisions are made. Actual land shape ratios, dimensions and irregularities encountered may require adjustments to the general forecasts above.
3) The software licensee shall take responsibility for the design specification values entered and any advice given or decisions made that are based on the forecast produced.

Exhibit 10.7 Waterford Apartment Forecast Record

Forecast Model RS1L

Development capacity forecast for ***APARTMENTS*** *using an* ***ADJACENT PARKING STRUCTURE*** *on the same premise.*
Given: *Gross land area available.* ***To Find:*** *Maximum dwelling unit capacity of the buildable land area given based on the design specification values entered below.*
Design Premise: *Building footprint adjacent to parking garage footprint within the core development area. All similar floors considered equal in area.*

DESIGN SPECIFICATION *Enter values in boxed areas where text is bold and blue. Express all fractions as decimals.*

Given:	**Gross land area**	GLA=	**0.658**	acres	28,662	SF
Land Variables:	Public/ private right-of-way & paved easements	W=	**0.000**	fraction of GLA	0	SF
	Net land area	NLA=	0.658	acres	28,662	SF
	Unbuildable and/or future expansion areas	U=	**0.000**	fraction of GLA	0	SF
	Gross land area reduction	X=	0.000	fraction of GLA	0	SF
	Buildable land area remaining	BLA=	**0.658**	acres	28,662	SF
Parking Variables:	Estimated net pkg. structure area per parking space	s =	**324**	SF		
	Parking spaces required per dwelling unit	u =	**0.78**			
	No. of parking levels contemplated	p=	**3.00**			
	Parking support as fraction of gross pkg. structure area (GPA)	Pu=	**0.098**			
	Net area for parking & circulation as fraction of GPA	Pe=	0.902	Pe+Pu must = 1		
	No. of loading spaces	l=	**1**			
	Gross area per loading space	b =	**0**	SF		
Site Variables:	**Project open space as fraction of BLA**	S=	**0.209**		5,990	SF
	Private driveways as fraction of BLA	R=	**0.155**		4,443	SF
	Misc. pavement as fraction of BLA	M=	**0.009**		258	SF
	Loading area as fraction of BLA	L=	0.000		0	SF
	Total site support areas as a fraction of BLA	Su=	0.373		10,691	SF
Core:	**Core development area as fraction of BLA**	C=	0.627	C+Su must = 1	17,971	SF
Building Variables:	Building efficiency as fraction of GBA	Be=				
	Building support as fraction of GBA	Bu=		Be+Bu must = 1		

NOTE: p=1 is a grade parking lot based on design premise. Increase the number of parking levels to increase the capacity forecast below. Other variables in blue, box and bold may also be changed.

Dwelling Unit Mix Table

DU dwelling unit type	GDA gross du area	CDA=GDA/Be comprehensive du area	MIX du mix	PDA = (CDA)MIX Pro-rated du area

Dwelling unit mix not available
Aggregate average dwelling unit area (AGG) found by dividing gross building area by number of dwelling units present

Aggregate avg. dwelling unit area	(AGG) =	**1,781**
Gross building area per parking space	(a)=	2,283

PLANNING FORECAST

no. of FLR bldg floors	min land area CORE for BCG & PLA	dwelling units **NDU**	pkg. spaces NPS	net pkg area NPA	gross pkg GPA struc area	pkg. struct cover **GPL** based on p above	gross bldg area GBA	footprint **BCA**	density per dBA bldable acre	gross bldg SF **SFAC** per buildable AC
1.00	17,971	9.6	7.5	2,423	2,686	895	17,076	17,076	14.57	25,951
2.00		18.3	14.2	4,616	5,118	1,706	32,531	16,266	27.76	49,439
3.00		26.2	20.4	6,610	7,329	2,443	46,586	15,529	39.75	70,799
4.00		33.4	26.0	8,432	9,348	3,116	59,422	14,855	50.71	90,306
5.00		40.0	31.2	10,102	11,199	3,733	71,191	14,238	60.75	108,193
6.00		46.1	35.9	11,639	12,903	4,301	82,022	13,670	69.99	124,653
7.00		51.7	40.3	13,058	14,476	4,825	92,022	13,146	78.52	139,850
8.00		56.9	44.4	14,372	15,933	5,311	101,282	12,660	86.43	153,925
9.00		61.7	48.1	15,592	17,286	5,762	109,884	12,209	93.77	166,996
10.00		66.2	51.6	16,729	18,546	6,182	117,893	11,789	100.60	179,168
11.00		70.4	54.9	17,790	19,722	6,574	125,369	11,397	106.98	190,531
12.00		74.3	58.0	18,782	20,823	6,941	132,365	11,030	112.95	201,162
15.00		84.7	66.1	21,411	23,737	7,912	150,887	10,059	128.75	229,312
19.00		**96.0**	**74.9**	**24,271**	**26,907**	**8,969**	**171,042**	**9,002**	**145.95**	**259,943**
20.00		98.5	76.8	24,894	27,599	9,200	175,436	8,772	149.70	266,620

WARNING: These are preliminary forecasts that must not be used to make final decisions.
1) These forecasts are not a substitute for the "due diligence" research that must be conducted to support the final definition of "unbuildable areas" above and the final decision to purchase land. This research includes, but is not limited to, verification of adequate subsurface soil, zoning, environmental clearance, access, title, utilities and water pressure, clearance from deed restriction, easement and right-of-way encumbrances, clearance from existing above and below ground facility conflicts, etc.
2) The most promising forecast(s) made on the basis of data entered in the design specification from "due diligence" research must be verified at the drawing board before funds are committed and land purchase decisions are made. Actual land shape ratios, dimensions and irregularities encountered may require adjustments to the general forecasts above.
3) The software licensee shall take responsibility for the design specification values entered and any advice given or decisions made that are based on the forecast produced.

Exhibit 10.8 Summit Chase Forecast When 70% Open Space Provided

Forecast Model RS2L

Development capacity forecast for ***APARTMENTS*** *based on the use of an* ***UNDERGROUND PARKING STRUCTURE.***
Given: *Gross land area.* ***To Find:*** *Dwelling unit capacity of the land area given based on the design specification values entered below.*
Design Premise: *Underground parking footprint may be larger, smaller, or equal to the building footprint above. All similar floors considered equal in area.*

DESIGN SPECIFICATION

Enter values in boxed areas where text is bold and blue. Express all fractions as decimals.

Given:	**Gross land area**	GLA=	**7.814**	acres	340,378	SF
Land Variables:	Public/ private right-of-way & paved easements	W=	**0.000**	fraction of GLA	0	SF
	Net land area	NLA=	7.814	acres	340,378	SF
	Future expansion and/or unbuildalbe areas	U=	**0.012**	fraction of GLA	3,982	SF
	Gross land area reduction	X=	0.012	fraction of GLA	3,982	SF
	Buildable land area remaining	BLA=	**7.723**	acres	336,395	SF
Parking Variables:	Est. net parking structure area per parking space	s =	**321.7**	SF		
*	Structure parking spaces provided per dwelling unit	u =	**1.056**			
	No. of loading spaces	l =	**1**			
	Gross area per loading space	b =	**1,700**		1,700	SF
Site Variables at Grade:	**Project open space as fraction of BLA**	S=	**0.700**		235,477	SF
	Private driveways as fraction of BLA	R=	**0.023**		7,737	SF
	Misc. pavement as fraction of BLA (includes 78 space surface parking area)	M=	**0.086**		28,930	SF
	Loading area as fraction of BLA	L=	0.005		1,700	SF
	Total site support areas at grade as a fraction of BLA	Su=	0.814		273,844	SF
Core:	**Core development area at grade as fraction of BLA**	C=	0.186	C+Su must = 1	62,552	SF
Below Grade:	Gross underground parking footprint (UNG) as fraction of BLA	G=	**0.100**		33,640	SF
*	Parking support within parking structure as fraction of UNG	Pu=	**0.100**			
	Net parking structure for pkg. & circulation as fraction of UNG	Pe=	0.900			
Building Variables:	Building efficiency as fraction of GBA	Be=				
	Building support as fraction of GBA	Bu=		Be+Bu must = 1		

Dwelling Unit Mix Table

DU dwelling unit type	GDA gross du area	CDA=GDA/Be comprehensive du area	MIX du mix	PDA = (CDA)MIX Prorated du area

Dwelling unit mix not available

Aggregate average dwelling unit area (AGG) found by dividing gross building area by number of dwelling units present

Aggregate avg. dwelling unit area (AGG) = **2,171**

PLANNING FORECAST

no. of bldg floors **FLR**	no. of pkg levels **p**	development area CORE at grade	density dBA per bldable acre	dwelling units **NDU**	pkg spaces NPS	net pkg area NPA	underground GPA incl BSU	gross bldg area **GBA**	gross bldg SF **SFAC** per buildable AC
1.00	0.32	62,552	3.7	28.8	30.4	9,788	10,876	62,552	8,100
2.00	0.65		7.5	57.6	60.9	19,576	21,751	125,103	16,200
3.00	0.97	footprint	11.2	86.4	91.3	29,364	32,627	187,655	24,299
4.00	1.29	BCA = CORE	14.9	115.2	121.7	39,152	43,502	250,206	32,399
5.00	1.62	62,552	18.7	144.1	152.1	48,940	54,378	312,758	40,499
6.00	1.94		22.4	172.9	182.6	58,728	65,253	375,309	48,599
7.00	2.26		26.1	201.7	213.0	68,516	76,129	437,861	56,699
8.00	2.59	pkg struc area / level	29.8	230.5	243.4	78,304	87,004	500,412	64,799
9.00	2.91	NPL	33.6	259.3	273.8	88,092	97,880	562,964	72,898
10.00	3.23	30,276	37.3	288.1	304.3	97,880	108,755	625,515	80,998
15.00	4.85		56.0	432.2	456.4	146,820	163,133	938,273	121,497
20.00	6.47		74.6	576.2	608.5	195,760	217,511	1,251,031	161,997
24.00	**7.76**		**89.5**	**691.5**	**730.2**	**234,912**	**261,013**	**1,501,237**	**194,396**
25.00	8.08		93.3	720.3	760.6	244,699	271,888	1,563,788	202,496
30.00	9.70		111.9	864.4	912.8	293,639	326,266	1,876,546	242,995

WARNING: These are preliminary forecasts that must not be used to make final decisions.
1) These forecasts are not a substitute for the "due diligence" research that must be conducted to support the final definition of "unbuildable aroas" above and the final decision to purchase land. This research includes, but is not limited to, verification of adequate subsurface soil, zoning, environmental clearance, access, title, utilities and water pressure, clearance from deed restriction, easement and right-of-way encumbrances, clearance from existing above and below ground facility conflicts, etc.
2) The most promising forecast(s) made on the basis of data entered in the design specification from "due diligence" research must be verified at the drawing board before funds are committed and land purchase decisions are made. Actual land shape ratios, dimensions and irregularities encountered may require adjustments to the general forecasts above.
3) The software licensee shall take responsibility for the design specification values entered and any advice given or decisions made that are based on the forecast produced.

Exhibit 10.9 Waterford Apartment Forecast when 30% Open Space Provided

Forecast Model RS1L

Development capacity forecast for ***APARTMENTS*** *using an* ***ADJACENT PARKING STRUCTURE*** *on the same premise.*
Given: *Gross land area available.* ***To Find:*** *Maximum dwelling unit capacity of the buildable land area given based on the design specification values entered below.*
Design Premise: *Building footprint adjacent to parking garage footprint within the core development area. All similar floors considered equal in area.*

DESIGN SPECIFICATION — *Enter values in boxed areas where text is bold and blue. Express all fractions as decimals.*

Given:	**Gross land area**	GLA=	**0.658**	acres	28,662	SF
Land Variables:	Public/ private right-of-way & paved easements	W=	**0.000**	fraction of GLA	0	SF
	Net land area	NLA=	0.658	acres	28,662	SF
	Unbuildable and/or future expansion areas	U=	**0.000**	fraction of GLA	0	SF
	Gross land area reduction	X=	0.000	fraction of GLA	0	SF
	Buildable land area remaining	BLA=	**0.658**	acres	28,662	SF
Parking Variables:	Estimated net pkg. structure area per parking space	s =	**324**	SF		
	Parking spaces required per dwelling unit	u =	**0.78**			
	No. of parking levels contemplated	p=	**3.00**			
	Pkg. support as fraction of gross pkg. structure area (GPA)	Pu=	**0.098**			
	Net area for parking & circulation as fraction of GPA	Pe=	0.902	Pe+Pu must = 1		
	No. of loading spaces	l=	**1**			
	Gross area per loading space	b =	**0**	SF		
Site Variables:	**Project open space as fraction of BLA**	S=	**0.300**		8,599	SF
	Private driveways as fraction of BLA	R=	**0.155**		4,443	SF
	Misc. pavement as fraction of BLA	M=	**0.009**		258	SF
	Loading area as fraction of BLA	L=	0.000		0	SF
	Total site support areas as a fraction of BLA	Su=	0.464		13,299	SF
Core:	**Core development area as fraction of BLA**	C=	0.536	C+Su must = 1	15,363	SF
Building Variables:	Building efficiency as fraction of GBA	Be=				
	Building support as fraction of GBA	Bu=		Be+Bu must = 1		

NOTE: p=1 is a grade parking lot based on design premise. Increase the number of parking levels to increase the capacity forecast below. Other variables in blue, box and bold may also be changed.

Dwelling Unit Mix Table

DU dwelling unit type	GDA gross du area	CDA=GDA/Be comprehensive du area	MIX du mix	PDA = (CDA)MIX Pro-rated du area

Dwelling unit mix not available

Aggregate average dwelling unit area (AGG) found by dividing gross building area by number of dwelling units present

Aggregate avg. dwelling unit area	(AGG) =	**1,781**
Gross building area planned, provided, or required per parking space	(a)=	2,283

PLANNING FORECAST

no. of FLR bldg floors	min land area CORE for BCG & PLA	dwelling units **NDU**	pkg spaces NPS	net pkg area NPA	gross pkg GPA struc area	pkg. struct cover **GPL** based on p above	gross bldg area GBA	footprint **BCA**	density per dBA bldable acre	gross bldg SF **SFAC** per buildable AC
1.00	15,363	8.2	6.4	2,071	2,296	765	14,598	14,598	12.46	22,185
2.00		15.6	12.2	3,946	4,375	1,458	27,810	13,905	23.73	42,264
3.00		22.4	17.4	5,651	6,265	2,088	39,824	13,275	33.98	60,523
4.00		28.5	22.2	7,208	7,991	2,664	50,797	12,699	43.35	77,200
5.00		34.2	26.7	8,636	9,574	3,191	60,859	12,172	51.93	92,491
6.00		39.4	30.7	9,950	11,031	3,677	70,118	11,686	59.83	106,562
7.00		44.2	34.5	11,163	12,375	4,125	78,666	11,238	67.13	119,553
8.00		48.6	37.9	12,286	13,621	4,540	86,583	10,823	73.88	131,585
9.00		52.7	41.1	13,329	14,777	4,926	93,936	10,437	80.16	142,759
10.00		56.6	44.1	14,301	15,855	5,285	100,782	10,078	86.00	153,165
11.00		60.2	46.9	15,208	16,860	5,620	107,174	9,743	91.45	162,878
12.00		63.5	49.6	16,056	17,801	5,934	113,154	9,429	96.56	171,967
15.00		72.4	56.5	18,303	20,292	6,764	128,988	8,599	110.07	196,030
19.00		**82.1**	**64.0**	**20,748**	**23,002**	**7,667**	**146,218**	**7,696**	**124.77**	**222,216**
20.00		84.2	65.7	21,281	23,593	7,864	149,974	7,499	127.98	227,924

WARNING: These are preliminary forecasts that must not be used to make final decisions.
1) These forecasts are not a substitute for the "due diligence" research that must be conducted to support the final definition of "unbuildable areas" above and the final decision to purchase land. This research includes, but is not limited to, verification of adequate subsurface soil, zoning, environmental clearance, access, title, utilities and water pressure, clearance from deed restriction, easement and right-of-way encumbrances, clearance from existing above and below ground facility conflicts, etc.
2) The most promising forecast(s) made on the basis of data entered in the design specification from "due diligence" research must be verified at the drawing board before funds are committed and land purchase decisions are made. Actual land shape ratios, dimensions and irregularities encountered may require adjustments to the general forecasts above.
3) The software licensee shall take responsibility for the design specification values entered and any advice given or decisions made that are based on the forecast produced.

forecast models could be used to reconcile these objectives by predicting the increased building height required to recover the development capacity lost when street-level context is improved with expanded open space.

Conclusion

The Summit Chase project used an underground parking structure to preserve open space and produced a very low density of 23.1 dBA. The Waterford project used an adjacent parking structure and produced a very high density of 145.95 dBA. This does not mean that underground parking structures produce low-density residential environments. Either structure could have been used to produce the opposite density. It does demonstrate the flexibility of parking structure systems to produce alternate lifestyles, and the ability of residential parking structure forecast models to compare the results that can be produced. This means that these forecast models can also be used to evaluate context plans for neighborhoods and districts before construction, and to express conclusions in terms capable of producing anticipated results. This method of design specification could protect the context integrity and lifestyle quality of every project eventually built within without dictating form and appearance.

Notes

1. The need for building cover and parking cover areas discourages this contribution.
2. Read this as 8 stories, no open space, and 100% development cover.
3. Read this as 2 stories and 33.8% development cover.
4. This includes the parking levels below.
5. There are many other objectives, but these are the objectives that are relevant to this discussion.

6. Additional parking levels could be introduced. Additional building floors could be constructed, other parking systems could be used, and revised design specifications could be written.

7. Aggregate average dwelling unit area AGG is equal to the gross building area divided by the total number of dwelling units provided. It is a function of the dwelling unit mix and floor plan area allocated per dwelling unit. The AGG value is a major factor affecting development capacity. It varies by development project and is a planning policy decision that directly affects capacity and intensity.

Zero-Parking Comparisons

Zero parking permits more shelter[1] to be built on less land and in taller buildings. It also produces increased development intensity that can result in overdevelopment when relief is not provided. Zero parking also allows shelter to be built in compact arrangements that can preserve more of the surrounding natural environment. If these urban arrangements are softened with places for people, the lifestyle becomes social. If places for people are ignored, the intensity produces a building-dominated universe traversed with paved movement systems that force people to crowd along the margins out of harm's way.

Development projects without parking on the premises are generally found in central business districts (CBDs) that provide public parking garages, commercial parking lots, and metered street parking within walking distance. They can also be found in older commercial areas of cities and villages that use a hitching-post method of parking cars in the public right-of-way. Cars have replaced horses, parking meters have replaced hitching posts, and parking garages have replaced stables to produce new urban cousins that appear quite different, but contain very similar design specification components. Figures 11.1 and 11.2 show that all

(a)

(b)

(c)

Figure 11.1 Gay Street Office—Example of hitching-post public parking with no private accommodations. This building represents 100% development cover and dependence on hitching-post parking in the public right-of-way. The nearest public parking is one block away. Building height was limited at the time of construction. (*a*) Overview; (*b*) site plan; (*c*) street view.

(Photos by [*a*] W. M. Hosack and [*c*] B. Higgins)

(a)

(b)

(c)

Figure 11.2 One Columbus—Example of modern office building with no private parking accommodations. (*a*) This dramatic building relies on the urban infrastructure of public parking to serve its population. (*b*) One such lot is located directly north of the building as shown on this site plan. (*c*) Street view.

(Photos by [*a*] W. M. Hosack and [*c*] B. Higgins)

rely on some form of remote supplemental parking to offset their on-site deficiencies and that all depend on public parking in the right-of-way for additional parking support.

Zero parking in the CBD forces someone else to address the issue. Zero parking in an older commercial center can easily produce deterioration when the issue is not addressed and customers are discouraged by the inconvenience. It amounts to a project subsidy that is absorbed within the urban fabric at a price, and the combined effect of zero parking and zero project open space is a recipe for intensity that can produce congestion and lifestyle disruption when collected into neighborhoods and districts without relief.[2] It is possible to define intensity, introduce open space at street level, and achieve equal or greater development capacity when building height is increased to compensate for the project open space provided. This is only common sense and is at the heart of the historic floor area ratio concept.[3]

The problem is that a floor area ratio does not say what it means, is most easily understood in zero-parking situations, and is subject to broad interpretation that can often produce unintended results. For instance, a floor area ratio of 1.0 means that a developer is permitted to multiply his or her gross land area by 1.0 to find the gross building area that he or she is permitted to construct. He or she may build this area as a 1-story building on 100% of the site, as a 2-story building on 50% of the site, and so forth. If the developer builds on 50%, half remains as open space, unless it is consumed by a parking lot. A 50% decision may result in permission to build three floors instead of two as a bonus for the open space that is provided.[4] The floor area ratio is intended to encourage height in return for project open space at street level, but it does not require minimum open space contributions. This means that it cannot be counted on to provide a consistently satisfactory streetscape context since the owner may often cover the entire site at his or her discretion.

Nonresidential Context

The Gay Street Office in Fig. 11.1 provides zero parking and represents a floor area ratio of 5.0. It is 5 stories tall and is built on 100% of the site. It provides no external places for people, and neither do its neighbors. The sidewalk is 12 ft wide, and pedestrians walk along this concrete channel between storefronts, auto grilles, and moving traffic. Trees are planted in the pavement to offer relief; street furniture reduces the active width of the channel; and some storefronts have been upgraded to improve their appearance. The building contains approximately 90,932 sq ft of area[5] and approximately 300 occupants, most of whom drive cars and park wherever they can in the surrounding city. It is a historic[6] hitching-post street presence that was limited to 5 stories for reasons that have since been forgotten. The building is a classic example found in every city that provides no external place for either people or cars. It was built in a time when travel distances had to be compact, and when the city was much smaller and easier to escape. Its context record is included as Table 11.1, and its site plan is included with Fig. 11.1. The forecast model using the measured context values is included as Exhibit 11.1. The amount of project open space specified in the model is zero, and the project built can be found in the row corresponding to a 5-story building. A glance at row 15 in Table 11.1 shows that the gross building area could have increased to 272,795 sq ft (approximately 850 people) if the building height had increased from 5 to 15 floors. This option may have been ignored for any number of reasons including, but not limited to, a lack of demand, a lack of capital, or a lack of elevator technology. As demand has increased and elevators have become commonplace, however, it is easy to see that the development potential of a zero-parking land area is enormous and that the temptation to redevelop can be irresistible.[7]

Table 11.1 Gay Street Office Context Record

Name	**Nitschke Building**
Address or Location	**35 E. Gay Street**
Land Use	**Office**

Parking System	*check box*
Surface parking around but not under building	
No Parking	**x**

Notes

		Description	Project Measurements		Context Summary SF	%	
Site							
		Gross land area in acres	**0.418**	GLA	18,186		
		Public/ private ROW and paved easements in SF	**0**	W		0.0%	of GLA
		Net lot area		NLA	18,186	100.0%	of GLA
		Total unbuildable area in SF	**0**	U		0.0%	of GLA
		Water area within unbuildable area in SF	**0**	WAT			of U
		Buildable land area		BLA	18,186	100.0%	of GLA
Development Cover							
		Gross building area in SF	**90,932**	GBA			
		Total building footprint in SF	**18,186**	B		100.0%	of BLA
		Total bldg. support in SF (stairs, corridors, elevators, etc.)		BSU			
		Building efficiency		Be			of GBA
	Residential only	Number of dwelling units		DU			
		Aggregate average dwelling unit area in SF		AGG			
		Number of building floors	**5.00**	FLR			
	G2 only	Unenclosed surface parking area under building in SF		PUB			
		Total surface parking area in SF (include internal landscape)		P or G		0.0%	of BLA
		Number of surface parking spaces		NPSs			
		Building cover over surface parking area		AIR			of GPA
		Gross surface parking area per parking space in SF		ss			
		Gross building area per surface parking space in SF		as			
		Surface parking spaces per dwelling unit		us	0.00		
	Pkg. struc. only	Total parking structure footprint in SF		P		0.0%	of BLA
	Pkg. struc. only	Number of parking structure floors		p			
	Pkg. struc. only	Total parking structure area in SF		GPA			
	Pkg. struc. only	Total parking structure spaces		NPSg			
	Pkg. struc. only	Total pkg. struc. support in SF (stairs, elevators, etc.)		PSS			
	Pkg. struc. only	Parking structure support percentage		Pu			of GPA
	Pkg. struc. only	Gross parking structure area per space in SF		sg	0		
	Pkg. struc. only	Gross building area per structure parking space in SF		ag	0		
	Pkg. struc. only	Parking structure spaces per dwelling unit		ug	0.00		
	Residential only	Total dwelling unit garage area in SF		GPA			
	Residential only	Dwelling unit garage under builiding in SF		GUB			
	Residential only	Number of dwelling unit garage parking spaces		GPS			
	Residential only	Garage parking spaces per dwelling		Gn	0.0		
	Residential only	Total dwelling unit garage area per space in SF		Ga			
	Residential only	Total dwelling unit garage cover in addition to building cover		Gc		0.0%	of BLA
		Total parking spaces			0.0		
	Residential only	Total parking spaces per dwelling unit		u			
		Gross building area per parking space total		a			
		Number of loading spaces		l			
		Total loading area in SF		LDA			
		Gross area per loading space in SF		b			
		Loading area percentage		L		0.0%	of BLA
		Driveway areas in SF		R		0.0%	of BLA
		Misc. pavement and social center pvmt. in SF		M		0.0%	of BLA
		TOTAL DEVELOPMENT COVER		D	18,186	**100.0%**	of BLA
Project Open Space							
		PROJECT OPEN SPACE		S	0	**0.0%**	of BLA
		Water area within project open space in SF		wat			of S
Summary			0.0%				
		Development balance index		BNX		1.000	
		Development intensity index		INX		6.000	
		Floor area ratio per buildable acre		BFAR		5.000	
		Density per buildable acre if applicable		dBA			
		Density per net acre if applicable		dNA			
		Density per gross acre if applicable		dGA			

Exhibit 11.1 Gay Street Office Design Specification and Forecast

Forecast Model CG1L

Development capacity forecast for ***NONRESIDENTIAL BUILDINGS*** *based on the use of an adjacent* ***GRADE PARKING LOT*** *located on the same premises. When s and a equal zero in the design specification below, the forecast pertains to conditions when* ***NO PARKING*** *is required.*

Given: *Gross land area.* ***To Find:*** *Maximum development capacity of the land area (gross building area potential) based on the design specification values entered below.* ***Premise:*** *All building floors considered equal in area.*

DESIGN SPECIFICATION

Enter values in boxed areas where text is bold and blue. Express all fractions as decimals.

Given:	**Gross land area**	**GLA=**	**0.4175**	acres	18,186	SF
Land Variables:	Public/ private right-of-way & paved easements	W=	**0.000**	fraction of GLA	0	SF
	Net land area	NLA=	0.418	acres	18,186	SF
	Unbuildable and/or future expansion areas	U=	**0.000**	fraction of GLA	0	SF
	Gross land area reduction	X=	0.000	fraction of GLA	0	SF
	Buildable land area remaining	**BLA=**	0.418	acres	18,186	SF
Parking Variables:	Est. gross pkg. lot area per space in SF	s =	**0**	ENTER ZERO IF NO PARKING REQUIRED		
	Building SF permitted per parking space	a =	**0**	ENTER ZERO IF NO PARKING REQUIRED		
	No. of loading spaces	l=	**0**			
	Gross area per loading space	b =	**0**	SF	0	SF
Site Variables:	**Project open space as fraction of BLA**	**S=**	**0.000**		0	SF
	Private driveways as fraction of BLA	R=	**0.000**		0	SF
	Misc. pavement as fraction of BLA	M=	**0.000**		0	SF
	Loading area as fraction of BLA	L=	0.000		0	SF
	Total site support areas as a fraction of BLA	Su=	0.000		0	SF
Core:	**Core development area as fraction of BLA**	**C=**	1.000	C+Su must = 1	18,186	SF

PLANNING FORECAST

no. of floors **FLR**	**CORE** minimum land area for BCG & PLA	gross building area **GBA**	parking lot area **PLA**	pkg spaces **NPS**	footprint **BCA**	bldg SF / acre **SFAC** function of BLA	flr area ratio **FAR** function of BLA
1.00	18,186	18,186	none	none	18,186	43,560	1.000
2.00		36,373	none	none	18,186	87,120	2.000
3.00		54,559	none	none	18,186	130,680	3.000
4.00		72,745	none	none	18,186	174,240	4.000
5.00		**90,932**	**none**	**none**	**18,186**	**217,800**	**5.000**
6.00		109,118	none	none	18,186	261,360	6.000
7.00		127,304	none	none	18,186	304,920	7.000
8.00	*NOTE: Be aware when BCA becomes too small to be feasible.*	145,490	none	none	18,186	348,480	8.000
9.00		163,677	none	none	18,186	392,040	9.000
10.00		181,863	none	none	18,186	435,600	10.000
11.00		200,049	none	none	18,186	479,160	11.000
12.00		218,236	none	none	18,186	522,720	12.000
13.00		236,422	none	none	18,186	566,280	13.000
14.00		254,608	none	none	18,186	609,840	14.000
15.00		272,795	none	none	18,186	653,400	15.000

WARNING: These are preliminary forecasts that must not be used to make final decisions.
1) These forecasts are not a substitute for the "due diligence" research that must be conducted to support the final definition of "unbuildable areas" above and the final decision to purchase land. This research includes, but is not limited to, verification of adequate subsurface soil, zoning, environmental clearance, access, title, utilities and water pressure, clearance from deed restriction, easement and right-of-way encumbrances, clearance from existing above and below ground facility conflicts, etc.
2) The most promising forecast(s) made on the basis of data entered in the design specification from "due diligence" research must be verified at the drawing board before funds are committed and land purchase decisions are made. Actual land shape ratios, dimensions and irregularities encountered may require adjustments to the general forecasts above.
3) The software licensee shall take responsibility for the design specification values entered and any advice given that is based on the forecast produced.

If this office had contributed 50% of its land area to streetscape amenity, Exhibit 11.2 shows that a 10-story building would have been required to produce the same gross building area. The floor area ratio is still 5.0, but the intensity index has changed from 5.0 to 10.5. This index indicates that the urban design context would change significantly since half of the lot has been provided as a place for people at street level. If the building height had been increased to 15 floors, Exhibit 11.2 shows that the development capacity of the site would have increased by 50% and the development intensity index would increase to 15.5, even though the land area built upon would have declined by 50% in both cases.

Table 11.2 illustrates the results of redevelopment when the intensity index increases to 27.0 on a 0.468-acre site. This land area is similar to the Gay Street area of 0.418 acres in Table 11.1, and the intensity level is easily feasible with today's building technology. The result is 407,472 sq ft of gross building area with a 19,934-sq-ft footprint covering 97.8% of the site. This is 4.5 times the Gay Street building area and illustrates the incentive to capture development capacity with redevelopment and new technology when zero parking requirements are involved. It also illustrates the tendency to cover this land with building cover in pursuit of the greatest amount of gross building area per acre of buildable land area.

Defining desirable relationships between open space and building height are beyond the scope of this book. It should be apparent from this simple example, however, that it is possible to quantitatively balance the financial objectives of a developer with the environmental need of an urban population and the public need for more shelter. Development forecast models can predict the options available, but context record research and design specifications are required to define limits. These limits can be used to produce an urban pattern and an urban quality of life that is capable of maintaining an attractive lifestyle for a growing population.

Exhibit 11.2 Gay Street Office with 50% Open Space

Forecast Model CG1L

Development capacity forecast for ***NONRESIDENTIAL BUILDINGS*** *based on the use of an adjacent* ***GRADE PARKING LOT*** *located on the same premises. When s and a equal zero in the design specification below, the forecast pertains to conditions when* ***NO PARKING*** *is required.*

Given: *Gross land area.* ***To Find:*** *Maximum development capacity of the land area (gross building area potential) based on the design specification values entered below.* ***Premise:*** *All building floors considered equal in area.*

DESIGN SPECIFICATION

Enter values in boxed areas where text is bold and blue. Express all fractions as decimals.

Given:	**Gross land area**	GLA=	0.4175	acres	18,186	SF
Land Variables:	Public/ private right-of-way & paved easements	W=	0.000	fraction of GLA	0	SF
	Net land area	NLA=	0.418	acres	18,186	SF
	Unbuildable and/or future expansion areas	U=	0.000	fraction of GLA	0	SF
	Gross land area reduction	X=	0.000	fraction of GLA	0	SF
	Buildable land area remaining	BLA=	0.418	acres	18,186	SF
Parking Variables:	Est. gross pkg. lot area per space in SF	s =	0	ENTER ZERO IF NO PARKING REQUIRED		
	Building SF permitted per parking space	a =	0	ENTER ZERO IF NO PARKING REQUIRED		
	No. of loading spaces	l=	0			
	Gross area per loading space	b =	0	SF	0	SF
Site Variables:	**Project open space as fraction of BLA**	S=	0.500		9,093	SF
	Private driveways as fraction of BLA	R=	0.000		0	SF
	Misc. pavement as fraction of BLA	M=	0.000		0	SF
	Loading area as fraction of BLA	L=	0.000		0	SF
	Total site support areas as a fraction of BLA	Su=	0.500		9,093	SF
Core:	**Core development area as fraction of BLA**	C=	0.500	C+Su must = 1	9,093	SF

PLANNING FORECAST

no. of floors **FLR**	**CORE** minimum land area for BCG & PLA	gross building area **GBA**	parking lot area **PLA**	pkg. spaces **NPS**	footprint **BCA**	bldg SF / acre **SFAC** function of BLA	flr area ratio **FAR** function of BLA
1.00	9,093	9,093	none	none	9,093	21,780	0.500
2.00		18,186	none	none	9,093	43,560	1.000
3.00		27,279	none	none	9,093	65,340	1.500
4.00	*NOTE:*	36,373	none	none	9,093	87,120	2.000
5.00	*Be aware when*	45,466	none	none	9,093	108,900	2.500
6.00	*BCA becomes too small to be*	54,559	none	none	9,093	130,680	3.000
7.00	*feasible.*	63,652	none	none	9,093	152,460	3.500
8.00		72,745	none	none	9,093	174,240	4.000
9.00		81,838	none	none	9,093	196,020	4.500
10.00		**90,932**	**none**	**none**	**9,093**	**217,800**	**5.000**
11.00		100,025	none	none	9,093	239,580	5.500
12.00		109,118	none	none	9,093	261,360	6.000
13.00		118,211	none	none	9,093	283,140	6.500
14.00		127,304	none	none	9,093	304,920	7.000
15.00		136,397	none	none	9,093	326,700	7.500

WARNING: These are preliminary forecasts that must not be used to make final decisions.

1) These forecasts are not a substitute for the "due diligence" research that must be conducted to support the final definition of "unbuildable areas" above and the final decision to purchase land. This research includes, but is not limited to, verification of adequate subsurface soil, zoning, environmental clearance, access, title, utilities and water pressure, clearance from deed restriction, easement and right-of-way encumbrances, clearance from existing above and below ground facility conflicts, etc.

2) The most promising forecast(s) made on the basis of data entered in the design specification from "due diligence" research must be verified at the drawing board before funds are committed and land purchase decisions are made. Actual land shape ratios, dimensions and irregularities encountered may require adjustments to the general forecasts above.

3) The software licensee shall take responsibility for the design specification values entered and any advice given that is based on the forecast produced.

Table 11.2 One Columbus Context Record

Name	One Columbus
Address or Location	10 W. Broad
Land Use	Office NP

Parking System	check box
Surface parking around but not under building	
No Parking	x
Notes	

	Description	Project Measurements		Context Summary SF	%	
Site						
	Gross land area in acres	0.468	GLA	20,386		
	Public/ private ROW and paved easements in SF	0	W		0.0%	of GLA
	Net lot area		NLA	20,386	100.0%	of GLA
	Total unbuildable area in SF	0	U		0.0%	of GLA
	Water area within unbuildable area in SF	0	WAT			of U
	Buildable land area		BLA	20,386	100.0%	of GLA
Development Cover						
	Gross building area in SF	407,472	GBA			
	Total building footprint in SF	19,934	B		97.8%	of BLA
	Total bldg. support in SF (stairs, corridors, elevators, etc.)		BSU			
	Building efficiency		Be			of GBA
Residential only	Number of dwelling units		DU			
	Aggregate average dwelling unit area in SF		AGG			
	Number of building floors	26.00	FLR			
G2 only	Unenclosed surface parking area under building in SF		PUB			
	Total surface parking area in SF (include internal landscape)		P or G		0.0%	of BLA
	Number of surface parking spaces		NPSs			
	Building cover over surface parking area		AIR			of GPA
	Gross surface parking area per parking space in SF		ss			
	Gross building area per surface parking space in SF		as			
	Surface parking spaces per dwelling unit		us	0.00		
Pkg. struc. only	Total parking structure footprint in SF		P		0.0%	of BLA
Pkg. struc. only	Parking structure area under building in SF					
Pkg. struc. only	Number of parking structure floors		p			
Pkg. struc. only	Total parking structure area in SF		GPA			
Pkg. struc. only	Total parking structure spaces		NPSg			
Pkg. struc. only	Total pkg. struc. support in SF (stairs, elevators, etc.)		PSS			
Pkg. struc. only	Parking structure support percentage		Pu			of GPA
Pkg. struc. only	Gross parking structure area per space in SF		sg	0		
Pkg. struc. only	Gross building area per structure parking space in SF		ag	0		
Pkg. struc. only	Parking structure spaces per dwelling unit		ug	0.00		
Residential only	Total dwelling unit garage area in SF		GPA			
Residential only	Dwelling unit garage under builiding in SF		GUB			
Residential only	Number of dwelling unit garage parking spaces		GPS			
Residential only	Garage parking spaces per dwelling		Gn	0.0		
Residential only	Total dwelling unit garage area per space in SF		Ga			
Residential only	Total dwelling unit garage cover in addition to building cover		Gc		0.0%	of BLA
	Total parking spaces			0.0		
Residential only	Total parking spaces per dwelling unit		u			
	Gross building area per parking space total		a			
	Number of loading spaces		l			
	Total loading area in SF		LDA			
	Gross area per loading space in SF		b			
	Loading area percentage		L		0.0%	of BLA
	Driveway areas in SF		R		0.0%	of BLA
	Misc. pavement and social center pvmt. in SF	452	M		2.2%	of BLA
	TOTAL DEVELOPMENT COVER		D	20,386	**100.0%**	of BLA
Project Open Space						
	PROJECT OPEN SPACE		S	0	**0.0%**	of BLA
	Water area within project open space in SF	0 0.0%	wat			of S
Summary						
	Development balance index		BNX		1.000	
	Development intensity index		INX		27.000	
	Floor area ratio per buildable acre		BFAR		19.988	
	Density per buildable acre if applicable		dBA			
	Density per net acre if applicable		dNA			
	Density per gross acre if applicable		dGA			

Streetscape open space dedication does not mean that the development potential of a zero-parking land area is restricted since building height can be used to compensate for the open space provided. There are reasonable limits to building height, however, that can be tested using the development capacity forecasting collection.

Residential Context

Zero parking can also be found in historic residential areas of cities that were located close to CBDs. After the introduction of the automobile, these homes became the residential equivalent of a zero-parking development. Figure 11.3 illustrates one of these urban houses, called City Park, that depends on the public street for parking. These homes can often be quite popular, but Fig. 11.3 does not convey the congested street parking and movement pattern that often results. City Park is one housing alternative that has the potential to improve our capacity to shelter an increasing population within more compact limits, but the parking issue must be resolved to reduce the competition between places for people and places for cars.

The context record for City Park is included as Table 11.3, and it shows that a density of 18.52 dwelling units per buildable acre was constructed. This density is generally associated with apartment projects and is much higher than most, if not all, suburb house (i.e., single-family, detached residential home) development. Unfortunately, the parking density is not commensurate with the housing density and congestion results. These homes represent a lesson from the past, however, that could easily be applied to the future need for more single-family housing should our parking, movement, and communication patterns change. In fact, Fig. 11.4 illustrates a modern variation of this urban house that accommodates the automobile but is not adaptable to historic homes. Its context record is included as Table 11.4.

(a)

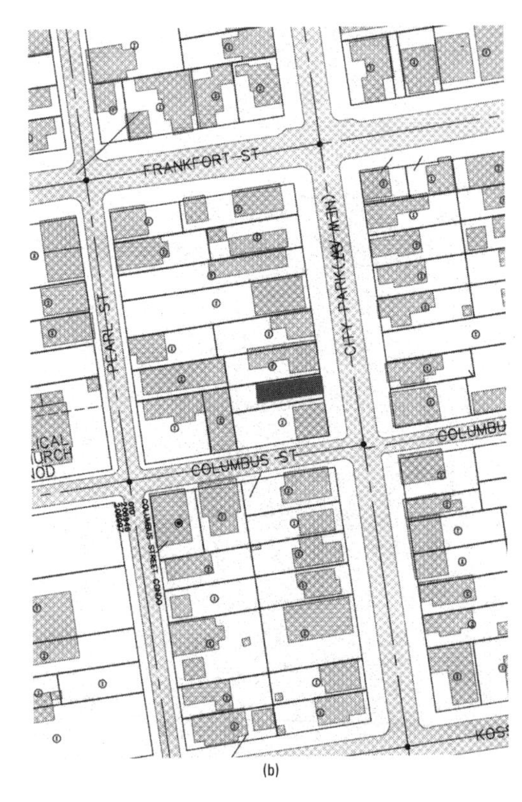

(b)

(c)

Figure 11.3 City Park Avenue—Example of an urban house with no private parking accommodations. (*a*) Street view. (*b*) Site plan illustrates narrow urban house lot with one side yard, a small rear yard, and dependence on the public right-of-way for parking. (*c*) Overview.

(Photos by W. M. Hosack)

Table 11.3 City Park Context Record

Name	
Address or Location	765 City Park
Land Use	Residential RNP

Parking System	check box
Surface parking around but not under building	
Surface parking around and under building	
No Parking	x
Notes	

	Description	Project Measurements		Context Summary SF	%	
Site						
	Gross land area in acres	0.054	GLA	2,352		
	Public/ private ROW and paved easements in SF	0	W		0.0%	of GLA
	Net lot area		NLA	2,352	100.0%	of GLA
	Total unbuildable area in SF	0	U		0.0%	of GLA
	Water area within unbuildable area in SF	0	WAT			of U
	Buildable land area		BLA	2,352	100.0%	of GLA
Development Cover						
	Gross building area in SF	1,972	GBA			
	Total building footprint in SF	1,188	B		50.5%	of BLA
	Total bldg. support in SF (stairs, corridors, elevators, etc.)		BSU			
	Building efficiency		Be			of GBA
Residential only	Number of dwelling units	1	DU			
	Aggregate average dwelling unit area in SF		AGG	1,972		
	Number of building floors	2.00	FLR			
G2 only	Unenclosed surface parking area under building in SF	0	PUB			
	Total surface parking area in SF (include internal landscape)	0	P or G		0.0%	of BLA
	Number of surface parking spaces	0	NPSs			
	Building cover over surface parking area		AIR			of GPA
	Gross surface parking area per parking space in SF		ss			
	Gross building area per surface parking space in SF		as			
	Surface parking spaces per dwelling unit		us	0.00		
Pkg. struc. only	Total parking structure footprint in SF		P		0.0%	of BLA
Pkg. struc. only	Parking structure area below grade in SF					
Pkg. struc. only	Parking structure area above grade under building in SF					
Pkg. struc. only	Number of parking structure floors		p			
Pkg. struc. only	Total parking structure area in SF		GPA			
Pkg. struc. only	Total parking structure spaces		NPSg			
Pkg. struc. only	Total pkg. struc. support in SF (stairs, elevators, etc.)		PSS			
Pkg. struc. only	Parking structure support percentage		Pu			of GPA
Pkg. struc. only	Gross parking structure area per space in SF		sg	0		
Pkg. struc. only	Gross building area per structure parking space in SF		ag	0		
Pkg. struc. only	Parking structure spaces per dwelling unit		ug	0.00		
Residential only	Total dwelling unit garage area in SF		GPA			
Residential only	Dwelling unit garage above grade under builiding in SF		GUB			
Residential only	Number of dwelling unit garage parking spaces		GPS			
Residential only	Garage parking spaces per dwelling		Gn	0.0		
Residential only	Total dwelling unit garage area per space in SF		Ga			
Residential only	Total dwelling unit garage cover in addition to building cover		Gc		0.0%	of BLA
	Total parking spaces			0.0		
Residential only	Total parking spaces per dwelling unit		u			
	Gross building area per parking space total		a			
	Number of loading spaces		l			
	Total loading area in SF		LDA			
	Gross area per loading space in SF		b			
	Loading area percentage		L		0.0%	of BLA
	Driveway areas in SF		R		0.0%	of BLA
	Misc. pavement and social center pvmt. in SF		M		0.0%	of BLA
	TOTAL DEVELOPMENT COVER		D	1,188	**50.5%**	of BLA
Project Open Space						
	PROJECT OPEN SPACE		S	1,164	**49.5%**	of BLA
	Water area within project open space in SF		wat			of S
Summary		0.0%				
	Development balance index		BNX		0.505	
	Development intensity index		INX		2.505	
	Floor area ratio per buildable acre		BFAR		0.838	
	Density per buildable acre if applicable		dBA		18.52	
	Density per net acre if applicable		dNA		18.52	
	Density per gross acre if applicable		dGA		18.52	

(a)

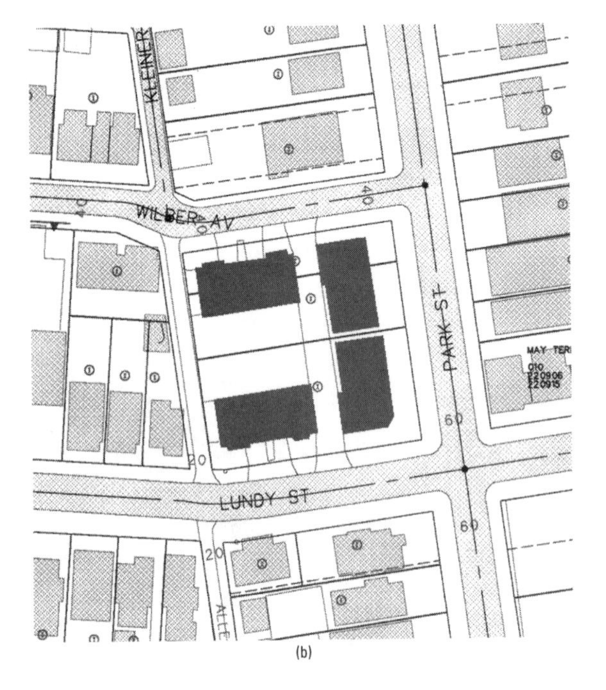

(b)

Figure 11.4 Park Street—Example of an urban house with private parking accommodations. (*a*) This building represents a modern urbanhouse variation with parking under the building. It accommodates the automobile on private property, and its context record shows that it exceeds the historic City Park urbanhouse density. (*b*) Site plan of urban house.

(Photo by B. Higgins)

Table 11.4: Park Street Context Record

Form CR

Name	Park Street
Address or Location	811-845 Park Street
Land Use	Residential RG2

Parking System	*check box*
Surface parking around but not under building	
Surface parking around and under building	x
Notes	

	Description	Project Measurements		Context Summary SF	Context Summary %	
Site						
	Gross land area in acres	0.744	GLA	32,409		
	Public/ private ROW and paved easements in SF	3,960	W		12.2%	of GLA
	Net lot area		NLA	28,449	87.8%	of GLA
	Total unbuildalbe area in SF	0	U		0.0%	of GLA
	Water area within unbuildable area in SF	0	WAT			of U
	Buildable land area		BLA	28,449	87.8%	of GLA
Development Cover						
	Gross building area in SF	24,644	GBA			
	Total building footprint in SF	12,322	B		43.3%	of BLA
	Total bldg. support in SF (stairs, corridors, elevators, etc.)		BSU			
	Building efficiency		Be			of GBA
residental only	Number of dwelling units	14	DU			
	Aggregate average dwelling unit area in SF		AGG	1,760		
	Number of building floors	2.00	FLR			
G2 only	Unenclosed surface parking area under building in SF		PUB			
	Total surface parking area in SF (include internal landscape)		P or G		0.0%	of BLA
	Number of surface parking spaces		NPSs			
	Building cover over surface parking area		AIR			of GPA
	Gross surface parking area per parking space in SF		ss			
	Gross building area per surface parking space in SF		as			
	Surface parking spaces per dwelling unit		us	0.00		
pkg. struc. only	Total parking structure footprint in SF		P		0.0%	of BLA
pkg. struc. only	Parking structure area under building in SF					
pkg. struc. only	Number of parking structure floors		p			
pkg. struc. only	Total parking structure area in SF		GPA			
pkg. struc. only	Total parking structure spaces		NPSg			
pkg. struc. only	Total pkg. struc. support in SF (stairs, elevators, etc.)		PSS			
pkg. struc. only	Parking structure support percentage		Pu			of GPA
pkg. struc. only	Gross parking structure area per space in SF		sg	0		
pkg. struc. only	Gross building area per structure parking space in SF		ag	0		
pkg. struc. only	Parking structure spaces per dwelling unit		ug	0.00		
residental only	Total dwelling unit garage area in SF	9,242	GPA			
residental only	Dwelling unit garage under builiding in SF	9,242	GUB			
residental only	Number of dwelling unit garage parking spaces	28	GPS			
residental only	Garage parking spaces per dwelling		Gn	2.0		
residental only	Total dwelling unit garage area per space in SF		Ga	330.1		
residental only	Total dwelling unit garage cover in addition to building cover		Gc		0.0%	of BLA
	Total parking spaces			28.0		
residental only	Total parking spaces per dwelling unit		u	2.0		
	Gross building area per parking space total		a	880.1		
	Number of loading spaces	0	l			
	Total loading area in SF	0	LDA			
	Gross area per loading space in SF		b			
	Loading area percentage		L		0.0%	of BLA
	Driveway areas in SF	1,216	R		4.3%	of BLA
	Misc. pavement and social center pvmt. in SF	252	M		0.9%	of BLA
	TOTAL DEVELOPMENT COVER		D	13,790	**48.5%**	of BLA
Project Open Space						
	PROJECT OPEN SPACE		S	14,659	**51.5%**	of BLA
	Water area within project open space in SF	0.0%	wat			of S
Summary						
	Development Balance Index		BNX		0.485	
	Development Intensity Index		INX		2.485	
	Floor area ratio per buildable acre		BFAR		0.866	
	Density per buildable acre if applicable		dBA		21.44	
	Density per net acre if applicable		dNA		21.44	
	Density per gross acre if applicable		dGA		18.82	

Conclusion

The CG1 and RG1 series of forecast models[8] apply to zero-parking design solutions when the design specification values that pertain to parking are entered as zero. The ultimate development capacity of land is reached when both parking and open space values are zero in one of these forecast models and when building height extends to its technological limit. When zero parking and zero open space are entered in a forecast model, however, the public need for these provisions does not disappear. It is either met by others, or a city's quality of life begins to imperceptibly deteriorate for many residents. When the public must contribute to offset these deficiencies and to ensure the success of a private real estate investment, the concept of development impact fees has been introduced to reduce this cost. This is not only true for zero parking and zero open space, but these two development approaches represent some of the greatest deficiencies that must be offset by public investment. A simple example of the ability of the development forecast collection to contribute to the evaluation of development impact fees follows.

If the Gay Street Office had provided a surface parking lot at a rather common rate of four spaces per 1000 gross sq ft of building area, Exhibit 11.3 shows that the land available could only support a gross building area of 10,698 sq ft. This is based on the current building height of five floors, no project open space, and a maximum footprint of 2140 sq ft. Since this footprint is unrealistically small, the practical limit under these circumstances is a 2-story building with a gross area of 9093 sq ft and a footprint of 4547 sq ft when no project open space is provided. This is a far cry from the unlimited development capacity of a zero-parking solution on the same lot, and it is 10% of the gross building area that presently occupies this lot. In other words, the original owner was able to construct 10 times

Exhibit 11.3 Gay Street Office with Surface Parking

Forecast Model CG1L

Development capacity forecast for ***NONRESIDENTIAL BUILDINGS*** *based on the use of an adjacent* ***GRADE PARKING LOT*** *located on the same premises. When s and a equal zero in the design specification below, the forecast pertains to conditions when* ***NO PARKING*** *is required.*

Given: *Gross land area.* ***To Find:*** *Maximum development capacity of the land area (gross building area potential) based on the design specification values entered below.* ***Premise:*** *All building floors considered equal in area.*

DESIGN SPECIFICATION

Enter values in boxed areas where text is bold and blue. Express all fractions as decimals.

Given:	**Gross land area**	**GLA=**	**0.4175**	acres	18,186	SF
Land Variables:	Public/ private right-of-way & paved easements	W=	**0.000**	fraction of GLA	0	SF
	Net land area	NLA=	0.418	acres	18,186	SF
	Unbuildable and/or future expansion areas	U=	**0.000**	fraction of GLA	0	SF
	Gross land area reduction	X=	0.000	fraction of GLA	0	SF
	Buildable land area remaining	**BLA=**	0.418	acres	18,186	SF
Parking Variables:	Est. gross pkg. lot area per space in SF	s =	**375**	ENTER ZERO IF NO PARKING REQUIRED		
	Building SF permitted per parking space	a =	**250**	ENTER ZERO IF NO PARKING REQUIRED		
	No. of loading spaces	l=	**0**			
	Gross area per loading space	b =	**0**	SF	0	SF
Site Variables:	**Project open space as fraction of BLA**	**S=**	**0.000**		0	SF
	Private driveways as fraction of BLA	R=	**0.000**		0	SF
	Misc. pavement as fraction of BLA	M=	**0.000**		0	SF
	Loading area as fraction of BLA	L=	0.000		0	SF
	Total site support areas as a fraction of BLA	Su=	0.000		0	SF
Core:	**Core development area as fraction of BLA**	**C=**	1.000	C+Su must = 1	18,186	SF

PLANNING FORECAST

no. of floors **FLR**	CORE minimum land area for BCG & PLA	gross building area **GBA**	parking lot area PLA	pkg spaces NPS	footprint **BCA**	bldg SF / acre SFAC function of BLA	flr area ratio FAR function of BLA
1.00	18,186	7,275	10,912	29.1	7,275	17,424	0.400
2.00		**9,093**	**13,640**	**36.4**	**4,547**	**21,780**	**0.500**
3.00		9,920	14,880	39.7	3,307	23,760	0.545
4.00		10,392	15,588	41.6	2,598	24,891	0.571
5.00		**10,698**	16,047	42.8	**2,140**	25,624	0.588
6.00		10,912	16,368	43.6	1,819	26,136	0.600
7.00	*NOTE: Be aware when BCA becomes too small to be feasible.*	11,070	16,605	44.3	1,581	26,515	0.609
8.00		11,192	16,787	44.8	1,399	26,806	0.615
9.00		11,288	16,932	45.2	1,254	27,037	0.621
10.00		11,366	17,050	45.5	1,137	27,225	0.625
11.00		11,431	17,147	45.7	1,039	27,381	0.629
12.00		11,486	17,229	45.9	957	27,512	0.632
13.00		11,533	17,299	46.1	887	27,623	0.634
14.00		11,573	17,360	46.3	827	27,720	0.636
15.00		11,608	17,412	46.4	774	27,804	0.638

WARNING: These are preliminary forecasts that must not be used to make final decisions.
1) These forecasts are not a substitute for the "due diligence" research that must be conducted to support the final definition of "unbuildable areas" above and the final decision to purchase land. This research includes, but is not limited to, verification of adequate subsurface soil, zoning, environmental clearance, access, title, utilities and water pressure, clearance from deed restriction, easement and right-of-way encumbrances, clearance from existing above and below ground facility conflicts, etc.
2) The most promising forecast(s) made on the basis of data entered in the design specification from "due diligence" research must be verified at the drawing board before funds are committed and land purchase decisions are made. Actual land shape ratios, dimensions and irregularities encountered may require adjustments to the general forecasts above.
3) The software licensee shall take responsibility for the design specification values entered and any advice given that is based on the forecast produced.

the surface parking lot capacity of the land available, and a new owner could easily build 10 times the current amount with sophisticated high-rise technology. Land developed to 100 times its surface parking capacity has a development impact that has been largely ignored because the missing parking has somehow been absorbed within the overall city context. This has produced an urban quilt of parking lots as an impromptu response to this increasing intensity, but it will not produce the urban quality of life that will become increasingly essential as populations expand.

It is also possible to calibrate urban intensity in multiples of surface parking capacity using the forecast model collection. The preceding example was produced by entering the Gay Street land area in Exhibit 11.3. This exhibit represents forecast model CG1L. The surface parking capacity of the site was then predicted based on no project open space being provided. This capacity was compared with that predicted by Exhibit 11.1 when no parking was required and the height limit was set at five floors. This method of defining intensity could be used to calibrate development impact fees that are a function of the parking omitted. The fees paid could then be dedicated to pay for supplemental parking garages and/or alternate forms of transportation to offset the intensity proposed.[9]

The entire concept of private development deficiencies and offsetting public contributions is beyond the scope of this book. The preceding simple example, however, was intended to demonstrate that the development forecast collection has capabilities that can be used to identify deficiencies and define solutions for offsetting private contributions that can help to balance the intensity introduced.

Notes

1. Shelter for all human activities that must be protected from an unstable environment.

2. Intensity has had to increase to the level of urban plague, public health hazard, human indignity, and social misery before the polemics of social reformers have been able to overcome those who demand proof in the face of the obvious—that there is a link between the use of land and our quality of life.

3. The floor area ratio and the skyplane limit were two early zoning attempts to limit the development intensity produced by zero-parking development plans. Both regulations have had limited success because they rely on the street right-of-way to produce the majority of public open space at the sidewalk level, and on the largesse of individual property owners to supplement these narrow avenues with expanded places for people. In an urban environment, this street level is often referred to as *streetscape,* and the quantity and quality of its presence play a major role in defining lifestyle and urban enjoyment. When the streetscape does not contain places for people, and when skyplane limits do not ensure that adequate light and air reach this street level, streetscape improvements become exercises in storefront beautification that attempt to substitute appearance for context. Both approaches have had limited success because of the following:
   - Floor area ratios and skyplane limits are primarily concerned with the effect of building height and permit the entire lot to be covered at the sidewalk level where we live.
   - Total lot coverage at sidewalk level overlooks places for people.
   - Right-of-way width can anticipate places for people, but local governments are poorly prepared to provide streetscape services that would benefit both the public and private enterprise.
   - Most rights-of-way are too narrow to serve both passive pedestrian and active motorist activity.
   - Many rights-of-way are subject to street widening, which reduces pedestrians' allocations.
   - Too little passive open space is provided by the public sector as a counterpoint to these active movement corridors.

4. The developer may have already received zero parking as a bonus, but this is often overlooked in the incentive process.

5. The actual area is somewhat less, given the subtleties of architectural floor plans, sections, and details. The area noted is a function of the footprint multiplied by the number of floors involved, and is close enough to

convey the building magnitude present on site. If a zoning regulation were being written, some form of building area reduction would have to be considered in relation to the height being proposed in order to ensure that natural light reaches street level.

6. Do not interpret this example as an argument to demolish historic buildings. This building is simply a convenient example of the issues surrounding zero parking for all urban buildings regardless of size and age.
7. The desire to convert historic nonelevator zero-parking buildings is heavily influenced by this simple arithmetic.
8. Includes forecast models CG1L, CG1B, RG1L, RG1B, and RG1D.
9. For instance, a parking garage space costs from $10,000 to $12,000. An impact fee could equal the number of spaces omitted multiplied by the parking garage cost per space.

Urban House Comparisons

All housing can be divided into three categories: (1) single-family detached suburb houses,[1] (2) single-family attached urban houses, and (3) apartments. These three groups can be distinguished by their typical dwelling unit relationship to the land, their method of dwelling unit separation, their average lot frontage per dwelling unit, and their building height potential.

An urban house is a single-family building that can contain twin-single, three-family, four-family, rowhouse, and townhouse[2] floor plan configurations. It can also include single-family detached homes with less than 40 ft of frontage, and often includes homes with less than 30 ft of frontage.[3] An urban house is directly entered from, and has a direct relationship with, the land. Each dwelling is attached to at least one other dwelling within the parent building, except for the narrow lots mentioned, and the parent building cannot extend beyond one dwelling unit in height because of its grade orientation.

Overview

Urban house development patterns track our transportation and economic evolution. During the colonial era, urban houses in villages like Williamsburg represented miniature farms. Main houses were placed on the streets, gardens were located in the middle, and stables and kitchens were entered from an alley. The industrial era squeezed the urban house garden to bring more people closer to work, but often kept the alley and stable. The automobile was placed in the stable and the alley became inconvenient. The street became clogged with parked cars seeking better maneuverability and front-door access. Children began darting between parked cars along socially oriented front yards, and the search for an accommodation became serious as more of the population entered the automobile age. Lots became wider. Alleys remained, but automobiles attempted to enter rear stables from the main street. These stables became detached garages and increased in size to accommodate larger vehicles. As a result, rear yards were squeezed again. Lots became larger in response and the suburb house multiplied in earnest. Sprawl emerged as populations grew; two automobiles became common; and more land was required to accommodate the increase. Mobility made it possible to disperse commercial activity from the urban core. More jobs were created closer to remote homes as a result. This encouraged more sprawl and allowed homeowners to mimic farms in new residential subdivisions with grass as their crop.[4] Urban house variations were introduced in the suburbs as buffers for single-family detached residential subdivisions, and the alley was eliminated. Populations continue to grow with similar suburban aspirations that further expand the total urban area, and we are still searching for an accommodation that will shelter increasing populations and limit sprawl by producing desirable housing at higher densities.

Our populations first clustered at higher densities because of the lack of mobility, the availability of work, and the need for social and economic organization within walking distance. Environmental preservation may become our second motivation to begin clustering at higher densities, even though our current mobility makes it possible to ignore this emerging awareness.

Density

The ability to shelter a given population is a function of the open space provided; the buildable area remaining for construction; and the height planned, permitted, or required. When every dwelling unit is grade oriented, most do not exceed 3.5 stories in height. The outer limit is generally 5 stories, and the common height is from 2 to 3 stories. This is true for both urban houses and suburb houses, since both are grade oriented. This has a distinctly limiting effect on the densities that can be achieved with two of the three housing groups that are available.

Density is traditionally expressed in terms of dwelling units constructed per acre of available land and is a benchmark that is often used as a design specification. It has been successful as a benchmark, but is too incomplete to perform as a specification capable of consistently producing successful development results. The components of the design specification panel in each residential forecast model present a more complete list of these essential ingredients. The forecast model collection has been created to predict the development capacity of land based on values assigned to these component ingredients, and the context record system has been created to measure and record the ingredient quantities used in existing development projects. This context record knowledge can then be used to evaluate the various ingredient quantities and combinations that are entered in the design specification panel of a forecast model for evaluation.

Gross building areas are subdivided to accommodate a variety of different land use activities within. (This subdivision is called a *building floor plan*. When combined with land subdivision and residential shelter, it produces density.) The capacity of land to accommodate building area is a function of the land use family, parking system, and design specification involved. The design components within a design specification are constant for each of the seven forecast model groups available.[5] The values assigned to these components change in relation to the intended use and design objectives of the project, but the component categories remain constant. The values assigned to these components establish the capacity of land to accommodate building area for an intended use. This building area capacity, generally referred to as *development capacity,* is then subdivided[6] to serve the activity planned for the gross building area within.

Dwelling units per acre is a traditional measure of residential intensity and a simplified benchmark of performance that cannot substitute for a design specification. It has been retained in all residential forecast models and context records as an index of results, but has been replaced by the design specification panel in order to produce more accurate development capacity forecasts. It can be a deceiving statistic, since average dwelling unit areas vary significantly and larger dwelling unit areas can be unfairly penalized by inflexible density limits even though the population, parking, and traffic intensity is unchanged. In fact, these density limits can produce circular negotiations over acceptable density levels that become stumbling blocks rather than measures of success. Density has been supplemented in all forecast models and context records with the SFAC index, a development capacity index that reports the gross square feet of residential building area produced per buildable acre. Neither index accounts for the broad range of dwelling unit areas and assortments that may be included in a building, and that directly affect achievable density, but the SFAC index allows residential shelter capacity to be compared with non-residential capacity. This more clearly illustrates the relationship between the capacity of land to support gross building area and the way that area is subdivided to serve unique activities sheltered within. The residential SFAC index can also be divided by an aggregate average dwelling unit area value AGG to produce a more accurate forecast or statement of the number of dwelling units that have been or can be constructed per acre. Neither index, however, is capable of leading the development design process nor substituting for a design specification, since both are designed to simply benchmark the performance of a complete set of these design specification values.

Case Studies

Nine urban houses have been selected for context record measurement and development capacity forecast modeling. They are located within the inner city, in a first ring suburb, in an edge city and in a remote planned community. Table 12.1 summarizes context records 12.1 through 12.9 and arranges each project in relation to the emphasis placed on private yard allocations (seventh column) within the total open space provided (third column). Northwest Court is at the top of this list and provides 54.1% common open space but no private or personal yard areas (sixth column).

Northwest Court

Northwest Court is a 2-story rowhouse rental complex near the inner city. It is illustrated in Fig. 12.1 and provides a residential level of total common open space, but no private yard areas and no personal association with the land or buildings beyond the grade-level entrances provided. Its aggregate average dwelling unit area is small, and the num-

Table 12.1 Urban House Data Summary*

Name	Context Record Table	Total Open Space, % of BLA *S*	Common Open Space, % of BLA COS	Private Yards, % of BLA Y	Private Yard Per Dwelling, in sq ft YRD	Private Yard Emphasis, % of OSA PYE	Avg. Habitable Footprint Plus Yard, in sq ft AFP	Aggregate Avg. Dwelling Unit Area, in sq ft AGG	Total Pkg. Spaces per Dwelling *p*	Parking Type	Dwellings per Buildable Acre dBA	Devel-opment Capacity Index, in sq ft SFAC	Devel-opment Intensity Index INX
Northwest Court	12.2	54.1%	54.1%	0.0%	0	0.0%	532	1064	1.52	Roadway, surface, and garage	14.83	17,110	2.459
Branford Commons	12.4	44.9%	31.8%	13.1%	563	29.2%	1448	1770	2.00	Surface and garage	10.16	20,216	2.551
Park Street	12.3	46.1%	28.2%	17.9%	364	38.9%	1244	1760	2.00	Garages under dwellings	21.40	51,886	2.539
Sherwood Villa	12.5	63.3%	35.6%	27.7%	2132	43.7%	3643	2267	2.68	Surface and garage	5.65	15,470	1.867
Strathmore Lane	12.6	59.8%	24.2%	35.6%	2380	59.4%	3880	3000	2.00	Garage	6.50	22,786	2.402
Northwest Boulevard Four-Family	12.7	67.8%	21.6%	46.2%	3000	68.1%	4200	1200	1.25	Garage	6.71	10,261	1.322
McKenzie Drive Two-Family	12.8	53.2%	12.4%	40.8%	2750	76.6%	4695	1945	2.00	Garage	6.45	15,413	1.468
Concord Village	12.9	19.6%	1.3%	18.3%	758	93.2%	2806	4097	2.00	Garage	10.50	47,653	2.804
City Park	12.10	44.8%	0.0%	44.8%	1054	100.0%	2242	1972	0.00	None	18.52	38,519	2.552

*Summarized from form GT context records.

(a)

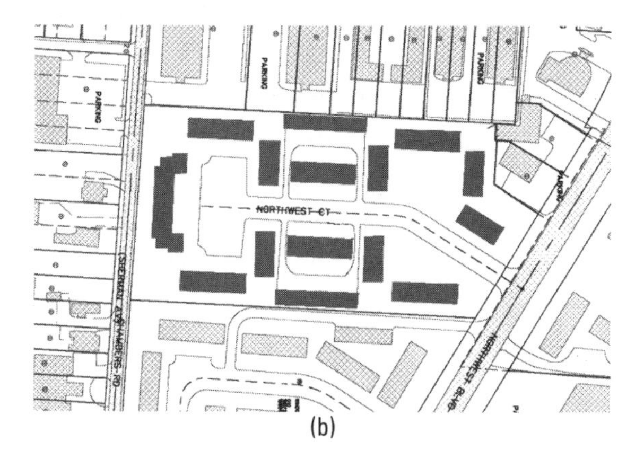

(b)

(c)

(d)

Figure 12.1 Northwest Court. Northwest Court provides a residential level of common open space but no private yard areas, and it depends on roadway parking to serve a large portion of its parking needs. When this approach is combined with the values measured in its context record, it produces a density of 14.83 dwelling units per buildable acre and a development capacity index of 17,110 sq ft of building area per acre of buildable land area. (*a*) Overview; (*b*) site plan; (*c*) and (*d*) street-level views.

(Photos by W. M. Hosack)

ber of parking spaces provided per dwelling unit is low since roadway parallel parking is used for part of this total. The density is relatively high based on the small dwelling unit areas, low parking ratio, and roadway parallel parking provided, but the SFAC value is modest because of the common open space quantity introduced and the surface parking provided. The development intensity index INX for this project is 2.459, which is well within residential expectations, but an impersonal quality is present since no open space is used for dwelling unit yard, patio, or deck areas, even though the total open space provided is adequate to accommodate this differentiation.

The record of Northwest Court context measurements is displayed in Table 12.2. A number of internal calculations based on the measurements entered are also displayed with this record. These values have been entered in the design specification panel of Exhibit 12.1 to demonstrate that the Northwest Court development results could have been predicted before construction using forecast model RGTL. This model can be found within the collection of development capacity forecast models attached to this book in CD-ROM format. (Refer to Fig. 1.1.) Exhibit 12.1 also shows that a 4% reduction in the common open space provided could have produced eight more dwelling units. It also shows that a 16% increase in the open space provided would have resulted in 30 less dwelling units, assuming all other specification values remain constant.[7]

DWELLING UNIT SPECIFICATION

Exhibit 12.1 has been somewhat simplified throughout this chapter. The model contains a panel, entitled Dwelling Unit Specification, that permits the user to investigate the development capacity implications of various dwelling unit types, mixes, areas, yards, and accessories. These options combine to produce an average habitable footprint and private yard area AFP per dwelling unit.[8] Dwelling unit breakdowns have

Table 12.2 Northwest Court Context Record

Name	**Northwest Court**
Address or Location	**1838 Northwest Court**
Land Use	**Residential RGT roadway parking**
Parking System	*Any combination of parking lot, roadway and/or garage parking*

Form GT

NOTE: An urban house is a single -family ATTACHED home that includes twin-singles, three-families, four-families, rowhouses, townhouses, garden apartments, etc. OR a single-family detached home on a lot less than 40 ft, wide. A townhouse implies separate ownership and defined private open space. A rowhouse implies rental status and less defined private open space.

	Description	Project Measurements		Context Summary	%	
Site						
	Gross land area in acres	4.450	GLA	193,842		
	Public/ private ROW and paved easements in SF	0	W		0.0%	of GLA
	Net lot area		NLA	193,842	100.0%	of GLA
	Total unbuildable and/or future expansion in SF	0	U		0.0%	of GLA
	Water area within unbuildable area in SF	0	WAT			of U
	Buildable land area		BLA	193,842	100.0%	of GLA
Development Cover						
	Gross habitable building area in SF	70,200	GBA			
	Total building footprint in SF	35,100	B		18.1%	of BLA
	Number of dwelling units	66	DU			
	Total private yard areas in SF	0				
	Avg. private yard area per dwelling unit			0		
	Aggregate average dwelling unit area in SF		AGG	1,064		
	Avg. habitable footprint plus yard in SF (no parking)		AFP	532		
	Number of building floors	2.00	FLR			
	Parking lot footprint in SF (include internal landscape)	12,000	P or G		6.2%	of BLA
	Parking lot footprint under building(s) in SF	0				
	Number of parking spaces	34	NPS			
	Parking lot spaces per dwelling unit		u	0.52		
	Gross parking lot area per parking space in SF		s	352.9		
	Private roadway parking spaces	40	RPS			
	Net roadway parking area in SF	15,840	RPA		8.2%	of BLA
	Roadway parking area under building(s) in SF	0				
	Private roadway parking spaces per dwelling unit		Rn	0.61		
	Net parking area per roadway space in SF		Rs	396.00		
	Total garage area in SF	5,940	GPA			
	Garage parking area under building(s) in SF	0				
	Number of garage parking spaces	26	GPS			
	Garage parking spaces per dwelling		Gn	0.39		
	Total garage area per space in SF		Ga	228.46		
	Total garage building cover percentage		Gc		3.1%	of BLA
	Total parking spaces per dwelling unit			1.52		
	Social and misc. building footprint(s) in SF	0	Sf			
	No. of social bldg. parking spaces	0	Sn			
	Social parking footprint (include internal landscape) in SF	0	Spla		0.0%	of BLA
	Driveway areas in SF	10,100	R		5.2%	of BLA
	Misc. pavement and social center pvmt. in SF	9,980	M		5.1%	of BLA
	TOTAL DEVELOPMENT COVER		D	88,960	**45.9%**	of BLA
Project Open Space						
	TOTAL PROJECT OPEN SPACE (OSA)		S	104,882	**54.1%**	of BLA
	Common open space area			104,882	54.1%	of BLA
	Private yard area			0	0.0%	of BLA
	Private yard area per dwelling unit			0		
	Private yard percentage of total project open space				0.0%	of OSA
	Water area within project open space in SF	0	wat	0	0.0%	of OSA
Summary						
	Development balance index		BNX		0.459	
	Development intensity index		INX		2.459	
	Floor area ratio per buildable acre		BFAR		0.362	
	Capacity index (gross building sq. ft. per buildable acre)		SFAC		17,110	
	Density per buildable acre if applicable		dBA		14.83	
	Density per net acre if applicable		dNA		14.83	
	Density per gross acre if applicable		dGA		14.83	

Exhibit 12.1 Northwest Court Design Specification and Forecast

Forecast Model RGTL

Development capacity forecast for ***TOWNHOUSES*** *based on the use of shared PARKING LOTS and / or PRIVATE GARAGES serving each dwelling unit.*
Given: *Gross land area.* ***To Find:*** *Maximum dwelling unit capacity of the land area given based on the design specification values entered below.*

DESIGN SPECIFICATION *Enter values in boxed areas where text is bold and blue. Express all fractions as decimals.*

			Value			
Given:	**Gross land area**	GLA=	**4.450**	acres	193,842	SF
Land Variables:	Public right-of-way as fraction of GLA	W=	**0.000**		0	SF
	Private right-of-way as fraction of GLA	Wp=	**0.000**		0	SF
	Net land area	NLA=	4.450	acres	193,842	SF
	Unbuildable and/or future expansion areas as fraction of GLA	U=	**0.000**		0	SF
	Total gross land area reduction as fraction of GLA	X=	0.000		0	SF
	Buildable land area remaining	BLA=	**4.450**	acres	193,842	SF
Parking Variables:	Parking lot spaces required or planned per DU	u=	**0.515**	enter zero if no shared parking lot		
	Est. gross pkg. area per pkg. lot space in SF	s =	**353**	enter zero if no shared parking lot		
	Parking lot area under bldg. per DU in SF		**0.0**			
	Roadway pkg. spaces per DU	Rn=	**0.606**	enter zero if no roadway parking	*NOTE: Express all fractions as decimals*	
	Est. net pkg area per roadway space in SF	Rs=	**396**	enter zero if no roadway parking		
	Roadway pkg. area under bldg. per DU in SF		**0.0**			
	Garage spaces per DU		**0.394**		1.52	total pkg per du
	Est. gross area per garage parking space in SF	Ga=	**228.5**	enter zero if no garage parking	511.8	pkg area per du
	Garage pkg. area under bldg. per DU in SF		**0.0**			
Support Variables:	Driveways as fraction of BLA	R=	**0.052**			
	Misc. pavement as fraction of BLA	M=	**0.051**		9,886	SF
	Outdoor social / entertainment area in SF	Sp=	**0**		0.000	fraction of BLA
	Social / service building(s) footprint in SF	Sf=	**0**		0.000	fraction of BLA
	Social / service building(s) gross area in SF	Sg=	**0**			
	Number of social / service parking spaces	Sn=	**0**	**400** gross SF / space	0.000	fraction of BLA
	Total support areas as a fraction of BLA	Su=	0.103		19,966	SF
	Core + open space + parking as fraction of BLA	CSR=	0.897	CS+Su must = 1	173,876	SF

DWELLING UNIT SPECIFICATION

dwelling unit type DU	no. of flrs. FLR above grade	dwelling unit MIX mix	total HAB habitable area	habitable area FTP footprint	private yard area YRD with townhouse	footprint and yard AFP no garage	du basement BSM area	du crawl CRW space area

532	NOTE: AFP override in effect
532	= AFP override

Enter zero in the adjacent box unless you wish to override the AFP value calculated above: **532**

DWELLING UNIT CAPACITY FORECAST *(based on common open space (COS) provided)*

common COS open space	number of NDU dwelling units	EFF	1BR	2BR	3BR	4BR	density per dBA buildable acre	parking lot NPS spaces	roadway RDS parking spaces	garage GAR parking spaces
		dwelling unit type (DU) quantities given mix above								
0.000	166.6						37.43	85.8	100.9	65.6
0.100	148.0						33.26	76.2	89.7	58.3
0.200	129.4	Dwelling unit data not forecast for this example					29.09	66.7	78.4	51.0
0.300	110.9						24.91	57.1	67.2	43.7
0.400	92.3						20.74	47.5	55.9	36.4
0.500	73.7						16.57	38.0	44.7	29.0
0.541	**66.1**						**14.86**	**34.0**	**40.1**	**26.0**
0.700	36.6						8.22	18.8	22.2	14.4
0.800	18.0						4.05	9.3	10.9	7.1

WARNING: These are preliminary forecasts that must not be used to make final decisions.
1) These forecasts are not a substitute for the "due diligence" research that must be conducted to support the final definition of "unbuildable areas" above and the final decision to purchase land. This research includes, but is not limited to, verification of adequate subsurface soil, zoning, environmental clearance, access, title, utilities and water pressure, clearance from deed restriction, easement and right-of-way encumbrances, clearance from existing above and below ground facility conflicts, etc.
2) The most promising forecast(s) made on the basis of data entered in the design specification from "due diligence" research must be verified at the drawing board before funds are committed and land purchase decisions are made. Actual land shape ratios, dimensions and irregularities encountered may require adjustments to the general forecasts above.
3) The software licensee shall take responsibility for the design specification values entered and any advice given that is based on the forecast produced.

been omitted from context record data collection forms since the AFP value is calculated within the context record from other data entered. The AFP value calculated in the context record has been used in each forecast model to produce the predictions shown, and the dwelling unit specification panel is blank in each exhibit as a result. The forecast models on the CD-ROM contain a data-override feature, discussed in Chap. 14, that permits the dwelling unit mix panel to be completed and still be overridden by a single alternate value to expedite potential comparisons. When writing a design specification for a new project, it would be wise to complete the dwelling unit specification panel rather than use the shortcut just described. This shortcut can be useful when preparing context records for existing projects, and during plan review; however, careful attention should be given to the dwelling unit specification when writing design specifications for new development projects, since these values directly affect the density that can be achieved.

Park Street

Park Street is a relatively new inner-city development project that devotes its first floor to private garages. The urban house occupies two floors above. Figure 12.2 illustrates a very compact site plan with a high SFAC development capacity index of 51,886 sq ft of building area per acre of buildable land area. Table 12.3 also shows that the total open space provided is 46.1% and that 38.9% of this open space is devoted to private yard emphasis. In this case, the private yards constitute second-level decks overlooking common open space below. This configuration also produces the highest density recorded at 21.40 dwelling units per buildable acre, and provides two parking spaces per dwelling unit. The driveways accommodate two more parking spaces, and the plan illustrates the land conservation that can result when the automobile is stored beneath the dwelling. This easily surpasses the City Park urban house SFAC develop-

(a)

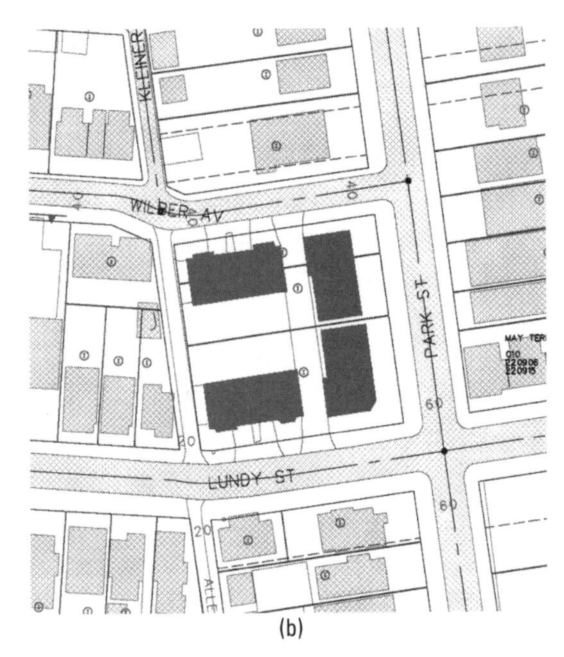

(b)

(c)

Figure 12.2 Park Street. Park Street places garage parking above grade beneath the building and surrounds cloistered common space with second-level private decks. When this approach is combined with the values measured in its context record, it produces a density of 21.40 dwelling units per buildable acre and a development capacity index of 51,886 sq ft of building area per acre of buildable land area. (*a*) and (*c*) Overviews; (*b*) site map.

(Photos by W. M. Hosack)

ment capacity of 38,519 sq ft of building area per acre of buildable land area and density of 18.52, even though the inner-city City Park example provides no on-site parking. The Park Street density is higher because the City Park example provides a larger habitable footprint and yard area per dwelling unit.

The preceding quoted data come from Table 12.3, which is the context record for Park Street. These data have been entered in Exhibit 12.2 to produce the development capac-

Table 12.3 Park Street Context Record

Name	Park Street
Address or Location	811-845 Park Street
Land Use	Urban house
	Subdwelling garage plan
Parking System	*Any combination of parking lot, roadway, and/or garage parking*

Form GT

NOTE: An urban house is a single -family ATTACHED home that includes twin-singles, three-families, four-families, rowhouses, townhouses, garden apartments, etc. OR a single-family detached home on a lot less than 40 ft. wide. A townhouse implies separate ownership and defined private open space. A rowhouse implies rental status and less defined private open space.

	Description	Project Measurements		Context Summary	%	
Site						
	Gross land area in acres	0.744	GLA	32,409		
	Public/ private ROW and paved easements in SF	3,960	W		12.2%	of GLA
	Net lot area		NLA	28,449	87.8%	of GLA
	Total unbuildable and/or future expansion in SF	0	U		0.0%	of GLA
	Water area within unbuildable area in SF	0	WAT			of U
	Buildable land area		BLA	28,449	87.8%	of GLA
Development Cover						
	Gross habitable building area in SF	24,644	GBA			
	Total building footprint in SF	12,322	B		43.3%	of BLA
	Number of dwelling units	14	DU			
	Total private yard areas in SF	5,100				
	Avg. private yard area per dwelling unit			364		
	Aggregate average dwelling unit area in SF		AGG	1,760		
	Avg. habitable footprint plus yard in SF (no parking)		AFP	1,244		
	Number of building floors	2.00	FLR			
	Parking lot footprint in SF (include internal landscape)	0	P or G		0.0%	of BLA
	Parking lot footprint under building(s) in SF	0				
	Number of parking spaces	0	NPS			
	Parking lot spaces per dwelling unit		u	0.00		
	Gross parking lot area per parking space in SF		s			
	Private roadway parking spaces	0	RPS			
	Net roadway parking area in SF	0	RPA		0.0%	of BLA
	Roadway parking area under building(s) in SF	0				
	Private roadway parking spaces per dwelling unit		Rn	0.00		
	Net parking area per roadway space in SF		Rs			
	Total garage area in SF	9,242	GPA			
	Garage parking area under building(s) in SF	9,242	Gu	330.07		
	Number of garage parking spaces	28	GPS			
	Garage parking spaces per dwelling		Gn	2.00		
	Total garage area per space in SF		Ga	330.07		
	Total garage building cover percentage		Gc		0.0%	of BLA
	Total parking spaces per dwelling unit			2.00		
	Social and misc. building footprint(s) in SF	0	Sf			
	No. of social bldg. parking spaces	0	Sn			
	Social parking footprint (include internal landscape) in SF	0	Spla		0.0%	of BLA
	Driveway areas in SF	1,600	R		5.6%	of BLA
	Misc. pavement and social center pvmt. in SF	1,400	M		4.9%	of BLA
	TOTAL DEVELOPMENT COVER		D	15,322	**53.9%**	of BLA
Project Open Space						
	TOTAL PROJECT OPEN SPACE (OSA)		S	13,127	**46.1%**	of BLA
	Common open space area			8,027	28.2%	of BLA
	Private yard area			5,100	17.9%	of BLA
	Private yard area per dwelling unit			364		
	Private yard percentage of total project open space				38.9%	of OSA
	Water area within project open space in SF	0	wat	0	0.0%	of OSA
Summary						
	Development balance index		BNX		0.539	
	Development intensity index		INX		2.539	
	Floor area ratio per buildable acre		BFAR		0.866	
	Capacity index (gross building sq. ft. per buildable acre)		SFAC		51,886	
	Density per buildable acre if applicable		dBA		21.44	
	Density per net acre if applicable		dNA		21.44	
	Density per gross acre if applicable		dGA		18.82	

Exhibit 12.2 Park Street Design Specification and Forecast

Forecast Model RGTL

Development capacity forecast for ***TOWNHOUSES*** *based on the use of shared PARKING LOTS and / or PRIVATE GARAGES serving each dwelling unit.*
Given: *Gross land area.* ***To Find:*** *Maximum dwelling unit capacity of the land area given based on the design specification values entered below.*

DESIGN SPECIFICATION

Enter values in boxed areas where text is bold and blue. Express all fractions as decimals.

Given:	**Gross land area**	GLA=	**0.744**	acres	32,409	SF
Land Variables:	Public right-of-way as fraction of GLA	W=	**0.122**	percent	3,954	SF
	Private right-of-way as fraction of GLA	Wp=	**0.000**	percent	0	SF
	Net land area	NLA=	0.653	acres	28,455	SF
	Unbuildable and/or future expansion areas as fraction of GLA	U=	**0.000**	percent	0	SF
	Total gross land area reduction as fraction of GLA	X=	0.122	percent	5,314	SF
	Buildable land area remaining	BLA=	**0.653**	acres	28,455	SF
Parking Variables:	Parking lot spaces required or planned per DU	u=	**0**	enter zero if no shared parking lot		
	Est. gross pkg. area per pkg. lot space in SF	s =	**0**	enter zero if no shared parking lot		
	Parking lot area under bldg. per DU in SF		**0.0**			
	Roadway pkg. spaces per DU	Rn=	**0**	enter zero if no roadway parking	*NOTE: Express all fractions as decimals*	
	Est. net pkg. area per roadway space in SF	Rs=	**0**	enter zero if no roadway parking		
	Roadway pkg. area under bldg. per DU in SF		**0.0**			
	Garage spaces per DU		**2**		2.00	total pkg per du
	Est. gross area per garage parking space in SF	Ga=	**330.1**	enter zero if no garage parking	660.14	pkg area per du
	Garage pkg. area under bldg. per DU in SF		**330.1**			
Support Variables:	Driveways as fraction of BLA	R=	**0.056**		1,593	SF
	Misc. pavement as fraction of BLA	M=	**0.049**		1,394	SF
	Outdoor social / entertainment area in SF	Sp=	**0**		0.000	fraction of BLA
	Social / service building(s) footprint in SF	Sf=	**0**		0.000	fraction of BLA
	Social / service building(s) gross area in SF	Sg=	**0**			
	Number of social / service parking spaces	Sn=	**0**	**400** gross SF / space	0.000	fraction of BLA
	Total support areas as a fraction of BLA	Su=	0.105		2,988	SF
	Core + open space + parking as fraction of BLA	CSR=	0.895	CS+Su must = 1	25,467	SF

DWELLING UNIT SPECIFICATION

dwelling unit type DU	no. of flrs. FLR above grade	dwelling unit MIX mix	total HAB habitable area	habitable area FTP footprint	private yard area YRD with townhouse	footprint and yard AFP no garage	du basement BSM area	du crawl CRW space area

	1,244	NOTE: AFP override in effect
Enter zero in the adjacent box unless you wish to override the AFP value calculated above:	**1,244**	= AFP override

DWELLING UNIT CAPACITY FORECAST

(based on common open space (COS) provided)

common COS open space	number of NDU dwelling units	EFF	1BR	2BR	3BR	4BR	density per dBA buildable acre	parking lot NPS spaces	roadway RDS parking spaces	garage GAR parking spaces
		dwelling unit type (DU) quantities given mix above								
0.000	20.5						31.34	0.0	0.0	40.9
0.100	18.2		Dwelling unit data not forecast for this example				27.84	0.0	0.0	36.4
0.200	15.9						24.34	0.0	0.0	31.8
0.282	**14.0**						**21.46**	**0.0**	**0.0**	**28.0**
0.300	13.6						20.83	0.0	0.0	27.2
0.500	9.0						13.83	0.0	0.0	18.1
0.600	6.7						10.33	0.0	0.0	13.5
0.700	4.5						6.83	0.0	0.0	8.9
0.800	2.2						3.33	0.0	0.0	4.3

WARNING: These are preliminary forecasts that must not be used to make final decisions.
1) These forecasts are not a substitute for the "due diligence" research that must be conducted to support the final definition of "unbuildable areas" above and the final decision to purchase land. This research includes, but is not limited to, verification of adequate subsurface soil, zoning, environmental clearance, access, title, utilities and water pressure, clearance from deed restriction, easement and right-of-way encumbrances, clearance from existing above and below ground facility conflicts, etc.
2) The most promising forecast(s) made on the basis of data entered in the design specification from "due diligence" research must be verified at the drawing board before funds are committed and land purchase decisions are made. Actual land shape ratios, dimensions and irregularities encountered may require adjustments to the general forecasts above.
3) The software licensee shall take responsibility for the design specification values entered and any advice given that is based on the forecast produced.

ity prediction for this project. This forecast shows what options were available during the planning phase of this project prior to contract document preparation and field construction. As usual, any change to the design specification values entered, to the common open space provided, or to the average dwelling unit specification value AFP anticipated would produce a new development capacity forecast.

Branford Commons, Sherwood Villa, and Strathmore Lane

These condominium urban homes have increasing amounts of common open space dedication within their subdivision site plans. The Branford Commons and Sherwood Villa projects are located within the jurisdiction of a first-ring suburb, but were constructed at a point in its expansion that brings them closer to typical suburban characteristics. The Strathmore Lane project is located in a remote planned community that is part of an edge-city jurisdiction. The Sherwood Villa project actually provides the greatest amount of total open space, but not the greatest amount of private yard or personal space per dwelling unit. This distinction belongs to the remote suburb, Strathmore Lane. (See Figs. 12.3 through 12.5.)

Table 12.1 shows that Strathmore Lane provides the least common open space and the most personal, or private, open space of the three. Since Strathmore Lane provided less total open space, however, it had a slightly higher density at 6.50. Neither project came close to the Branford Commons density of 10.16, however, since both provided significantly greater amounts of total open space, had much larger habitable footprint areas AFP, and constructed larger aggregate average dwelling unit areas AGG.

The measurements that produced these statistics are recorded in the context records in Tables 12.4 through 12.6. The forecast models that predict these results from the values measured are included as Exhibits 12.3 through 12.5.

(a)

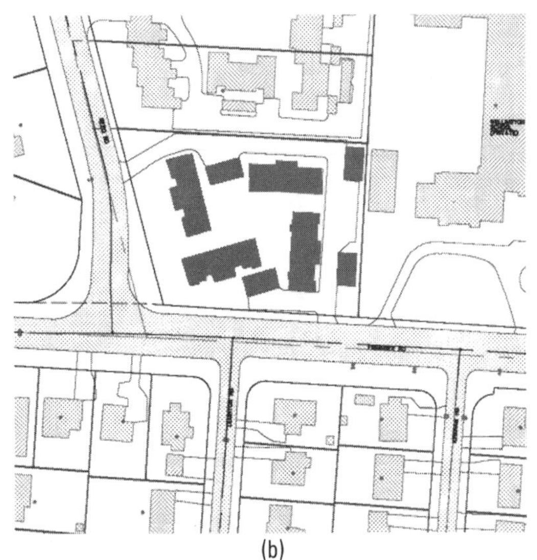

(b)

(c)

(d)

Figure 12.3 Branford Commons. (*a*) Branford Commons provides cloistered common open space with parking support arranged along the perimeter. Selected units have small private yards, but many are limited to the individual space associated with their front-entry doors. This produces some private yard emphasis, but privacy depends primarily on the cloistered open space that is shared. When this approach is combined with the values measured in its context record, it produces a density of 10.16 dwelling units per buildable acre and a development capacity index of 20,216 sq ft of building area per acre of buildable land area. (*b*) Site map. (*c*) Courtyard orientation. (*d*) Parking orientation.

(Photos by W. M. Hosack)

(a)

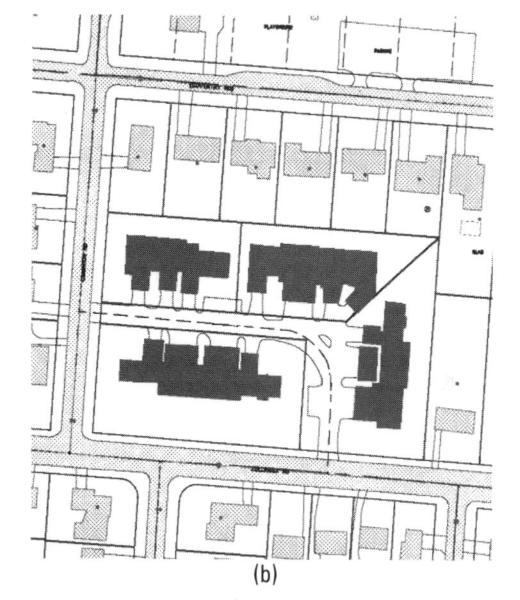

(b)

(c)

Figure 12.4 Sherwood Villa. Sherwood Villa places parking in the center of the complex and private yard areas along the perimeter to increase private yard emphasis. This produces a service entry impression to visitors, walls along the perimeter, and higher noise levels adjacent to yard areas abutting traffic movement. When this approach is combined with the values measured in its context record, it produces a density of 5.65 dwelling units per buildable acre and a development capacity index of 15,470 sq ft of building area per acre of buildable land area. (*a*) Overview; (*b*) site plan; (*c*) street-level view.

(Photos by W. M. Hosack)

(a)

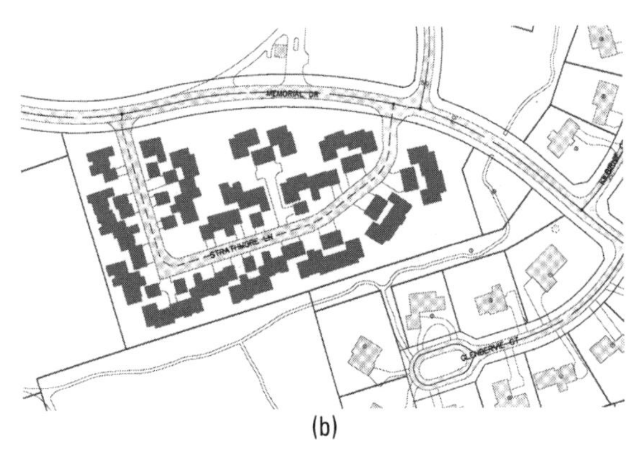

(b)

(c)

(d)

Figure 12.5 Strathmore Lane. Strathmore Lane is an edge-city version of the Sherwood Villa planning concept that benefits from more interesting topography, more surrounding open space, and lower traffic volumes. When this approach is combined with the values measured in its context record, it produces a density of 6.5 dwelling units per buildable acre and a development capacity index of 22,786 sq ft of building area per acre of buildable land area. (*a*), (*c*), and (*d*) Street-level views; (*b*) site plan.

(Photos by W. M. Hosack)

Table 12.4 Branford Context Record

Name	**Branford Condominiums**
Address or Location	**NE corner Reed and Fishinger Roads**
Land Use	**Urban house**
	Surface parking and remote garage plan
Parking System	*Any combination of parking lot, roadway, and/or garage parking*

Form GT

NOTE: An urban house is a single family ATTACHED home that includes twin-singles, three-families, four-families, rowhouses, townhouses, garden apartments, etc. OR a single-family detached home on a lot less than 40 ft. wide. A townhouse implies separate ownership and defined private open space. A rowhouse implies rental status and less defined private open space.

	Description	Project Measurements		Context Summary	%	
Site						
	Gross land area in acres	1.575	GLA	68,607		
	Public/ private ROW and paved easements in SF	0	W		0.0%	of GLA
	Net lot area		NLA	68,607	100.0%	of GLA
	Total unbuildable and/or future expansion in SF	0	U		0.0%	of GLA
	Water area within unbuildable area in SF	0	WAT			of U
	Buildable land area		BLA	68,607	100.0%	of GLA
Development Cover						
	Gross habitable building area in SF	28,320	GBA			
	Total building footprint in SF	14,160	B		20.6%	of BLA
	Number of dwelling units	16	DU			
	Total private yard areas in SF	9,000				
	Avg. private yard area per dwelling unit			563		
	Aggregate average dwelling unit area in SF		AGG	1,770		
	Avg. habitable footprint plus yard in SF (no parking)		AFP	1,448		
	Number of building floors	2.00	FLR			
	Parking lot footprint in SF (include internal landscape)	7,500	P or G		10.9%	of BLA
	Parking lot footprint under building(s) in SF	0				
	Number of parking spaces	16	NPS			
	Parking lot spaces per dwelling unit		u	1.00		
	Gross parking lot area per parking space in SF		s	468.8		
	Private roadway parking spaces	0	RPS			
	Net roadway parking area in SF	0	RPA		0.0%	of BLA
	Roadway parking area under building(s) in SF	0				
	Private roadway parking spaces per dwelling unit		Rn	0.00		
	Net parking area per roadway space in SF		Rs			
	Total garage area in SF	3,520	GPA			
	Garage parking area under building(s) in SF	0				
	Number of garage parking spaces	16	GPS			
	Garage parking spaces per dwelling		Gn	1.00		
	Total garage area per space in SF		Ga	220.00		
	Total garage building cover percentage		Gc		5.1%	of BLA
	Total parking spaces per dwelling unit			2.00		
	Social and misc. building footprint(s) in SF	0	Sf			
	No. of social bldg. parking spaces	0	Sn			
	Social parking footprint (include internal landscape) in SF	0	Spla		0.0%	of BLA
	Driveway areas in SF	10,450	R		15.2%	of BLA
	Misc. pavement and social center pvmt. in SF	2,160	M		3.1%	of BLA
	TOTAL DEVELOPMENT COVER		D	37,790	**55.1%**	of BLA
Project Open Space						
	TOTAL PROJECT OPEN SPACE (OSA)		S	30,817	**44.9%**	of BLA
	Common open space area			21,817	31.8%	of BLA
	Private yard area			9,000	13.1%	of BLA
	Private yard area per dwelling unit			563		
	Private yard percentage of total project open space				29.2%	of OSA
	Water area within project open space in SF	0	wat	0	0.0%	of OSA
Summary						
	Development balance index		BNX		0.551	
	Development intensity index		INX		2.551	
	Floor area ratio per buildable acre		BFAR		0.413	
	Capacity index (gross building sq. ft. per buildable acre)		SFAC		20,216	
	Density per buildable acre if applicable		dBA		10.16	
	Density per net acre if applicable		dNA		10.16	
	Density per gross acre if applicable		dGA		10.16	

Table 12.5 *Sherwood Villa Context Record*

Name	Sherwood Villa
Address or Location	Zollinger at Cimmarron
Land Use	Urban house
	Adjacent garage parking plan
Parking System	*Any combination of parking lot, roadway and/or garage parking*

Form GT

NOTE: An urban house is a single -family ATTACHED home that includes twin-singles, three-families, four-families, rowhouses, townhouses, garden apartments, etc. OR a single-family detached home on a lot less than 40 ft. wide. A townhouse implies separate ownership and defined private open space. A rowhouse implies rental status and less defined private open space.

	Description	Project Measurements		Context Summary	%	
Site						
	Gross land area in acres	3.941	GLA	171,670		
	Public/ private ROW and paved easements in SF	25,300	W		14.7%	of GLA
	Net lot area		NLA	146,370	85.3%	of GLA
	Total unbuildable and/or future expansion in SF	0	U		0.0%	of GLA
	Water area within unbuildable area in SF	0	WAT			of U
	Buildable land area		BLA	146,370	85.3%	of GLA
Development Cover						
	Gross habitable building area in SF	43,072	GBA			
	Total building footprint in SF	28,715	B		19.6%	of BLA
	Number of dwelling units	19	DU			
	Total private yard areas in SF	40,500				
	Avg. private yard area per dwelling unit			2,132		
	Aggregate average dwelling unit area in SF		AGG	2,267		
	Avg. habitable footprint plus yard in SF (no parking)		AFP	3,643		
	Number of building floors	1.50	FLR			
	Parking lot footprint in SF (include internal landscape)	3,910	P or G		2.7%	of BLA
	Parking lot footprint under building(s) in SF	0				
	Number of parking spaces	13	NPS			
	Parking lot spaces per dwelling unit		u	0.68		
	Gross parking lot area per parking space in SF		s	300.8		
	Private roadway parking spaces	0	RPS			
	Net roadway parking area in SF	0	RPA		0.0%	of BLA
	Roadway parking area under building(s) in SF	0				
	Private roadway parking spaces per dwelling unit		Rn	0.00		
	Net parking area per roadway space in SF		Rs			
	Total garage area in SF	8,910	GPA			
	Garage parking area under building(s) in SF	0				
	Number of garage parking spaces	38	GPS			
	Garage parking spaces per dwelling		Gn	2.00		
	Total garage area per space in SF		Ga	234.47		
	Total garage building cover percentage		Gc		6.1%	of BLA
	Total parking spaces per dwelling unit			2.68		
	Social and misc. building footprint(s) in SF	0	Sf			
	No. of social bldg. parking spaces	0	Sn			
	Social parking footprint (include internal landscape) in SF	0	Spla		0.0%	of BLA
	Driveway areas in SF	7,400	R		5.1%	of BLA
	Misc. pavement and social center pvmt. in SF	4,800	M		3.3%	of BLA
	TOTAL DEVELOPMENT COVER		D	53,735	**36.7%**	of BLA
Project Open Space						
	TOTAL PROJECT OPEN SPACE (OSA)		S	92,635	**63.3%**	of BLA
	Common open space area			52,135	35.6%	of BLA
	Private yard area			40,500	27.7%	of BLA
	Private yard area per dwelling unit			2,132		
	Private yard percentage of total project open space				43.7%	of OSA
	Water area within project open space in SF	0	wat	0	0.0%	of OSA
Summary						
	Development balance index		BNX		0.367	
	Development intensity index		INX		1.867	
	Floor area ratio per buildable acre		BFAR		0.294	
	Capacity index (gross building sq. ft. per buildable acre)		SFAC		15,470	
	Density per buildable acre if applicable		dBA		5.65	
	Density per net acre if applicable		dNA		5.65	
	Density per gross acre if applicable		dGA		4.82	

Table 12.6 Strathmore Context Record

Name	**Strathmore Lane**
Address or Location	**Strathmore Lane**
Land Use	**Urban house**
	Adjacent garage parking plan
Parking System	*Any combination of parking lot, roadway, and/or garage parking*

Form GT

NOTE: An urban house is a single family ATTACHED home that includes twin-singles, three-families, four-families, rowhouses, townhouses, garden apartments, etc. OR a single-family detached home on a lot less than 40 ft. wide. A townhouse implies separate ownership and defined private open space. A rowhouse implies rental status and less defined private open space.

	Description	Project Measurements		Context Summary	%	
Site						
	Gross land area in acres	5.520	GLA	240,451		
	Public/ private ROW and paved easements in SF	26,100	W		10.9%	of GLA
	Net lot area		NLA	214,351	89.1%	of GLA
	Total unbuildalbe and/or future expansion in SF	0	U		0.0%	of GLA
	Water area within unbuildable area in SF	0	WAT			of U
	Buildable land area		BLA	214,351	89.1%	of GLA
Development Cover						
	Gross habitable building area in SF	96,000	GBA			
	Total building footprint in SF	48,000	B		22.4%	of BLA
	Number of dwelling units	32	DU			
	Total private yard areas in SF	76,150				
	Avg. private yard area per dwelling unit			2,380		
	Aggregate average dwelling unit area in SF		AGG	3,000		
	Avg. habitable footprint plus yard in SF (no parking)		AFP	3,880		
	Number of building floors	2.00	FLR			
	Parking lot footprint in SF (include internal landscape)	0	P or G		0.0%	of BLA
	Parking lot footprint under building(s) in SF	0				
	Number of parking spaces	0	NPS			
	Parking lot spaces per dwelling unit		u	0.00		
	Gross parking lot area per parking space in SF		s			
	Private roadway parking spaces	0	RPS			
	Net roadway parking area in SF	0	RPA		0.0%	of BLA
	Roadway parking area under building(s) in SF	0				
	Private roadway parking spaces per dwelling unit		Rn	0.00		
	Net parking area per roadway space in SF		Rs			
	Total garage area in SF	16,128	GPA			
	Garage parking area under building(s) in SF	0				
	Number of garage parking spaces	64	GPS			
	Garage parking spaces per dwelling		Gn	2.00		
	Total garage area per space in SF		Ga	252.00		
	Total garage building cover percentage		Gc		7.5%	of BLA
	Total parking spaces per dwelling unit			2.00		
	Social and misc. building footprint(s) in SF	0	Sf			
	No. of social bldg. parking spaces	0	Sn			
	Social parking footprint (include internal landscape) in SF	0	Spla		0.0%	of BLA
	Driveway areas in SF	12,500	R		5.8%	of BLA
	Misc. pavement and social center pvmt. in SF	9,600	M		4.5%	of BLA
	TOTAL DEVELOPMENT COVER		D	86,228	**40.2%**	of BLA
Project Open Space						
	TOTAL PROJECT OPEN SPACE (OSA)		S	128,123	**59.8%**	of BLA
	Common open space area			51,973	24.2%	of BLA
	Private yard area			76,150	35.5%	of BLA
	Private yard area per dwelling unit			2,380		
	Private yard percentage of total project open space				59.4%	of OSA
	Water area within project open space in SF	0	wat	0	0.0%	of OSA
Summary						
	Development balance index		BNX		0.402	
	Development intensity index		INX		2.402	
	Floor area ratio per buildable acre		BFAR		0.448	
	Capacity index (gross building sq. ft. per buildable acre)		SFAC		22,786	
	Density per buildable acre if applicable		dBA		6.50	
	Density per net acre if applicable		dNA		6.50	
	Density per gross acre if applicable		dGA		5.80	

Exhibit 12.3 Branford Commons Design Specification and Forecast

Forecast Model RGTL

Development capacity forecast for ***TOWNHOUSES*** *based on the use of shared PARKING LOTS and / or PRIVATE GARAGES serving each dwelling unit.*
Given: *Gross land area.* ***To Find:*** *Maximum dwelling unit capacity of the land area given based on the design specification values entered below.*

DESIGN SPECIFICATION

Enter values in boxed areas where text is bold and blue. Express all fractions as decimals.

			Value			
Given:	**Gross land area**	GLA=	**1.575**	acres	68,607	SF
Land Variables:	Public right-of-way as fraction of GLA	W=	**0.000**		0	SF
	Private right-of-way as fraction of GLA	Wp=	**0.000**		0	SF
	Net land area	NLA=	1.575	acres	68,607	SF
	Unbuildable and/or future expansion areas as fraction of GLA	U=	**0.000**		0	SF
	Total gross land area reduction as fraction of GLA	X=	0.000		0	SF
	Buildable land area remaining	**BLA=**	**1.575**	acres	68,607	SF
Parking Variables:	Parking lot spaces required or planned per DU	u=	**1**	enter zero if no shared parking lot		
	Est. gross pkg. area per pkg lot space in SF	s =	**468.8**	enter zero if no shared parking lot		
	Pkg. lot area under bldg. per DU in SF		**0.0**			
	Roadway pkg. spaces per DU	Rn=	**0**	enter zero if no roadway parking	*NOTE: Express all fractions as decimals*	
	Est. net pkg. area per roadway space in SF	Rs=	**0**	enter zero if no roadway parking		
	Roadway pkg. area under bldg. per DU in SF		**0.0**			
	Garage spaces per DU		**1**		2.00	total pkg per du
	Est. gross area per garage parking space in SF	Ga=	**220.0**	enter zero if no garage parking	688.8	pkg area per du
	Garage pkg. area under bldg. per DU in SF		**0.0**			
Support Variables:	Driveways as fraction of BLA	R=	**0.152**			
	Misc. pavement as fraction of BLA	M=	**0.031**		2,127	SF
	Outdoor social / entertainment area in SF	Sp=	**0**		0.000	fraction of BLA
	Social / service building(s) footprint in SF	Sf=	**0**		0.000	fraction of BLA
	Social / service building(s) gross area in SF	Sg=	**0**			
	Number of social / service parking spaces	Sn=	**0**	**400** gross SF / space	0.000	fraction of BLA
	Total support areas as a fraction of BLA	Su=	0.183		12,555	SF
	Core + open space + parking as fraction of BLA	**CSR=**	0.817	CS+Su must = 1	56,052	SF

DWELLING UNIT SPECIFICATION

dwelling unit type DU	no. of flrs. FLR above grade	dwelling unit MIX mix	total HAB habitable area	habitable area FTP footprint	private yard area YRD with townhouse	footprint and yard AFP no garage	du basement BSM area	du crawl CRW space area

	1,448	NOTE: AFP override in effect
Enter zero in the adjacent box unless you wish to override the AFP value calculated above:	**1,448**	= AFP override

DWELLING UNIT CAPACITY FORECAST

(based on common open space (COS) provided)

common COS open space	number of NDU dwelling units	EFF	1BR	2BR	3BR	4BR	density per dBA buildable acre	parking lot NPS spaces	roadway RDS parking spaces	garage GAR parking spaces
0.000	26.2						16.66	26.2	0.0	26.2
0.100	23.0						14.62	23.0	0.0	23.0
0.200	19.8			Dwelling unit data not forecast for this example			12.58	19.8	0.0	19.8
0.300	16.6						10.54	16.6	0.0	16.6
0.318	**16.0**						**10.17**	**16.0**	**0.0**	**16.0**
0.500	10.2						6.46	10.2	0.0	10.2
0.600	7.0						4.42	7.0	0.0	7.0
0.700	3.8						2.39	3.8	0.0	3.8
0.800	0.5						0.35	0.5	0.0	0.5

(EFF–4BR: dwelling unit type (DU) quantities given mix above)

WARNING: These are preliminary forecasts that must not be used to make final decisions.
1) These forecasts are not a substitute for the "due diligence" research that must be conducted to support the final definition of "unbuildable areas" above and the final decision to purchase land. This research includes, but is not limited to, verification of adequate subsurface soil, zoning, environmental clearance, access, title, utilities and water pressure, clearance from deed restriction, easement and right-of-way encumbrances, clearance from existing above and below ground facility conflicts, etc.
2) The most promising forecast(s) made on the basis of data entered in the design specification from "due diligence" research must be verified at the drawing board before funds are committed and land purchase decisions are made. Actual land shape ratios, dimensions and irregularities encountered may require adjustments to the general forecasts above.
3) The software licensee shall take responsibility for the design specification values entered and any advice given that is based on the forecast produced.

Exhibit 12.4 Sherwood Villa Design Specification and Forecast

Forecast Model RGTL

*Development capacity forecast for **TOWNHOUSES** based on the use of shared PARKING LOTS and / or PRIVATE GARAGES serving each dwelling unit.*
***Given:** Gross land area. **To Find:** Maximum dwelling unit capacity of the land area given based on the design specification values entered below.*

DESIGN SPECIFICATION

Enter values in boxed areas where text is bold and blue. Express all fractions as decimals.

Given:	**Gross land area**	GLA=	**3.941**	acres	171,670	SF
Land Variables:	Public right-of-way as fraction of GLA	W=	**0.147**	percent	25,235	SF
	Private right-of-way as fraction of GLA	Wp=	**0.000**	percent	0	SF
	Net land area	NLA=	3.362	acres	146,434	SF
	Unbuildable and/or future expansion areas as fraction of GLA	U=	**0.000**	percent	0	SF
	Total gross land area reduction as fraction of GLA	X=	0.147	percent	6,403	SF
	Buildable land area remaining	BLA=	**3.362**	acres	146,434	SF
Parking Variables:	Parking lot spaces required or planned per DU	u=	**0.68**	enter zero if no shared parking lot		
	Est. gross pkg. area per pkg. lot space in SF	s =	**300.8**	enter zero if no shared parking lot		
	Parking lot area under bldg. per DU in SF		**0.0**			
	Roadway pkg. spaces per DU	Rn=	**0**	enter zero if no roadway parking	*NOTE: Express all fractions as decimals*	
	Est. net pkg. area per roadway space in SF	Rs=	**0**	enter zero if no roadway parking		
	Roadway pkg. area under bldg. per DU in SF		**0.0**			
	Garage spaces per DU		**2**		2.68	total pkg per du
	Est. gross area per garage parking space in SF	Ga=	**234.5**	enter zero if no garage parking	673.484	pkg area per du
	Garage pkg. area under bldg. per DU in SF		**0.0**			
Support Variables:	Driveways as fraction of BLA	R=	**0.051**		7,468	SF
	Misc. pavement as fraction of BLA	M=	**0.033**		4,832	SF
	Outdoor social / entertainment area in SF	Sp=	**0**		0.000	fraction of BLA
	Social / service building(s) footprint in SF	Sf=	**0**		0.000	fraction of BLA
	Social / service building(s) gross area in SF	Sg=	**0**			
	Number of social / service parking spaces	Sn=	**0**	**400** gross SF / space	0.000	fraction of BLA
	Total support areas as a fraction of BLA	Su=	0.084		12,300	SF
	Core + open space + parking as fraction of BLA	CSR=	0.916	CS+Su must = 1	134,134	SF

DWELLING UNIT SPECIFICATION

dwelling unit type DU	no. of flrs. FLR above grade	dwelling unit MIX mix	total HAB habitable area	habitable area FTP footprint	private yard area YRD with townhouse	footprint and yard AFP no garage	du basement BSM area	du crawl CRW space area

Enter zero in the adjacent box unless you wish to override the AFP value calculated above: | 3,643 | NOTE: AFP override in effect
3,643 = AFP override

DWELLING UNIT CAPACITY FORECAST

(based on common open space (COS) provided)

common COS open space	number of NDU dwelling units	EFF	1BR	2BR	3BR	4BR	density per dBA buildable acre	parking lot NPS spaces	roadway RDS parking spaces	garage GAR parking spaces
0.000	31.1						9.24	21.1	0.0	62.1
0.100	27.7						8.23	18.8	0.0	55.4
0.200	24.3		Dwelling unit data not forecast for this example				7.23	16.5	0.0	48.6
0.282	21.5						6.40	14.6	0.0	43.0
0.356	**19.0**						**5.65**	**12.9**	**0.0**	**38.0**
0.500	14.1						4.20	9.6	0.0	28.2
0.600	10.7						3.19	7.3	0.0	21.4
0.700	7.3						2.18	5.0	0.0	14.7
0.800	3.9						1.17	2.7	0.0	7.9

(dwelling unit type (DU) quantities given mix above)

WARNING: These are preliminary forecasts that must not be used to make final decisions.
1) These forecasts are not a substitute for the "due diligence" research that must be conducted to support the final definition of "unbuildable areas" above and the final decision to purchase land. This research includes, but is not limited to, verification of adequate subsurface soil, zoning, environmental clearance, access, title, utilities and water pressure, clearance from deed restriction, easement and right-of-way encumbrances, clearance from existing above and below ground facility conflicts, etc.
2) The most promising forecast(s) made on the basis of data entered in the design specification from "due diligence" research must be verified at the drawing board before funds are committed and land purchase decisions are made. Actual land shape ratios, dimensions and irregularities encountered may require adjustments to the general forecasts above.
3) The software licensee shall take responsibility for the design specification values entered and any advice given that is based on the forecast produced.

Exhibit 12.5 Strathmore Lane Design Specification and Forecast

Forecast Model RGTL

Development capacity forecast for ***TOWNHOUSES*** *based on the use of shared PARKING LOTS and / or PRIVATE GARAGES serving each dwelling unit.*
Given: *Gross land area.* ***To Find:*** *Maximum dwelling unit capacity of the land area given based on the design specification values entered below.*

DESIGN SPECIFICATION

Enter values in boxed areas where text is bold and blue. Express all fractions as decimals.

Given:	**Gross land area**	GLA=	**5.520**	acres	240,451	SF
Land Variables:	Public right-of-way as fraction of GLA	W=	**0.000**	percent	0	SF
	Private right-of-way as fraction of GLA	Wp=	**0.109**	percent	26,209	SF
	Net land area	NLA=	4.918	acres	214,242	SF
	Unbuildable and/or future expansion areas as fraction of GLA	U=	**0.000**	percent	0	SF
	Total gross land area reduction as fraction of GLA	X=	0.109	percent	4,748	SF
	Buildable land area remaining	BLA=	**4.918**	acres	214,242	SF
Parking Variables:	Parking lot spaces required or planned per DU	u=	**0**	enter zero if no shared parking lot		
	Est. gross pkg. area per pkg. lot space in SF	s =	**0**	enter zero if no shared parking lot		
	Parking lot area under bldg. per DU in SF		**0.0**			
	Roadway pkg. spaces per DU	Rn=	**0**	enter zero if no roadway parking		
	Est. net pkg. area per roadway space in SF	Rs=	**0**	enter zero if no roadway parking		
	Roadway pkg. area under bldg. per DU in SF		**0.0**			
	Garage spaces per DU		**2**		2.00	total pkg per du
	Est. gross area per garage parking space in SF	Ga=	**252.0**	enter zero if no garage parking	504	pkg area per du
	Garage pkg. area under bldg. per DU in SF		**0.0**			
Support Variables:	Driveways as fraction of BLA	R=	**0.058**		12,426	SF
	Misc. pavement as fraction of BLA	M=	**0.045**		9,641	SF
	Outdoor social / entertainment area in SF	Sp=	**0**		0.000	fraction of BLA
	Social / service building(s) footprint in SF	Sf=	**0**		0.000	fraction of BLA
	Social / service building(s) gross area in SF	Sg=	**0**			
	Number of social / service parking spaces	Sn=	**0**	**400** gross SF / space	0.000	fraction of BLA
	Total support areas as a fraction of BLA	Su=	0.103		22,067	SF
	Core + open space + parking as fraction of BLA	CSR=	0.897	CS+Su must = 1	192,175	SF

NOTE: Express all fractions as decimals

DWELLING UNIT SPECIFICATION

dwelling unit type DU	no. of flrs. FLR above grade	dwelling unit MIX mix	total HAB habitable area	habitable area FTP footprint	private yard area YRD with townhouse	footprint and yard AFP no garage	du basement BSM area	du crawl CRW space area

	3,880	NOTE: AFP override in effect
Enter zero in the adjacent box unless you wish to override the AFP value calculated above:	**3,880**	= AFP override

DWELLING UNIT CAPACITY FORECAST

(based on common open space (COS) provided)

common COS open space	number of NDU dwelling units	EFF	1BR	2BR	3BR	4BR	density per dBA buildable acre	parking lot NPS spaces	roadway RDS parking spaces	garage GAR parking spaces
		dwelling unit type (DU) quantities given mix above								
0.000	43.8						8.91	0.0	0.0	87.7
0.100	38.9						7.92	0.0	0.0	77.9
0.200	34.1						6.93	0.0	0.0	68.1
0.242	**32.0**						**6.51**	**0.0**	**0.0**	**64.0**
0.400	24.3						4.94	0.0	0.0	48.6
0.500	19.4		Dwelling unit data not forecast for this example				3.94	0.0	0.0	38.8
0.600	14.5						2.95	0.0	0.0	29.0
0.700	9.6						1.96	0.0	0.0	19.3
0.800	4.7						0.96	0.0	0.0	9.5

WARNING: These are preliminary forecasts that must not be used to make final decisions.
1) These forecasts are not a substitute for the "due diligence" research that must be conducted to support the final definition of "unbuildable areas" above and the final decision to purchase land. This research includes, but is not limited to, verification of adequate subsurface soil, zoning, environmental clearance, access, title, utilities and water pressure, clearance from deed restriction, easement and right-of-way encumbrances, clearance from existing above and below ground facility conflicts, etc.
2) The most promising forecast(s) made on the basis of data entered in the design specification from "due diligence" research must be verified at the drawing board before funds are committed and land purchase decisions are made. Actual land shape ratios, dimensions and irregularities encountered may require adjustments to the general forecasts above.
3) The software licensee shall take responsibility for the design specification values entered and any advice given that is based on the forecast produced.

Northwest Boulevard Four-Family and McKenzie Drive Two-Family Urban Homes

The urban house transition to suburb house begins with four-family homes and steps down to two-family homes that are variously called *doubles, twin-singles,* or *duplexes.* A 2-story, 4-family home is still in the townhouse genre of building types, but a 1-story, 4-family home begins to take on the visual characteristics of a single-family detached home. Two-family homes are generally seen as transitional under any building height circumstance, but remain attached to each other and are classified as urban homes. A four-family home from a first-ring suburb is illustrated in Fig. 12.6, and a two-family home in a middle-ring area of the city's evolution is illustrated in Fig. 12.7.

The densities achieved by these two examples are similar, but the development capacity indexes are quite different. The two-family home provides less total open space per dwelling and larger dwelling unit areas. This results in an SFAC of 15,413 sq ft of gross building area produced per buildable acre and an AFP of 4695 sq ft, including private yard area. The four-family home provides more total open space and smaller dwelling unit areas to produce an AFP of 10,261 sq ft of gross building area per buildable acre and an AFP of 4200 sq ft. The SFAC value indicates the total amount of shelter area produced per acre of available land. A dwelling unit's AFP value indicates the amount of land allocated to the private interior and exterior living areas of a dwelling unit. As the AFP value becomes smaller, the dwelling unit becomes less grade oriented. As the SFAC becomes smaller, the average dwelling unit shrinks in total area. In the case of these four-family and two-family examples, the AFP value increases from 4200 sq ft per dwelling to 4695 sq ft per dwelling as the number of dwelling units declines from four to two. No common open space is provided, so this is the total average amount of land area devoted to the habitable interior and exterior activities of each dwelling unit.[9]

(a)

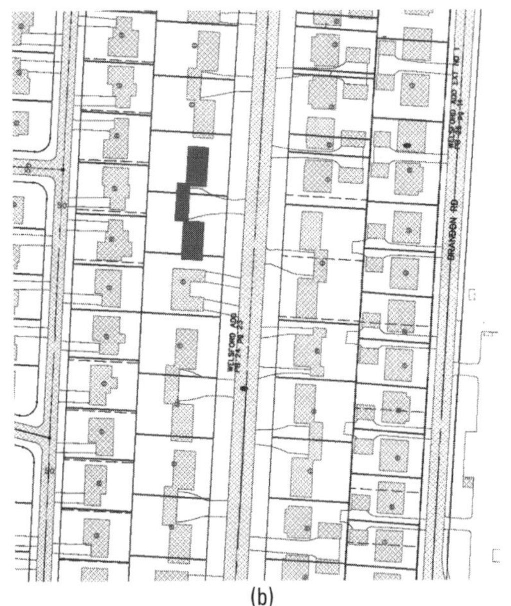

(b)

(c)

Figure 12.6 Northwest Boulevard four-family home. This four-family dwelling makes parking directly accessible from the public street, places private yard areas in the rear, directly associates the living room with the front door, and symbolizes our confusion over the automobile. At the turn of the century, the work areas of the home were in the rear and the social areas were along the public street and associated with the front door. The front yard is now associated with appearance, traffic, and automobile access. The rear yard is associated with privacy and entertainment. To this day, however, the tradition of a living room along the public street remains, and a family room is placed to the rear where much social activity actually occurs. When this approach is combined with the values measured in its context record, it produces a density of 6.71 dwelling units per buildable acre and a development capacity index of 10,261 sq ft of building area per acre of buildable land area. (*a*) Street-level view; (*b*) site plan; (*c*) overview.

(Photos by W. M. Hosack)

(a)

(b) (c)

Figure 12.7 McKenzie Drive two-family home. This is a smaller version of the Northwest Boulevard planning concept built at a later period in the twentieth century. The prominence of the garage location forces social activity to the rear and the front yard becomes a photo opportunity. It has been used as a transition from office land use to single-family neighborhood, and the total open space provided is 53.2% of the total buildable area. Of this open space, 76.6% is devoted to private yard area, or what has been referred to as *private yard emphasis.* When this approach is combined with the values measured in its context record, it produces a density of 6.45 dwelling units per buildable acre and a development capacity index of 15,413 sq ft of building area per acre of buildable land area. (*a*) Street-level view; (*b*) site plan; (*c*) overview.

(Photos by W. M. Hosack)

The context records used to measure and record these projects are included as Tables 12.7 and 12.8. Site plans and illustrations of these projects are included as Figs. 12.6 and 12.7. The forecast models that could have predicted these results based on the values entered in each design specification are included as Exhibits 12.6 and 12.7. By now, the concepts of context record measurement and forecast model predictions should be familiar, so these examples are included as patterns that can be used to duplicate these results and examine other options with the software attached to this book.

Concord Village

This project can represent a confusing set of statistics without some background explanation. It is the most expensive and exclusive of the condominium projects included and is illustrated in Fig. 12.8. It has one of the smallest yard area allocations per dwelling unit, however. The sixth column in Table 12.1 shows that an average of 758 sq ft per dwelling unit is provided. The total open space provided is only 19.6% of the buildable lot area, but 93.2% of this allocation represents private yards and courtyards. The revealing statistic for this project is found in the ninth column. The small amount of private exterior open space provided has been counterbalanced by the highest amount of gross building area per dwelling unit. The ninth column shows that an average of 4097 gross sq ft is provided, and the thirteenth column shows that this represents a development capacity of 47,655 sq ft per acre. This area is often spread over three floors, and when combined with the average yard area provided, produces a very modest habitable footprint area AFP of 2806 sq ft. This, in turn, permits a density of 10.50 dwelling units per acre to be reached, even though two garage parking spaces are provided per dwelling unit. In this case then, exterior space is traded for interior; building height is held to 2.5 stories plus finished basements;

Table 12.7 Four-Family Context Record

Name	**Northwest Four-Family**
Address or Location	**Northwest Boulevard**
Land Use	**Urban house**
	Adjacent garage parking plan
Parking System	*Any combination of parking lot, roadway, and/or garage parking*

Form GT

NOTE: An urban house is a single -family ATTACHED home that includes twin-singles, three-families, four-families, rowhouses, townhouses, garden apartments, etc. OR a single-family detached home on a lot less than f40 ft. wide. A townhouse implies separate ownership and defined private open space. A rowhouse implies rental status and less defined private open space.

	Description	Project Measurements		Context Summary	%	
Site						
	Gross land area in acres	**0.826**	GLA	35,981		
	Public/ private ROW and paved easements in SF	**10,000**	W		27.8%	of GLA
	Net lot area		NLA	25,981	72.2%	of GLA
	Total unbuildable and/or future expansion in SF	**0**	U		0.0%	of GLA
	Water area within unbuildable area in SF	**0**	WAT			of U
	Buildable land area		BLA	25,981	72.2%	of GLA
Development Cover						
	Gross habitable building area in SF	**4,800**	GBA			
	Total building footprint in SF	**4,800**	B		18.5%	of BLA
	Number of dwelling units	**4**	DU			
	Total private yard areas in SF	**12,000**				
	Avg. private yard area per dwelling unit			3,000		
	Aggregate average dwelling unit area in SF		AGG	1,200		
	Avg. habitable footprint plus yard in SF (no parking)		AFP	4,200		
	Number of building floors	**1.00**	FLR			
	Parking lot footprint in SF (include internal landscape)	**0**	P or G		0.0%	of BLA
	Parking lot footprint under building(s) in SF	**0**				
	Number of parking spaces	**0**	NPS			
	Parking lot spaces per dwelling unit		u	0.00		
	Gross parking lot area per parking space in SF		s			
	Private roadway parking spaces	**0**	RPS			
	Net roadway parking area in SF	**0**	RPA		0.0%	of BLA
	Roadway parking area under building(s) in SF	**0**				
	Private roadway parking spaces per dwelling unit		Rn	0.00		
	Net parking area per roadway space in SF		Rs			
	Total garage area in SF	**1,320**	GPA			
	Garage parking area under building(s) in SF	**0**				
	Number of garage parking spaces	**5**	GPS			
	Garage parking spaces per dwelling		Gn	1.25		
	Total garage area per space in SF		Ga	264.00		
	Total garage building cover percentage		Gc		5.1%	of BLA
	Total parking spaces per dwelling unit			1.25		
	Social and misc. building footprint(s) in SF	**0**	Sf			
	No. of social bldg. parking spaces	**0**	Sn			
	Social parking footprint (include internal landscape) in SF	**0**	Spla		0.0%	of BLA
	Driveway areas in SF	**2,062**	R		7.9%	of BLA
	Misc. pavement and social center pvmt. in SF	**180**	M		0.7%	of BLA
	TOTAL DEVELOPMENT COVER		D	8,362	**32.2%**	of BLA
Project Open Space						
	TOTAL PROJECT OPEN SPACE (OSA)		S	17,619	**67.8%**	of BLA
	Common open space area			5,619	21.6%	of BLA
	Private yard area			12,000	46.2%	of BLA
	Private yard area per dwelling unit			3,000		
	Private yard percentage of total project open space				68.1%	of OSA
	Water area within project open space in SF	**0**	wat	0	0.0%	of OSA
Summary						
	Development balance index		BNX		0.322	
	Development intensity index		INX		1.322	
	Floor area ratio per buildable acre		BFAR		0.185	
	Capacity index (gross building sq. ft. per buildable acre)		SFAC		10,261	
	Density per buildable acre if applicable		dBA		6.71	
	Density per net acre if applicable		dNA		6.71	
	Density per gross acre if applicable		dGA		4.84	

Table 12.8 Two-Family Context Record

Name	**McKenzie Two-Family**
Address or Location	**McKenzie Drive**
Land Use	**Urban house**
	Adjacent garage parking plan
Parking System	*Any combination of parking lot, roadway, and/or garage parking*

Form GT

NOTE: An urban house is a single -family ATTACHED home that includes twin-singles, three-families, four-families, rowhouses, townhouses, garden apartments, etc. OR a single-family detached home on a lot less than 40 ft.wide. A townhouse implies separate ownership and defined private open space. A rowhouse implies rental status and less defined private open space.

	Description	Project Measurements		Context Summary	%	
Site						
	Gross land area in acres	**0.379**	GLA	16,509		
	Public/ private ROW and paved easements in SF	**3,000**	W		18.2%	of GLA
	Net lot area		NLA	13,509	81.8%	of GLA
	Total unbuildable and/or future expansion in SF	**0**	U		0.0%	of GLA
	Water area within unbuildable area in SF	**0**	WAT			of U
	Buildable land area		BLA	13,509	81.8%	of GLA
Development Cover						
	Gross habitable building area in SF	**3,890**	GBA			
	Total building footprint in SF	**3,890**	B		28.8%	of BLA
	Number of dwelling units	**2**	DU			
	Total private yard areas in SF	**5,500**				
	Avg. private yard area per dwelling unit			2,750		
	Aggregate average dwelling unit area in SF		AGG	1,945		
	Avg. habitable footprint plus yard in SF (no parking)		AFP	4,695		
	Number of building floors	**1.00**	FLR			
	Parking lot footprint in SF (include internal landscape)	**0**	P or G		0.0%	of BLA
	Parking lot footprint under building(s) in SF	**0**				
	Number of parking spaces	**0**	NPS			
	Parking lot spaces per dwelling unit		u	0.00		
	Gross parking lot area per parking space in SF		s			
	Private roadway parking spaces	**0**	RPS			
	Net roadway parking area in SF	**0**	RPA		0.0%	of BLA
	Roadway parking area under building(s) in SF	**0**				
	Private roadway parking spaces per dwelling unit		Rn	0.00		
	Net parking area per roadway space in SF		Rs			
	Total garage area in SF	**890**	GPA			
	Garage parking area under building(s) in SF	**0**				
	Number of garage parking spaces	**4**	GPS			
	Garage parking spaces per dwelling		Gn	2.00		
	Total garage area per space in SF		Ga	222.50		
	Total garage building cover percentage		Gc		6.6%	of BLA
	Total parking spaces per dwelling unit			2.00		
	Social and misc. building footprint(s) in SF	**0**	Sf			
	No. of social bldg. parking spaces	**0**	Sn			
	Social parking footprint (include internal landscape) in SF	**0**	Spla		0.0%	of BLA
	Driveway areas in SF	**1,368**	R		10.1%	of BLA
	Misc. pavement and social center pvmt. in SF	**180**	M		1.3%	of BLA
	TOTAL DEVELOPMENT COVER		D	6,328	**46.8%**	of BLA
Project Open Space						
	TOTAL PROJECT OPEN SPACE (OSA)		S	7,181	**53.2%**	of BLA
	Common open space area			1,681	12.4%	of BLA
	Private yard area			5,500	40.7%	of BLA
	Private yard area per dwelling unit			2,750		
	Private yard percentage of total project open space				76.6%	of OSA
	Water area within project open space in SF	**0**	wat	0	0.0%	of OSA
Summary						
	Development balance index		BNX		0.468	
	Development intensity index		INX		1.468	
	Floor area ratio per buildable acre		BFAR		0.288	
	Capacity index (gross building sq. ft. per buildable acre)		SFAC		15,413	
	Density per buildable acre if applicable		dBA		6.45	
	Density per net acre if applicable		dNA		6.45	
	Density per gross acre if applicable		dGA		5.28	

Exhibit 12.6 Northwest Boulevard Four-Family Design Specification and Forecast

Forecast Model RGTL

*Development capacity forecast for **TOWNHOUSES** based on the use of shared PARKING LOTS and / or PRIVATE GARAGES serving each dwelling unit.*
***Given:** Gross land area. **To Find:** Maximum dwelling unit capacity of the land area given based on the design specification values entered below.*

DESIGN SPECIFICATION

Enter values in boxed areas where text is bold and blue. Express all fractions as decimals.

			Value			
Given:	**Gross land area**	**GLA=**	**0.826**	acres	35,981	SF
Land Variables:	Public right-of-way as fraction of GLA	W=	**0.278**	percent	10,003	SF
	Private right-of-way as fraction of GLA	Wp=	**0.000**	percent	0	SF
	Net land area	NLA=	0.596	acres	25,978	SF
	Unbuildable and/or future expansion areas as fraction of GLA	U=	**0.000**	percent	0	SF
	Total gross land area reduction as fraction of GLA	X=	0.278	percent	12,110	SF
	Buildable land area remaining	**BLA=**	**0.596**	acres	25,978	SF
Parking Variables:	Parking lot spaces required or planned per DU	u=	**0**	enter zero if no shared parking lot		
	Est. gross pkg. area per pkg. lot space in SF	s =	**0**	enter zero if no shared parking lot		
	Parking lot area under bldg. per DU in SF		**0.0**			
	Roadway pkg. spaces per DU	Rn=	**0**	enter zero if no roadway parking	*NOTE: Express all fractions as decimals*	
	Est. net pkg area per roadway space in SF	Rs=	**0**	enter zero if no roadway parking		
	Roadway pkg. area under bldg. per DU in SF		**0.0**			
	Garage spaces per DU		**1.25**		1.25	total pkg per du
	Est. gross area per garage parking space in SF	Ga=	**264.0**	enter zero if no garage parking	330	pkg area per du
	Garage pkg. area under bldg. per DU in SF		**0.0**			
Support Variables:	Driveways as fraction of BLA	R=	**0.079**		2,052	SF
	Misc. pavement as fraction of BLA	M=	**0.007**		182	SF
	Outdoor social / entertainment area in SF	Sp=	**0**		0.000	fraction of BLA
	Social / service building(s) footprint in SF	Sf=	**0**		0.000	fraction of BLA
	Social / service building(s) gross area in SF	Sg=	**0**			
	Number of social / service parking spaces	Sn=	**0**	**400** gross SF / space	0.000	fraction of BLA
	Total support areas as a fraction of BLA	Su=	0.086		2,234	SF
	Core + open space + parking as fraction of BLA	**CSR=**	0.914	CS+Su must = 1	23,744	SF

DWELLING UNIT SPECIFICATION

dwelling unit type DU	no. of flrs. FLR above grade	dwelling unit MIX mix	total HAB habitable area	habitable area FTP footprint	private yard area YRD with townhouse	footprint and yard AFP no garage	du basement BSM area	du crawl CRW space area

	4,200	NOTE: AFP override in effect
Enter zero in the adjacent box unless you wish to override the AFP value calculated above:	**4,200**	= AFP override

DWELLING UNIT CAPACITY FORECAST

(based on common open space (COS) provided)

common COS open space	number of NDU dwelling units	EFF	1BR	2BR	3BR	4BR	density per dBA buildable acre	parking lot NPS spaces	roadway RDS parking spaces	garage GAR parking spaces
		dwelling unit type (DU) quantities given mix above								
0.000	5.2						8.79	0.0	0.0	6.6
0.100	4.7						7.83	0.0	0.0	5.8
0.200	4.1						6.87	0.0	0.0	5.1
0.216	**4.0**						**6.71**	**0.0**	**0.0**	**5.0**
0.400	2.9						4.94	0.0	0.0	3.7
0.500	2.4			Dwelling unit data not forecast for this example			3.98	0.0	0.0	3.0
0.600	1.8						3.02	0.0	0.0	2.3
0.700	1.2						2.06	0.0	0.0	1.5
0.800	0.7						1.10	0.0	0.0	0.8

WARNING: These are preliminary forecasts that must not be used to make final decisions.
1) These forecasts are not a substitute for the "due diligence" research that must be conducted to support the final definition of "unbuildable areas" above and the final decision to purchase land. This research includes, but is not limited to, verification of adequate subsurface soil, zoning, environmental clearance, access, title, utilities and water pressure, clearance from deed restriction, easement and right-of-way encumbrances, clearance from existing above and below ground facility conflicts, etc.
2) The most promising forecast(s) made on the basis of data entered in the design specification from "due diligence" research must be verified at the drawing board before funds are committed and land purchase decisions are made. Actual land shape ratios, dimensions and irregularities encountered may require adjustments to the general forecasts above.
3) The software licensee shall take responsibility for the design specification values entered and any advice given that is based on the forecast produced.

Exhibit 12.7 McKenzie Drive Two-Family Design Specification and Forecast

Forecast Model RGTL

Development capacity forecast for ***TOWNHOUSES*** *based on the use of shared PARKING LOTS and / or PRIVATE GARAGES serving each dwelling unit.*
Given: *Gross land area.* ***To Find:*** *Maximum dwelling unit capacity of the land area given based on the design specification values entered below.*

DESIGN SPECIFICATION

Enter values in boxed areas where text is bold and blue. Express all fractions as decimals.

			Value			
Given:	**Gross land area**	GLA=	**0.379**	acres	16,509	SF
Land Variables:	Public right-of-way as fraction of GLA	W=	**0.182**	percent	3,005	SF
	Private right-of-way as fraction of GLA	Wp=	**0.000**	percent	0	SF
	Net land area	NLA=	0.310	acres	13,505	SF
	Unbuildable and/or future expansion areas as fraction of GLA	U=	**0.000**	percent	0	SF
	Total gross land area reduction as fraction of GLA	X=	0.182	percent	7,928	SF
	Buildable land area remaining	BLA=	**0.310**	acres	13,505	SF
Parking Variables:	Parking lot spaces required or planned per DU	u=	**0**	enter zero if no shared parking lot		
	Est. gross pkg. area per pkg. lot space in SF	s =	**0**	enter zero if no shared parking lot		
	Parking lot area under bldg. per DU in SF		**0.0**			
	Roadway pkg. spaces per DU	Rn=	**0**	enter zero if no roadway parking		
	Est. net pkg. area per roadway space in SF	Rs=	**0**	enter zero if no roadway parking		
	Roadway pkg. area under bldg. per DU in SF		**0.0**			
	Garage spaces per DU		**2**		2.00	total pkg per du
	Est. gross area per garage parking space in SF	Ga=	**222.5**	enter zero if no garage parking	445	pkg area per du
	Garage pkg. area under bldg. per DU in SF		**0.0**			
Support Variables:	Driveways as fraction of BLA	R=	**0.101**		1,364	SF
	Misc. pavement as fraction of BLA	M=	**0.013**		176	SF
	Outdoor social / entertainment area in SF	Sp=	**0**		0.000	fraction of BLA
	Social / service building(s) footprint in SF	Sf=	**0**		0.000	fraction of BLA
	Social / service building(s) gross area in SF	Sg=	**0**			
	Number of social / service parking spaces	Sn=	**0**	**0** gross SF / space	0.000	fraction of BLA
	Total support areas as a fraction of BLA	Su=	0.114		1,540	SF
	Core + open space + parking as fraction of BLA	CSR=	0.886	CS+Su must = 1	11,965	SF

NOTE: Express all fractions as decimals

DWELLING UNIT SPECIFICATION

dwelling unit type DU	no. of flrs. FLR above grade	dwelling unit MIX mix	total HAB habitable area	habitable area FTP footprint	private yard area YRD with townhouse	footprint and yard AFP no garage	du basement BSM area	du crawl CRW space area

	4,695	NOTE: AFP override in effect
Enter zero in the adjacent box unless you wish to override the AFP value calculated above:	**4,695**	= AFP override

DWELLING UNIT CAPACITY FORECAST

(based on common open space (COS) provided)

common COS open space	number of NDU dwelling units	EFF	1BR	2BR	3BR	4BR	density per dBA buildable acre	parking lot NPS spaces	roadway RDS parking spaces	garage GAR parking spaces
0.000	2.3						7.51	0.0	0.0	4.7
0.100	2.1						6.66	0.0	0.0	4.1
0.124	**2.0**						**6.46**	**0.0**	**0.0**	**4.0**
0.300	1.5						4.97	0.0	0.0	3.1
0.400	1.3						4.12	0.0	0.0	2.6
0.500	1.0			Dwelling unit data not forecast for this example			3.27	0.0	0.0	2.0
0.600	0.8						2.42	0.0	0.0	1.5
0.700	0.5						1.58	0.0	0.0	1.0
0.800	0.2						0.73	0.0	0.0	0.5

(EFF–4BR: dwelling unit type (DU) quantities given mix above)

WARNING: These are preliminary forecasts that must not be used to make final decisions.
1) These forecasts are not a substitute for the "due diligence" research that must be conducted to support the final definition of "unbuildable areas" above and the final decision to purchase land. This research includes, but is not limited to, verification of adequate subsurface soil, zoning, environmental clearance, access, title, utilities and water pressure, clearance from deed restriction, easement and right-of-way encumbrances, clearance from existing above and below ground facility conflicts, etc.
2) The most promising forecast(s) made on the basis of data entered in the design specification from "due diligence" research must be verified at the drawing board before funds are committed and land purchase decisions are made. Actual land shape ratios, dimensions and irregularities encountered may require adjustments to the general forecasts above.
3) The software licensee shall take responsibility for the design specification values entered and any advice given that is based on the forecast produced.

(a)

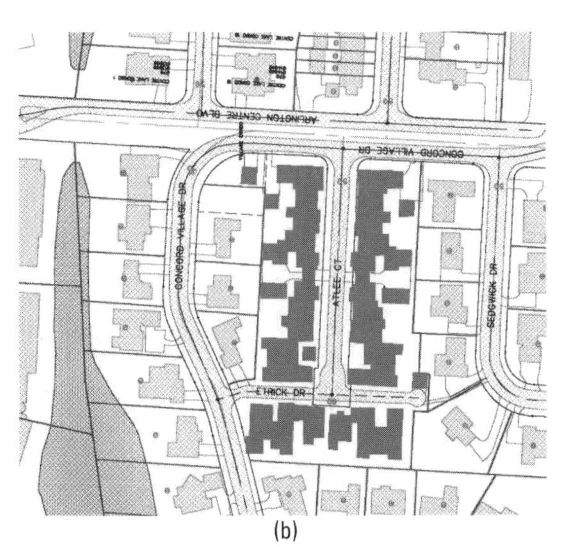

(b)

(c)

Figure 12.8 Concord Village. Concord Village is a late-twentieth-century development that eliminates the front yard and ensures that the public street carries limited traffic associated with the neighborhood. The public street is a hammerhead version of a traditional cul-de-sac, but isolation is avoided by connecting this cul-de-sac to surrounding streets with much narrower private alley streets. The width of these alley streets ensures that they are only useful for residential service, but their front-door location makes them more convenient and less isolated than early-twentieth-century alleys. Garage doors are much less prominent with this approach, but the total open space provided by this project is much less than traditional residential housing projects. However, 93.2% of that provided is devoted to private yard emphasis. When this approach is combined with the values measured in its context record, it produces a density of 10.5 dwelling units per buildable acre and a development capacity index of 47,653 sq ft of building area per acre of buildable land area. (*a*) Street-level view; (*b*) site plan; (*c*) overview.

(Photos by W. M. Hosack)

smaller private exterior areas are retained; garages are provided; and common open space is held to a minimum to permit the density of 10 dwelling units per acre to be reached. The complete context record for this project is included as Table 12.9, and the corresponding forecast model is included as Exhibit 12.8.

This project was an infill project built in the 1980s when most construction activity was occurring in the suburban fringe. It is reminiscent of the early urban houses of Williamsburg and seems to indicate that at least part of the housing market is willing to accept much less yard in return for larger interior areas. Even though exterior areas are held to a minimum, however, each unit is associated with its own personal exterior space, individual identification, and condominium ownership.

City Park

City Park is a classic urban home left over from the Industrial Revolution. It has not solved its parking problem, and the surrounding public streets are still crowded with excess demand. Its intrinsic value is being rediscovered throughout the country, however, and its SFAC development capacity index of 38,519 sq ft of gross building area per acre of buildable land area represents an opportunity to shelter many more people within the grade-oriented format favored by most. It represents land and home ownership, private exterior space, and materials and methods of construction meant to endure. The Park Street project demonstrates one method of solving the parking problem for new construction in these areas, but the lesson cannot be applied to existing historic homes like City Park. (See Fig. 12.9.) The context record and forecast model for City Park are included as Table 12.10 and Exhibit 12.9, respectively. They are interesting as points of historic reference, but the latter half of the twentieth century has been engaged in attempted modifications to this specification that can accommodate the

Table 12.9 Concord Village Context Record

Name	**Concord Village**
Address or Location	**Atlee Court**
Land Use	**Urban house**
	Attached and detached garage parking
Parking System	*Any combination of parking lot, roadway, and/or garage parking*

Form GT

NOTE: An urban house is a single-family ATTACHED home that includes twin-singles, three-families, four-families, rowhouses, townhouses, garden apartments, etc. OR a single-family detached home on a lot less than 40 ft. wide. A townhouse implies separate ownership and defined private open space. A rowhouse implies rental status and less defined private open space.

	Description	Project Measurements		Context Summary	%	
Site						
	Gross land area in acres	**3.210**	GLA	139,828		
	Public/ private ROW and paved easements in SF	**32,000**	W		22.9%	of GLA
	Net lot area		NLA	107,828	77.1%	of GLA
	Total unbuildalbe and/or future expansion in SF	**0**	U		0.0%	of GLA
	Water area within unbuildable area in SF	**0**	WAT			of U
	Buildable land area		BLA	107,828	77.1%	of GLA
Development Cover						
	Gross habitable building area in SF	**106,520**	GBA			
	Total building footprint in SF	**53,260**	B		49.4%	of BLA
	Number of dwelling units	**26**	DU			
	Total private yard areas in SF	**19,700**				
	Avg. private yard area per dwelling unit			758		
	Aggregate average dwelling unit area in SF		AGG	4,097		
	Avg. habitable footprint plus yard in SF (no parking)		AFP	2,806		
	Number of building floors	**2.00**	FLR			
	Parking lot footprint in SF (include internal landscape)	**0**	P or G		0.0%	of BLA
	Parking lot footprint under building(s) in SF	**0**				
	Number of parking spaces	**0**	NPS			
	Parking lot spaces per dwelling unit		u	0.00		
	Gross parking lot area per parking space in SF		s			
	Private roadway parking spaces	**0**	RPS			
	Net roadway parking area in SF	**0**	RPA		0.0%	of BLA
	Roadway parking area under building(s) in SF	**0**				
	Private roadway parking spaces per dwelling unit		Rn	0.00		
	Net parking area per roadway space in SF		Rs			
	Total garage area in SF	**11,440**	GPA			
	Garage parking area under building(s) in SF	**0**				
	Number of garage parking spaces	**52**	GPS			
	Garage parking spaces per dwelling		Gn	2.00		
	Total garage area per space in SF		Ga	220.00		
	Total garage building cover percentage		Gc		10.6%	of BLA
	Total parking spaces per dwelling unit			2.00		
	Social and misc. building footprint(s) in SF	**0**	Sf			
	No. of social bldg. parking spaces	**0**	Sn			
	Social parking footprint (include internal landscape) in SF	**0**	Spla		0.0%	of BLA
	Driveway areas in SF	**7,200**	R		6.7%	of BLA
	Misc. pavement and social center pvmt. in SF	**14,800**	M		13.7%	of BLA
	TOTAL DEVELOPMENT COVER		D	86,700	**80.4%**	of BLA
Project Open Space						
	TOTAL PROJECT OPEN SPACE (OSA)		S	21,128	**19.6%**	of BLA
	Common open space area			1,428	1.3%	of BLA
	Private yard area			19,700	18.3%	of BLA
	Private yard area per dwelling unit			758		
	Private yard percentage of total project open space				93.2%	of OSA
	Water area within project open space in SF	**0**	wat	0	0.0%	of OSA
Summary						
	Development balance index		BNX		0.804	
	Development intensity index		INX		2.804	
	Floor area ratio per buildable acre		BFAR		0.988	
	Capacity index (gross building sq. ft. per buildable acre)		SFAC		47,653	
	Density per buildable acre if applicable		dBA		10.50	
	Density per net acre if applicable		dNA		10.50	
	Density per gross acre if applicable		dGA		8.10	

Exhibit 12.8 Concord Village Design Specification and Forecast

Forecast Model RGTL

Development capacity forecast for ***TOWNHOUSES*** *based on the use of shared PARKING LOTS and / or PRIVATE GARAGES serving each dwelling unit.*
Given: *Gross land area.* ***To Find:*** *Maximum dwelling unit capacity of the land area given based on the design specification values entered below.*

DESIGN SPECIFICATION

Enter values in boxed areas where text is bold and blue. Express all fractions as decimals.

Given:	**Gross land area**	**GLA=**	**3.210**	acres	139,828	SF
Land Variables:	Public right-of-way as fraction of GLA	W=	**0.229**	percent	32,021	SF
	Private right-of-way as fraction of GLA	Wp=	**0.000**	percent	0	SF
	Net land area	NLA=	2.475	acres	107,807	SF
	Unbuildable and/or future expansion areas as fraction of GLA	U=	**0.000**	percent	0	SF
	Total gross land area reduction as fraction of GLA	X=	0.229	percent	9,975	SF
	Buildable land area remaining	**BLA=**	**2.475**	acres	107,807	SF
Parking Variables:	Parking lot spaces required or planned per DU	u=	**0**	enter zero if no shared parking lot		
	Est. gross pkg. area per pkg. lot space in SF	s =	**0**	enter zero if no shared parking lot		
	Parking lot area under bldg. per DU in SF		**0.0**			
	Roadway pkg. spaces per DU	Rn=	**0**	enter zero if no roadway parking	*NOTE: Express all fractions as decimals*	
	Est. net pkg area per roadway space in SF	Rs=	**0**	enter zero if no roadway parking		
	Roadway pkg. area under bldg. per DU in SF		**0.0**			
	Garage spaces per DU		**2**		2.00	total pkg per du
	Est. gross area per garage parking space in SF	Ga=	**220.0**	enter zero if no garage parking	440	pkg area per du
	Garage pkg. area under bldg. per DU in SF		**0.0**			
Support Variables:	Driveways as fraction of BLA	R=	**0.067**		7,223	SF
	Misc. pavement as fraction of BLA	M=	**0.137**		14,770	SF
	Outdoor social / entertainment area in SF	Sp=	**0**		0.000	fraction of BLA
	Social / service building(s) footprint in SF	Sf=	**0**		0.000	fraction of BLA
	Social / service building(s) gross area in SF	Sg=	**0**			
	Number of social / service parking spaces	Sn=	**0**	**0** gross SF / space	0.000	fraction of BLA
	Total support areas as a fraction of BLA	Su=	0.204		21,993	SF
	Core + open space + parking as fraction of BLA	**CSR=**	0.796	CS+Su must = 1	85,814	SF

DWELLING UNIT SPECIFICATION

dwelling unit type DU	no. of flrs. FLR above grade	dwelling unit MIX mix	total HAB habitable area	habitable area FTP footprint	private yard area YRD with townhouse	footprint and yard AFP no garage	du basement BSM area	du crawl CRW space area

	2,806	NOTE: AFP override in effect
Enter zero in the adjacent box unless you wish to override the AFP value calculated above:	**2,806**	= AFP override

DWELLING UNIT CAPACITY FORECAST

(based on common open space (COS) provided)

common COS open space	number of NDU dwelling units	EFF	1BR	2BR	3BR	4BR	density per dBA buildable acre	parking lot NPS spaces	roadway RDS parking spaces	garage GAR parking spaces
		dwelling unit type (DU) quantities given mix above								
0.0000	26.4						10.68	0.0	0.0	52.9
0.0126	**26.0**						**10.51**	**0.0**	**0.0**	**52.0**
0.2000	19.8						8.00	0.0	0.0	39.6
0.3000	16.5						6.66	0.0	0.0	32.9
0.4000	13.2						5.31	0.0	0.0	26.3
0.5000	9.8	Dwelling unit data not forecast for this example					3.97	0.0	0.0	19.7
0.6000	6.5						2.63	0.0	0.0	13.0
0.7000	3.2						1.29	0.0	0.0	6.4
0.8000	-0.1						-0.05	0.0	0.0	-0.3

WARNING: These are preliminary forecasts that must not be used to make final decisions.
1) These forecasts are not a substitute for the "due diligence" research that must be conducted to support the final definition of "unbuildable areas" above and the final decision to purchase land. This research includes, but is not limited to, verification of adequate subsurface soil, zoning, environmental clearance, access, title, utilities and water pressure, clearance from deed restriction, easement and right-of-way encumbrances, clearance from existing above and below ground facility conflicts, etc.
2) The most promising forecast(s) made on the basis of data entered in the design specification from "due diligence" research must be verified at the drawing board before funds are committed and land purchase decisions are made. Actual land shape ratios, dimensions and irregularities encountered may require adjustments to the general forecasts above.
3) The software licensee shall take responsibility for the design specification values entered and any advice given that is based on the forecast produced.

(a)

(b)

(c)

Figure 12.9 City Park Avenue. City Park was built in the nineteenth century and provides no private parking accommodation, nor is the lot size adequate to add this amenity. It provides 44.8% of its total lot area as open space and 100% of this is devoted to private yard area. The building is placed on the street, and the street accommodates both parked cars and traffic generated by the adjacent central business district. Sidewalk activity is above suburban levels, and the neighborhood has gone through a cycle of original development and decline that is typical of many cities. It is currently enjoying a period of popularity as lifestyles have adjusted to its lack of automobile accommodation and its central location attributes are recognized. When this approach is combined with the values measured in its context record, it produces a density of 18.52 dwelling units per buildable acre and a development capacity index of 38,519 sq ft of building area per acre of buildable land area. (*a*) Street-level view; (*b*) site plan; (*c*) overview.

(Photos by W. M. Hosack)

Table 12.10 City Park Context Record

Name	**City Park**
Address or Location	**765 City Park, German Village**
Land Use	**Urbanhouse**
	Street parking plan
Parking System	*Any combination of parking lot, roadway, and/or garage parking*

Form GT

NOTE: An urban house is a single -family ATTACHED home that includes twin-singles, three-families, four-families, rowhouses, townhouses, garden apartments, etc. OR a single-family detached home on a lot less than 40 ft. wide. A townhouse implies separate ownership and defined private open space. A rowhouse implies rental status and less defined private open space.

	Description	Project Measurements		Context Summary	%	
Site						
	Gross land area in acres	**0.054**	GLA	2,352		
	Public/ private ROW and paved easements in SF	**0**	W		0.0%	of GLA
	Net lot area		NLA	2,352	100.0%	of GLA
	Total unbuildable and/or future expansion in SF	**0**	U		0.0%	of GLA
	Water area within unbuildable area in SF	**0**	WAT			of U
	Buildable land area		BLA	2,352	100.0%	of GLA
Development Cover						
	Gross habitable building area in SF	**1,972**	GBA			
	Total building footprint in SF	**1,188**	B		50.5%	of BLA
	Number of dwelling units	**1**	DU			
	Total private yard areas in SF	**1,054**				
	Avg. private yard area per dwelling unit			1,054		
	Aggregate average dwelling unit area in SF		AGG	1,972		
	Avg. habitable footprint plus yard in SF (no parking)		AFP	2,242		
	Number of building floors	**2.00**	FLR			
	Parking lot footprint in SF (include internal landscape)	**0**	P or G		0.0%	of BLA
	Parking lot footprint under building(s) in SF	**0**				
	Number of parking spaces	**0**	NPS			
	Parking lot spaces per dwelling unit		u	0.00		
	Gross parking lot area per parking space in SF		s			
	Private roadway parking spaces	**0**	RPS			
	Net roadway parking area in SF	**0**	RPA		0.0%	of BLA
	Roadway parking area under building(s) in SF	**0**				
	Private roadway parking spaces per dwelling unit		Rn	0.00		
	Net parking area per roadway space in SF		Rs			
	Total garage area in SF	**0**	GPA			
	Garage parking area under building(s) in SF	**0**				
	Number of garage parking spaces	**0**	GPS			
	Garage parking spaces per dwelling		Gn	0.00		
	Total garage area per space in SF		Ga			
	Total garage building cover percentage		Gc			of BLA
	Total parking spaces per dwelling unit			0.00		
	Social and misc. building footprint(s) in SF	**0**	Sf			
	No. of social bldg. parking spaces	**0**	Sn			
	Social parking footprint (include internal landscape) in SF	**0**	Spla		0.0%	of BLA
	Driveway areas in SF	**0**	R		0.0%	of BLA
	Misc. pavement and social center pvmt. in SF	**110**	M		4.7%	of BLA
	TOTAL DEVELOPMENT COVER		D	1,298	**55.2%**	of BLA
Project Open Space						
	TOTAL PROJECT OPEN SPACE (OSA)		S	1,054	**44.8%**	of BLA
	Common open space area			0	0.0%	of BLA
	Private yard area			1,054	44.8%	of BLA
	Private yard area per dwelling unit			1,054		
	Private yard percentage of total project open space				100.0%	of OSA
	Water area within project open space in SF	**0**	wat	0	0.0%	of OSA
Summary						
	Development balance index		BNX		0.552	
	Development intensity index		INX		2.552	
	Floor area ratio per buildable acre		BFAR		0.838	
	Capacity index (gross building sq. ft. per buildable acre)		SFAC		36,519	
	Density per buildable acre if applicable		dBA		18.52	
	Density per net acre if applicable		dNA		18.52	
	Density per gross acre if applicable		dGA		18.52	

Exhibit 12.9 City Park Design Specification and Forecast

Forecast Model RGTL

Development capacity forecast for ***TOWNHOUSES*** *based on the use of shared PARKING LOTS and / or PRIVATE GARAGES serving each dwelling unit.*
Given: *Gross land area.* ***To Find:*** *Maximum dwelling unit capacity of the land area given based on the design specification values entered below.*

DESIGN SPECIFICATION

Enter values in boxed areas where text is bold and blue. Express all fractions as decimals.

			Value			
Given:	**Gross land area**	GLA=	**0.073**	acres	3,162	SF
Land Variables:	Public right-of-way as fraction of GLA	W=	**0.257**	percent	812	SF
	Private right-of-way as fraction of GLA	Wp=	**0.000**	percent	0	SF
	Net land area	NLA=	0.054	acres	2,351	SF
	Unbuildable and/or future expansion areas as fraction of GLA	U=	**0.000**	percent	0	SF
	Total gross land area reduction as fraction of GLA	X=	0.257	percent	11,182	SF
	Buildable land area remaining	BLA=	**0.054**	acres	2,351	SF
Parking Variables:	Parking lot spaces required or planned per DU	u=	**0**	enter zero if no shared parking lot		
	Est. gross pkg. area per pkg. lot space in SF	s =	**0**	enter zero if no shared parking lot		
	Pkg. lot area under bldg. per DU in SF		**0.0**			
	Roadway pkg. spaces per DU	Rn=	**0**	enter zero if no roadway parking	*NOTE: Express all fractions as decimals*	
	Est. net pkg. area per roadway space in SF	Rs=	**0**	enter zero if no roadway parking		
	Roadway pkg. area under bldg. per DU in SF		**0.0**			
	Garage spaces per DU		**0**		0.00	total pkg per du
	Est. gross area per garage parking space in SF	Ga=	**0.0**	enter zero if no garage parking	0	pkg area per du
	Garage pkg. area under bldg. per DU in SF		**0.0**			
Support Variables:	Driveways as fraction of BLA	R=	**0.000**		0	SF
	Misc. pavement as fraction of BLA	M=	**0.047**		110	SF
	Outdoor social / entertainment area in SF	Sp=	**0**		0.000	fraction of BLA
	Social / service building(s) footprint in SF	Sf=	**0**		0.000	fraction of BLA
	Social / service building(s) gross area in SF	Sg=	**0**			
	Number of social / service parking spaces	Sn=	**0**	**0** gross SF / space	0.000	fraction of BLA
	Total support areas as a fraction of BLA	Su=	0.047		110	SF
	Core + open space + parking as fraction of BLA	CSR=	0.953	CS+Su must = 1	2,240	SF

DWELLING UNIT SPECIFICATION

dwelling unit type DU	no. of flrs. FLR above grade	dwelling unit MIX mix	total HAB habitable area	habitable area FTP footprint	private yard area YRD with townhouse	footprint and yard AFP no garage	du basement BSM area	du crawl CRW space area

	2,242	NOTE: AFP override in effect
Enter zero in the adjacent box unless you wish to override the AFP value calculated above:	**2,242**	= AFP override

DWELLING UNIT CAPACITY FORECAST

(based on common open space (COS) provided)

common COS open space	number of NDU dwelling units	EFF	1BR	2BR	3BR	4BR	density per dBA buildable acre	parking lot NPS spaces	roadway RDS parking spaces	garage GAR parking spaces
0.000	**1.0**						**18.52**	**0.0**	**0.0**	**0.0**
0.100	0.9						16.57	0.0	0.0	0.0
0.200	0.8						14.63	0.0	0.0	0.0
0.300	0.7						12.69	0.0	0.0	0.0
0.400	0.6	Dwelling unit data not forecast for this example					10.74	0.0	0.0	0.0
0.500	0.5						8.80	0.0	0.0	0.0
0.600	0.4						6.86	0.0	0.0	0.0
0.700	0.3						4.92	0.0	0.0	0.0
0.800	0.2						2.97	0.0	0.0	0.0

(EFF, 1BR, 2BR, 3BR, 4BR: dwelling unit type (DU) quantities given mix above)

WARNING: These are preliminary forecasts that must not be used to make final decisions.
1) These forecasts are not a substitute for the "due diligence" research that must be conducted to support the final definition of "unbuildable areas" above and the final decision to purchase land. This research includes, but is not limited to, verification of adequate subsurface soil, zoning, environmental clearance, access, title, utilities and water pressure, clearance from deed restriction, easement and right-of-way encumbrances, clearance from existing above and below ground facility conflicts, etc.
2) The most promising forecast(s) made on the basis of data entered in the design specification from "due diligence" research must be verified at the drawing board before funds are committed and land purchase decisions are made. Actual land shape ratios, dimensions and irregularities encountered may require adjustments to the general forecasts above.
3) The software licensee shall take responsibility for the design specification values entered and any advice given that is based on the forecast produced.

automobile and our evolving lifestyles. Unfortunately, these homes will remain restoration projects without duplication until some method of automobile accommodation or substitution can be associated with what is otherwise a very human, pedestrian, mixed-use urban pattern that encourages social interaction.

Notes

1. The choice was between *farmhouse* and *suburb house*. A farmhouse probably hates to be referred to as a suburb house since a suburb house is a mere imitation, but it is in the minority and shares the same characteristics in different quantities. The term *suburb house* was chosen to avoid confusion with apologies to the farmhouses of the world, but the user can easily adopt his/her own preference.
2. A townhouse implies separate ownership and defined private open space. A rowhouse implies rental status and less defined private open space. Both can have a similar appearance, but townhouses generally receive more elaborate treatment.
3. Homes on such narrow lots are generally accompanied by street parking. In some cases, alleys and detached garages are included.
4. The early industrial revolution and its lack of mobility had preempted this attachment.
5. CG: nonresidential surface parking group; CS: nonresidential structure parking group; RG: residential apartment surface parking group; RS: residential apartment structure parking group; RGT: residential urban house group; RSF: residential suburb house group; MX: mixed-use group.
6. This subdivision is called a *building floor plan*.
7. For those wishing to experiment, forecast model RGTL can be called to the screen, and the specification values shown in Exhibit 12.1 can be entered to produce the same page. These values can then be altered to explore other alternatives that were available at the time of initial project design.
8. Garage and surface parking areas are calculated separately.
9. Remember that garage areas are not considered habitable areas.

Suburb House Comparisons

Suburb houses[1] represent one of two residential housing classes that are able to maintain an association with the land for each dwelling unit and family sheltered within. They do not conform to the standard classification system used to organize much of the work in this book, however, and the system in Table 13.1 has been created to identify the case studies included in this chapter.[2]

A suburb house is distinguished more by the lot area, open space, and street system that surround it than by its own building area and style. In other words, a suburb house is referred to as a single-family detached dwelling, and the amount of detachment establishes the context in which it resides.

Lot area and lot frontage are often regulated subdivision characteristics. Building cover and building height may or may not be regulated, but become subdivision characteristics once constructed. Building cover and miscellaneous pavement cover combine to produce development cover. Development cover is balanced by open space, expands over time, and establishes the balance and stormwater runoff characteristics of the suburb house and subdivision. Specifying the total net land area allocation per dwelling, the lot

Table 13.1 Suburb House Divisions of the Built Environment

Kingdom	Division	Shelter
Phylum	Category	Residential
Class	Building type	Suburb house
Order	Land organization	Standard plan, cluster plan
Family	Lot area range*	Farm, estate, garden, large, medium, small, very small
Genus	Lot frontage	50 ft, 60 ft, 75 ft, 90 ft, etc.
Species	Building cover and development cover	10%, 20%, etc.
Subspecies	Building height	2.5 stories or 35 ft
	Street pattern†	Grid, modified grid, cul-de-sac, curvilinear, hybrid
	Specific lot area	Area of lot in acres
	Building appearance	Style, materials, setbacks

*See Table 13.3 for a more detailed explanation.
†Ibid.

area per dwelling, lot frontage, building cover, building height, and development cover firmly establishes the suburb house and subdivision context that will evolve.[3]

The organization of suburb house yard areas and open space within a subdivision also affects the context in which the suburb house resides. This book refers to the two primary forms of organization as *standard plan* and *cluster plan* arrangements. *Standard plans* are created when a subdivision allocates all or nearly all available land area to private suburb house lots and street rights-of-way. The context of these neighborhoods is influenced by the size of the lots platted and the street pattern introduced. *Cluster plans* are created when suburb house standard lots are reduced in size within the same area in order to assemble the remaining land as shared neighborhood open space. The context of these neighborhoods is influenced by the cluster plan organization of common open space, lot sizes, and street pattern. Building height is a secondary issue in both cases since it is generally limited to 2.5 stories, but glaring inconsistencies can produce conflict.

A lot area range can indicate the neighborhood context of the suburb house when standard plan organization is used. For instance, Table 13.2 suggests that a standard plan estate lot extends from 2 acres to 5 acres. However, subdivision context is not necessarily indicated by the lot areas involved when cluster plan organization is used, since cluster plans reduce lot areas in order to assemble common open space. In this case, the common open space allocation per dwelling combines with the lot area allocation to describe the neighborhood context of the suburb house.

Land organization and lot area are the first of six subdivision characteristics listed in Table 13.1 that can be used to classify subdivisions and the suburb houses within, and have been used in this book.

If the classification system in Table 13.1 were used to define a suburb house, for example, the definition would read, "Standard plan context, estate lot range, 200-ft frontage, 10% building cover, 15% total development cover, and 2.5-story building height." Real estate definitions, such as four-car garage, five bedrooms, five baths, Tudor style, beautiful view, wonderful schools, and such are useful when defining a commodity to be sold, but have little relevance when defining the environmental context that a suburb house and its subdivision plan will produce. The case studies included in this chapter will explain these classification

Table 13.2 Lot Area Ranges and Context Categories

Context Category	Equal to/More Than	Less Than
Farm lots	5 acres	No limit
Estate lots	2 acres	5 acres
Garden lots	1 acre	2 acres
Picture lots	15,000 sq ft	43,560 sq ft
Lawn lots	9,000 sq ft	15,000 sq ft
Yard lots	6,000 sq ft	9,000 sq ft
Narrow lots	40-ft frontage	6,000 sq ft
Urban houses	0 sq ft	40-ft frontage

characteristics in more detail, but the following is a brief overview.

Lot Organization

Standard plan and cluster plan lot arrangements have been previously defined, but many subdivisions now contain entry walls, landscaped areas, guardhouses, and the like that identify the subdivision name and are maintained through common subdivision assessment. This is one form of common open space, but is small in relation to the private lots created and is often included with standard plan subdivisions. It does not represent the open space implied by the term *cluster plan*. In a standard plan subdivision most, if not all, land is allocated to either road right-of-way, roadway easements, or private lots. In this same area, a cluster plan will make the lots smaller and allocate the remaining land to common open space in the form of parks, ponds, bikeways, nature preserves, and so on. The intent is to remove some of the open space amenity from private yards and place the savings in more visible and socially accessible locations that can be shared within the subdivision. The number of dwelling units provided per acre may be the same and housing capacity may not be improved, but streetscape appearance can be less monotonous and more socially oriented.[4] If the lots are too small, however, they may struggle to accommodate the automobile and have significantly less building expansion potential over time.

Rigid subdivision standards can make it difficult to use the cluster plan approach for suburb house development, however, since the concept is to reduce lot areas from established minimums. These reductions contribute to shared open space within the same total area that would be occupied by a standard plan subdivision. Therefore, cluster plans represent a different method of organizing open space within a subdivision. They do not necessarily increase housing capacity nor conserve agricultural land and are not environmentally significant unless the open space preserved has this status.

Lot Area Ranges

Suburb houses are generally distinguished by the amount of lot area allocated for the private use of each family. Lot sizes are often regulated by zoning district and may or may not be grouped into broader context categories. The context categories listed in Table 13.2 are based on lot area ranges and have been used to simplify the referencing system in this chapter.

Lot Frontage

Lot frontage is a primary subdivision characteristic that affects the curbside appearance of a suburb house and its ability to accommodate the automobile. Inadequate lot frontage makes it difficult for a suburb house to include both a two-car garage and an inviting front-door entry. Lot frontage affects the usefulness of land, site plan, and floor plan composition, convenience, and accessibility. It is included in the suburb house classification system for this reason.

Development Cover

Development cover includes building cover and miscellaneous pavement cover and does as much to establish neighborhood context as the total amount of land area provided per lot. For instance, a 4-acre estate lot is expected to provide substantial amounts of open space around the suburb house constructed. If 80% of this lot is covered with house and pavement, it contradicts the context expected, but the zoning setback lines in place may permit this magnitude of development cover.

Development cover is a statistic that has context, amenity, lifestyle, and public health implications. It can also represent the design limit for each suburb house in a subdivision, since the open space remaining establishes the visual context and stormwater protection for that neighborhood. Public health is involved since excess development cover produces runoff that exceeds the design capacity of the storm sewer system. This can produce street flooding, contamination, and basement flooding when combined sewers are involved. For new construction, however, it is not difficult to regulate the maximum amount of development cover permitted in relation to the storm sewer capacity included with the subdivision, and to ensure that this development cover limit anticipates a reasonable amount of future expansion for the suburb house.

In many older subdivisions, home sizes have been expanded, patios have been added, driveways have been widened, garages have been enlarged, sheds have been built, and so forth. These additional roofs and pavement add to the stormwater volume that the older sewer system is expected to receive, and the development cover limits implied by the original storm sewer capacity constructed may be exceeded.[5]

Building Height

Building height combines with building cover to produce three-dimensional building form.[6] This form is called *massing,* and a massing envelope represents the maximum building volume that can be constructed on a lot. Lot frontage can influence building height, since a narrow lot may force an owner to produce a multifloor plan to meet his or her needs. If suburb house building cover, building height, and total development cover are specified for a given lot, the maximum development capacity of the lot and the maximum three-dimensional volume potential of the lot are defined.

Street Pattern

Suburb house appearance, privacy, function, and social involvement are affected by the street pattern that ties it to the community. These patterns are often referred to as *grid systems, modified grid systems, curvilinear systems, cul-de-sac systems,* and *hybrid systems.* Their function is often referred to as *neighborhood, local, collector, arterial,* or *limited access,* and their level of service or congestion is generally graded like a school report card. A suburb house is affected by the function and pattern of the street system serving it. For this reason, the classification of a suburb house includes the nature of the street system that links it to the neighborhood and community.

Specific Lot Area

Specific lot areas are not as important as the lot area context range that encompasses them. The specific area of a suburb house lot, therefore, has been included as a subspecies characteristic that serves to place the suburb house within a private lot context range. Suburb houses are generally allocated from 6000 sq ft to 5 acres of private lot area per dwelling. These allocations shelter from 7.26 to 0.20 families per acre. 12,000- to 15,000-sq-ft lawn lots seem to be a popular, though not always affordable, lot size that easily accommodates the automobile in either front-loading or side-loading garage configurations.[7] These statistics make it apparent, however, that there is a clear limit to the housing capacity of single-family subdivisions and the suburb houses that use them.

Performance

The index of housing capacity (also referred to as *development capacity*) or performance for single-family subdivisions is lots per acre, but the term *dwelling units per acre* is often used since *lot* and *dwelling unit* are synonymous for this shelter class. The greatest suburb house capacity is produced with the smallest lots. The yard lot range and the urban house lot range are subdivided in the following list to illustrate the upper limits of suburb house development capacity. When platted, they produce the following densities:

Housing Type	Area, sq ft	Lot Frontage, ft	Density, Lots per Buildable Acre
Yard lots	7,200	60	6.05
Yard lots	6,000	50	7.26
Urban house lots	4,800	40	9.07
Urban house lots	3,600	30	12.1

These densities and those calculated in Table 13.3 should illustrate that the window of broadly marketable and affordable single-family suburb house densities is relatively narrow and generally ranges from two to seven lots per net acre. This would be roughly 1.5 to 5.25 lots per gross acre, depending on the street pattern and right-of-way planned or present.

Case Studies

Eight suburb house case studies are included in this chapter to illustrate some of the classification and capacity characteristics of this shelter category. The SF context record form is used to record suburb house characteristics, and the RSFL forecast model is used to predict and evaluate other suburb house development options with and without cluster plan organization and common open space provisions.

Table 13.3 Suburb House Data Summary*

Context Record Summary of Suburb House and Subdivision Averages

Subdivision Name	Context record table	Subdivision Characteristics	Total Buildable Area in Acres BLA	No. of Lots	No. of Lots per Net Acre Lba	Common Open Space as Percentage of Net Land Area COS	Average Private Lot Area, in sq ft (Excludes Common Open Space) ALA	Maximum Development Cover as a Percentage of Average Lot Area $D = B + M$	Maximum Building Cover as Percentage of Average Lot Area B	Maximum Miscellaneous Pavement as Percentage of Average Lot Area M	Minimum Private Yard Remaining at Build-Out as % of Average Lot Area $S = 1 - D$	Total Development Cover Area in sq ft DCA	Total Private Yard Area in sq ft YRD	Private Yard as a Multiple of Maximum Development Cover Potential Y = YRD/DCA	Footprint Present or Planned in sq ft FTP	Maximum Footprint Potential in sq ft BCG	Footprint Expansion Potential in sq ft FEX = BCG − FTP	Footprint Expansion Potential as Percentage of Planned Footprint FEX/FTP	Current Private, Common, and Expansion Open Space per Average Lot
Doone/Essex	13.1	MYG	5.58	27	4.84	0.0%	9,003	41.5%	29.5%	12.0%	58.5%	3,736	5,267	1.41	1,855	2,656	801	43.2%	58.5%
Fisher Place	13.2	MLC	1.77	7	3.96	0.0%	10,995	35.0%	24.6%	10.4%	65.0%	3,848	7,147	1.86	2,000	2,705	705	35.3%	65.0%
Donegal Cliffs	13.3	MPSc	4.64	13	2.80	0.0%	15,551	40.0%	35.0%	5.0%	60.0%	6,220	9,331	1.50	2,400	5,443	3,043	126.8%	60.0%
Jefferson Twp. cluster	13.4	CLM	29.41	76	2.58	40.5%	10,028	40.0%	35.0%	5.0%	60.0%	4,011	6,017	1.50	2,500	3,510	1,010	40.4%	82.2%
Jefferson Twp. traditional	13.5	MPM	29.75	76	2.55	0.0%	17,052	35.0%	30.0%	5.0%	65.0%	5,968	11,084	1.86	2,500	5,116	2,616	104.6%	80.3%
Glenbervie	13.6	CPC	4.63	10	2.16	3.5%	19,475	42.0%	37.0%	5.0%	58.0%	8,180	11,296	1.38	2,300	7,206	4,906	213.3%	59.5%
Invergordon	13.7	CLC	5.22	5	0.96	73.4%	12,108	35.0%	32.0%	3.0%	65.0%	4,238	7,870	1.86	2,800	3,875	1,075	38.4%	93.0%
Squirrel Bend	13.8	MEC	30.33	15	0.49	0.0%	87,980	25.0%	13.0%	12.0%	75.0%	21,995	65,985	3.00	7,550	11,437	3,887	51.5%	75.0%

Subdivision Characteristics

Lot Pattern	Lot Area Range	Equal to/ More Than	Less Than	Street Pattern	Street Function
	F: farm lot	5 acres	no limit		
M: standard lot	E: estate lot	2 acres	5 acres	G: grid system	N: neighborhood access
C: cluster lot	G: garden lot	1 acre	2 acres	M: modified grid system	L: local traffic
	P: picture lot	15,000 sq ft	43,560 sq ft	S: curvilinear system	C: collector traffic
	L: lawn lot	9,000 sq ft	15,000 sq ft	C: cul-de-sac system	A: arterial traffic
	Y: yard lot	6,000 sq ft	9,000 sq ft	Gc: grid with cul-de-sacs	F: limited access traffic
	N: narrow lot	40-ft frontage	6,000 sq ft	Sc: curvilinear with cul-de-sacs	
	U: urban lot	0 sq ft	40-ft frontage	Mc: modified grid with cul-de-sacs	

*Taken from SF context records.

Standard Plan with Yard Lot Context

The historic Doone/Essex subdivision is a standard plan subdivision with a grid street pattern. Its average lot area slightly exceeds the yard lot range by 3 sq ft, but its yard lot character has placed it in this classification.

The Doone/Essex subdivision is served by a collector street system with an unknown service level. Its site plan is included with Fig. 13.1. Many detached garages are present because the lot frontages are relatively narrow and the floor plans are wide, leaving room for a driveway to a wider detached garage behind. Some homes have single-car attached garages that are entered from the street, and a few own a lot and a half or two lots, which makes it feasible to include an attached two-car garage.

(a)

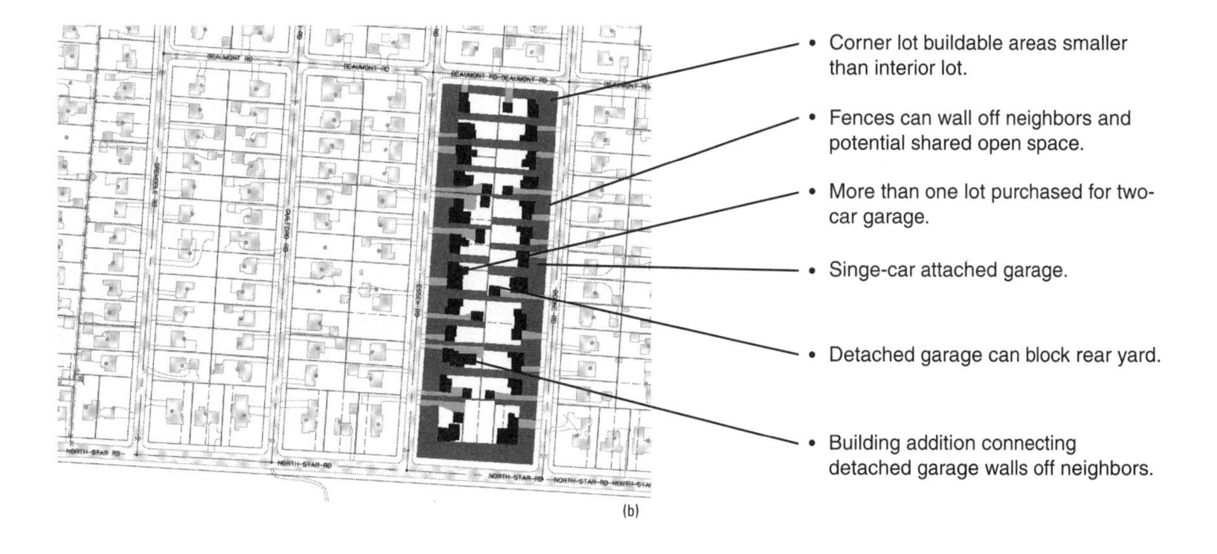

(b)

(c)

(d)

(e)

Figure 13.1 Doone/Essex suburb houses—standard plan arrangement with yard lot context, 60-ft frontage, and 41.5% development cover. Doone/Essex is within a first-ring suburb and includes yard lot frontages, detached garages, 2-story homes, sidewalks, and streets with curbs and gutters as typical features. (*a*) Overview; (*b*) site plan; (*c*)–(*e*) frontal views.

(Photos by W. M. Hosack)

The Doone/Essex subdivision context measurements are recorded in Table 13.4 and show an average lot size of 9003 sq ft. These records also show an average footprint of 1855 sq ft, including garage; an average building footprint expansion capability of 801 sq ft; and 58.5% minimum project open space if the total development cover[8] allocation is ever constructed.

The forecast model for Doone/Essex is included as Exhibit 13.1. The values entered in the design specification panel are found in the project context record of measured values included in Table 13.4. The lot specification panel does not break down each lot area and type present in the subdivision. It simply uses the average lot characteristics that represent the entire subdivision. Corner lot types have been given special mention in the text to follow, however. Based on these data, the lot capacity forecast panel within Exhibit 13.1 predicts that 27 lots can be created when no common open space is provided, and that each lot will be 9006 sq ft in area. The actual average lot area measured was 9003 sq ft, and the dwelling unit count was 27.[9]

COMMON OPEN SPACE

The lot forecast panel further predicts that only 16 lots of the same average area could have been created if 40% common open space had been provided in a cluster plan setting. However, Table 13.4 notes that *if* the average lot area had been reduced to 5404 sq ft, the same 27 lots could have been created along with 40% common open space. 5000 sq ft is equal to a 50-by-100-ft lot. A 9000-sq-ft lot is equal to a 75-by-120-ft lot. A 5000-sq-ft lot is very close to the urban house category and represents a neighborhood context that would be quite different from that presently found along Doone/Essex. Both would produce the same housing capacity of 4.84 families per buildable acre, but each would produce a different environment because of the lot plan organization differences involved.

Table 13.4 Doone / Essex Context Record

Form SF

	Form SF
Name	**Doone / Essex**
Address or Location	**First block W. of N. Star**
Land Use	**Suburb house**
	Standard plan, yard lot context
Parking System	*Interior parking in garage or carport or exterior parking on site*

NOTE: Single-family means a DETACHED dwelling unit (home) intended for the use of only one family unit.

	Description	Project Measurements		Context Summary SF	%	
Site						
	Gross land area in acres	7.013	GLA	305,486		
	Public/ private ROW and paved easements in SF	62,400	W		20.43%	of GLA
	Net lot area		NLA	243,086	79.6%	of GLA
	Total unbuildable area in SF not common open space	0	U		0.0%	of GLA
	Water area within unbuildable area in SF	0	WAT			of U
	Buildable land area in acres and SF	5.58	BLA	**243,086**	79.6%	of GLA
Common Open Space						
	Common open space as a % of BLA	0.00%	COS	0		of BLA
	Water area within common open space in SF	0 0.0%				
Subdivision Development						
	Total number of lots	27	NLT			
	Average private lot area in SF		ALA	**9,003**		
	Avg. private lot area plus prorated common open space			**9,003**		
	Avg. lot frontage in feet	60.0	FRN			
	Avg. lot depth to width ratio			2.50		
	Total dev. cover planned per lot as a % of BLA	41.50%	D	3,736		
	Avg. total private yard area per lot in SF		YRD**	5,267	58.5%	of LOT
	Private yard as multiple of total dev. cover area		Y	1.41		
	Misc. pavement incl. driveways as a % of lot	12.00%	M	1,080		
	Total potential building cover area per lot		BCG	2,656	29.5%	of LOT
	Misc. pavement as % of total bldg. cover potential		Mbc		40.7%	of BCG
	Total avg. building cover present per lot in SF	1,855	FTP			
	Building footprint expansion capacity remaining		FXA	801	43.2%	of FTP
	Current private and common open space including bldg. expansion areas		Sn	163,831	67.4%	of BCG
	Anticipated private and common open space minimum permanent provision		S	142,205	58.5%	of BCG
	No. of lots with 1-story homes					
	No. of lots with 1.5-story homes					
	No. of lots with 2-story homes	10				
	No. of lots with 2.5-story homes					
	No. of lots with 3.0-story homes					
	No. of lots with 3.5-story homes					
	Average building height in stories				2.00	
	Max. permitted building height in stories	2.5				
Summary						
	Lots per buildable acre				4.84	
	Lots per net acre				4.84	
	Lots per gross acre				3.85	
	Current development intensity index		CINX		**2.326**	
	Anticipated max. development intensity index		AINX		**2.915**	

private yard areas
gross housing area
side yards
net housing area

Typical subdivision. Gross buildable area within dashed lines with emphasis on front yards. Subtract side yards to find net buildable area. Divide by number of lots to find average per lot.

common open space

Common open space. Separate lot ownership. Front yards de-emphasized. When there is no separate lot ownership, buildable areas can be defined within a condominium arrangement or not.

***This can be a deceiving statistic since many zoning ordinances permit detached garages, swimming pools, sheds, fences, etc. to be placed in side and rear yards as additional development cover. This displaces the open space required by setback lines and in the case of fences and garages disrupts the continuity of this rear yard amenity. In older neighborhoods where detached garages are common, the combination of fences, garages , and sheds reduces rear and side yards to postage stamp amenities that bear no resemblance to the open space implied by the setback statistic quoted.*

Exhibit 13.1 Doone / Essex Design Specification and Forecast

Forecast Model RSFL

Development capacity forecast for ***SINGLE-FAMILY DETACHED HOUSING*** *with no shared or common parking lots serving each dwelling unit.*

Given: *Gross land area.* ***To Find:*** *(1) Maximum number of lots that can be created from the land area given based on the design specification values entered below. (2) Maximum gross dwelling unit area that can be built on each lot created.*

DESIGN SPECIFICATION *Enter values in boxed areas where text is bold and blue. Express all fractions as decimals.*

Given:	**Gross land area**	GLA=	**7.013**	acres	305,486 SF
Land Variables:	Est. right-of-way dedication as fraction of GLA	W=	**0.204**		62,411 SF
	Private roadway as fraction of GLA	R=	**0.000**		
	School land donation as fraction of GLA	Ds=	**0.000**		
	Park land donation as fraction of GLA	Dp=	**0.000**		
	Other land donation as fraction of GLA	Do=	**0.000**		
	Net land area	NLA=	5.580	acres	243,075 SF
	Unbuildable and/or future expansion areas as fraction of GLA	U=	**0.000**		0 SF
	Total gross land area reduction as fraction of GLA	X=	0.204		62,411 SF
	Buildable land area remaining	**BLA=**	**5.580**	acres	243,075 SF
Core:	**Core + open space available**	**CS=**	**5.580**	acres	243,075 SF

LOT SPECIFICATION

	total bldg cover potential BCG incl expansion & garage	garage GAR footprint	misc. pavement & drive Mbc as multiple of BCG	total development DCA cover area	private yards Y as multiple of DCA	total lot LOT area	lot LTX mix	aggregate avg. AAL lot areas in mix	total lot TLA areas in mix
Type 1 Interior Lot									
Type 2 Interior Lot									
Type 3 Interior Lot	**2,656**	**220**	**0.41**	3,737	**1.41**	9,006	**1.000**	9,006	9,006
Type 4 Interior Lot									
Type 5 Interior Lot									
Type 6 Interior Lot									
Type 7 Interior Lot									
Total Average Development Cover				3,737					
Mix Total (do not exceed 1.0)							1.00		
Aggregate Avg. Lot Size Based on Mix:				41.49%	= DCA		AAL =	**9,006**	
Enter zero in the adjacent box unless you wish to override the AAL value calculated above:								**0**	= AAL override

LOT CAPACITY FORECAST *(based on common open space value (COS) entered below.*

DEFINITION of COMMON OPEN SPACE (COS): Open space provided for common benefit and massing separation in addition to that provided for private use and setbacks on individual lots. On a given land area, private lot sizes decrease as common open space increases when all other development factors are constant.
NOTE: Common open space values in bold & blue in the table below may be changed to examine other implications. When COS = 0.0 all open space is provided on private lots (OSA) in the form of yards and a typical subdivision is implied.

COS:	**0.00**	**0.10**	**0.20**	**0.30**	**0.40**	**0.50**	**0.60**	**0.70**	**0.80**
No. of lots:	27.0	24.3	21.6	18.9	16.2	13.5	10.8	8.1	5.4
Net density (dNA):	4.84	4.35	3.87	3.39	2.90	2.42	1.93	1.45	0.97
BLA density (dBA):	4.84	4.35	3.87	3.39	2.90	2.42	1.93	1.45	0.97
Lot size compensation	9,006	8,106	7,205	6,304	5,404	4,503	3,602	2,702	1,801
No. of lots desired:	27.0	27.0	27.0	27.0	27.0	27.0	27.0	27.0	27.0

GROSS BUILDING AREA per LOT FORECAST

	basement as fraction of BSM habitable footprint	basement BSA area	crawl space CRW area	NET POTENTIAL BUILDING AREA INCLUDING GARAGE but EXCLUDING BASEMENT and POTENTIAL EXPANSION — number of building floors: 1	1.5	2	2.5	3	3.5
Type 1 Interior Lot	**1.000**								
Type 2 Interior Lot	**1.000**								
Type 3 Interior Lot	**1.000**	2,436	0	2,656	3,984	5,312	6,640	7,968	9,296
Type 4 Interior Lot	**1.000**								
Type 5 Interior Lot	**1.000**								
Type 6 Interior Lot	**1.000**								
Type 7 Interior Lot	**1.000**								

WARNING: These are preliminary forecasts that must not be used to make final decisions.
1) These forecasts are not a substitute for the "due diligence" research that must be conducted to support the final definition of "unbuildable areas" above and the final decision to purchase land. This research includes, but is not limited to, verification of adequate subsurface soil, zoning, environmental clearance, access, title, utilities and water pressure, clearance from deed restriction, easement and right-of-way encumbrances, clearance from existing above and below ground facility conflicts, etc.
2) The most promising forecast(s) made on the basis of data entered in the design specification from "due diligence" research must be verified at the drawing board before funds are committed and land purchase decisions are made. Actual land shape ratios, dimensions and irregularities encountered may require adjustments to the general forecasts above.
3) The software licensee shall take responsibility for the design specification values entered and any advice given that is based on the forecast produced.

GROSS BUILDING AREA CAPACITY

The gross building area forecast panel in Exhibit 13.1 predicts the total building area capacity of the 9006-sq-ft lot, based on the lot specification values entered above. This forecast applies to the private lot areas provided and does not take common open space into account. The gross building area forecast panel shows that complete build-out[10] of all permitted development cover on the 9006-sq-ft lot forecast with a 1-story building could produce 2656 sq ft of gross building area including garage. A 2-story building would produce 5312 gross sq ft. For those familiar with 9000-sq-ft lots, this "ultimate development capacity" potential is well above the investment generally made on lots of this size.

A building area forecast serves to verify that the yard area multiple Y of development cover will not adversely affect the development capacity of the lot. In this case, Exhibit 13.1 shows that this multiple is 1.41, but detached garages, sheds, swimming pools, and so on may be constructed in this yard area.

DEVELOPMENT COVER LIMITS

The Doone/Essex average development cover percentage of 41.5%, shown in Table 13.3 (column labeled "$D + B = M$") may be higher than the impervious cover assumption originally used to design this historic subdivision's storm sewer sizes and capacities. As a result, neighborhoods like these that continue to add development cover may experience storm water problems over time. This is particularly true if the older storm system is combined with sanitary sewer responsibilities.

When forecasting new subdivision opportunities, it is always helpful to correlate the planned building footprint and its potential expansion over time with the total development cover anticipated and the storm sewer capacity contemplated. Exhibit 13.1 is based on forecast model RSFL and predicts the Doone/Essex development cover percentage DCA in its lot specification panel. This prediction is based on the total building cover, garage cover, and miscellaneous pavement cover planned or present per lot. Forecast model RSFL is included with the forecast model collection in this book and can be used to predict other development implications based on the criteria entered in its design specification.

CORNER LOTS

The corner lots in this subdivision deserve mention since they represent a common issue. North on the site plan in Fig. 13.1 is at the top of the page and the eastern corner lots have been assembled from smaller platted lots to produce the three large lots shown. Two of these lots have homes and both benefit from the open middle lot since they face a high-volume collector street. The corner lots at the western end of the block are larger than their interior lot neighbors, but not as large as their eastern cousins. They are enlarged to offset the fact that the front yard setback requirements for the two adjoining streets and the rear yard requirement combine to reduce the permitted housing footprint area within the lot. This area, however, is still not as large as its interior lot neighbors, but the pressure to expand is the same. This can, and often does, produce excessive requests for corner lot variances. It is an issue that can be resolved easily, however, by requiring that the corner lot area within front, side, and rear setback lines of new subdivisions be no less than the area provided for its interior lot neighbors.[11]

Standard Plan with Lawn Lot Context and Cluster Plan Context with Picture Lots

The suburb homes at Fisher Place and Glenbervie Court are both arranged around short cul-de-sac streets and have similar floor plan areas and appearances, but they have widely different land area allocations per dwelling.

Fisher Place is illustrated by Fig. 13.2 and is an infill project of seven lots that are served by a collector street carrying much higher traffic volumes within a first-ring suburb. It is isolated from the surrounding grid-pattern neighborhood by its dead-end street arrangement and its much higher housing values. Table 13.3 also notes that the average lot area is approximately 10,995 sq ft (column labeled "ALA"). The development cover limit is 35% (column labeled "$D + B = M$") and the average expansion potential is 705 sq ft of footprint area (column labeled "FEX = BCG − FTP").

Glenbervie is illustrated by Fig. 13.3, and its average lot is almost twice as large, but its average suburb house footprint is only 300 sq ft greater than Fisher Place. This is shown in Table 13.3. Glenbervie produces a much greater apparent lawn area, but the potential building expansion area on a Glenbervie lot could actually result in less open space over time. The minimum lawn area on the Glenbervie lot is 1.38 times the maximum potential development cover, while the Fisher Place multiple is 1.86.

The biggest difference between these two subdivisions, however, is that Fisher Place is an isolated infill project, while Glenbervie is part of an arrangement of cul-de-sac clusters served by limited-volume collector streets that are separated by common open space and connected by multipurpose pathways.[12] This open space system increases the Glenbervie land allocation per dwelling and reduces the housing capacity to 2.16 lots per acre, while Fisher Place is able to provide 3.96 lots per acre with greater assurance that its open space yard areas will remain over time. Neither is designed to solve the world's housing problem, but the first is simply a practical response to an older neighborhood with sidewalks, while the other is part of a planned community that contains an integrated pathway system and a protected hierarchy of street functions.

The context records for Fisher Place and Glenbervie are presented as Tables 13.5 and 13.6 and are summarized in Table 13.3. Exhibits 13.2 and 13.3 show that these results could have been predicted by forecast model RSFL. The lot capacity forecast panel in the Fisher Place model, Exhibit 13.2, also shows the implications of a 30% common open space provision. If introduced, it would have either reduced the number of lots to 4.9 or would have required the seven lots to be reduced to an average of 7706 sq ft. This is equivalent to a 60- by 120-ft lot and could not have accommodated the suburb house floor plans built in this subdivision. Since the common open space produced would only be approximately 15,000 sq ft, and would not be connected to any larger open space plan, its use in this infill application was never seriously considered. It is only mentioned to show how the RSFL forecast model can be used to evaluate the question.

(a)

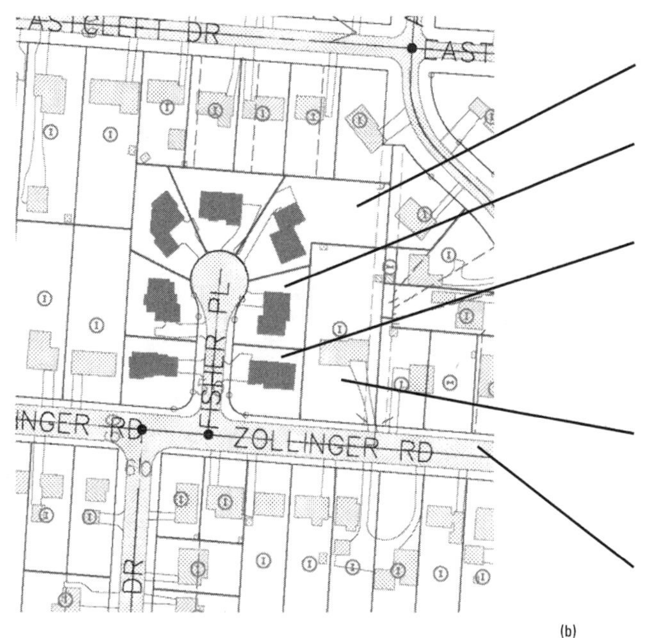

- This area artificially increases average lot areas.
- Large homes on smaller lots with less expansion potential, similar to Invergordon.
- Setbacks affect placement.
- Building cover and development cover regulations limit intensity and protect stormwater system.
- Infill lots are quite different from lot sizes and building characteristics of surrounding neighborhood.
- Four-lane collector road serves community shopping center.

(b)

(c)

Figure 13.2 Fisher Place infill suburb houses—standard plan arrangement with lawn lot context, 90-ft average frontage, and 35% development cover. Fisher Place is an infill subdivision within a first-ring suburb. It has attached garages, 2-story homes, sidewalks, and streets with curbs and gutters as typical features. Three of the seven lots have side-loading garages. (*a*) Overview; (*b*) site plan; (*c*) frontal view.

(Photos by W. M. Hosack)

(a)

(b)

(c)

(d)

Figure 13.3 Glenbervie suburb houses—cluster plan context with picture lots, 105-ft average frontage, and 42% development cover. Glenbervie is within the outer limits of an edge city and provides picture lot frontage; attached garages; 2-story homes; and no sidewalk, curb, or gutter construction. Additional features include wide frontage, 19,475-sq-ft-average lot area, 60% side-loading garages, common open space along one side and rear of subdivision, cul-de-sac pavement impact reduced with landscaping emphasis, and a potential development cover expansion that could reduce private rear yard areas. (*a*) Overview; (*b*) site plan; (*c*) and (*d*) frontal views.

Table 13.5 Fisher Place Context Record

Form SF

Name	**Fisher Place**
Address or Location	**Fisher Place**
Land Use	**Suburb house**
	Standard plan, lawn lot context
Parking System	*Interior parking in garage or carport or exterior parking on site*

NOTE: Single-family means a DETACHED dwelling unit (home) intended for the use of only one family unit.

	Description	Project Measurements		Context Summary SF	%	
Site						
	Gross land area in acres	2.020	GLA	87,991		
	Public/ private ROW and paved easements in SF	11,026	W		12.53%	of GLA
	Net lot area		NLA	76,965	87.5%	of GLA
	Total unbuildable area in SF not common open space	0	U		0.0%	of GLA
	Water area within unbuildable area in SF	0	WAT			of U
	Buildable land area in acres and SF	1.77	BLA	**76,965**	87.5%	of GLA
Common Open Space						
	Common open space as a % of BLA	0.00%	COS	0		of BLA
	Water area within common open space in SF	0 0.0%				
Subdivision Development						
	Total number of lots	7	NLT			
	Average private lot area in SF		ALA	**10,995**		
	Avg. private lot area plus prorated common open space			**10,995**		
	Avg. lot frontage in feet	90.0	FRN			
	Avg. lot depth to width ratio			1.36		
	Total dev. cover planned per lot as a % of BLA	35.00%	D	3,848		
	Avg. total private yard area per lot in SF		YRD**	7,147	65.0%	of LOT
	Private yard as multiple of total dev. cover area		Y	1.86		
	Misc. pavement incl. driveways as a % of lot	10.40%	M	1,143		
	Total potential building cover area per lot		BCG	2,705	24.6%	of LOT
	Misc. pavement as % of total bldg. cover potential		Mbc		42.3%	of BCG
	Total avg. building cover present per lot in SF	2,000	FTP			
	Building footprint expansion capacity remaining		FXA	705	35.2%	of FTP
	Current private and common open space including bldg. expansion areas		Sn	54,961	71.4%	of BCG
	Anticipated private and common open space minimum permanent provision		S	50,027	65.0%	of BCG
	No. of lots with 1-story homes					
	No. of lots with 1.5-story homes	4				
	No. of lots with 2-story homes	3				
	No. of lots with 2.5-story homes					
	No. of lots with 3.0-story homes					
	No. of lots with 3.5-story homes					
	Average building height in stories				1.71	
	Max. permitted building height in stories	2.5				
Summary						
	Lots per buildable acre				3.96	
	Lots per net acre				3.96	
	Lots per gross acre				3.47	
	Current development intensity index		CINX		**2.000**	
	Anticipated max. development intensity index		AINX		**2.850**	

Typical subdivision. Gross buildable area within dashed lines with emphasis on front yards. Subtract side yards to find net buildable area. Divide by number of lots to find average per lot.

Common open space. Separate lot ownership. Front yards de-emphasized. When there is no separate lot ownership, buildable areas can be defined within a condominium arrangement or not.

***This can be a deceiving statistic since many zoning ordinances permit detached garages, swimming pools, sheds, fences etc. to be placed in side and rear yards as additional development cover. This displaces the open space required by setback lines and in the case of fences and garages disrupts the continuity of this rear yard amenity. In older neighborhoods where detached garages are common, the combination of fences, garages, and sheds reduces rear and side yards to postage stamp amenities that bear no resemblance to the open space implied by the setback statistic quoted.*

Table 13.6 Glenbervie Context Record

Form SF

Name	**Glenbervie**
Address or Location	**Glenbervie Drive**
Land Use	**Suburb house**
	Cluster plan context, picture lots
Parking System	*Interior parking in garage or carport or exterior parking on site*

NOTE: Single-family means a DETACHED dwelling unit (home) intended for the use of only one family unit.

	Description	Project Measurements		Context Summary SF	%	
Site						
	Gross land area in acres	**5.370**	GLA	233,917		
	Public/ private ROW and paved easements in SF	**32,100**	W		13.72%	of GLA
	Net lot area		NLA	201,817	86.3%	of GLA
	Total unbuildable area in SF not common open space	**0**	U		0.0%	of GLA
	Water area within unbuildable area in SF	**0**	WAT			of U
	Buildable land area in acres and SF	4.63	BLA	**201,817**	86.3%	of GLA
Common Open Space						
	Common open space as a % of BLA	**3.50%**	COS	7,064		of BLA
	Water area within common open space in SF	**0**				
		0.0%				
Subdivision Development						
	Total number of lots	**10**	NLT			
	Average private lot area in SF		ALA	**19,475**		
	Avg. private lot area plus prorated common open space			**20,182**		
	Avg. lot frontage in feet	**105.0**	FRN			
	Avg. lot depth to width ratio			1.77		
	Total dev. cover planned per lot as a % of BLA	**42.00%**	D	8,180		
	Avg. total private yard area per lot in SF		YRD**	11,296	58.0%	of LOT
	Private yard as multiple of total dev. cover area		Y	1.38		
	Misc. pavement incl. driveways as a % of lot	**5.00%**	M	974		
	Total potential building cover area per lot		BCG	7,206	37.0%	of LOT
	Misc. pavement as % of total bldg. cover potential		Mbc		13.5%	of BCG
	Total avg. building cover present per lot in SF	**2,300**	FTP			
	Building footprint expansion capacity remaining		FXA	4,906	213.3%	of FTP
	Current private and common open space including bldg. expansion areas		Sn	169,080	83.8%	of BCG
	Anticipated private and common open space minimum permanent provision		S	120,021	59.5%	of BCG
	No. of lots with 1-story homes					
	No. of lots with 1.5-story homes					
	No. of lots with 2-story homes	**10**				
	No. of lots with 2.5-story homes					
	No. of lots with 3.0-story homes					
	No. of lots with 3.5-story homes					
	Average building height in stories				2.00	
	Max. permitted building height in stories	**2.5**				
Summary						
	Lots per buildable acre				2.16	
	Lots per net acre				2.16	
	Lots per gross acre				1.86	
	Current development intensity index		CINX		**2.162**	
	Anticipated max. development intensity index		AINX		**2.905**	

private yard areas
gross housing area
side yards
net housing area

Typical subdivision. Gross buildable area within dashed lines with emphasis on front yards. Subtract side yards to find net buildable area. Divide by number of lots to find average per lot.

common open space

Common open space. Separate lot ownership. Front yards de-emphasized. When there is no separate lot ownership, buildable areas can be defined within a condominium arrangement.

***This can be a deceiving statistic since many zoning ordinances permit detached garages, swimming pools, sheds, fences,etc. to be placed in side and rear yards as additional development cover. This displaces the open space required by setback lines and in the case of fences and garages disrupts the continuity of this rear yard amenity. In older neighborhoods where detached garages are common, the combination of fences, garages, and sheds reduces rear and side yards to postage stamp amenities that bear no resemblance to the open space implied by the setback statistic quoted.*

Exhibit 13.2 Fisher Place Design Specification and Forecast

Forecast Model RSFL

Development capacity forecast for ***SINGLE-FAMILY DETACHED HOUSING*** *with no shared or common parking lots serving each dwelling unit.*

Given: *Gross land area.* ***To Find:*** *(1) Maximum number of lots that can be created from the land area given based on the design specification values entered below. (2) Maximum gross dwelling unit area that can be built on each lot created.*

DESIGN SPECIFICATION

Enter values in boxed areas where text is bold and blue. Express all fractions as decimals.

Given:	**Gross land area**	GLA=	**2.020**	acres	87,991	SF
Land Variables:	Est. right-of-way dedication as fraction of GLA	W=	**0.125**		11,025	SF
	Private roadway as fraction of GLA	R=	**0.000**			
	School land donation as fraction of GLA	Ds=	**0.000**			
	Park land donation as fraction of GLA	Dp=	**0.000**			
	Other land donation as fraction of GLA	Do=	**0.000**			
	Net land area	NLA=	1.767	acres	76,966	SF
	Unbuildable and/or future expansion areas as fraction of GLA	U=	**0.000**		0	SF
	Total gross land area reduction as fraction of GLA	X=	0.125		11,025	SF
	Buildable land area remaining	**BLA=**	**1.767**	acres	76,966	SF
Core:	**Core + open space available**	**CS=**	**1.767**	acres	76,966	SF

LOT SPECIFICATION

	total bldg cover potential BCG incl expansion & garage	garage GAR footprint	misc. pavement & drive Mbc as multiple of BCG	total development DCA cover area	private yards Y as multiple of DCA	total lot LOT area	lot LTX mix	aggregate avg. AAL lot areas in mix	total lot TLA areas in mix
Type 1 Interior Lot									
Type 2 Interior Lot									
Type 3 Interior Lot	**2,705**	**484**	**0.42**	3,849	**1.86**	11,009	**1.000**	11,009	11,009
Type 4 Interior Lot									
Type 5 Interior Lot									
Type 6 Interior Lot									
Type 7 Interior Lot									
Total Average Development Cover				3,849					
Mix Total (do not exceed 1.0)							1.00		
Aggregate Avg. Lot Size Based on Mix:				34.97%	= DCA		AAL =	**11,009**	
Enter zero in the adjacent box unless you wish to override the AAL value calculated above:								**0**	= AAL override

LOT CAPACITY FORECAST

(based on common open space value (COS) entered below.

DEFINITION of COMMON OPEN SPACE (COS): Open space provided for common benefit and massing separation in addition to that provided for private use and setbacks on individual lots. On a given land area, private lot sizes decrease as common open space increases when all other development factors are constant.

NOTE: Common open space values in bold & blue in the table below may be changed to examine other implications. When COS = 0.0 all open space is provided on private lots (OSA) in the form of yards and a typical subdivision is implied.

COS:	**0.00**	**0.10**	**0.20**	**0.30**	**0.40**	**0.50**	**0.60**	**0.70**	**0.80**
No. of lots:	7.0	6.3	5.6	4.9	4.2	3.5	2.8	2.1	1.4
Net density (dNA):	3.96	3.56	3.17	2.77	2.37	1.98	1.58	1.19	0.79
BLA density (dBA):	3.96	3.56	3.17	2.77	2.37	1.98	1.58	1.19	0.79
Lot size compensation	11,009	9,908	8,807	7,706	6,605	5,504	4,404	3,303	2,202
No. of lots:	7.0	7.0	7.0	7.0	7.0	7.0	7.0	7.0	7.0

GROSS BUILDING AREA per LOT FORECAST

	basement as fraction of BSM habitable footprint	basement BSA area	crawl space CRW area	NET POTENTIAL BUILDING AREA INCLUDING GARAGE but EXCLUDING BASEMENT and POTENTIAL EXPANSION — number of building floors: 1	1.5	2	2.5	3	3.5
Type 1 Interior Lot	**1.000**								
Type 2 Interior Lot	**1.000**								
Type 3 Interior Lot	**1.000**	2,221	0	2,705	4,058	5,410	6,763	8,115	9,468
Type 4 Interior Lot	**1.000**								
Type 5 Interior Lot	**1.000**								
Type 6 Interior Lot	**1.000**								
Type 7 Interior Lot	**1.000**								

WARNING: These are preliminary forecasts that must not be used to make final decisions.

1) These forecasts are not a substitute for the "due diligence" research that must be conducted to support the final definition of "unbuildable areas" above and the final decision to purchase land. This research includes, but is not limited to, verification of adequate subsurface soil, zoning, environmental clearance, access, title, utilities and water pressure, clearance from deed restriction, easement and right-of-way encumbrances, clearance from existing above and below ground facility conflicts, etc.

2) The most promising forecast(s) made on the basis of data entered in the design specification from "due diligence" research must be verified at the drawing board before funds are committed and land purchase decisions are made. Actual land shape ratios, dimensions and irregularities encountered may require adjustments to the general forecasts above.

3) The software licensee shall take responsibility for the design specification values entered and any advice given that is based on the forecast produced.

Exhibit 13.3 Glenbervie Design Specification and Forecast

Forecast Model RSFL

Development capacity forecast for ***SINGLE-FAMILY DETACHED HOUSING*** *with no shared or common parking lots serving each dwelling unit.*

Given: *Gross land area.* ***To Find:*** *(1) Maximum number of lots that can be created from the land area given based on the design specification values entered below. (2) Maximum gross dwelling unit area that can be built on each lot created.*

DESIGN SPECIFICATION

Enter values in boxed areas where text is bold and blue. Express all fractions as decimals.

Given:	**Gross land area**	GLA=	**5.370**	acres	233,917 SF
Land Variables:	Est. right-of-way dedication as fraction of GLA	W=	**0.137**		32,093 SF
	Private roadway as fraction of GLA	R=	**0.000**		
	School land donation as fraction of GLA	Ds=	**0.000**		
	Park land donation as fraction of GLA	Dp=	**0.000**		
	Other land donation as fraction of GLA	Do=	**0.000**		
	Net land area	NLA=	4.633	acres	201,824 SF
	Unbuildable and/or future expansion areas as fraction of GLA	U=	**0.000**		0 SF
	Total gross land area reduction as fraction of GLA	X=	0.137		32,093 SF
	Buildable land area remaining	**BLA=**	**4.633**	acres	201,824 SF
Core:	**Core + open space available**	**CS=**	**4.633**	acres	201,824 SF

LOT SPECIFICATION

	total bldg cover potential BCG incl expansion & garage	garage GAR footprint	misc. pavement & drive Mbc as multiple of BCG	total development DCA cover area	private yards Y as multiple of DCA	total lot LOT area	lot LTX mix	aggregate avg. AAL lot areas in mix	total lot TLA areas in mix
Type 1 Interior Lot									
Type 2 Interior Lot									
Type 3 Interior Lot	**7,206**	**528**	**0.135**	8,179	**1.381**	19,474	**1.000**	19,474	19,474
Type 4 Interior Lot									
Type 5 Interior Lot									
Type 6 Interior Lot									
Type 7 Interior Lot									
Total Average Development Cover				8,179					
Mix Total (do not exceed 1.0)							1.00		
Aggregate Avg. Lot Size Based on Mix:				42.00%	= DCA		AAL =	**19,474**	

Enter zero in the adjacent box unless you wish to override the AAL value calculated above: **0** = AAL override

LOT CAPACITY FORECAST

(based on common open space value (COS) entered below)

DEFINITION of COMMON OPEN SPACE (COS): Open space provided for common benefit and massing separation in addition to that provided for private use and setbacks on individual lots. On a given land area, private lot sizes decrease as common open space increases when all other development factors are constant.
NOTE: Common open space values in bold & blue in the table below may be changed to examine other implications. When COS = 0.0 all open space is provided on private lots (OSA) in the form of yards and a typical subdivision is implied.

COS:	**0.035**	**0.10**	**0.20**	**0.30**	**0.40**	**0.50**	**0.60**	**0.70**	**0.80**
No. of lots:	10.0	9.3	8.3	7.3	6.2	5.2	4.1	3.1	2.1
Net density (dNA):	2.16	2.01	1.79	1.57	1.34	1.12	0.89	0.67	0.45
BLA density (dBA):	2.16	2.01	1.79	1.57	1.34	1.12	0.89	0.67	0.45
Lot size compensation	19,474	18,162	16,144	14,126	12,108	10,090	8,072	6,054	4,036
No. of lots desired:	10.0	10.0	10.0	10.0	10.0	10.0	10.0	10.0	10.0

GROSS BUILDING AREA per LOT FORECAST

	basement as fraction of BSM habitable footprint	basement BSA area	crawl space CRW area	NET POTENTIAL BUILDING AREA INCLUDING GARAGE but EXCLUDING BASEMENT and POTENTIAL EXPANSION — number of building floors: 1	1.5	2	2.5	3	3.5
Type 1 Interior Lot	**1.000**								
Type 2 Interior Lot	**1.000**								
Type 3 Interior Lot	**1.000**	6,678	0	7,206	10,809	14,412	18,015	21,618	25,221
Type 4 Interior Lot	**1.000**								
Type 5 Interior Lot	**1.000**								
Type 6 Interior Lot	**1.000**								
Type 7 Interior Lot	**1.000**								

WARNING: These are preliminary forecasts that must not be used to make final decisions.
1) These forecasts are not a substitute for the "due diligence" research that must be conducted to support the final definition of "unbuildable areas" above and the final decision to purchase land. This research includes, but is not limited to, verification of adequate subsurface soil, zoning, environmental clearance, access, title, utilities and water pressure, clearance from deed restriction, easement and right-of-way encumbrances, clearance from existing above and below ground facility conflicts, etc.
2) The most promising forecast(s) made on the basis of data entered in the design specification from "due diligence" research must be verified at the drawing board before funds are committed and land purchase decisions are made. Actual land shape ratios, dimensions and irregularities encountered may require adjustments to the general forecasts above.
3) The software licensee shall take responsibility for the design specification values entered and any advice given that is based on the forecast produced.

Standard Plan with Picture Lot Context

The Donegal Cliffs subdivision offers an average lot area of 15,551 sq ft, which falls within the picture lot range. Donegal Cliffs is illustrated by Fig. 13.4 and is distinguished by a curvilinear street and modified grid pattern combined with cul-de-sac branches. The advantage of the Donegal Cliffs plan is that it produces less monotony. It disguises front-loading garages and parked cars with its curved streets while retaining the neighborhood integration that is a characteristic of a grid street pattern.

Table 13.3 (column labeled "Lba") shows that Donegal Cliffs actually produces a higher density than the Jefferson Township cluster plan that will be discussed later, even though the Jefferson Township lot size is smaller. The Jefferson Township cluster plan provides 10,028 sq ft of private lot area per family and 40.5% common open space. The Donegal Cliffs plan provides 15,551 sq ft of private lot area per family and no common open space. If Donegal Cliffs provided 40% common open space, Exhibit 13.4 shows that Donegal Cliffs would only be able to allocate an average of 9332 sq ft to each private lot. This is because Jefferson Township actually provides more total land area per dwelling unit. The 0% common open space column in the lot capacity forecast panel of Exhibit 13.5 explains this point. It shows that the Jefferson cluster plan provides 20,931 sq ft of total land area per lot when no common open space is provided. Therefore, the Donegal Cliffs lot starts with less total area per lot and would produce less private area if an amount of common open space roughly equal to the Jefferson Township cluster were subtracted. This illustrates a typical cluster plan relationship.

Table 13.3, in the column labeled "FEX/FTP," shows that the Donegal Cliffs building expansion potential is 126.8% of its current building cover area and the Glenbervie potential is 213.3%. It is unlikely that the Donegal

(a)

Figure 13.4 Donegal Cliffs suburb houses—standard plan arrangement with picture lot context, 100-ft average frontage, and 40% development cover. Donegal is within an edge city with attached garages, 2-story homes, sidewalks, and streets with curbs and gutters as typical features. 12 of 15 lots have front-loading garages. A curvilinear street pattern adds variety and conceals garage door monotony. (*a*) Overview; (*b*) site plan; (*c*) and (*d*) two shots of the same home to illustrate garage door concealment.

(Photos by W. M. Hosack)

(b)

(c)

(d)

Exhibit 13.4 Donegal Cliffs Design Specification and Forecast

Forecast Model RSFL

Development capacity forecast for ***SINGLE-FAMILY DETACHED HOUSING*** *with no shared or common parking lots serving each dwelling unit.*

Given: *Gross land area.* ***To Find:*** *(1) Maximum number of lots that can be created from the land area given based on the design specification values entered below. (2) Maximum gross dwelling unit area that can be built on each lot created.*

DESIGN SPECIFICATION *Enter values in boxed areas where text is bold and blue. Express all fractions as decimals.*

Given:	**Gross land area**	GLA=	**5.519**	acres	240,408	SF
Land Variables:	Est. right-of-way dedication as fraction of GLA	W=	**0.159**		38,225	SF
	Private roadway as fraction of GLA	R=	**0.000**			
	School land donation as fraction of GLA	Ds=	**0.000**			
	Park land donation as fraction of GLA	Dp=	**0.000**			
	Other land donation as fraction of GLA	Do=	**0.000**			
	Net land area	NLA=	4.641	acres	202,183	SF
	Unbuildable and/or future expansion areas as fraction of GLA	U=	**0.000**		0	SF
	Total gross land area reduction as fraction of GLA	X=	0.159		38,225	SF
	Buildable land area remaining	**BLA=**	**4.641**	acres	202,183	SF
Core:	**Core + open space available**	**CS=**	**4.641**	acres	202,183	SF

LOT SPECIFICATION

	total bldg cover potential BCG incl expansion & garage	garage GAR footprint	misc. pavement & drive Mbc as multiple of BCG	total development DCA cover area	private yards Y as multiple of DCA	total lot LOT area	lot LTX mix	aggregate avg. AAL lot areas in mix	total lot TLA areas in mix
Type 1 Interior Lot									
Type 2 Interior Lot									
Type 3 Interior Lot	**5,443**	**528**	**0.14**	6,221	**1.50**	15,553	**1.000**	15,553	15,553
Type 4 Interior Lot									
Type 5 Interior Lot									
Type 6 Interior Lot									
Type 7 Interior Lot									
Total Average Development Cover				6,221					
Mix Total (do not exceed 1.0)							1.00		
Aggregate Avg. Lot Size Based on Mix:				40.00%	= DCA		AAL =	**15,553**	
Enter zero in the adjacent box unless you wish to override the AAL value calculated above:								**0**	= AAL override

LOT CAPACITY FORECAST *(based on common open space value (COS) entered below)*

DEFINITION of COMMON OPEN SPACE (COS): Open space provided for common benefit and massing separation in addition to that provided for private use and setbacks on individual lots. On a given land area, private lot sizes decrease as common open space increases when all other development factors are constant.
NOTE: Common open space values in bold & blue in the table below may be changed to examine other implications. When COS = 0.0 all open space is provided on private lots (OSA) in the form of yards and a typical subdivision is implied.

COS:	**0.00**	**0.10**	**0.20**	**0.30**	**0.40**	**0.50**	**0.60**	**0.70**	**0.80**
No. of lots:	13.0	11.7	10.4	9.1	7.8	6.5	5.2	3.9	2.6
Net density (dNA):	2.80	2.52	2.24	1.96	1.68	1.40	1.12	0.84	0.56
BLA density (dBA):	2.80	2.52	2.24	1.96	1.68	1.40	1.12	0.84	0.56
Lot size compensation	15,553	13,998	12,443	10,887	9,332	7,777	6,221	4,666	3,111
No. of lots:	13.0	13.0	13.0	13.0	13.0	13.0	13.0	13.0	13.0

GROSS BUILDING AREA per LOT FORECAST

	basement as fraction of BSM habitable footprint	basement BSA area	crawl space CRW area	NET POTENTIAL BUILDING AREA INCLUDING GARAGE but EXCLUDING BASEMENT and POTENTIAL EXPANSION — number of building floors: 1	1.5	2	2.5	3	3.5
Type 1 Interior Lot	**1.000**								
Type 2 Interior Lot	**1.000**								
Type 3 Interior Lot	**1.000**	4,915	0	5,443	8,165	10,886	13,608	16,329	19,051
Type 4 Interior Lot	**1.000**								
Type 5 Interior Lot	**1.000**								
Type 6 Interior Lot	**1.000**								
Type 7 Interior Lot	**1.000**								

WARNING: These are preliminary forecasts that must not be used to make final decisions.
1) These forecasts are not a substitute for the "due diligence" research that must be conducted to support the final definition of "unbuildable areas" above and the final decision to purchase land. This research includes, but is not limited to, verification of adequate subsurface soil, zoning, environmental clearance, access, title, utilities and water pressure, clearance from deed restriction, easement and right-of-way encumbrances, clearance from existing above and below ground facility conflicts, etc.
2) The most promising forecast(s) made on the basis of data entered in the design specification from "due diligence" research must be verified at the drawing board before funds are committed and land purchase decisions are made. Actual land shape ratios, dimensions and irregularities encountered may require adjustments to the general forecasts above.
3) The software licensee shall take responsibility for the design specification values entered and any advice given that is based on the forecast produced.

Exhibit 13.5 Jefferson Township Cluster Lot Design Specification and Forecast

Forecast Model RSFL

Development capacity forecast for ***SINGLE-FAMILY DETACHED HOUSING*** *with no shared or common parking lots serving each dwelling unit.*

Given: *Gross land area.* ***To Find:*** *(1) Maximum number of lots that can be created from the land area given based on the design specification values entered below. (2) Maximum gross dwelling unit area that can be built on each lot created.*

DESIGN SPECIFICATION *Enter values in boxed areas where text is bold and blue. Express all fractions as decimals.*

Given:	**Gross land area**	GLA=	**34.870**	acres	1,518,937	SF
Land Variables:	Est. right-of-way dedication as fraction of GLA	W=	**0.157**		238,017	SF
	Private roadway as fraction of GLA	R=	**0.000**			
	School land donation as fraction of GLA	Ds=	**0.000**			
	Park land donation as fraction of GLA	Dp=	**0.000**			
	Other land donation as fraction of GLA	Do=	**0.000**			
	Net land area	NLA=	29.406	acres	1,280,920	SF
	Unbuildable and/or future expansion areas as fraction of GLA	U=	**0.000**		0	SF
	Total gross land area reduction as fraction of GLA	X=	0.157		238,017	SF
	Buildable land area remaining	BLA=	**29.406**	acres	1,280,920	SF
Core:	**Core + open space available**	CS=	**29.406**	acres	1,280,920	SF

LOT SPECIFICATION

	total bldg cover potential BCG incl expansion & garage	garage GAR footprint	misc. pavement & drive Mbc as multiple of BCG	total development DCA cover area	private yards Y as multiple of DCA	total lot LOT area	lot LTX mix	aggregate avg. AAL lot areas in mix	total lot TLA areas in mix
Type 1 Interior Lot									
Type 2 Interior Lot									
Type 3 Interior Lot	**3,510**	**484**	**0.14**	4,012	**1.50**	10,030	**1.000**	10,030	10,030
Type 4 Interior Lot									
Type 5 Interior Lot									
Type 6 Interior Lot									
Type 7 Interior Lot									
Total Average Development Cover				4,012					
Mix Total (do not exceed 1.0)							1.00		
Aggregate Avg. Lot Size Based on Mix:				40.00%	= DCA		AAL =	**10,030**	

Enter zero in the adjacent box unless you wish to override the AAL value calculated above: **0** = AAL override

LOT CAPACITY FORECAST *(based on common open space value (COS) entered below)*

DEFINITION of COMMON OPEN SPACE (COS): Open space provided for common benefit and massing separation in addition to that provided for private use and setbacks on individual lots. On a given land area, private lot sizes decrease as common open space increases when all other development factors are constant.
NOTE: Common open space values in bold & blue in the table below may be changed to examine other implications. When COS = 0.0 all open space is provided on private lots (OSA) in the form of yards and a typical subdivision is implied.

COS:	**0.00**	**0.10**	**0.20**	**0.30**	**0.405**	**0.50**	**0.60**	**0.70**	**0.80**
No. of lots:	127.7	114.9	102.2	89.4	76.0	63.9	51.1	38.3	25.5
Net density (dNA):	4.34	3.91	3.47	3.04	2.58	2.17	1.74	1.30	0.87
BLA density (dBA):	4.34	3.91	3.47	3.04	2.58	2.17	1.74	1.30	0.87
Lot size compensation	16,854	15,169	13,483	11,798	10,028	8,427	6,742	5,056	3,371
No. of lots desired:	**76.0**	76.0	76.0	76.0	76.0	76.0	76.0	76.0	76.0

GROSS BUILDING AREA per LOT FORECAST

	basement as fraction of BSM habitable footprint	basement BSA area	crawl space CRW area	NET POTENTIAL BUILDING AREA INCLUDING GARAGE but EXCLUDING BASEMENT and POTENTIAL EXPANSION — number of building floors: 1	1.5	2	2.5	3	3.5
Type 1 Interior Lot	**1.000**								
Type 2 Interior Lot	**1.000**								
Type 3 Interior Lot	**1.000**	3,026	0	3,510	5,265	7,020	8,775	10,530	12,285
Type 4 Interior Lot	**1.000**								
Type 5 Interior Lot	**1.000**								
Type 6 Interior Lot	**1.000**								
Type 7 Interior Lot	**1.000**								

WARNING: These are preliminary forecasts that must not be used to make final decisions.
1) These forecasts are not a substitute for the "due diligence" research that must be conducted to support the final definition of "unbuildable areas" above and the final decision to purchase land. This research includes, but is not limited to, verification of adequate subsurface soil, zoning, environmental clearance, access, title, utilities and water pressure, clearance from deed restriction, easement and right-of-way encumbrances, clearance from existing above and below ground facility conflicts, etc.
2) The most promising forecast(s) made on the basis of data entered in the design specification from "due diligence" research must be verified at the drawing board before funds are committed and land purchase decisions are made. Actual land shape ratios, dimensions and irregularities encountered may require adjustments to the general forecasts above.
3) The software licensee shall take responsibility for the design specification values entered and any advice given that is based on the forecast produced.

Cliffs and Glenbervie suburb houses will be developed to these limits. If they were, their storm sewer capacity would come under extreme pressure. What results is larger lawn areas. Therefore, zoning ordinances do not necessarily protect these spacious neighborhoods. The impracticality and expense of achieving the development cover permitted on such large lots often, but not always, protects them. Since sprawl is an issue and housing capacity a need, the question is how much expansion capacity is required and what yard multiple (in the column labeled "Total Private Yard Area in sq ft") will produce an adequate balance between existing development cover, future expansion potential, and open space in single-family neighborhoods. The context record for Donegal Cliffs is shown in Table 13.7.

Cluster Plan Context with Lawn Lots and Standard Plan with Picture Lot Context

A Jefferson Township cluster plan proposal is shown in Fig. 13.5, and its context record is included as Table 13.8. A more typical standard plan for the same area is shown in Fig. 13.6, and its context record is included as Table 13.9. Both are summarized in Table 13.3 under the subdivision names, "Jefferson Township cluster" and "Jefferson Township traditional." Table 13.3 shows that the Jefferson Township traditional plan provides 17,052 sq ft of land area per dwelling unit and no common open space. The cluster plan reduces this lot size to 10,028 sq ft of land area per dwelling unit and provides 40.5% of the land area as common open space. The same building footprint can be placed on both lots (column labeled "FTP"), but the building expansion potential on the standard lot is 1.04 times the original footprint area (column labeled "FEX/FTP"), while the expansion potential on the cluster lot is 0.40 times the original footprint area. Since both total land areas are equal, this example shows that the cluster plan open space arrangement comes at the expense of future building expansion

Table 13.7 Donegal Cliffs Context Record

Form SF

Name	**Donegal Cliffs**
Address or Location	**Donegal Cliffs Drive**
Land Use	**Suburb house**
	Standard plan, picture lot context
Parking System	*Interior parking in garage or carport or exterior parking on site*

NOTE: Single-family means a DETACHED dwelling unit (home) intended for the use of only one family unit.

	Description	Project Measurements		Context Summary SF	%	
Site						
	Gross land area in acres	**5.519**	GLA	240,408		
	Public/ private ROW and paved easements in SF	**38,250**	W		15.91%	of GLA
	Net lot area		NLA	202,158	84.1%	of GLA
	Total unbuildable area in SF not common open space	**0**	U		0.0%	of GLA
	Water area within unbuildable area in SF	**0**	WAT			of U
	Buildable land area in acres and SF	4.64	BLA	**202,158**	84.1%	of GLA
Common Open Space						
	Common open space as a % of BLA	**0.00%**	COS	0		of BLA
	Water area within common open space in SF	**0**				
		0.0%				
Subdivision Development						
	Total number of lots	**13**	NLT			
	Average private lot area in SF		ALA	**15,551**		
	Avg. private lot area plus prorated common open space			**15,551**		
	Avg. lot frontage in feet	**40.00%**	D	6,220		
	Avg. lot depth to width ratio	**100.0**	FRN			
	Total dev. cover planned per lot as a % of BLA			1.56		
	Avg. total private yard area per lot in SF		YRD**	9,330	60.0%	of LOT
	Private yard as multiple of total dev. cover area		Y	1.50		
	Misc. pavement incl. driveways as a % of lot	**5.00%**	MSP	778		
	Total potential building cover area per lot		BCG	5,443	35.0%	of LOT
	Misc. pavement as % of total bldg. cover potential		Mbc		14.3%	of BCG
	Total avg. building cover present per lot in SF	**2,400**	FTP			
	Building footprint expansion capacity remaining		FXA	3,043	126.8%	of FTP
	Current private and common open space including bldg. expansion areas		Sn	160,850	79.6%	of BCG
	Anticipated private and common open space minimum permanent provision		S	121,295	60.0%	of BCG
	No. of lots with 1-story homes					
	No. of lots with 1.5-story homes					
	No. of lots with 2-story homes	**13**				
	No. of lots with 2.5-story homes					
	No. of lots with 3.0-story homes					
	No. of lots with 3.5-story homes					
	Average building height in stories				2.00	
	Max. permitted building height in stories	**2.5**				
Summary						
	Lots per buildable acre				2.80	
	Lots per net acre				2.80	
	Lots per gross acre				2.36	
	Current development intensity index		CINX		**2.204**	
	Anticipated max. development intensity index		AINX		**2.900**	

Typical subdivision. Gross buildable area within dashed lines with emphasis on front yards. Subtract side yards to find net buildable area. Divide by number of lots to find average per lot.

Common open space. Separate lot ownership. Front yards de-emphasized. When there is no separate lot ownership, buildable areas can be defined within a condominium arrangement or not.

***This can be a deceiving statistic since many zoning ordinances permit detached garages, swimming pools, sheds, fences,etc. to be placed in side and rear yards as additional development cover. This displaces the open space required by setback lines and in the case of fences and garages disrupts the continuity of this rear yard amenity. In older neighborhoods where detached garages are common, the combination of fences, garages, and sheds reduces rear and side yards to postage stamp amenities that bear no resemblance to the open space implied by the setback statistic quoted.*

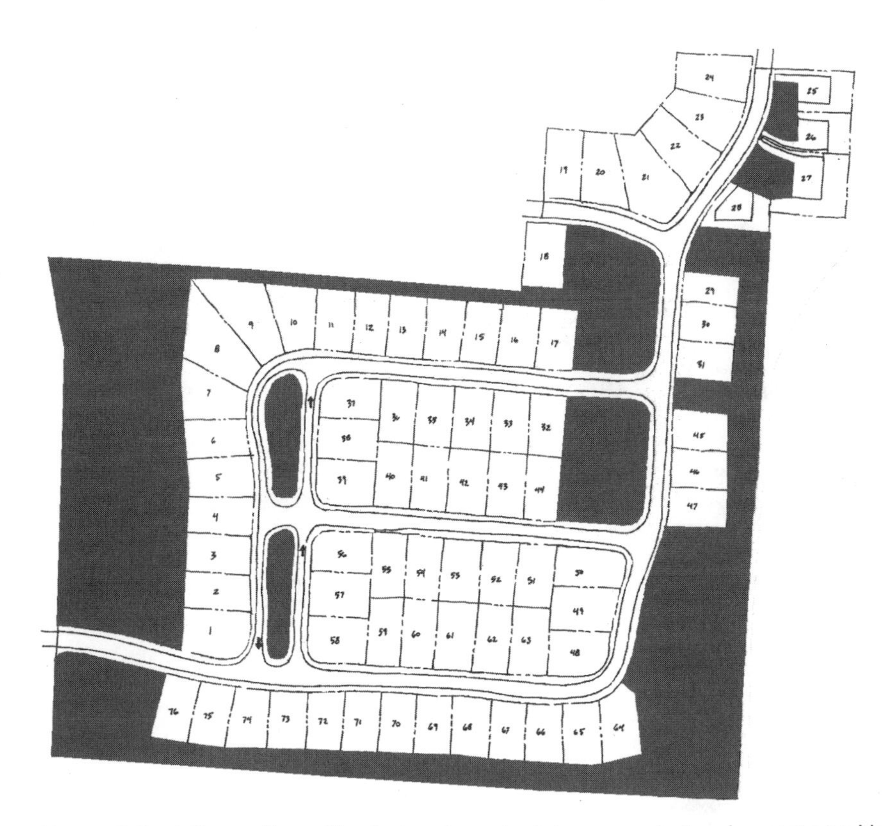

Figure 13.5 Jefferson Township cluster plan suburb houses—cluster plan context with lawn lots, 80-ft average frontage, and 40% development cover within a rural township. Garages, building heights, and roadway construction are undetermined. Included among the features are a modified grid street pattern that weaves the neighborhood together, a boulevard that creates a central green, and a common open space that requires association management.

(Sketch by Lane Kendig, Inc., Mundelein, Illinois; courtesy of Jefferson Township, Ohio)

potential on the individual lots. Table 13.3, in the column labeled "Current Private, Common, and Expansion Open Space per Average Lot," shows that the percentage of open space allocations provided by the two proposals are roughly the same, but the location, privacy, and accessibility of this open space is dramatically different. The cluster plan results in a total private plus common open space allocation of 82.2%, while the traditional subdivision provides 80.3% of the total land as private lawn area.

Table 13.8 Jefferson Township Cluster

Form SF

Name	**Jefferson Township**
Address or Location	**Design study**
Land Use	**Suburb house**
	Cluster plan context, lawn lots
Parking System	*Interior parking in garage or carport or exterior parking on site*

NOTE: Single-family means a DETACHED dwelling unit (home) intended for the use of only one family unit.

	Description	Project Measurements		Context Summary SF	%	
Site						
	Gross land area in acres	**34.870**	GLA	1,518,937		
	Public/ private ROW and paved easements in SF	**238,000**	W		15.67%	of GLA
	Net lot area		NLA	1,280,937	84.3%	of GLA
	Total unbuildable area in SF not common open space	**0**	U		0.0%	of GLA
	Water area within unbuildable area in SF	**0**	WAT			of U
	Buildable land area in acres and SF	29.41	BLA	**1,280,937**	84.3%	of GLA
Common Open Space						
	Common open space as a % of BLA	**40.50%**	COS	518,780		of BLA
	Water area within common open space in SF	**0** 0.0%				
Subdivision Development						
	Total number of lots	**76**	NLT			
	Average private lot area in SF		ALA	**10,028**		
	Avg. private lot area plus prorated common open space			**16,854**		
	Avg. lot frontage in feet	**80.0**	FRN			
	Avg. lot depth to width ratio			1.57		
	Total dev. cover planned per lot as a % of BLA	**40.00%**	D	4,011		
	Avg. total private yard area per lot in SF		YRD**	6,017	60.0%	of LOT
	Private yard as multiple of total dev. cover area		Y	1.50		
	Misc. pavement incl. driveways as a % of lot	**5.00%**	M	501		
	Total potential building cover area per lot		BCG	3,510	35.0%	of LOT
	Misc. pavement as % of total bldg. cover potential		Mbc		14.3%	of BCG
	Total avg. building cover present per lot in SF	**2,500**	FTP			
	Building footprint expansion capacity remaining		FXA	1,010	40.4%	of FTP
	Current private and common open space including bldg. expansion areas		Sn	1,052,829	82.2%	of BCG
	Anticipated private and common open space minimum permanent provision		S	976,074	76.2%	of BCG
	No. of lots with 1-story homes	**20**				
	No. of lots with 1.5-story homes					
	No. of lots with 2-story homes	**56**				
	No. of lots with 2.5-story homes					
	No. of lots with 3.0-story homes					
	No. of lots with 3.5-story homes					
	Average building height in stories				1.74	
	Max. permitted building height in stories	**2.5**				
Summary						
	Lots per buildable acre				2.58	
	Lots per net acre				2.58	
	Lots per gross acre				2.18	
	Current development intensity index		CINX		**1.915**	
	Anticipated max. development intensity index		AINX		**2.738**	

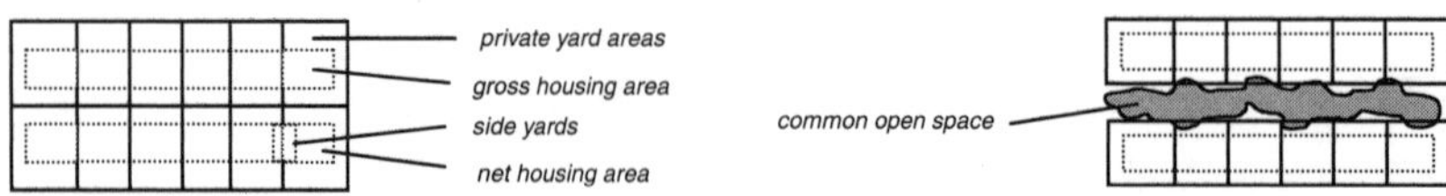

Typical subdivision. Gross buildable area within dashed lines with emphasis on front yards. Subtract side yards to find net buildable area. Divide by number of lots to find average per lot.

Common open space. Separate lot ownership. Front yards de-emphasized. When there is no separate lot ownership, buildable areas can be defined within a condominium arrangement.

***This can be a deceiving statistic since many zoning ordinances permit detached garages, swimming pools, sheds, fences, etc. to be placed in side and rear yards as additional development cover. This displaces the open space required by setback lines and in the case of fences and garages disrupts the continuity of this rear yard amenity. In older neighborhoods where detached garages are common, the combination of fences, garages, and sheds reduces rear and side yards to postage stamp amenities that bear no resemblance to the open space implied by the setback statistic quoted.*

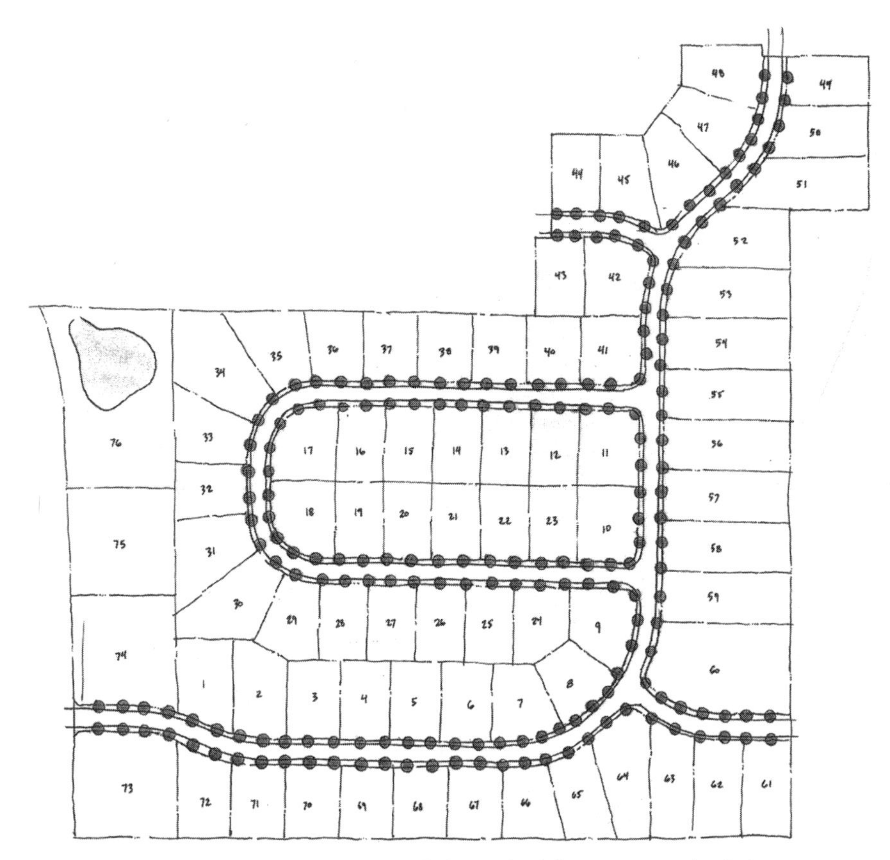

Figure 13.6 Jefferson Township standard plan suburb houses—standard plan arrangement with picture lot context, 112.9-ft average frontage, and 35% development cover. Garages, building heights, and roadway construction are undetermined. Included among the features are a loop street in a modified grid street pattern that creates a consolidated neighborhood context. The lack of a boulevard and shared open space eliminates neighborhood symbolism; repetitive street trees are used to create a common neighborhood identity; and two lots (75 and 76) are completely isolated.

(Sketch by Lane Kendig, Inc., Mundelein, Illinois; courtesy of Jefferson Township, Ohio)

Neither arrangement represents a problem for suburb house development in this example since both the cluster plan and the standard plan provide ample land areas. In fact, the standard plan and the cluster plan produce roughly the same density.[13] This occurs when the cluster plan reduces individual lot areas to compensate for the common

Table 13.9 Jefferson Township Traditional

Form SF

Name	**Jefferson Township**
Address or Location	**Design study**
Land Use	**Suburb house**
	Standard plan, picture lot context
Parking System	*Interior parking in garage or carport or exterior parking on site*

NOTE: Single-family means a DETACHED dwelling unit (home) intended for the use of only one family unit.

	Description	Project Measurements		Context Summary SF	%	
Site						
	Gross land area in acres	**34.870**	GLA	1,518,937		
	Public/ private ROW and paved easements in SF	**223,000**	W		14.68%	of GLA
	Net lot area		NLA	1,295,937	85.3%	of GLA
	Total unbuildable area in SF not common open space	**0**	U		0.0%	of GLA
	Water area within unbuildable area in SF	**0**	WAT			of U
	Buildable land area in acres and SF	29.75	BLA	**1,295,937**	85.3%	of GLA
Common Open Space						
	Common open space as a % of BLA	**0.00%**	COS	0		of BLA
	Water area within common open space in SF	**0** 0.0%				
Subdivision Development						
	Total number of lots	**76**	NLT			
	Average private lot area in SF		ALA	**17,052**		
	Avg. private lot area plus prorated common open space			**17,052**		
	Avg. lot frontage in feet	**112.9**	FRN			
	Avg. lot depth to width ratio			1.34		
	Total dev. cover planned per lot as a % of BLA	**35.00%**	D	5,968		
	Avg. total private yard area per lot in SF		YRD**	11,084	65.0%	of LOT
	Private yard as multiple of total dev. cover area		Y	1.86		
	Misc. pavement incl. driveways as a % of lot	**5.00%**	M	853		
	Total potential building cover area per lot		BCG	5,116	30.0%	of LOT
	Misc. pavement as % of total bldg. cover potential		Mbc		16.7%	of BCG
	Total avg. building cover present per lot in SF	**2,500**	FTP			
	Building footprint expansion capacity remaining		FXA	2,616	104.6%	of FTP
	Current private and common open space including bldg. expansion areas		Sn	1,041,140	80.3%	of BCG
	Anticipated private and common open space minimum permanent provision		S	842,359	65.0%	of BCG
	No. of lots with 1-story homes	**20**				
	No. of lots with 1.5-story homes					
	No. of lots with 2-story homes	**56**				
	No. of lots with 2.5-story homes					
	No. of lots with 3.0-story homes					
	No. of lots with 3.5-story homes					
	Average building height in stories				1.74	
	Max. permitted building height in stories	**2.5**				
Summary						
	Lots per buildable acre				2.55	
	Lots per net acre				2.55	
	Lots per gross acre				2.18	
	Current development intensity index		CINX		**1.933**	
	Anticipated max. development intensity index		AINX		**2.850**	

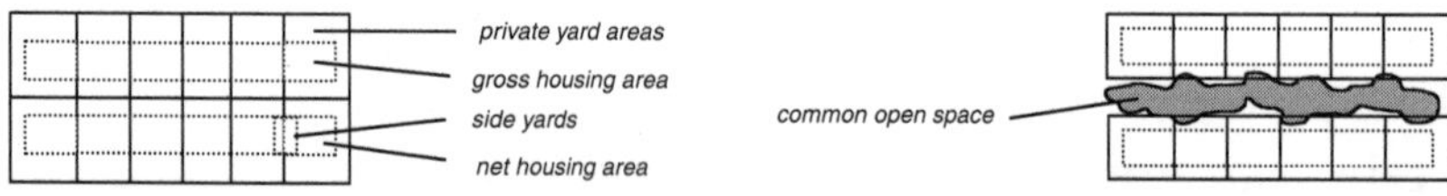

Typical subdivision. Gross buildable area within dashed lines with emphasis on front yards. Subtract side yards to find net buildable area. Divide by number of lots to find average per lot.

Common open space. Separate lot ownership. Front yards de-emphasized. When there is no separate lot ownership, buildable areas can be defined within a condominium arrangement.

***This can be a deceiving statistic since many zoning ordinances permit detached garages, swimming pools, sheds, fences ,etc. to be placed in side and rear yards as additional development cover. This displaces the open space required by setback lines and in the case of fences and garages disrupts the continuity of this rear yard amenity. In older neighborhoods where detached garages are common, the combination of fences, garages, and sheds reduces rear and side yards to postage stamp amenities that bear no resemblance to the open space implied by the setback statistic quoted.*

open space provided in order to produce the same number of dwelling units.

Exhibit 13.5 explains the subtleties of this story with the least amount of confusing detail. The design specification panel in Exhibit 13.5 recites the essential context information for the cluster plan project. The lot specification panel defines the average lot area to be used in the cluster plan, and the lot capacity forecast panel predicts that seventy-six 10,030-sq-ft lots can be created on the buildable land area available when 40.5% common open space is provided. Exhibit 13.6 also shows that an equal number of larger standard lots can be created when no common open space is provided. This is the essence of the cluster plan concept. A planner-developer has two options that are illustrated in the lot capacity forecast panel. In the first two rows of this panel, common open space COS increases and the number of lots decreases because the average cluster lot area AAL is held constant. The net density dNA in the third row declines under this scenario because the number of lots declines with increasing common open space. In the bottom two rows, the average cluster lot declines in area to compensate for the common open space introduced and the number of lots remains constant. This also means that the net density remains constant, but that lot size and buildable area decline from the standard lot benchmark in the 0% COS column. This means that the lot area in a cluster plan can be held constant while the common open space introduced reduces the number of potential lots, or the lot area can be reduced in relation to increasing common open space in order to maintain the same number of potential lots. The first option reduces the housing capacity of the site. The second maintains the capacity but reduces the lot area allocation per family.

The density of the Jefferson Township proposal was between 2.55 and 2.58 lots per net acre depending on the plan involved. The proposal was designed as a traditional

Exhibit 13.6 Jefferson Township Standard Lot Design Specification and Forecast

Forecast Model RSFL

Development capacity forecast for ***SINGLE-FAMILY DETACHED HOUSING*** *with no shared or common parking lots serving each dwelling unit.*

Given: *Gross land area.* ***To Find:*** *(1) Maximum number of lots that can be created from the land area given based on the design specification values entered below. (2) Maximum gross dwelling unit area that can be built on each lot created.*

DESIGN SPECIFICATION

Enter values in boxed areas where text is bold and blue. Express all fractions as decimals.

Given:	**Gross land area**	GLA=	**34.870**	acres	1,518,937	SF
Land Variables:	Est. right-of-way dedication as fraction of GLA	W=	**0.147**		222,980	SF
	Private roadway as fraction of GLA	R=	**0.000**			
	School land donation as fraction of GLA	Ds=	**0.000**			
	Park land donation as fraction of GLA	Dp=	**0.000**			
	Other land donation as fraction of GLA	Do=	**0.000**			
	Net land area	NLA=	29.751	acres	1,295,957	SF
	Unbuildable and/or future expansion areas as fraction of GLA	U=	**0.000**		0	SF
	Total gross land area reduction as fraction of GLA	X=	0.147		222,980	SF
	Buildable land area remaining	**BLA=**	**29.751**	acres	1,295,957	SF
Core:	**Core + open space available**	**CS=**	**29.751**	acres	1,295,957	SF

LOT SPECIFICATION

	total bldg cover potential BCG incl expansion & garage	garage GAR footprint	misc. pavement & drive Mbc as multiple of BCG	total development DCA cover area	private yards Y as multiple of DCA	total lot LOT area	lot LTX mix	aggregate avg. AAL lot areas in mix	total lot TLA areas in mix
Type 1 Interior Lot									
Type 2 Interior Lot									
Type 3 Interior Lot	**5,116**	**484**	**0.17**	5,965	**1.86**	17,061	**1.000**	17,061	17,061
Type 4 Interior Lot									
Type 5 Interior Lot									
Type 6 Interior Lot									
Type 7 Interior Lot									
Total Average Development Cover				5,965					
Mix Total (do not exceed 1.0)							1.00		
Aggregate Avg. Lot Size Based on Mix:				34.97%	= DCA		AAL =	**17,061**	
Enter zero in the adjacent box unless you wish to override the AAL value calculated above:								**0**	= AAL override

LOT CAPACITY FORECAST

(based on common open space value (COS) entered below)

DEFINITION of COMMON OPEN SPACE (COS): Open space provided for common benefit and massing separation in addition to that provided for private use and setbacks on individual lots. On a given land area, private lot sizes decrease as common open space increases when all other development factors are constant.
NOTE: Common open space values in bold & blue in the table below may be changed to examine other implications. When COS = 0.0 all open space is provided on private lots (OSA) in the form of yards and a typical subdivision is implied.

COS:	**0.00**	**0.10**	**0.20**	**0.30**	**0.40**	**0.50**	**0.60**	**0.70**	**0.80**
No. of lots:	76.0	68.4	60.8	53.2	45.6	38.0	30.4	22.8	15.2
Net density (dNA):	2.55	2.30	2.04	1.79	1.53	1.28	1.02	0.77	0.51
BLA density (dBA):	2.55	2.30	2.04	1.79	1.53	1.28	1.02	0.77	0.51
Lot size compensation	17,061	15,355	13,649	11,942	10,236	8,530	6,824	5,118	3,412
No. of lots desired:	76.0	76.0	76.0	76.0	76.0	76.0	76.0	76.0	76.0

GROSS BUILDING AREA per LOT FORECAST

	basement as fraction of BSM habitable footprint	basement BSA area	crawl space CRW area	NET POTENTIAL BUILDING AREA INCLUDING GARAGE but EXCLUDING BASEMENT and POTENTIAL EXPANSION number of building floors: 1	1.5	2	2.5	3	3.5
Type 1 Interior Lot	**1.000**								
Type 2 Interior Lot	**1.000**								
Type 3 Interior Lot	**1.000**	4,632	0	5,116	7,674	10,232	12,790	15,348	17,906
Type 4 Interior Lot	**1.000**								
Type 5 Interior Lot	**1.000**								
Type 6 Interior Lot	**1.000**								
Type 7 Interior Lot	**1.000**								

WARNING: These are preliminary forecasts that must not be used to make final decisions.
1) These forecasts are not a substitute for the "due diligence" research that must be conducted to support the final definition of "unbuildable areas" above and the final decision to purchase land. This research includes, but is not limited to, verification of adequate subsurface soil, zoning, environmental clearance, access, title, utilities and water pressure, clearance from deed restriction, easement and right-of-way encumbrances, clearance from existing above and below ground facility conflicts, etc.
2) The most promising forecast(s) made on the basis of data entered in the design specification from "due diligence" research must be verified at the drawing board before funds are committed and land purchase decisions are made. Actual land shape ratios, dimensions and irregularities encountered may require adjustments to the general forecasts above.
3) The software licensee shall take responsibility for the design specification values entered and any advice given that is based on the forecast produced.

17,052-sq-ft standard lot subdivision in Fig. 13.6 and as a 10,028-sq-ft cluster lot subdivision in Fig. 13.5. Even the reduced lot areas in the cluster plan represent 83.5-by-120-ft lots. This is a generous area by most standards, and the reduction represents more of a landscape design change than a lifestyle change at this density.

Cluster Plan Context with Lawn Lots

Invergordon is a very deceptive cluster plan within a golf course community. Its private lot area statistic places it in the lawn lot range, but the total private and common open space allocated per unit would place it in the garden lot range. The community has average housing costs, lot sizes, and association fees that are beyond the reach of most in the larger regional area. It represents, however, a context and lifestyle that are desired by many.

Invergordon is illustrated by Fig. 13.7 and represents a classic cul-de-sac cluster plan arrangement reached from a low-volume collector street and surrounded by common open space. Each protected cluster within the community is connected with a pathway system that is used for recreation, exercise, and childhood enjoyment. The pathway does not produce broad social interaction since it is removed from each house and most residents use cars to reach other homes and clusters. No sidewalks are provided, which reinforces the "home as a castle" arrangement, and the front-door impression of each home is inviting but remote. Both front and rear yards have little functional purpose, except for the privacy, occasional gathering, and building setting they provide.

This cluster plan context, lawn lot community substitutes several golf courses for the farm fields that once separated villages. It also substitutes asphalt bikeways for the dirt paths that connected homes, common open space for intervening woods, and greater intensity on an increasing number of paved streets. It represents our desire to retain an

(a)

KIRKALDY CT
KILBIRNIE CT
GLENALMOND CT
INVERGORDON CT
MEMORIAL DR
DUNNIKER PARK DR
GLENBERVIE CT

(b)

(c)

(d)

Figure 13.7 Invergordon suburb houses—cluster plan context with lawn lots, 86-ft average frontage, and 35% development cover. Invergordon is within the outer limits of an edge city and provides lawn lot frontages, attached garages, 2-story homes, and no sidewalk, curb, or gutter construction. It includes 20% side-loading garages, pedestrian paths, pedestrian path tunnels under roadways, and common open space. (*a*) Overview; (*b*) site plan; (*c*) frontal view; (*d*) pathway that runs under the road through a pedestrian tunnel.

(Photos by W. M. Hosack)

attachment to the land, however, and is at the low-density end of a spectrum of suburban housing that has attempted to miniaturize this historic farmland pattern to fit an increasing urban demand.

The context record for Invergordon is presented as Table 13.10 and is summarized in Table 13.3. This context record is a recipe of affluence that does not provide large private lots, since the average is 12,108 sq ft, but it provides an additional 73.4% of the net land area as common open space. When total private yard areas and building expansion areas are added to the common open space allocation, 93% of the net land area is devoted to these different open space functions.

Table 13.10 shows that this is actually a cluster plan subdivision that provides 45,434 sq ft of land area per family, but allocates most to common open space managed by

Table 13.10 Invergordon Context Record

Form SF

Name	**Invergordon**
Address or Location	**Invergordon Drive**
Land Use	**Suburb house**
	Cluster plan context, lawn lots
Parking System	*Interior parking in garage or carport or exterior parking on site*

NOTE: Single-family means a DETACHED dwelling unit (home) intended for the use of only one family unit.

	Description	Project Measurements		Context Summary SF	%	
Site						
	Gross land area in acres	**5.366**	GLA	233,743		
	Public/ private ROW and paved easements in SF	**6,573**	W		2.81%	of GLA
	Net lot area		NLA	227,170	97.2%	of GLA
	Total unbuildable area in SF not common open space	**0**	U		0.0%	of GLA
	Water area within unbuildable area in SF	**0**	WAT			of U
	Buildable land area in acres and SF	5.22	BLA	**227,170**	97.2%	of GLA
Common Open Space						
	Common open space as a % of BLA	**73.4%**	COS	166,629		of BLA
	Water area within common open space in SF	**0** 0.0%				
Subdivision Development						
	Total number of lots	**5**	NLT			
	Average private lot area in SF		ALA	**12,108**		
	Avg. private lot area plus prorated common open space			**45,434**		
	Avg. lot frontage in feet	**86.0**	FRN			
	Avg. lot depth to width ratio			1.64		
	Total dev. cover planned per lot as a % of BLA	**35.00%**	D	4,238		
	Avg. total private yard area per lot in SF		YRD**	7,870	65.0%	of LOT
	Private yard as multiple of total dev. cover area		Y	1.86		
	Misc. pavement incl. driveways as a % of lot	**3.00%**	M	363		
	Total potential building cover area per lot		BCG	3,875	32.0%	of LOT
	Misc. pavement as % of total bldg. cover potential		Mbc		9.4%	of BCG
	Total avg. building cover present per lot in SF	**2,800**	FTP			
	Building footprint expansion capacity remaining		FXA	1,075	38.4%	of FTP
	Current private and common open space including bldg. expansion areas		Sn	211,354	93.0%	of BLA
	Anticipated private and common open space minimum permanent provision		S	205,981	90.7%	of BLA
	No. of lots with 1-story homes					
	No. of lots with 1.5-story homes					
	No. of lots with 2-story homes	**5**				
	No. of lots with 2.5-story homes					
	No. of lots with 3.0-story homes					
	No. of lots with 3.5-story homes					
	Average building height in stories				2.00	
	Max. permitted building height in stories	**2.5**				
Summary						
	Lots per buildable acre				0.96	
	Lots per net acre				0.96	
	Lots per gross acre				0.93	
	Current development intensity index		CINX		**2.070**	
	Anticipated max. development intensity index		AINX		**2.593**	

private yard areas
gross housing area
side yards
net housing area

Typical subdivision. Gross buildable area within dashed lines with emphasis on front yards. Subtract side yards to find net buildable area. Divide by number of lots to find average per lot.

common open space

Common open space. Separate lot ownership. Front yards de-emphasized. When there is no separate lot ownership, buildable areas can be defined within a condominium arrangement.

***This can be a deceiving statistic since many zoning ordinances permit detached garages, swimming pools, sheds, fences, etc. to be placed in side and rear yards as additional development cover. This displaces the open space required by setback lines and in the case of fences and garages disrupts the continuity of this rear yard amenity. In older neighborhoods where detached garages are common, the combination of fences, garages, and sheds reduces rear and side yards to postage stamp amenities that bear no resemblance to the open space implied by the setback statistic quoted.*

association assessment. What may look like lawn lots, therefore, are actually garden lot land allocations that share common open space to create small village identities. This all fits easily because the Invergordon density is 0.96 lots per net acre, which is more than adequate to accommodate the automobile and most urban lifestyles.

The forecast model for Invergordon is presented as Exhibit 13.7. The design specification reflects the subdivision's context record measurements, and the lot capacity forecast panel predicts that five lots will be created when 73% of the net land area is provided as common open space. It also shows that an average lot of 45,435 sq ft could have been created if no common open space had been provided. This baseline comparison is found in the 0% common open space column of the lot forecast table. This column shows that the Invergordon cluster plan is comparable to a traditional garden lot subdivision providing 1.04-acre parcels for sale with no common open space allocation.

Standard Plan with Estate Lot Context

The Squirrel Bend subdivision is one step away from farm status and falls within the estate lot category. It is illustrated in Fig. 13.8 and represents large homes with footprints greater than 5000 sq ft and estate lots greater than 2 acres.[14] If these homes were placed on farm lots greater than 5 acres, the sense of privacy and remoteness would be enhanced by greater removal from the road and greater surrounding open space, but the essential characteristics of the estate lot category would simply be magnified on the farm lot. These homes are at one end of the density spectrum and represent a housing alternative that is available to a very small fraction of the population.[15]

The automobile has given the urban dweller an opportunity to return to the farm, and the context record for Squirrel Bend adds definition to the pattern. This record is included as Table 13.11 and is summarized in Table 13.3.

Exhibit 13.7 Invergordon Design Specification and Forecast

Forecast Model RSFL

Development capacity forecast for ***SINGLE-FAMILY DETACHED HOUSING*** *with no shared or common parking lots serving each dwelling unit.*

Given: *Gross land area.* ***To Find:*** *(1) Maximum number of lots that can be created from the land area given based on the design specification values entered below. (2) Maximum gross dwelling unit area that can be built on each lot created.*

DESIGN SPECIFICATION

Enter values in boxed areas where text is bold and blue. Express all fractions as decimals.

Given:	**Gross land area**	GLA=	**5.366**	acres	233,743	SF
Land Variables:	Est. right-of-way dedication as fraction of GLA	W=	**0.028**		6,568	SF
	Private roadway as fraction of GLA	R=	**0.000**			
	School land donation as fraction of GLA	Ds=	**0.000**			
	Park land donation as fraction of GLA	Dp=	**0.000**			
	Other land donation as fraction of GLA	Do=	**0.000**			
	Net land area	NLA=	5.215	acres	227,175	SF
	Unbuildable and/or future expansion areas as fraction of GLA	U=	**0.000**		0	SF
	Total gross land area reduction as fraction of GLA	X=	0.028		6,568	SF
	Buildable land area remaining	**BLA=**	**5.215**	acres	227,175	SF
Core:	**Core + open space available**	**CS=**	**5.215**	acres	227,175	SF

LOT SPECIFICATION

	total bldg cover potential BCG incl expansion & garage	garage GAR footprint	misc. pavement & drive Mbc as multiple of BCG	total development DCA cover area	private yards Y as multiple of DCA	total lot LOT area	lot LTX mix	aggregate avg. AAL lot areas in mix	total lot TLA areas in mix
Type 1 Interior Lot									
Type 2 Interior Lot									
Type 3 Interior Lot	**3,875**	**528**	**0.09**	4,237	**1.86**	12,119	**1.000**	12,119	12,119
Type 4 Interior Lot									
Type 5 Interior Lot									
Type 6 Interior Lot									
Type 7 Interior Lot									
Total Average Development Cover				4,237					
Mix Total (do not exceed 1.0)							1.00		
Aggregate Avg. Lot Size Based on Mix:				34.97%	= DCA		AAL =	**12,119**	

Enter zero in the adjacent box unless you wish to override the AAL value calculated above: **0** = AAL override

LOT CAPACITY FORECAST

(based on common open space value (COS) entered below)

DEFINITION of COMMON OPEN SPACE (COS): Open space provided for common benefit and massing separation in addition to that provided for private use and setbacks on individual lots. On a given land area, private lot sizes decrease as common open space increases when all other development factors are constant.
NOTE: Common open space values in bold & blue in the table below may be changed to examine other implications. When COS = 0.0 all open space is provided on private lots (OSA) in the form of yards and a typical subdivision is implied.

COS:	**0.00**	**0.10**	**0.20**	**0.30**	**0.40**	**0.50**	**0.60**	**0.70**	**0.73**
No. of lots:	18.7	16.9	15.0	13.1	11.2	9.4	7.5	5.6	5.0
Net density (dNA):	3.59	3.23	2.88	2.52	2.16	1.80	1.44	1.08	0.96
BLA density (dBA):	3.59	3.23	2.88	2.52	2.16	1.80	1.44	1.08	0.96
Lot size compensation	45,435	40,891	36,348	31,804	27,261	22,717	18,174	13,630	12,086
No. of lots desired:	**5.0**	5.0	5.0	5.0	5.0	5.0	5.0	5.0	5.0

GROSS BUILDING AREA per LOT FORECAST

	basement as fraction of BSM habitable footprint	basement BSA area	crawl space CRW area	NET POTENTIAL BUILDING AREA INCLUDING GARAGE but EXCLUDING BASEMENT and POTENTIAL EXPANSION — number of building floors: 1	1.5	2	2.5	3	3.5
Type 1 Interior Lot	**1.000**								
Type 2 Interior Lot	**1.000**								
Type 3 Interior Lot	**1.000**	3,347	0	3,875	5,813	7,750	9,688	11,625	13,563
Type 4 Interior Lot	**1.000**								
Type 5 Interior Lot	**1.000**								
Type 6 Interior Lot	**1.000**								
Type 7 Interior Lot	**1.000**								

WARNING: These are preliminary forecasts that must not be used to make final decisions.
1) These forecasts are not a substitute for the "due diligence" research that must be conducted to support the final definition of "unbuildable areas" above and the final decision to purchase land. This research includes, but is not limited to, verification of adequate subsurface soil, zoning, environmental clearance, access, title, utilities and water pressure, clearance from deed restriction, easement and right-of-way encumbrances, clearance from existing above and below ground facility conflicts, etc.
2) The most promising forecast(s) made on the basis of data entered in the design specification from "due diligence" research must be verified at the drawing board before funds are committed and land purchase decisions are made. Actual land shape ratios, dimensions and irregularities encountered may require adjustments to the general forecasts above.
3) The software licensee shall take responsibility for the design specification values entered and any advice given that is based on the forecast produced.

The summary shows that Squirrel Bend is not a cluster plan subdivision. It allocates all buildable land area to standard lots that are an average of 87,980 sq ft (2.02 acres) in area. This produces a density of one-half lot per acre. Its development cover area is more than adequate to cover future expansion, and its yard areas are three times the maximum development cover potential. Since the current average footprint area constructed is a fraction of the total building cover permitted, the yard areas are larger than the minimums anticipated. This adds to the estate lot impression, which this chapter has suggested will commence at or near 2 acres and 75% open space. Invergordon, by contrast, does not have an estate lot appearance. The lots are smaller and much open space is shared in a cluster-housing pattern that reduces individual maintenance responsibilities. The cluster approach provides a combined private and common open space allocation of 73.4%, but the pattern produces protected villages rather than isolated mansions. This is because Invergordon begins with total land area allocations per dwelling unit that are one-half of the Squirrel Bend allocations. Even though the percentage allocations are similar, they are a function of two entirely different baseline areas. This produces the different context photographs in Figs. 13.7 and 13.8.

Squirrel Bend also has an average miscellaneous pavement allocation of 12%. This is the same as the Doone/Essex allocation percentage, but has a different function. The Doone/Essex pavement percentage is high because the lot areas are small and driveways to detached garages in the rear of the lot are a necessity. The Squirrel Bend percentage is high because it uses pavement for circular drives, entertainment, and access roads to remote structures. The percentages may be similar, but the context and purposes represented can be quite different when the baseline land area allocation per dwelling unit is taken into account.

(a)

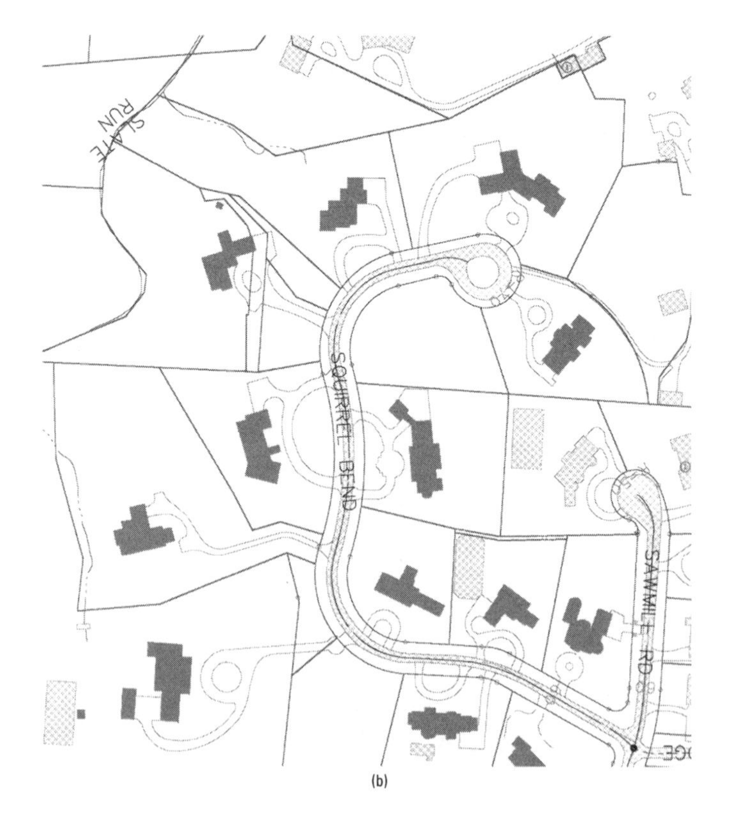

(b)

(c)

Figure 13.8 Squirrel Bend suburb houses—standard plan arrangement with estate lot context, 214-ft average frontage, and 25% development cover. Squirrel Bend is within the outer edge of a first-ring suburb and provides estate lot frontages, attached garages, 2-story homes, and no sidewalk, curb, or gutter construction. 85% of these homes have side-loading garages, and 85% have circular drives. (*a*) Overview; (*b*) site plan; (*c*) frontal view.
(Photos by W. M. Hosack)

The Squirrel Bend building cover percentage is the lowest shown in Table 13.3, even though the average building footprint area is the largest. This is because the lot area is large and the greater building footprint is still a smaller percentage of the total area. Setback lines that do not take this into account can contradict the purpose of estate lot zoning, which is to provide the largest amount of open space and the largest housing area opportunities. Setback lines at the picture lot, garden lot, estate lot, and farm lot levels often produce large footprint potential and small yard area requirements as a percentage of the total area. This is a contradiction that can be easily resolved when present.

The Squirrel Bend forecast model is shown as Exhibit 13.8. The lot capacity forecast panel predicts that fifteen 2-acre lots can be created when no common open space is provided. This represents the Squirrel Bend neighborhood photographed. The model also predicts that fifteen 1-acre lots can be created when 50% of the land area is devoted to common open space. This approach could produce an Inver-

Table 13.11 Squirrel Bend Context Record

Form SF

Name	**Squirrel Bend Subdivision**
Address or Location	**Squirrel Bend Drive**
Land Use	**Suburb house**
	Standard plan, estate lot context
Parking System	*Interior parking in garage or carport or exterior parking on site*

NOTE: Single-family means a DETACHED dwelling unit (home) intended for the use of only one family unit.

	Description	Project Measurements		Context Summary SF	%	
Site						
	Gross land area in acres	**32.574**	GLA	1,418,923		
	Public/ private ROW and paved easements in SF	**99,223**	W		6.99%	of GLA
	Net lot area		NLA	1,319,700	93.0%	of GLA
	Total unbuildable area in SF not common open space	**0**	U		0.0%	of GLA
	Water area within unbuildable area in SF	**0**	WAT			of U
	Buildable land area in acres and SF	30.30	BLA	**1,319,700**	93.0%	of GLA
Common Open Space						
	Common open space as a % of BLA	**0.00%**	COS	0		of BLA
	Water area within common open space in SF	**0** 0.0%	WAT			
Subdivision Development						
	Total number of lots	**15**	NLT			
	Average private lot area in SF		ALA	**87,980**		
	Avg. private lot area plus prorated common open space			**87,980**		
	Avg. lot frontage in feet	**214.0**	FRN			
	Avg. lot depth to width ratio			1.92		
	Total dev. cover planned per lot as a % of BLA	**25.00%**	D	21,995		
	Avg. total private yard area per lot in SF		YRD**	65,985	75.0%	of LOT
	Private yard as multiple of total dev. cover area		Y	3.00		
	Misc. pavement incl. driveways as a % of lot	**12.00%**	M	10,558		
	Total potential building cover area per lot		BCG	11,437	13.0%	of LOT
	Misc. pavement as % of total bldg. cover potential		Mbc		92.3%	of BCG
	Total avg. building cover present per lot in SF	**7,550**	FTP			
	Building footprint expansion capacity remaining		FXA	3,887	51.5%	of FTP
	Current private and common open space including bldg. expansion areas		Sn	1,048,086	79.4%	of BCG
	Anticipated private and common open space minimum permanent provision		S	989,775	75.0%	of BCG
	No. of lots with 1-story homes					
	No. of lots with 1.5-story homes					
	No. of lots with 2-story homes	**15**				
	No. of lots with 2.5-story homes					
	No. of lots with 3.0-story homes					
	No. of lots with 3.5-story homes					
	Average building height in stories				2.00	
	Max. permitted building height in stories	**2.5**				
Summary						
	Lots per buildable acre				0.50	
	Lots per net acre				0.50	
	Lots per gross acre				0.46	
	Current development intensity index		CINX		**2.206**	
	Anticipated max. development intensity index		AINX		**2.750**	

private yard areas
gross housing area
side yards
net housing area

Typical subdivision. Gross buildable area within dashed lines with emphasis on front yards. Subtract side yards to find net buildable area. Divide by number of lots to find average per lot.

common open space

Common open space. Separate lot ownership. Front yards de-emphasized. When there is no separate lot ownership, buildable areas can be defined within a condominium arrangement.

***This can be a deceiving statistic since many zoning ordinances permit detached garages, swimming pools, sheds, fences, etc. to be placed in side and rear yards as additional development cover. This displaces the open space required by setback lines and in the case of fences and garages disrupts the continuity of this rear yard amenity. In older neighborhoods where detached garages are common, the combination of fences, garages, and sheds reduces rear and side yards to postage stamp amenities that bear no resemblance to the open space implied by the setback statistic quoted.*

Exhibit 13.8 Squirrel Bend Design Specification and Forecast

Forecast Model RSFL

Development capacity forecast for ***SINGLE-FAMILY DETACHED HOUSING*** *with no shared or common parking lots serving each dwelling unit.*
Given: *Gross land area.* ***To Find:*** *(1) Maximum number of lots that can be created from the land area given based on the design specification values entered below. (2) Maximum gross dwelling unit area that can be built on each lot created.*

DESIGN SPECIFICATION

Enter values in boxed areas where text is bold and blue. Express all fractions as decimals.

Given:	**Gross land area**	GLA=	**32.574**	acres	1,418,923	SF
Land Variables:	Est. right-of-way dedication as fraction of GLA	W=	**0.069**		97,906	SF
	Private roadway as fraction of GLA	R=	**0.000**			
	School land donation as fraction of GLA	Ds=	**0.000**			
	Park land donation as fraction of GLA	Dp=	**0.000**			
	Other land donation as fraction of GLA	Do=	**0.000**			
	Net land area	NLA=	30.326	acres	1,321,018	SF
	Unbuildable and/or future expansion areas as fraction of GLA	U=	**0.000**		0	SF
	Total gross land area reduction as fraction of GLA	X=	0.069		97,906	SF
	Buildable land area remaining	**BLA=**	**30.326**	acres	1,321,018	SF
Core:	**Core + open space available**	**CS=**	**30.326**	acres	1,321,018	SF

LOT SPECIFICATION

	total bldg cover incl BCG expansion & garage	garage GAR footprint	misc. pavement & drive Mbc as multiple of BCG	total development DCA cover area	private yards Y as multiple of DCA	total lot LOT area	lot LTX mix	aggregate avg. AAL lot areas in mix	total lot TLA areas in mix
Type 1 Interior Lot									
Type 2 Interior Lot									
Type 3 Interior Lot	**11,437**	**900**	**0.92**	21,993	**3.00**	87,973	**1.000**	87,973	87,973
Type 4 Interior Lot									
Type 5 Interior Lot									
Type 6 Interior Lot									
Type 7 Interior Lot									
Total Average Development Cover				21,993					
Mix Total (do not exceed 1.0)							1.00		
Aggregate Avg. Lot Size Based on Mix:				25.00%	= DCA		AAL =	**87,973**	
Enter zero in the adjacent box unless you wish to override the AAL value calculated above:								**0**	= AAL override

LOT CAPACITY FORECAST

(based on common open space value (COS) entered below)

DEFINITION of COMMON OPEN SPACE (COS): Open space provided for common benefit and massing separation in addition to that provided for private use and setbacks on individual lots. On a given land area, private lot sizes decrease as common open space increases when all other development factors are constant.
NOTE: Common open space values in bold & blue in the table below may be changed to examine other implications. When COS = 0.0 all open space is provided on private lots (OSA) in the form of yards and a typical subdivision is implied.

COS:	**0.00**	**0.10**	**0.20**	**0.30**	**0.40**	**0.50**	**0.60**	**0.70**	**0.80**
No. of lots:	15.0	13.5	12.0	10.5	9.0	7.5	6.0	4.5	3.0
Net density (dNA):	0.50	0.45	0.40	0.35	0.30	0.25	0.20	0.15	0.10
BLA density (dBA):	0.50	0.45	0.40	0.35	0.30	0.25	0.20	0.15	0.10
Lot size compensation	87,973	79,176	70,379	61,581	52,784	43,987	35,189	26,392	17,595
No. of lots desired:	15.0	15.0	15.0	15.0	15.0	15.0	15.0	15.0	15.0

GROSS BUILDING AREA per LOT FORECAST

	basement as fraction of BSM habitable footprint	basement BSA area	crawl space CRW area	NET POTENTIAL BUILDING AREA INCLUDING GARAGE but EXCLUDING BASEMENT and POTENTIAL EXPANSION — number of building floors: 1	1.5	2	2.5	3	3.5
Type 1 Interior Lot	**1.000**								
Type 2 Interior Lot	**1.000**								
Type 3 Interior Lot	**1.000**	10,537	0	11,437	17,156	22,874	28,593	34,311	40,030
Type 4 Interior Lot	**1.000**								
Type 5 Interior Lot	**1.000**								
Type 6 Interior Lot	**1.000**								
Type 7 Interior Lot	**1.000**								

WARNING: These are preliminary forecasts that must not be used to make final decisions.
1) These forecasts are not a substitute for the "due diligence" research that must be conducted to support the final definition of "unbuildable areas" above and the final decision to purchase land. This research includes, but is not limited to, verification of adequate subsurface soil, zoning, environmental clearance, access, title, utilities and water pressure, clearance from deed restriction, easement and right-of-way encumbrances, clearance from existing above and below ground facility conflicts, etc.
2) The most promising forecast(s) made on the basis of data entered in the design specification from "due diligence" research must be verified at the drawing board before funds are committed and land purchase decisions are made. Actual land shape ratios, dimensions and irregularities encountered may require adjustments to the general forecasts above.
3) The software licensee shall take responsibility for the design specification values entered and any advice given that is based on the forecast produced.

gordon or Jefferson Township cluster lot arrangement, but the private lot areas would be at least twice as large. The density and housing capacity would be the same as the Squirrel Bend standard plan, but the cluster plan organization would produce a different neighborhood context for the suburb houses involved.

Summary

The column labeled "Lba" in Table 13.3 has been arranged in order of decreasing density in order to compare it with the column labeled "ALA," which lists the private land area provided per suburb house. This comparison shows that density produces erratic private land areas per dwelling unit that are a function of the lot organization decision made (standard plan versus cluster plan). Many zoning ordinances do not take the cluster plan option into account when specifying minimum lot areas per zone. When cluster plan options are included, the lot arrangement decision determines how that total land area is to be allocated within the subdivision. Therefore, cluster plan organization generally reduces the amount of private land area allocated per suburb house while maintaining the same density by providing shared common open space.

Most of us have a subconscious desire for farmland space on suburban lots as our urban populations increase, but are becoming aware that balance must be maintained with our natural and agricultural world. We need a tool to accurately forecast development capacity so that we can compare this capacity with the context, lifestyle, and quality of life that it implies. Without this tool, we cannot begin to produce the amounts of shelter we need to protect an increasing population within development limits that respect the balance we must maintain with our natural and agricultural worlds.[16]

The collection of 30 forecast models (35 including the mixed-use series) presented in Chapter 1 is the tool offered by this book to help us learn to live within limits. The collection covers the four classes of shelter suggested in Appendix A. Context record measurement forms have been introduced to help define existing projects and can be used by many to compare the relative merits of existing "recipes." These comparisons will help refine future design specifications that can be used in forecast models to predict and plan housing capacity and context. This comparative approach and forecasting system are intended to advance the lifestyle and quality of life represented by the original projects measured, and to help us focus on the development capacity of land in relation to the shelter we need to construct for an increasing population.

The forecast model RSFL has been used with each of the context record definitions in this chapter to demonstrate the process of development capacity analysis. It has also been used to explain some of the development capacity and context alternatives that were available when the original subdivision design specification was intuitively written at the drawing board.

Subdivisions and suburb houses are directly related and could easily be classified by the same subdivision context terminology, for instance: narrow houses, yardhouses, lawnhouses, picture houses, garden houses, estate houses, and farmhouses. All of these detached housing variations are based on the private lot area provided per dwelling, however, and have been collectively referred to as *suburb houses* in this book since they are a function of this lot area allocation.

The architectural style of a suburb house is not as significant as the way it consumes land. The essential function it provides is shelter, and the essential relationship it maintains is with the natural environment that sustains it. Suburb house balance has been referred to as the relationship between development cover and open space on a given land area. This is both a micro- and macrolevel definition. The balance between suburb house development cover and open space on an individual lot or land area represents the microlevel. The macrolevel is represented by the relationship between development cover of the built environment and open space of the natural and agricultural world at neighborhood, community, regional, national, and international levels of scale. Balance is not the only consideration, however, since it may be achieved with development intensity that negatively affects our quality of life. The development forecast collection and the context record measurement system are offered as tools to those who may be interested in pursuing these fundamental issues of shelter, intensity, and survival at both the micro- and macrolevels of our environment.

Notes

1. Suburb houses are farmhouses on smaller land areas with more stylistic treatment and less productive purpose. Many farms have become lots. Many villages have become subdivisions, and the ranch house and freeway have become national symbols of our ability to expand. All housing consumes land, withdraws energy, and discharges waste. An increasing population with improved mobility has produced a sprawling housing pattern that has simply magnified these characteristics.
2. See App. A for other suggested divisions of the built environment.
3. Additional suburb house specifications often address permitted location, appearance, and social impact with such topics as setbacks, building materials, roof shape, landscape features, yard accessories, and performance standards.
4. This approach has many variations. In some, the common open space is placed in the rear and all homes surround a shared parklike setting. In others, common open space is placed along the street to interrupt the repetition present. In others, common open space fingers separate cul-de-sac clusters and are used for bikeways and increased separation. All approaches have their critics, but the intent is to provide shared open space that benefits everyone in addition to private yards that benefit a single family and are often much larger than necessary to provide family privacy and building expansion potential.

5. The purpose of yard areas and common open space in single-family subdivisions is not only separation and visual amenity; it is also needed by the storm sewer system to control runoff and protect discharge capacity.
6. Building form in development capacity evaluation should be visualized as a simple square or rectangular volume produced by multiplying building cover by building height. The volume produced is referred to as *massing,* but should not be confused with architectural form and mass. A simple development capacity forecast is given architectural definition with specific floor plans, sections, and elevations that do not exceed the original area and volume forecast, but that sculpt its appearance in response to an owner's requirements. This sculpture is given detail and definition that transform the forecast from simple geometry to complex architecture that is capable of meeting today's shelter requirements.
7. When lot frontage is not excessively narrow.
8. All development cover statistics represent estimates of the development cover permitted by each local zoning ordinance and have not been verified with each local government.
9. These slight aberrations are primarily due to the rounding characteristics of the software.
10. The term *build-out* is often used to mean finishing tenant spaces in a speculative office building; however, in this context, it means complete use of all permitted development cover on a lot.
11. The area within these setback lines is commonly referred to as the *buildable area* of the lot but could more accurately be referred to as *housing area. Buildable area* is a misnomer since many garages, sheds, pools, patios, driveways, sidewalks, and the like may be constructed outside of this buildable area. This book changes the use of the term *buildable area* to mean gross land area minus naturally unbuildable areas and road rights-of-way. *Buildable area* in this context means the entire land area that is capable of accepting construction, minus land devoted to public rights-of-way and private roadway easements. In this context, required setback areas are included within the total private open space or yard area provided.
12. Glenbervie actually has little common open space amenity and is similar to a traditional subdivision plan, but is connected to a larger common open space plan.

13. The density could be the same, but the cluster lot plan has changed the street plan and land consumption slightly.

14. These size and area characteristics are supplemented by the fact that 85% of these lots have circular drives and side-loading garages. This is not necessarily a characteristic but a rather common preference among these landowners. The area per lot is more than ample to accommodate these design features, and cost is not a limiting factor.

15. The entire single-family detached housing spectrum has a relatively narrow band of density alternatives as well, but wide appeal. The characteristics of this entire shelter class are not self-limiting, however, and too much reliance on this housing alternative will simply add to the sprawl that is already a cause for concern.

16. The kingdoms of the built environment that must be balanced with the kingdoms of our natural world are suggested as shelter, movement, open space, and life support in App. A.

Apartment House Comparisons

Apartments reduce the project area consumed by each dwelling unit and increase the housing capacity of land as a result. For instance, if an apartment building is 2 stories in height, the gross building footprint area is 10,000 sq ft, there are 10 dwelling units per floor, and the average footprint per dwelling unit is 500 gross sq ft (10,000/20). If the building is 20 stories in height with 200 dwelling units and the same footprint area, the land consumed per dwelling unit is 50 sq ft, and the land area required per dwelling unit is 10% of the 2-story building requirement. This is a simplified example that does not take roadway, unbuildable land, parking, open space, and miscellaneous pavement per dwelling unit into account, but it illustrates the advantage of building height when land conservation and housing capacity are considerations.

Apartment houses contain single-family dwelling units without direct and immediate access to the land[1] and represent the greatest departure from our agrarian past. The term *apartment* is generally used to imply a rental or lease arrangement, while the term *condominium* is used to imply ownership. These legal relationships may affect commitment, respect, responsibility, and affordability, but have no bearing on the physical capacity of land to accommodate shelter. Since the objective of the context record system and development forecast collection is to predict this capacity, the term *apartment house* as used here disregards ownership status for all such building types.[2]

Building height and *project open space* are two primary characteristics of apartment houses. In fact, building height and project open space are two pressure points of development. *Development cover* counterbalances project open space and represents the land left after open space is provided. *Building cover* remains after *parking cover* and *miscellaneous pavement areas* are subtracted from development cover. *Gross building area* is produced when available building cover is multiplied by the building height planned or permitted. Therefore, the scope of development capacity possibilities for a given site is limited by the building height and project open space planned or permitted. A great range of development capacity possibilities exist within each set of open space and building height limits. This range is a function of the parking system options and design specification value options chosen, but any given set of open space and building height limits establishes the environmental parameters of the development capacity options available. This may seem to have more relevance to planners than to developers and investors, but the combination of building height and open space requirements limits the context format, infrastructure demand, development capacity, tax yield, and return-on-investment options for any given site and affects all concerned with the construction of our built environment.

Building height is generally referred to as *low-rise, mid-rise,* and *high-rise,* but the ranges are broad, imprecise, and subject to individual interpretation. Building height combines with building cover to produce building mass, and the arrangement of this mass in space produces urban form and contributes to urban intensity. The intermediate height categories in Table 14.1 are suggested as a way to expand the definition and visualization of the building mass, form, and

Table 14.1 Building Height Context Categories

Height, ft		
Greater Than	**Less Than or Equal To**	**Context Category**
0	2.5	Low-rise
2.5	5	High low-rise
5	10	Low mid-rise
10	15	Mid-rise
15	20	High mid-rise
20	30	Low high-rise
30	50	High-rise
50	No limit	Sky-rise

intensity that is created when building height is combined with building cover.

Open space counterbalances development cover and offsets building height to decrease development intensity. Open space categories are suggested in Chap. 13 to convey an impression of the suburb house environments implied when these allocations increase. A similar system is suggested in Table 14.2 to convey an impression of the apartment house environments implied by various open space allocation ranges.

The characteristics of eight apartment house case studies are summarized in Table 14.3. Four are new and four have been drawn from other sections of this book and are not dis-

Table 14.2 Apartment House Open Space Context Categories

Open Space, %		
Greater Than	**Less Than or Equal To**	**Context Category**
0	25	CBD
25	35	Transition
35	45	Courtyard
45	55	Yard
55	65	Lawn
65	75	Garden
75	85	Estate
85	95	Park
95	100	Farm

Table 14.3 Apartment House Data Summary from SF Context Records

Land and Building Data

Apartment Name	Location	Type	Context	Gross Land Area GLA	Buildable Land Area BLA	Gross Building Area GBA	Total Building Footprint Area BCA	Number of Dwelling Units NDU	Number of Building Floors FLR	Average Gross Dwelling Unit Area per Building Floor AFP	Density per Buildable Acre dBA	Building sq ft per Dwelling Unit ADU	Building sq ft per Buildable Acre SFAC	Gross Building Area per Parking Space BPS
Britton	Edge city	Low-rise	Garden	28.33	27.26	448,200	149,400	353	3	423	**12.95**	1269.7	16,442	689.5
Summit Chase	First ring	Low high-rise	Estate	7.81	7.722	386,400	16,100	178	24	90	**23.05**	2170.8	50,038	1452.6
Turnberry	Suburb	Low-rise	Yard	0.50	0.5	16,184	8,092	12	2	674	**24.00**	1348.7	32,368	809.2
Jefferson	Suburb	Low-rise	Transition	1.76	1.756	62,400	20,800	72	3	289	**41.00**	866.7	35,535	611.0
Lane Crest	Suburb	Low-rise	CBD	0.53	0.525	23,856	7,952	36	3	221	**68.57**	662.7	45,440	994.0
Townley Court	CBD	High low-rise	Courtyard	1.12	0.958	83,320	20,830	102	4	204	**106.40**	816.9	86,975	
Library Park	CBD	High low-rise	Courtyard	1.94	1.262	119,040	29,760	135	4	220	**107.00**	881.7	94,347	
Waterford	CBD	High mid-rise	CBD	0.66	0.658	171,000	9,000	96	19	94	**145.90**	1781.3	259,878	2280.0

Parking and Summary Data

Apartment Name	Location	Street Parking Only	Surface Parking Spaces NPS	Average Gross Area per Surface Parking Space *s*	Structure Parking Spaces NPS	Average Gross Area per Structure Parking Space *s*	Garage Parking Spaces NPS	Average Gross Area per Garage Parking Space *s*	Building Cover Cover as % of Buildable Area *B*	Develop-ment Cover as % of Buildable Area *D*	Project Open Spaces as % of Buildable Area *S*	Total Parking Spaces Provided NPS	Develop-ment Capacity Index INX	Total Parking Spaces per Dwelling Unit PDU
Britton	Edge city		500	376					0.126	0.341	0.659	500	3.341	1.800
Summit Chase	First ring		78	354	188	357			0.048	0.162	0.838	266	24.162	1.500
Turnberry	Suburb		5	390			150	211	0.372	0.501	0.499	155	3.501	1.670
Jefferson	Suburb		102	311			75	360	0.272	0.696	0.304	177	3.696	1.420
Lane Crest	Suburb		24	375					0.348	0.772	0.228	24	3.772	0.700
Townley Court	CBD	Yes							0.499	0.547	0.453	0	4.547	
Library Park	CBD	Yes							0.541	0.591	0.409	0	4.591	
Waterford	CBD				75	360			0.314	0.791	0.209	75	19.791	0.780

Classification Criteria

Stories: Greater Than	Stories: Less Than or Equal To	Height Classification	Index
1	2.5	Low-rise	LO
2.5	5	High low-rise	HL
5	10	Low mid-rise	LM
10	15	Mid-rise	MD
15	20	High mid-rise	HM
20	30	Low high-rise	LH
30	50	High-rise	HI
50	No limit	Sky-rise	SK

Percent Open Space: Greater Than	Percent Open Space: Less Than or Equal To	Context	Index
0	25	CBD	C
25	35	Transition	T
35	45	Courtyard	D
45	55	Yard	Y
55	65	Lawn	L
65	75	Picture	G
75	85	Garden	E
85	95	Estate	P
95	100	Farm	F

cussed in detail. All eight records are included in this chapter for ease of reference, however.

Table 14.4 isolates elements within Table 14.3 in a search for patterns that make some apartment houses more desirable than others. No one factor represents a secret ingredient, but it may be helpful to look at the relationships measured and contrast them with a given set of preferences in order to look for desirable patterns that may exist.[3]

Table 14.4 shows that decreasing open space does not produce a consistent decrease in the preference rating. Project open space below 50% of the buildable area results in a decreasing preference rating, but 21% open space in the central business district produces a preferred rating.[4] Table 14.5 shows that increasing the number of floors FLR has no direct relationship to the average dwelling unit area ADU provided.[5] Table 14.6 shows that increasing density dBA has no relationship to the average dwelling unit area provided, even though it might be expected to decline.[6] Table 14.7 shows that the gross building area constructed per acre SFAC has no direct relationship to the average dwelling unit area provided, the number of floors constructed, or the open space *S* contributed. Table 14.8 produces a surprise.

Table 14.4 Project Open Space Comparison to Preference

Project Open Space as % of Buildable Area *S*	Building sq ft per Dwelling Unit ADU	Number of Building Floors FLR	Total Parking Spaces per Dwelling Unit PDU	Apartment Name	Location	Context	Preference (1 = Preferred)
0.838	2170.8	24	1.500	Summit Chase	First ring	Estate	**1**
0.659	1269.7	3	1.800	Britton	Edge city	Garden	**1**
0.499	1348.7	2	1.670	Turnberry	Suburb	Yard	**1**
0.453	816.9	4	None	Townley Court	CBD	Courtyard	**2**
0.409	881.7	4	None	Library Park	CBD	Courtyard	**2**
0.304	866.7	3	1.420	Jefferson	Suburb	Transition	**3**
0.228	662.7	3	0.700	Lane Crest	Suburb	CBD	**3**
0.209	1781.3	19	0.780	Waterford	CBD	CBD	**1**

Table 14.5 Average Dwelling Unit Area Compared to Density and Preference

Gross Building sq ft per Dwelling Unit ADU	Density per Buildable Acre dBA	Apartment Name	Location	Context	Preference (1 = Preferred)
2170.8	23.05	Summit Chase	First ring	Estate	**1**
1781.3	145.90	Waterford	CBD	CBD	**1**
1348.7	24.00	Turnberry	Suburb	Yard	**1**
1269.7	12.95	Britton	Edge city	Garden	**1**
881.7	107.00	Library Park	CBD	Courtyard	**2**
866.7	41.00	Jefferson	Suburb	Transition	**3**
816.9	106.40	Townley Court	CBD	Courtyard	**2**
662.7	68.57	Lane Crest	Suburb	CBD	**3**

It shows that the order of preference roughly parallels the average dwelling unit area provided. This is a surprise because average dwelling unit area was not a factor in the author's preference ranking, even though it is a significant influence on the quality of life provided.[7] These dwelling unit areas did not produce the context encountered, but they were a part of the design specification that did, and seem to be reflections of the complete package of decisions that combined to produce each environment.[8]

Tables 14.4 through 14.8 serve to indicate that there may be no single criteria that can produce the sustainable

Table 14.6 Density Compared to Building Height, Average Dwelling Unit Area, and Preference

Density per Buildable Acre dBA	Number of Building Floors FLR	Building sq ft per Dwelling Unit ADU	Apartment Name	Location	Context	Preference (1 = Preferred)
12.95	3	1269.7	Britton	Edge city	Garden	**1**
23.05	24	2170.8	Summit Chase	First ring	Yard	**1**
24.00	2	1348.7	Turnberry	Suburb	Transition	**1**
41.00	3	866.7	Jefferson	Suburb	CBD	**3**
68.57	3	662.7	Lane Crest	Suburb	Courtyard	**3**
106.40	4	816.9	Townley Court	CBD	Courtyard	**2**
107.00	4	881.7	Library Park	CBD	CBD	**2**
145.90	19	1781.3	Waterford	CBD	Estate	**1**

Table 14.7 Development Capacity Index SFAC Compared to Average Dwelling Unit Area ADU, Number of Floors FLR, Project Open Space *S*, and Preference

Building sq ft per Buildable Acre SFAC	Building sq ft per Dwelling Unit ADU	Number of Building Floors FLR	Project Open Space as % of Buildable Area *S*	Total Parking Spaces Provided NPS	Gross Building Area per Parking Space BPS	Total Parking Spaces per Dwelling Unit PDU	Apartment Name	Location	Context	Preference (1 = Preferred)
16,442	**1269.7**	**3**	**0.659**	500	689.5	1.800	Britton	Edge city	Garden	**1**
32,368	**1348.7**	**2**	**0.499**	155	809.2	1.670	Turnberry	Suburb	Yard	**1**
35,535	**866.7**	**3**	**0.304**	177	611.0	1.420	Jefferson	Suburb	Transition	**3**
45,440	**662.7**	**2**	**0.228**	24	994.0	0.700	Lane Crest	Suburb	CBD	**3**
50,038	**2170.8**	**24**	**0.838**	266	1,452.6	1.500	Summit Chase	First ring	Estate	**1**
86,975	**816.9**	**4**	**0.453**	0			Townley Court	CBD	Courtyard	**2**
94,347	**881.7**	**4**	**0.409**	0			Library Park	CBD	Courtyard	**2**
259,878	**1781.3**	**19**	**0.209**	75	2,280.0	0.780	Waterford	CBD	CBD	**1**

Table 14.8 Average Dwelling Unit Area Compared to Order of Preference

Building sq ft per Dwelling Unit ADU	Preference (1 = Preferred)	Apartment Name	Location	Context	Type	Gross Land Area GLA	Buildable Land Area BLA	Gross Building Area GBA	Total Building Footprint Area BCA	Number of Dwelling Units NDU	Number of Building Floors FLR	Average Gross Dwelling Unit Area per Building Floor AFP	Density per Buildable Acre dBA	Building sq ft per Buildable Acre SFAC	Gross Building Area per Parking Space BPS
2170.8	1	Summit Chase	First ring	Estate	Low high-rise	7.81	7.722	386,400	16,100	178	24	90	**23.05**	50,038	1453
1781.3	1	Waterford	CBD	CBD	High mid-rise	0.66	0.658	171,000	9,000	96	19	94	**145.90**	259,878	2280
1348.7	1	Turnberry	Suburb	Yard	Low-rise	0.50	0.5	16,184	8,092	12	2	674	**24.00**	32,368	809
1269.7	1	Britton	Edge city	Garden	Low-rise	28.33	27.26	448,200	149,400	353	3	423	**12.95**	16,442	690
881.7	2	Library Park	CBD	Courtyard	High low-rise	1.94	1.262	119,040	29,760	135	4	220	**107.00**	94,347	
866.7	3	Jefferson	Suburb	Transition	Low-rise	1.76	1.756	62,400	20,800	72	3	289	**41.00**	35,535	611
816.9	2	Townley Court	CBD	Courtyard	High low-rise	1.12	0.958	83,320	20,830	102	4	204	**106.40**	86,975	
662.7	3	Lane Crest	Suburb	CBD	Low-rise	0.53	0.525	23,856	7,952	36	3	221	**68.57**	45,440	994

Parking and Summary Data

Building sq ft per Dwelling Unit ADU	Preference (1 = Preferred)	Apartment Name	Location	Street Parking Only	Surface Parking Spaces NPS	Average Gross Area per Parking Space *s*	Structure Parking Spaces NPS	Average Gross Area per Parking Space *s*	Garage Parking Spaces NPS	Average Gross Area per Parking Space *s*	Building Cover as % of Buildable Area *B*	Devel-opment Cover as % of Buildable Area *D*	Project Open Space as % of Buildable Area *S*	Total Parking Spaces Provided NPS	Devel-opment Capacity Index INX	Total Parking Spaces per Dwelling Unit PDU
2170.8	1	Summit Chase	First ring		78	354	188	357			0.048	0.162	0.838	266	24.162	1.500
1781.3	1	Waterford	CBD				75	360			0.314	0.791	0.209	75	19.791	0.780
1348.7	1	Turnberry	Suburb		5	390			150	211	0.372	0.501	0.499	155	3.501	1.670
1269.7	1	Britton	Edge city		500	376					0.126	0.341	0.659	500	3.341	1.800
881.7	2	Library Park	CBD	Yes							0.541	0.591	0.409	0	4.591	
866.7	3	Jefferson	Suburb		102	311			75	360	0.272	0.696	0.304	177	3.696	1.420
816.9	2	Townley Court	CBD	Yes							0.499	0.547	0.453	0	4.547	
662.7	3	Lane Crest	Suburb		24	375					0.348	0.772	0.228	24	3.772	0.700

shelter and quality of life that we seek. There is, however, a group of criteria that combine to produce these results. They are listed in the design specification panel of each forecast model in the development forecast collection. The values assigned to these criteria, or components, can be entered to produce an outline for final drawing-board evaluation that can produce desired results consistently. It is also possible to enter values that will produce negative results.

Data Override

This feature has been introduced to simplify data collection and entry. The purpose of data override is to eliminate the need to gather and enter detailed dwelling unit information in a context record and to permit quick comparison of dwelling unit mix alternatives in a forecast model. For this reason, a context record form requests only gross building area and total dwelling unit information. The form automatically divides these values to calculate an aggregate average dwelling unit area AGG to represent the project. This value can also be entered in the data-override box of any residential forecast model. The box is located in the lower right-hand corner of the dwelling unit mix panel in Exhibit 14.1 and overrides the need to complete the DU, GDA, and MIX columns in this panel. Data in these columns will be considered only if the value in the data-override box is 0.

The override feature can be a useful comparative tool. When the dwelling unit mix panel contains a complete set of values, the override box can be used to compare the development implications of this dwelling unit mix with an alternate AGG. For instance, Exhibit 14.1 predicts that 353 dwelling units NDU can be constructed in a collection of 3-story buildings.[9] This forecast contains an AGG of 1270 sq ft in the data-override box. A detailed dwelling unit mix is compared in Exhibit 14.2. The override box contains a null value, and an aggregate average value of 1365 sq ft is cal-

Exhibit 14.1 Example of Data Override Feature

Forecast Model RG1L

Development capacity forecast for ***APARTMENTS*** *based on the use of an adjacent* ***GRADE PARKING LOT*** *located on the same premises. When s and u equal zero in the design specification below, the forecast pertains to conditions when* ***NO PARKING*** *is required.*

Given: *Gross land area.* ***To Find:*** *Maximum dwelling unit capacity of the land area given based on the design specification values entered below.* ***Premise:*** *All building floors considered equal in area.*

DESIGN SPECIFICATION

Enter values in boxed areas where text is bold and blue. Express all fractions as decimals.

Given:	**Gross land area**	GLA=	**28.330**	acres	1,234,055	SF
Land Variables:	Public/ private right-of-way & paved easements	W=	**0.038**	fraction of GLA	46,647	SF
	Net land area	NLA=	27.259	acres	1,187,408	SF
	Unbuildable and/or future expansion areas	U=	**0.000**	fraction of GLA	0	SF
	Gross land area reduction	X=	0.038	fraction of GLA	46,647	SF
	Buildable land area remaining	BLA=	**27.259**	acres	1,187,408	SF
Parking Variables:	Est. gross pkg. lot area per pkg. space in SF	s =	**376**	ENTER ZERO IF NO PARKING REQUIRED		
	Parking lot spaces planned or required per dwelling unit	u=	**1.42**	ENTER ZERO IF NO PARKING REQUIRED		
	Garage parking spaces planned or required per dwelling unit	Gn=	**0.424**	ENTER ZERO IF NO PARKING REQUIRED		
	Gross building area per garage space	Ga=	**211.2**	ENTER ZERO IF NO PARKING REQUIRED		
	No. of loading spaces	l =	**0**			
	Gross area per loading space	b =	**0**	SF	0	SF
Site Variables:	**Project open space as fraction of BLA**	S=	**0.659**		782,502	SF
	Private driveways as fraction of BLA	R=	**0.000**		0	SF
	Misc. pavement as fraction of BLA	M=	**0.030**		35,622	SF
	Loading area as fraction of BLA	L=	0.000		0	SF
	Total site support areas as a fraction of BLA	Su=	0.689		818,124	SF
Core:	**Core development area as fraction of BLA**	C=	0.311	C+Su must = 1	369,284	SF
Building Variables:	Building efficiency as percentage of GBA	Be=	**0.850**	must have a value entered		
	Building support as fraction of GBA	Bu=	0.150	Be + Bu must = 1		

Dwelling Unit Mix Table:

DU dwelling unit type	GDA gross du area	CDA=GDA/Be comprehensive du area	MIX du mix	PDA = (CDA)MIX Prorated du area
EFF	**500**		**10%**	
1 BR	**750**		**20%**	
2 BR	**1,200**	data override	**40%**	data override
3 BR	**1,500**		**20%**	
4 BR	**1,800**		**10%**	

Aggregate avg. dwelling unit area	(AGG) =	**1,270**
GBA sf per parking space	a=	689
Enter zero in the adjacent box unless you wish to override the AGG value calculated above		**1,270** **override**

PLANNING FORECAST

no. of floors FLR	CORE minimum lot area for BCG & PLA	density per net acre dNA	dwelling units **NDU**	pkg. lot spaces NPS	parking lot area **PLA**	garage spaces GPS	garage area **GAR**	gross bldg area GBA no garages	footprint **BCA**	density per dBA bldable acre
1.00	369,284	7.15	195	276.9	104,131	82.7	17,465	247,688	247,688	7.15
2.00		10.76	293	416.7	156,673	124.4	26,277	372,667	186,334	10.76
3.00		**12.94**	**353**	**500.9**	**188,353**	**149.6**	**31,590**	**448,022**	**149,341**	**12.94**
4.00		14.40	392	557.3	209,537	166.4	35,143	498,412	124,603	14.40
5.00		15.44	421	597.6	224,701	178.4	37,687	534,481	106,896	15.44
6.00		16.22	442	627.9	236,091	187.5	39,597	561,574	93,596	16.22
7.00	*NOTE: Be aware when BCA becomes too small to be feasible.*	16.83	459	651.5	244,960	194.5	41,085	582,671	83,239	16.83
8.00		17.32	472	670.4	252,062	200.2	42,276	599,564	74,946	17.32
9.00		17.72	483	685.8	257,878	204.8	43,251	613,396	68,155	17.72
10.00		18.05	492	698.7	262,726	208.6	44,064	624,930	62,493	18.05
11.00		18.33	500	709.7	266,831	211.9	44,753	634,694	57,699	18.33
12.00		18.58	506	719.0	270,352	214.7	45,343	643,067	53,589	18.58
13.00		18.79	512	727.1	273,404	217.1	45,855	650,327	50,025	18.79
14.00		18.97	517	734.2	276,075	219.2	46,303	656,681	46,906	18.97

WARNING: These are preliminary forecasts that must not be used to make final decisions.

1) These forecasts are not a substitute for the "due diligence" research that must be conducted to support the final definition of "unbuildable areas" above and the final decision to purchase land. This research includes, but is not limited to, verification of adequate subsurface soil, zoning, environmental clearance, access, title, utilities and water pressure, clearance from deed restriction, easement and right-of-way encumbrances, clearance from existing above and below ground facility conflicts, etc.

2) The most promising forecast(s) made on the basis of data entered in the design specification from "due diligence" research must be verified at the drawing board before funds are committed and land purchase decisions are made. Actual land shape ratios, dimensions and irregularities encountered may require adjustments to the general forecasts above.

3) The software licensee shall take responsibility for the design specification values entered and any advice given that is based on the forecast produced.

Exhibit 14.2 Example of Disengaged Data Override Feature

Forecast Model RG1L

Development capacity forecast for ***APARTMENTS*** *based on the use of an adjacent* ***GRADE PARKING LOT*** *located on the same premises. When s and u equal zero in the design specification below, the forecast pertains to conditions when* ***NO PARKING*** *is required.*

Given: *Gross land area.* ***To Find:*** *Maximum dwelling unit capacity of the land area given based on the design specification values entered below.* ***Premise:*** *All building floors considered equal in area.*

DESIGN SPECIFICATION

Enter values in boxed areas where text is bold and blue. Express all fractions as decimals.

Given:	**Gross land area**	GLA=	**28.330**	acres	1,234,055	SF
Land Variables:	Public/ private right-of-way & paved easements	W=	**0.038**	fraction of GLA	46,647	SF
	Net land area	NLA=	27.259	acres	1,187,408	SF
	Unbuildable and/or future expansion areas	U=	**0.000**	fraction of GLA	0	SF
	Gross land area reduction	X=	0.038	fraction of GLA	46,647	SF
	Buildable land area remaining	BLA=	**27.259**	acres	1,187,408	SF
Parking Variables:	Est. gross pkg. lot area per pkg. space in SF	s =	**376**	ENTER ZERO IF NO PARKING REQUIRED		
	Parking lot spaces planned or required per dwelling unit	u=	**1.42**	ENTER ZERO IF NO PARKING REQUIRED		
	Garage parking spaces planned or required per dwelling unit	Gn=	**0.424**	ENTER ZERO IF NO PARKING REQUIRED		
	Gross building area per garage space	Ga=	**211.2**	ENTER ZERO IF NO PARKING REQUIRED		
	No. of loading spaces	l =	**0**			
	Gross area per loading space	b =	**0**	SF	0	SF
Site Variables:	**Project open space as fraction of BLA**	S=	**0.659**		782,502	SF
	Private driveways as fraction of BLA	R=	**0.000**		0	SF
	Misc. pavement as fraction of BLA	M=	**0.030**		35,622	SF
	Loading area as fraction of BLA	L=	0.000		0	SF
	Total site support areas as a fraction of BLA	Su=	0.689		818,124	SF
Core:	**Core development area as fraction of BLA**	C=	0.311	C+Su must = 1	369,284	SF
Building Variables:	Building efficiency as percentage of GBA	Be=	**0.850**	must have a value entered		
	Building support as fraction of GBA	Bu=	0.150	Be + Bu must = 1		

Dwelling Unit Mix Table:

DU dwelling unit type	GDA gross du area	CDA=GDA/Be comprehensive du area	MIX du mix	PDA = (CDA)MIX Prorated du area
EFF	**500**	588	**10%**	59
1 BR	**750**	882	**20%**	176
2 BR	**1,200**	1,412	**40%**	565
3 BR	**1,500**	1,765	**20%**	353
4 BR	**1,800**	2,118	**10%**	212
			Aggregate avg. dwelling unit area (AGG) =	**1,365**
			GBA sf per parking space a=	740
			Enter zero in the adjacent box unless you wish to override the AGG value calculated above	0

PLANNING FORECAST

no. of floors FLR	CORE minimum lot area for BCG & PLA	density per net acre dNA	dwelling units **NDU**	pkg. lot spaces NPS	parking lot area **PLA**	garage spaces GPS	garage area **GAR**	gross bldg area GBA no garages	footprint **BCA**	density per dBA bldable acre
1.00	369,284	6.81	186	263.8	99,170	78.8	16,633	253,481	253,481	6.81
2.00		10.37	283	401.6	150,991	119.9	25,324	385,936	192,968	10.37
3.00		**12.56**	**342**	**486.3**	**182,839**	**145.2**	**30,666**	**467,338**	**155,779**	**12.56**
4.00		14.04	383	543.6	204,394	162.3	34,281	522,434	130,609	14.04
5.00		15.11	412	585.0	219,953	174.7	36,890	562,202	112,440	15.11
6.00		15.92	434	616.3	231,712	184.0	38,863	592,258	98,710	15.92
7.00	*NOTE: Be aware when BCA becomes too small to be feasible.*	16.55	451	640.7	240,911	191.3	40,405	615,771	87,967	16.55
8.00		17.06	465	660.4	248,305	197.2	41,646	634,669	79,334	17.06
9.00		17.48	476	676.5	254,377	202.0	42,664	650,190	72,243	17.48
10.00		17.83	486	690.0	259,452	206.0	43,515	663,163	66,316	17.83
11.00		18.12	494	701.5	263,758	209.5	44,237	674,169	61,288	18.12
12.00		18.38	501	711.3	267,457	212.4	44,858	683,624	56,969	18.38
13.00		18.60	507	719.9	270,669	214.9	45,397	691,834	53,218	18.60
14.00		18.79	512	727.4	273,484	217.2	45,869	699,029	49,931	18.79

WARNING: These are preliminary forecasts that must not be used to make final decisions.

1) These forecasts are not a substitute for the "due diligence" research that must be conducted to support the final definition of "unbuildable areas" above and the final decision to purchase land. This research includes, but is not limited to, verification of adequate subsurface soil, zoning, environmental clearance, access, title, utilities and water pressure, clearance from deed restriction, easement and right-of-way encumbrances, clearance from existing above and below ground facility conflicts, etc.

2) The most promising forecast(s) made on the basis of data entered in the design specification from "due diligence" research must be verified at the drawing board before funds are committed and land purchase decisions are made. Actual land shape ratios, dimensions and irregularities encountered may require adjustments to the general forecasts above.

3) The software licensee shall take responsibility for the design specification values entered and any advice given that is based on the forecast produced.

culated from the values entered. All other values are equal to those in Exhibit 14.1. The resulting forecast predicts that only 342 dwelling units can be constructed when the AGG value increases as shown in Exhibit 14.2.

The project open space provided in both examples is 65.9% of the buildable area. Reducing this open space provision would increase the core development area and the potential number of larger dwelling units that could be accommodated. This open space could be reduced to bring the total number of units back to the benchmark of 353. A number of other design specification values could also be adjusted, but resulting context must be kept in mind.

This ability to quickly compare dwelling unit areas and mix allocations with their impact on the development capacity and context of a site can help to evaluate the following factors: (1) the open space provided and the intensity proposed, (2) the relationship of building height to the dwelling unit area provided, (3) the external project massing and context that results, (4) the impact this context will have on the neighborhood context, (4) the return on investment this development capacity represents, (5) the municipal tax yield and infrastructure burden that is implied, (6) the quality of shelter provided, (7) the housing capacity represented, and (8) the land conservation that results.

Context Records and Forecast Models

Three context record forms (CR, general purpose; GT, urban houses; and SF, suburb houses) have been designed to standardize the method of collecting relevant development capacity information. They mathematically convert project measurements to design specification values that define each project. These values can be entered in a forecast model to predict future results for other project areas based on the recipe derived, or they can be altered to test other options. The following collection of apartment house case studies uses context form CR to illustrate this process of data collection. Form CR has been used in previous chapters with a different identification panel. This panel has been expanded in these examples to illustrate how projects can be classified by class, order, family, genus, species, and location. For those less interested in classification and more interested in practical forecasts, the information panel in each context record format on the CD-ROM included with this book can be customized to reflect past or preferred formats. Any identification that differentiates one set of context record measurements from another is adequate from this perspective, since the shelter class options listed in App. A and the intensity index found are adequate for indexing development capacity records and recipes.

For those interested in broader applications, the address portion of form CR has been expanded to include a general indication of the growth location implied by the address provided. This geographic location can be indexed by the ring designations in Table 14.9. The shelter classification section of the identification panel has also been expanded to permit project identification by the hierarchy listed in App. A.

The ring designations used should not be considered homogeneous two-dimensional waves that flow evenly from an historic core. The term *ring* in this text is a title of convenience that has more to do with time, awareness, and accomplishment than geometry. Urban growth rings plotted over time can have very irregular and discontinuous shapes. These shapes include many connected elements that combine over rings and time to create urban form. This form is heavily influenced by the external forces and accomplishments of a given time period, and by the natural constraints of the surrounding environment. (It is also organic and alive, since we drive it from within.) The location of a project area within the growth rings of this form can often help to visualize its larger context.

Table 14.9 Irregular Rings of Urban Growth

Ring Arrangement	Ring Composition
Historic core	Shelter Matrix
Transition ring	Core duplicates
First ring	Strip substitutes
Inner rings	Land use clusters*
Central rings	Isolated contradictions
Previous ring	Edge conflicts
Cambrian ring	Disease
Edge cities	Movement paths
	Life-support systems
	Open space relief

*A land use cluster is typically referred to as a land use area, but the term *area* is amorphous and conveys a sense of sprawl rather than a sense of place. A cryptic zoning designation reinforces this abstract relationship and social disconnection. Subdivision names are an informal attempt to reestablish a sense of place within these zones. When a land use area is seen as a neighborhood, district, cluster, or village, a sense of place is reinforced and edge conflicts are more easily recognized as serious issues. Isolated contradictions, random change, and spreading interference within a land use cluster can also be more easily recognized as activity that can disrupt the cluster lifestyle, and as evidence that long-range evolutionary plans are needed to guide the change when justified. Random change may be legally justifiable at the present time based on microlevel relationships and political intuition, but it can also represent disease within the organism that is often taken far too casually.

Case Studies

Eight case studies are included in Table 14.3, but four have been summarized in previous chapters. The following expanded description of the remaining four includes several photographs, a site plan, a context record, and at least one forecast model for each. These packages represent previous recipes that have been constructed. These recipes can be examined for success, they can be used to predict future results, and they can be modified to examine alternative solutions.

Garden Context, High Low-Rise with Surface Parking in a Cambrian Ring Location

Britton Woods is a 3-story edge city apartment complex that uses surface parking and detached garages to serve its residents. Figure 14.1 illustrates the project, which was built

Figure 14.1 Britton Woods apartment houses—garden context, high low-rise, G1 surface parking system, lease; Cambrian-ring edge city location. Britton Woods produces a low housing capacity of 12.95 units per buildable acre because of the low height, substantial open space, and adequate parking provided. It is an example of an introverted site plan with internal amenity surrounded by building mass, perimeter parking lots, and garages that attempt to shelter its activity and identity from the higher-intensity nonresidential land uses that abut. It serves as a buffer for low-density residential neighborhoods beyond. (*a*) Overview; (*b*) site plan; (*c*) fountain; (*d*) parking area.

(Photos by W. M. Hosack)

with currently popular amenities that include entry drive image, ponds, fountains, tennis courts, social center, pool, detached garages, sidewalks, landscaping, and connection to a municipal bikeway system. Table 14.10 displays its context record. It is first in this discussion simply because it has the lowest density. Most Britton Woods apartment buildings are arranged on the site to surround a shared amenity such as a pond or lawn and to turn their backs on entry from perimeter parking lots and garages. Several are located as freestanding entry features. The Britton Woods site plan is *introverted* because it surrounds interior amenities with perimeter buildings and parking lots. A site plan is *extroverted* when apartments surround an interior service core of parking lots and garages and face out to a perimeter amenity. Asherton is included as Figure 14.2 to illustrate a typical extroverted arrangement, and its context record is included as Table 14.11, but it is not included in the four case studies provided. Other terms, such as *linear, freestanding,* and *courtyard,* are also informally used to describe site plan arrangement themes. These themes influence appearance, but also affect the amount of land consumed for roadway, parking, miscellaneous pavement, and building cover. This affects the amount of land left for open space, the final context produced, and the ultimate development capacity of the land available.

Britton Woods provides 65.9% open space, and 10.7% of this is pond area. The surface parking lots provided are very evident, but the open space provided is generous. The development capacity of the 28.330 acres is much greater than that constructed because of the height and open space introduced. Capacity can also be a function of zoning, however, and the building height adopted by Britton Woods is clearly a limiting factor. Both height and open space combine in this project to produce a low-density residential apartment project that is compatible with its edge city suburban neighbors and the market that is drawn to these areas.

Table 14.10 Britton Woods Context Record

Form CR

Name	**Windsor at Britton Woods**
Address	**5489 Crescent Ridge Drive, Dublin, OH**
Location	**Cambrian Ring Edge City**
Class and Order	**Apartment house, garden context**
Family, Genus, Species	**High low-rise, G1 parking, lease**
Parking System	*check box*
Surface parking around but not under building	x
Surface parking around and under building	
No on-site parking	
Notes	

		Description	Project Measurements		Context Summary SF	%	
Site					SF	%	
		Gross land area in acres	28.330	GLA	1,234,055		
		Public/ private ROW and paved easements in SF	46,640	W		3.8%	of GLA
		Net lot area		NLA	1,187,415	96.2%	of GLA
		Total unbuildable area in SF		U		0.0%	of GLA
		Water area within unbuildable area in SF		WAT			of U
		Buildable land area		BLA	1,187,415	96.2%	of GLA
Development Cover							
		Gross building area in SF	448,200	GBA			
		Total building footprint in SF	149,400	B		12.6%	of BLA
		Total bldg. support in SF (stairs, corridors, elevators, etc.)		BSU			
		Building efficiency		Be			of GBA
	Residential only	Number of dwelling units	353	DU			
		Aggregate average dwelling unit area in SF		AGG	1,270		
		Number of building floors	3.00	FLR			
	G2 only	Unenclosed surface parking area under building in SF		PUB			
		Total surface parking area in SF**	187,880	P or G		15.8%	of BLA
		Number of surface parking spaces on premise	500	NPSs			
		Building cover over surface parking area		AIR			of GPA
		Gross surface parking area per parking space in SF		ss	376		
		Gross building area per surface parking space in SF		as	896		
		Surface parking spaces per dwelling unit		us	1.42		
	Pkg. struc. only	Total parking structure footprint in SF		P		0.0%	of BLA
	Pkg. struc. only	Parking structure area below grade in SF					
	Pkg. struc. only	Parking structure area above grade under building in SF					
	Pkg. struc. only	Number of parking structure floors		p			
	Pkg. struc. only	Total parking structure area in SF		GPA			
	Pkg. struc. only	Total parking structure spaces		NPSg			
	Pkg. struc. only	Total pkg. struc. support in SF (stairs, elevators, etc.)		PSS			
	Pkg. struc. only	Parking structure support percentage		Pu			of GPA
	Pkg. struc. only	Gross parking structure area per space in SF		sg	0		
	Pkg. struc. only	Gross building area per structure parking space in SF		ag	0		
	Pkg. struc. only	Parking structure spaces per dwelling unit		ug	0.00		
	Residential only	Total dwelling unit garage area(s) in SF	31,680	GPA			
	Residential only	Dwelling unit garage above grade under builiding in SF		GUB			
	Residential only	Number of dwelling unit garage parking spaces	150	GPS			
	Residential only	Garage parking spaces per dwelling		Gn	0.4		
	Residential only	Total dwelling unit garage area per space in SF		Ga	211.2		
	Residential only	Total dwelling unit garage cover in addition to building cover		Gc		2.7%	of BLA
		Total parking spaces			650.0		
	Residential only	Total parking spaces per dwelling unit		u	1.8		
		Gross building area per parking space total		a	689.5		
		Number of loading spaces		l			
		Total loading area in SF		LDA			
		Gross area per loading space in SF		b			
		Loading area percentage		L		0.0%	of BLA
		Driveway areas in SF		R		0.0%	of BLA
		Misc. pavement and social center pvmt. in SF	35,600	M		3.0%	of BLA
		TOTAL DEVELOPMENT COVER		D	404,560	**34.1%**	of BLA
Project Open Space							
		PROJECT OPEN SPACE		S	782,855	**65.9%**	of BLA
		Water area within project open space in SF	83,500 (10.7%)	wat			of S
Summary							
		Development balance index		BNX		0.341	
		Development intensity index		INX		3.341	
		Floor area ratio per buildable acre		BFAR		0.377	
		Capacity index (gross building sq. ft. per buildable acre)		SFAC		16,442	
		Density per buildable acre if applicable		dBA		12.95	
		Density per net acre if applicable		dNA		12.95	
		Density per gross acre if applicable		dGA		12.46	

** Include internal parking lot landscaping, circulation aisles, and private roadway parking space areas. Express private roads separately as driveway areas.

(a)

(b)

(c)

Britton Woods could easily have increased its development capacity by increasing its average building height or by reducing the open space context provided. Exhibit 14.3 displays the design specification for this project and the planning forecast based on this data. The forecast is equal to the context measured for a 3-story building complex. Other building height options are included in the forecast, however, and the remainder of the forecast panel displays these options. For instance, if a 2-story building height had been used, the panel predicts that the number of dwelling units would decline to 293. If a 5-story building height had been used, the panel predicts that the number of dwelling units would have increased to 421. All other design specification values remain constant for these predictions.

Open space and parking requirements also have a significant effect on development capacity. Exhibit 14.4 shows that an open space reduction to 50% of the buildable land area would increase the 3-story housing capacity to 533 dwelling units. This increase occurs because more land is

(d)

Figure 14.2 Asherton apartment houses—garden context, low-rise buildings, G1 surface parking, lease; Cambrian-ring edge city location. Asherton is located in an affluent edge city and is an example of an extroverted site plan with central parking service cores at each housing cluster and perimeter amenity. It provides suburb house shelter capacity of 5.02 dwelling units per buildable acre in spite of its apartment house format. This results because of the extensive amounts of open space, generous aggregate average dwelling unit areas, and low-rise building height constructed. (See context record in Table 14.11). (*a*) Overview; (*b*) site plan; (*c*) street view; (*d*) parking area.

(Photos by W. M. Hosack)

Table 14.11 Asherton Context Record

Name	**Asherton**
Address	**Earlington Parkway**
Location	**Edge City**
Class and Order	**Apartment house, garden context**
Family, Genus, Species	**Low-rise, G1 parking, lease**
Parking System	*check box*
Surface parking around but not under building	x
Surface parking around and under building	
Surface and/or garage parking	

		Description	Project Measurements		Context Summary SF	%	
Site					SF	%	
		Gross land area in acres	21.330	GLA	929,135		
		Public/ private ROW and paved easements in SF	35,750	W		3.8%	of GLA
		Net lot area		NLA	893,385	96.2%	of GLA
		Total unbuildable area in SF		U		0.0%	of GLA
		Water area within unbuildable area in SF		WAT			of U
		Buildable land area		BLA	893,385	96.2%	of GLA
Development Cover							
		Gross building area in SF	178,700	GBA			
		Total building footprint in SF	89,350	B		10.0%	of BLA
		Total bldg. support in SF (stairs, corridors, elevators, etc.)		BSU			
		Building efficiency		Be			of GBA
	Residential only	Number of dwelling units	103	DU			
		Aggregate average dwelling unit area in SF		AGG	1,735		
		Number of building floors	2.00	FLR			
	G2 only	Unenclosed surface parking area under building in SF		PUB			
		Total surface parking area in SF**	60,780	P or G		6.8%	of BLA
		Number of surface parking spaces on premise	146	NPSs			
		Building cover over surface parking area		AIR			of GPA
		Gross surface parking area per parking space in SF		ss	416		
		Gross building area per surface parking space in SF		as	1,224		
		Surface parking spaces per dwelling unit		us	1.42		
	Pkg. struc. only	Total parking structure footprint in SF		P		0.0%	of BLA
	Pkg. struc. only	Parking structure area below grade in SF		UNG			of BLA
	Pkg. struc. only	Parking structure area above grade under building in SF		PSC			
	Pkg. struc. only	Number of parking structure floors		p			
	Pkg. struc. only	Total parking structure area in SF		GPA			
	Pkg. struc. only	Total parking structure spaces		NPSg			
	Pkg. struc. only	Total pkg. struc. support % (stairs, elevators, etc.)		PSS			
	Pkg. struc. only	Parking structure support SF		Pu			of GPA
	Pkg. struc. only	Gross parking structure area per space in SF		sg	0		
	Pkg. struc. only	Net parking structure area per space in SF		ag			
	Pkg. struc. only	Gross building area per structure parking space in SF			0		
	Pkg. struc. only	Parking structure spaces per dwelling unit		ug	0.00		
	Residential only	Total dwelling unit garage area(s) in SF	10,000	GPA			
	Residential only	Dwelling unit garage above grade under builiding in SF	0	GUB			
	Residential only	Number of dwelling unit garage parking spaces	40	GPS			
	Residential only	Garage parking spaces per dwelling		Gn	0.4		
	Residential only	Total dwelling unit garage area per space in SF		Ga	250.0		
	Residential only	Total dwelling unit garage cover in addition to building cover		Gc		1.1%	of BLA
		Total parking spaces		NPS	186.0		
	Residential only	Total parking spaces per dwelling unit		u	1.8		
		Gross building area per parking space total		a	960.8		
		Number of loading spaces	0	l			
		Total loading area in SF	0	LDA			
		Gross area per loading space in SF		b			
		Loading area percentage		L		0.0%	of BLA
		Driveway areas in SF	6,500	R		0.7%	of BLA
		Misc. pavement and social center pvmt. in SF	17,200	M		1.9%	of BLA
		TOTAL DEVELOPMENT COVER		D	183,830	**20.6%**	of BLA
Project Open Space							
		PROJECT OPEN SPACE		S	709,555	**79.4%**	of BLA
		Water area within project open space in SF	179,500	wat			of S
			25.3%				
Summary							
		Development balance index		BNX		0.206	
		Development intensity index		INX		2.206	
		Floor area ratio per buildable acre		BFAR		0.200	
		Capacity index (gross building sq. ft. per buildable acre)		SFAC		8,713	
		Density per buildable acre if applicable		dBA		5.02	
		Density per net acre if applicable		dNA		5.02	
		Density per gross acre if applicable		dGA		4.83	

** Include internal parking lot landscaping, circulation aisles, and private roadway parking space areas. Express private roads separately as driveway areas.

Exhibit 14.3 Britton Woods Forecast Record

Forecast Model RG1L

Development capacity forecast for ***APARTMENTS*** *based on the use of an adjacent* ***GRADE PARKING LOT*** *located on the same premises. When s and u equal zero in the design specification below, the forecast pertains to conditions when* ***NO PARKING*** *is required.*

Given: *Gross land area.* ***To Find:*** *Maximum dwelling unit capacity of the land area given based on the design specification values entered below.* ***Premise:*** *All building floors considered equal in area.*

DESIGN SPECIFICATION

Enter values in boxed areas where text is bold and blue. Express all fractions as decimals.

Given:	**Gross land area**	**GLA=**	**28.330**	acres	1,234,055	SF
Land Variables:	Public/ private right-of-way & paved easements	W=	**0.038**	fraction of GLA	46,647	SF
	Net land area	NLA=	27.259	acres	1,187,408	SF
	Unbuildable and/or future expansion areas	U=	**0.000**	fraction of GLA	0	SF
	Gross land area reduction	X=	0.038	fraction of GLA	46,647	SF
	Buildable land area remaining	**BLA=**	**27.259**	acres	1,187,408	SF
Parking Variables:	Est. gross pkg. lot area per pkg. space in SF	s =	**376**	ENTER ZERO IF NO PARKING REQUIRED		
	Parking lot spaces planned or required per dwelling unit	u=	**1.42**	ENTER ZERO IF NO PARKING REQUIRED		
	Garage parking spaces planned or required per dwelling unit	Gn=	**0.424**	ENTER ZERO IF NO PARKING REQUIRED		
	Gross building area per garage space	Ga=	**211.2**	ENTER ZERO IF NO PARKING REQUIRED		
	No. of loading spaces	l =	**0**			
	Gross area per loading space	b =	**0**	SF	0	SF
Site Variables:	**Project open space as fraction of BLA**	**S=**	**0.659**		782,502	SF
	Private driveways as fraction of BLA	R=	**0.000**		0	SF
	Misc. pavement as fraction of BLA	M=	**0.030**		35,622	SF
	Loading area as fraction of BLA	L=	0.000		0	SF
	Total site support areas as a fraction of BLA	Su=	0.689		818,124	SF
Core:	**Core development area as fraction of BLA**	**C=**	0.311	C+Su must = 1	369,284	SF
Building Variables:	Building efficiency as percentage of GBA	Be=	**0.850**	must have a value entered		
	Building support as fraction of GBA	Bu=	0.150	Be + Bu must = 1		

Dwelling Unit Mix Table:

DU dwelling unit type	GDA gross du area	CDA=GDA/Be comprehensive du area	MIX du mix	PDA = (CDA)MIX Prorated du area
EFF	**500**		**10%**	
1 BR	**750**		**20%**	
2 BR	**1,200**	data override	**40%**	data override
3 BR	**1,500**		**20%**	
4 BR	**1,800**		**10%**	

Aggregate avg. dwelling unit area (AGG) = **1,270**

GBA sf per parking space a= 689

Enter zero in the adjacent box unless you wish to override the AGG value calculated above **1,270** **override**

PLANNING FORECAST

no. of floors FLR	CORE minimum lot area for BCG & PLA	density per net acre dNA	dwelling units **NDU**	pkg. lot spaces NPS	parking lot area **PLA**	garage spaces GPS	garage area **GAR**	gross bldg area GBA no garages	footprint **BCA**	density per dBA bldable acre
1.00	369,284	7.15	195	276.9	104,131	82.7	17,465	247,688	247,688	7.15
2.00		10.76	293	416.7	156,673	124.4	26,277	372,667	186,334	10.76
3.00		**12.94**	**353**	**500.9**	**188,353**	**149.6**	**31,590**	**448,022**	**149,341**	**12.94**
4.00		14.40	392	557.3	209,537	166.4	35,143	498,412	124,603	14.40
5.00		15.44	421	597.6	224,701	178.4	37,687	534,481	106,896	15.44
6.00		16.22	442	627.9	236,091	187.5	39,597	561,574	93,596	16.22
7.00	*NOTE: Be aware when BCA becomes too small to be feasible.*	16.83	459	651.5	244,960	194.5	41,085	582,671	83,239	16.83
8.00		17.32	472	670.4	252,062	200.2	42,276	599,564	74,946	17.32
9.00		17.72	483	685.8	257,878	204.8	43,251	613,396	68,155	17.72
10.00		18.05	492	698.7	262,726	208.6	44,064	624,930	62,493	18.05
11.00		18.33	500	709.7	266,831	211.9	44,753	634,694	57,699	18.33
12.00		18.58	506	719.0	270,352	214.7	45,343	643,067	53,589	18.58
13.00		18.79	512	727.1	273,404	217.1	45,855	650,327	50,025	18.79
14.00		18.97	517	734.2	276,075	219.2	46,303	656,681	46,906	18.97

WARNING: These are preliminary forecasts that must not be used to make final decisions.

1) These forecasts are not a substitute for the "due diligence" research that must be conducted to support the final definition of "unbuildable areas" above and the final decision to purchase land. This research includes, but is not limited to, verification of adequate subsurface soil, zoning, environmental clearance, access, title, utilities and water pressure, clearance from deed restriction, easement and right-of-way encumbrances, clearance from existing above and below ground facility conflicts, etc.

2) The most promising forecast(s) made on the basis of data entered in the design specification from "due diligence" research must be verified at the drawing board before funds are committed and land purchase decisions are made. Actual land shape ratios, dimensions and irregularities encountered may require adjustments to the general forecasts above.

3) The software licensee shall take responsibility for the design specification values entered and any advice given that is based on the forecast produced.

Exhibit 14.4 Britton Woods Forecast with 50% Open Space

Forecast Model RG1L

*Development capacity forecast for **APARTMENTS** based on the use of an adjacent **GRADE PARKING LOT** located on the same premises. When s and u equal zero in the design specification below, the forecast pertains to conditions when **NO PARKING** is required.*

***Given:** Gross land area. **To Find:** Maximum dwelling unit capacity of the land area given based on the design specification values entered below. **Premise:** All building floors considered equal in area.*

DESIGN SPECIFICATION

Enter values in boxed areas where text is bold and blue. Express all fractions as decimals.

Given:	**Gross land area**	GLA=	**28.330**	acres	1,234,055	SF
Land Variables:	Public/ private right-of-way & paved easements	W=	**0.038**	fraction of GLA	46,647	SF
	Net land area	NLA=	27.259	acres	1,187,408	SF
	Unbuildable and/or future expansion areas	U=	**0.000**	fraction of GLA	0	SF
	Gross land area reduction	X=	0.038	fraction of GLA	46,647	SF
	Buildable land area remaining	BLA=	**27.259**	acres	1,187,408	SF
Parking Variables:	Est. gross pkg. lot area per pkg. space in SF	s =	**376**	ENTER ZERO IF NO PARKING REQUIRED		
	Parking lot spaces planned or required per dwelling unit	u=	**1.42**	ENTER ZERO IF NO PARKING REQUIRED		
	Garage parking spaces planned or required per dwelling unit	Gn=	**0.424**	ENTER ZERO IF NO PARKING REQUIRED		
	Gross building area per garage space	Ga=	**211.2**	ENTER ZERO IF NO PARKING REQUIRED		
	No. of loading spaces	l =	**0**			
	Gross area per loading space	b =	**0**	SF	0	SF
Site Variables:	**Project open space as fraction of BLA**	S=	**0.500**		593,704	SF
	Private driveways as fraction of BLA	R=	**0.000**		0	SF
	Misc. pavement as fraction of BLA	M=	**0.030**		35,622	SF
	Loading area as fraction of BLA	L=	0.000		0	SF
	Total site support areas as a fraction of BLA	Su=	0.530		629,326	SF
Core:	**Core development area as fraction of BLA**	C=	0.470	C+Su must = 1	558,082	SF
Building Variables:	Building efficiency as percentage of GBA	Be=	**0.850**	must have a value entered		
	Building support as fraction of GBA	Bu=	0.150	Be + Bu must = 1		

Dwelling Unit Mix Table:

DU dwelling unit type	GDA gross du area	CDA=GDA/Be comprehensive du area	MIX du mix	PDA = (CDA)MIX Pro-rated du area
EFF	**500**		**10%**	
1 BR	**750**		**20%**	
2 BR	**1,200**	data override	**40%**	data override
3 BR	**1,500**		**20%**	
4 BR	**1,800**		**10%**	

Aggregate avg. dwelling unit area	(AGG) =	**1,270**
GBA sf per parking space	a=	689
Enter zero in the adjacent box unless you wish to override the AGG value calculated above		**1,270** **override**

PLANNING FORECAST

no. of floors FLR	CORE minimum lot area for BCG & PLA	density per net acre dNA	dwelling units **NDU**	pkg lot spaces NPS	parking lot area **PLA**	garage spaces GPS	garage area **GAR**	gross bldg area GBA no garages	footprint **BCA**	density per dBA bldable acre
1.00	558,082	10.81	295	418.5	157,368	125.0	26,394	374,320	374,320	10.81
2.00		16.27	443	629.7	236,773	188.0	39,711	563,195	281,598	16.27
3.00		**19.56**	**533**	**757.0**	**284,649**	**226.0**	**47,741**	**677,075**	**225,692**	**19.56**
4.00		21.76	593	842.2	316,664	251.5	53,111	753,227	188,307	21.76
5.00		23.33	636	903.1	339,580	269.7	56,954	807,736	161,547	23.33
6.00		24.51	668	948.9	356,793	283.3	59,841	848,681	141,447	24.51
7.00	*NOTE: Be aware when BCA becomes too small to be feasible.*	25.44	693	984.6	370,197	294.0	62,089	880,564	125,795	25.44
8.00		26.17	713	1,013.1	380,930	302.5	63,889	906,094	113,262	26.17
9.00		26.78	730	1,036.5	389,718	309.5	65,363	926,997	103,000	26.78
10.00		27.28	744	1,056.0	397,046	315.3	66,592	944,428	94,443	27.28
11.00		27.71	755	1,072.5	403,250	320.2	67,633	959,184	87,199	27.71
12.00		28.07	765	1,086.6	408,570	324.5	68,525	971,838	80,987	28.07
13.00		28.39	774	1,098.9	413,182	328.1	69,299	982,809	75,601	28.39
14.00		28.67	781	1,109.6	417,219	331.3	69,976	992,412	70,887	28.67

WARNING: These are preliminary forecasts that must not be used to make final decisions.

1) These forecasts are not a substitute for the "due diligence" research that must be conducted to support the final definition of "unbuildable areas" above and the final decision to purchase land. This research includes, but is not limited to, verification of adequate subsurface soil, zoning, environmental clearance, access, title, utilities and water pressure, clearance from deed restriction, easement and right-of-way encumbrances, clearance from existing above and below ground facility conflicts, etc.

2) The most promising forecast(s) made on the basis of data entered in the design specification from "due diligence" research must be verified at the drawing board before funds are committed and land purchase decisions are made. Actual land shape ratios, dimensions and irregularities encountered may require adjustments to the general forecasts above.

3) The software licensee shall take responsibility for the design specification values entered and any advice given that is based on the forecast produced.

available for the core development area and less is provided as open space. This reduction is a serious decision, however, since it implies a context change and a potential impact on market attraction. Therefore, housing capacity increases when project open space is reduced and all other design specification values remain constant, but the increase has context implications.

It should be fairly evident that if parking requirements increase and all other design specification values remain constant, housing capacity declines. Exhibit 14.5 simply confirms this. Britton Woods currently provides an average of 1.42 surface parking spaces and 0.424 garage spaces per dwelling unit. The number of surface parking spaces is increased to an average of 2 per dwelling unit in Exhibit 14.5. If the open provision is held constant, this parking increase consumes more core development area and reduces the dwelling unit potential of the complex to 292. Therefore, parking is also a significant decision, since too much reduces housing capacity and unnecessarily increases pavement, glare, cost, and runoff, while too little increases congestion, inconvenience, and conflict while also reducing market attraction. This focus on parking, open space, and building height should not be interpreted to mean that other values in a design specification can be discounted, since all combine to influence development capacity. It simply means that building height, open space, and parking are three of the most prominent.

Yard Context, Low-Rise with Underground Parking in an Inner-Ring Suburb Location

Turnberry Condominium, shown in Fig. 14.3, is a 2-story apartment building that uses a 1-story underground parking garage in combination with supplemental surface parking to serve residents within a suburban environment. It has been chosen to illustrate the use of forecast model RS2L, which is designed to address apartment houses and underground

Exhibit 14.5 Britton Woods Forecast with 2 Parking Spaces per Dwelling

Forecast Model RG1L

Development capacity forecast for ***APARTMENTS*** *based on the use of an adjacent* ***GRADE PARKING LOT*** *located on the same premises. When s and u equal zero in the design specification below, the forecast pertains to conditions when* ***NO PARKING*** *is required.*

Given: *Gross land area.* ***To Find:*** *Maximum dwelling unit capacity of the land area given based on the design specification values entered below.* ***Premise:*** *All building floors considered equal in area.*

DESIGN SPECIFICATION *Enter values in boxed areas where text is bold and blue. Express all fractions as decimals.*

Given:	**Gross land area**	**GLA=**	**28.330**	acres	1,234,055	SF
Land Variables:	Public/ private right-of-way & paved easements	W=	**0.038**	fraction of GLA	46,647	SF
	Net land area	NLA=	27.259	acres	1,187,408	SF
	Unbuildable and/or future expansion areas	U=	**0.000**	fraction of GLA	0	SF
	Gross land area reduction	X=	0.038	fraction of GLA	46,647	SF
	Buildable land area remaining	**BLA=**	**27.259**	acres	1,187,408	SF
Parking Variables:	Est. gross pkg. lot area per pkg. space in SF	s =	**376**	ENTER ZERO IF NO PARKING REQUIRED		
	Parking lot spaces planned or required per dwelling unit	u=	**2.0**	ENTER ZERO IF NO PARKING REQUIRED		
	Garage parking spaces planned or required per dwelling unit	Gn=	**0.424**	ENTER ZERO IF NO PARKING REQUIRED		
	Gross building area per garage space	Ga=	**211.2**	ENTER ZERO IF NO PARKING REQUIRED		
	No. of loading spaces	l =	**0**			
	Gross area per loading space	b =	**0**	SF	0	SF
Site Variables:	**Project open space as fraction of BLA**	**S=**	**0.659**		782,502	SF
	Private driveways as fraction of BLA	R=	**0.000**		0	SF
	Misc. pavement as fraction of BLA	M=	**0.030**		35,622	SF
	Loading area as fraction of BLA	L=	0.000		0	SF
	Total site support areas as a fraction of BLA	Su=	0.689		818,124	SF
Core:	**Core development area as fraction of BLA**	**C=**	0.311	C+Su must = 1	369,284	SF
Building Variables:	Building efficiency as percentage of GBA	Be=	**0.850**	must have a value entered		
	Building support as fraction of GBA	Bu=	0.150	Be + Bu must = 1		

Dwelling Unit Mix Table:

DU dwelling unit type	GDA gross du area	CDA=GDA/Be comprehensive du area	MIX du mix	PDA = (CDA)MIX Pro-rated du area
EFF	**500**	data override	**10%**	data override
1 BR	**750**		**20%**	
2 BR	**1,200**		**40%**	
3 BR	**1,500**		**20%**	
4 BR	**1,800**		**10%**	

Aggregate avg. dwelling unit area (AGG) = **1,270**

GBA sf per parking space a= 524

PLANNING FORECAST Enter zero in the adjacent box unless you wish to override the AGG value calculated above **1,270** **override**

no. of floors FLR	CORE minimum lot area for BCG & PLA	density per net acre dNA	dwelling units **NDU**	pkg. lot spaces NPS	parking lot area **PLA**	garage spaces GPS	garage area **GAR**	gross bldg area GBA no garages	footprint **BCA**	density per dBA bldable acre
1.00	369,284	6.42	175	349.8	131,515	74.2	15,661	222,107	222,107	6.42
2.00		9.17	250	500.2	188,075	106.0	22,396	317,626	158,813	9.17
3.00		**10.71**	**292**	**583.9**	**219,547**	**123.8**	**26,144**	**370,778**	**123,593**	**10.71**
4.00		11.69	319	637.2	239,594	135.1	28,531	404,634	101,158	11.69
5.00		12.37	337	674.2	253,482	142.9	30,185	428,087	85,617	12.37
6.00		12.86	351	701.3	263,670	148.7	31,398	445,294	74,216	12.86
7.00	*NOTE: Be aware when BCA becomes too small to be feasible.*	13.24	361	722.0	271,464	153.1	32,326	458,456	65,494	13.24
8.00		13.54	369	738.3	277,618	156.5	33,059	468,850	58,606	13.54
9.00		13.79	376	751.6	282,602	159.3	33,652	477,266	53,030	13.79
10.00		13.99	381	762.6	286,719	161.7	34,143	484,220	48,422	13.99
11.00		14.16	386	771.8	290,178	163.6	34,555	490,061	44,551	14.16
12.00		14.30	390	779.6	293,125	165.3	34,906	495,038	41,253	14.30
13.00		14.42	393	786.3	295,666	166.7	35,208	499,329	38,410	14.42
14.00		14.53	396	792.2	297,879	168.0	35,472	503,067	35,933	14.53

WARNING: These are preliminary forecasts that must not be used to make final decisions.
1) These forecasts are not a substitute for the "due diligence" research that must be conducted to support the final definition of "unbuildable areas" above and the final decision to purchase land. This research includes, but is not limited to, verification of adequate subsurface soil, zoning, environmental clearance, access, title, utilities and water pressure, clearance from deed restriction, easement and right-of-way encumbrances, clearance from existing above and below ground facility conflicts, etc.
2) The most promising forecast(s) made on the basis of data entered in the design specification from "due diligence" research must be verified at the drawing board before funds are committed and land purchase decisions are made. Actual land shape ratios, dimensions and irregularities encountered may require adjustments to the general forecasts above.
3) The software licensee shall take responsibility for the design specification values entered and any advice given that is based on the forecast produced.

(a)

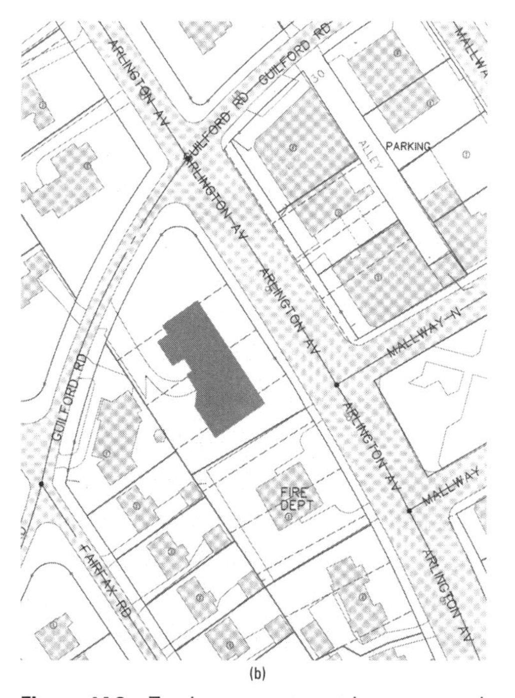

(b)

(c)

Figure 14.3 Turnberry apartment houses—yard context, low-rise buildings, S2 underground parking, condominium; inner-ring suburb. Turnberry produces a relatively high density of 24 units per buildable acre in spite of the low height and open space provided because of the underground parking level used. It abuts nonresidential land use on two sides and serves as a buffer for single-family residential housing abutting the third side. (*a*) Street view; (*b*) site plan; (*c*) overview.

(Photos by W. M. Hosack)

parking systems when the land area is known. Table 14.12 displays the context record for Turnberry. This record shows that Turnberry introduces 12 dwelling units on 0.5 acre of land with 15 underground and 5 surface parking spaces. The aggregate average area per dwelling unit is 1349 sq ft, and 49.9% open space is provided. The processed data from this record has been entered in the design specification panel of Exhibit 14.6. This model predicts the actual project measured, and these predictions can be found in the second row of the planning forecast panel. As usual, a greater number of building floors produces a greater number of dwelling units, but the number of underground parking levels p increases as well to provide the increased parking required. The underground garage, however, is currently contained within the building footprint above. (Note that the UNG value and the BCA value in row 2 are approximately equal.) If the underground parking footprint became larger than the building footprint, the number of parking levels could be reduced. In this example, if increased height were used to increase the number of dwelling units produced, the single underground parking surface could expand beyond the building footprint to accommodate the greater number of cars required to serve the increased number of dwelling units. For instance, the planning forecast panel in Exhibit 14.6 shows that a 4-story building will produce a NDU of 20 dwelling units and will require a UNG of 20,073 gross sq ft of underground parking area. Since the net buildable land area BLA is 21,780 sq ft, a one-level parking garage can fit below grade within these property lines. This assumes that the property shape is adequate.[10] The underground parking footprint value G in the design specification can be changed from 0.372 to 0.922 of the buildable land area to produce a one-level underground parking structure forecast for this four-floor building concept. The building footprint would be 6758 sq ft per floor. The surface parking area would increase to 3282 sq ft on top of the garage. The open space

Table 14.12 Turnberry Condominiums Context Record

Form CR

Name	**Turnberry Condominiums**	
Address	**2121 Arlington**	
Location	**Inner ring suburb**	
Class and Order	**Apartment house, yard context**	
Family, Genus, Species	**Low-rise, S2 parking, condominium**	
Parking System		*check box*
Structure parking adjacent to building		
Structure parking underground		x
Structure parking above grade under building		
Notes	**Limited G1 surface parking**	

	Description	Project Measurements		Context Summary SF	%	
Site						
	Gross land area in acres	0.500	GLA	21,780		
	Public/ private ROW and paved easements in SF	0	W		0.0%	of GLA
	Net lot area		NLA	21,780	100.0%	of GLA
	Total unbuildable area in SF	0	U		0.0%	of GLA
	Water area within unbuildable area in SF	0	WAT			of U
	Buildable land area		BLA	21,780	100.0%	of GLA
Development Cover						
	Gross building area in SF (exclude attached garages)	16,184	GBA			
	Total building footprint in SF (exclude attached garages)	8,092	B		37.15%	of BLA
	Total bldg. support in SF (stairs, corridors, elevators, etc.)		BSU			
	Building efficiency		Be			of GBA
Residential only	Number of dwelling units	12	DU			
	Aggregate average dwelling unit area in SF		AGG	1,349		
	Number of building floors	2.00	FLR			
G2 only	Unenclosed surface parking area under building in SF	0	PUB			
	Total surface parking area in SF**	1,950	P or G		9.0%	of BLA
	Number of surface parking spaces on premise	5	NPSs			
	Building cover over surface parking area		AIR			of GPA
	Gross surface parking area per parking space in SF		ss	390		
	Gross building area per surface parking space in SF		as	3,237		
	Surface parking spaces per dwelling unit		us	0.42		
Pkg. struc. only	Total parking structure footprint in SF	8,092	P		0.0%	of BLA
Pkg. struc. only	Parking structure area below grade in SF	8,092	UNG		37.15%	of BLA
Pkg. struc. only	Parking structure area above grade under building in SF	0	PSC			
Pkg. struc. only	Number of parking structure floors	1	p			
Pkg. struc. only	Number of parking structure floors beneath building	1				
Pkg. struc. only	Total parking structure area in SF	8,092	GPA			
Pkg. struc. only	Total parking structure spaces	15	NPSg			
Pkg. struc. only	Parking structure support percentage	0.200	Pu			of GPA
Pkg. struc. only	Total pkg. struc. support in SF (stairs, elevators, etc.)		PSS	1,618		
Pkg. struc. only	Gross parking structure area per space in SF		sg	539.5		
Pkg. struc. only	Net parking structure area per space in SF			431.6		
Pkg. struc. only	Gross building area per structure parking space in SF		ag	1,079		
Pkg. struc. only	Parking structure spaces per dwelling unit		ug	1.25		
Residential only	Total dwelling unit garage area(s) in SF		GPA			
Residential only	Dwelling unit garage above grade under builiding in SF		GUB			
Residential only	Number of dwelling unit garage parking spaces		GPS			
Residential only	Garage parking spaces per dwelling		Gn	0.0		
Residential only	Total dwelling unit garage area per space in SF		Ga			
Residential only	Total dwelling unit garage cover in addition to building cover		Gc		0.0%	of BLA
	Total parking spaces		NPS	20.0		
Residential only	Total parking spaces per dwelling unit		u	1.67		
	Gross building area per parking space total		a	809.2		
	Number of loading spaces		l			
	Total loading area in SF		LDA			
	Gross area per loading space in SF		b			
	Loading area percentage		L		0.0%	of BLA
	Driveway areas in SF	600	R		2.8%	of BLA
	Misc. pavement and social center pvmt. in SF	270	M		1.2%	of BLA
	TOTAL DEVELOPMENT COVER		D	10,912	**50.1%**	of BLA
Project Open Space						
	PROJECT OPEN SPACE		S	10,868	**49.9%**	of BLA
	Water area within project open space in SF		wat			of S
Summary		0.0%				
	Development balance index		BNX		50.1%	
	Development intensity index		INX		3.501	
	Floor area ratio per buildable acre		BFAR		0.743	
	Capacity index (gross building sq. ft. per buildable acre)		SFAC		32,368	
	Density per buildable acre if applicable		dBA		24.00	
	Density per net acre if applicable		dNA		24.00	
	Density per gross acre if applicable		dGA		24.00	

** Include internal parking lot landscaping, circulation aisles, and private roadway parking space areas. Express private roads separately as driveway areas.

Exhibit 14.6 Turnberry Condominium Forecast Record

Forecast Model RS2L

Development capacity forecast for ***APARTMENTS*** *based on the use of an* ***UNDERGROUND PARKING STRUCTURE.***
Given: *Gross land area.* ***To Find:*** *Dwelling unit capacity of the land area given based on the design specification values entered below.*
Design Premise: *Underground parking footprint may be larger, smaller, or equal to the building footprint above. All similar floors considered equal in area.*

DESIGN SPECIFICATION *Enter values in boxed areas where text is bold and blue. Express all fractions as decimals.*

Given:	**Gross land area**	GLA=	**0.500**	acres		21,780	SF
Land Variables:	Public/ private right-of-way & paved easements	W=	**0.000**	fraction of GLA		0	SF
	Net land area	NLA=	0.500	acres		21,780	SF
	Future expansion and/or unbuildalbe areas	U=	**0.000**	fraction of GLA		0	SF
	Gross land area reduction	X=	0.000	fraction of GLA		0	SF
	Buildable land area remaining	**BLA=**	**0.500**	acres		21,780	SF
Parking Variables:	Parking structure spaces planned, provided or required per dwelling unit	u =	**1.25**	**0.42**	surface parking		
	Est. net parking structure area per parking space	s =	**431.5**	**390**	surface parking		
	No. of loading spaces	l =	**0**				
	Gross area per loading space	b =	**0**			0	SF
Site Variables at Grade:	**Project open space as fraction of BLA**	**S=**	**0.499**	⇦		10,868	SF
	Private driveways as fraction of BLA	R=	**0.028**			610	SF
	Misc. pavement as fraction of BLA	M=	**0.012**			261	SF
	Loading area as fraction of BLA	L=	0.000			0	SF
	Total site support areas at grade as a fraction of BLA	Su=	0.539			11,739	SF
Core:	**Core development area at grade as fraction of BLA**	**C=**	0.461	C+Su must = 1		10,041	SF
Below Grade:	Gross underground parking area (UNG) as fraction of BLA	G=	**0.372**			8,091	SF
	Parking support within parking structure as fraction of UNG	Pu=	**0.200**				
	Net parking structure for pkg. & circulation as fraction of UNG	Pe=	0.800				
Building Variables:	Building efficiency as fraction of GBA	Be=	**0.850**	must have a value >0 entered			
	Building support as fraction of GBA	Bu=	0.150	Be+Bu must = 1			

Dwelling Unit Mix Table

DU dwelling unit type	GDA gross du area	CDA=GDA/Be comprehensive du area	MIX du mix	PDA = (CDA)MIX Pro-rated du area
EFF	**500**		**10%**	
1 BR	**750**		**20%**	
2 BR	**1,200**	data override	**40%**	data override
3 BR	**1,500**		**20%**	
4 BR	**1,800**		**10%**	

Aggregate avg. dwelling unit area (AGG) = **1,349**

PLANNING FORECAST Enter zero in the adjacent box unless you wish to override the AGG value calculated above **1,349** **override**

no. of bldg. floors **FLR**	no. of pkg. levels p	CORE	number of **NDU** dwelling units	underground NPS pkg. spaces	total TPS pkg. spaces	gross pkg. area UNG underground	surface PLA parking area	gross **GBA** building area	building BCA footprint area	density dBA per bldable acre
1.00	0.55		6.6	8.3	11.1	4,475	1,087	8,953	8,953	13.3
2.00	**1.00**		**12.0**	**15.0**	**20.0**	**8,075**	**1,962**	**16,157**	**8,079**	**24.0**
3.00	1.36		16.4	20.5	27.3	11,035	2,681	22,079	7,360	32.7
4.00	2.48	BCA+PLA	20.0	25.0	33.5	20,073	3,282	27,033	6,758	40.1
5.00	3.10	10,041	23.2	28.9	38.7	25,091	3,793	31,238	6,248	46.3
6.00	3.72		25.8	32.3	43.1	30,109	4,232	34,852	5,809	51.7
7.00	4.34	gross pkg struc area / level	28.2	35.2	47.0	35,127	4,613	37,992	5,427	56.3
8.00	4.96	GPL	30.2	37.8	50.4	40,146	4,947	40,745	5,093	60.4
9.00	5.58	8,091	32.0	40.0	53.5	45,164	5,243	43,179	4,798	64.0
10.00	6.20		33.6	42.0	56.1	50,182	5,506	45,346	4,535	67.2
11.00	6.82	net pkg struc area / level	35.1	43.8	58.5	55,200	5,742	47,287	4,299	70.1
12.00	7.44	NPL	36.4	45.4	60.7	60,218	5,954	49,037	4,086	72.7
13.00	8.06	6,473	37.5	46.9	62.7	65,237	6,147	50,621	3,894	75.1
14.00	8.68		38.6	48.2	64.5	70,255	6,322	52,064	3,719	77.2
15.00	9.30		39.6	49.5	66.1	75,273	6,482	53,382	3,559	79.1

WARNING: These are preliminary forecasts that must not be used to make final decisions.
1) These forecasts are not a substitute for the "due diligence" research that must be conducted to support the final definition of "unbuildable areas" above and the final decision to purchase land. This research includes, but is not limited to, verification of adequate subsurface soil, zoning, environmental clearance, access, title, utilities and water pressure, clearance from deed restriction, easement and right-of-way encumbrances, clearance from existing above and below ground facility conflicts, etc.
2) The most promising forecast(s) made on the basis of data entered in the design specification from "due diligence" research must be verified at the drawing board before funds are committed and land purchase decisions are made. Actual land shape ratios, dimensions and irregularities encountered may require adjustments to the general forecasts above.
3) The software licensee shall take responsibility for the design specification values entered and any advice given or decisions made that are based on the forecast produced.

would remain at 10,868 sq ft, and the 1-story underground parking footprint would be 20,073 sq ft. Exhibit 14.7 illustrates this revised forecast when the value *G* is adjusted to 0.922 in the design specification panel.

Please note that the second row in Exhibit 14.7 still shows that a 2-story building will produce 12 dwelling units and that its underground parking level UNG will equal the building cover area BCA required. However, the number of parking levels *p* predicted is 0.40. This is because the levels predicted are a function of the underground area available for parking. In this example, the area *G* allocated is equal to 92.2% of the buildable land area. The forecast table is predicting that only 40% of this area is required when a 2-story apartment house with 12 dwelling units is constructed. A comparison of the UNG and BCA columns, however, shows that this 40% is equal to the building footprint predicted, and that the underground garage is equal to the building basement when a 2-story apartment house is constructed.

Underground parking structures provide a great degree of flexibility, the ability to magnify housing capacity, and the ability to preserve open space on a given land area. They also represent additional complexity. The relationship between underground parking and surface building cover has just been discussed. This case also uses surface parking in addition to the underground parking that is provided. This parking is recorded in the project context record and entered in the design specification panel of Exhibit 14.6. The planning forecast panel predicts the surface parking area required in the PLA column and the total number of parking spaces provided in the TPS column. Subtracting TPS from the number of underground parking spaces NPS predicted reveals that five surface parking spaces are provided. As the number of floors increases in the planning forecast panel, the number of dwelling units NDU, the total number of parking spaces TPS, and the surface parking area PLA also increase. The building footprint area BCA decreases in rela-

Exhibit 14.7 Turnberry Condominium with Expanded Underground Parking

Forecast Model RS2L

Development capacity forecast for ***APARTMENTS*** *based on the use of an* ***UNDERGROUND PARKING STRUCTURE.***
Given: *Gross land area.* ***To Find:*** *Dwelling unit capacity of the land area given based on the design specification values entered below.*
Design Premise: *Underground parking footprint may be larger, smaller, or equal to the building footprint above. All similar floors considered equal in area.*

DESIGN SPECIFICATION *Enter values in boxed areas where text is bold and blue. Express all fractions as decimals.*

Given:	**Gross land aArea**	GLA=	**0.500**	acres	21,780	SF
Land Variables:	Public/ private right-of-way & paved easements	W=	**0.000**	fraction of GLA	0	SF
	Net land area	NLA=	0.500	acres	21,780	SF
	Future expansion and/or unbuildalbe areas	U=	**0.000**	fraction of GLA	0	SF
	Gross land area reduction	X=	0.000	fraction of GLA	0	SF
	Buildable land area remaining	BLA=	**0.500**	acres	21,780	SF
Parking Variables:	Parking structure spaces planned, provided or required per dwelling unit	u =	**1.25**	**0.42**	surface parking	
	Est. net parking structure area per parking space	s =	**431.5**	**390**	surface parking	
	No. of loading spaces	l =	**0**			
	Gross area per loading space	b =	**0**		0	SF
Site Variables at Grade:	**Project open space as fraction of BLA**	S=	**0.499**	⇦	10,868	SF
	Private driveways as fraction of BLA	R=	**0.028**		610	SF
	Misc. pavement as fraction of BLA	M=	**0.012**		261	SF
	Loading area as fraction of BLA	L=	0.000		0	SF
	Total site support areas at grade as a fraction of BLA	Su=	0.539		11,739	SF
Core:	**Core development area at grade as fraction of BLA**	C=	0.461	C+Su must = 1	10,041	SF
Below Grade:	Gross underground parking area (UNG) as fraction of BLA	G=	**0.922**		20,081	SF
	Parking support within parking structure as fraction of UNG	Pu=	**0.200**			
	Net parking structure for pkg. & circulation as fraction of UNG	Pe=	0.800			
Building Variables:	Building efficiency as fraction of GBA	Be=	**0.850**	must have a value >0 entered		
	Building support as fraction of GBA	Bu=	0.150	Be+Bu must = 1		

Dwelling Unit Mix Table

DU dwelling unit type	GDA gross du area	CDA=GDA/Be comprehensive du area	MIX du mix	PDA = (CDA)MIX Pro-rated du area
EFF	**500**		**10%**	
1 BR	**750**		**20%**	
2 BR	**1,200**	data override	**40%**	data override
3 BR	**1,500**		**20%**	
4 BR	**1,800**		**10%**	

Aggregate avg. dwelling unit area (AGG) = **1,349**

PLANNING FORECAST Enter zero in the adjacent box unless you wish to override the AGG value calculated above **1,349** **override**

no. of bldg. floors **FLR**	no. of pkg. levels **p**	CORE	number of **NDU** dwelling units	underground NPS pkg. spaces	total TPS pkg. spaces	gross pkg. area UNG underground	surface PLA parking area	gross **GBA** building area	building BCA footprint area	density dBA per bldable acre
1.00	0.22	BCA+PLA	6.6	8.3	11.1	4,475	1,087	8,953	8,953	13.3
2.00	0.40	10,041	12.0	15.0	20.0	8,075	1,962	16,157	8,079	24.0
3.00	0.55		16.4	20.5	27.3	11,035	2,681	22,079	7,360	32.7
4.00	**1.00**		**20.0**	**25.0**	**33.5**	**20,073**	**3,282**	**27,033**	**6,758**	**40.1**
5.00	1.25		23.2	28.9	38.7	25,091	3,793	31,238	6,248	46.3
6.00	1.50		25.8	32.3	43.1	30,109	4,232	34,852	5,809	51.7
7.00	1.75	gross pkg struc area / level	28.2	35.2	47.0	35,127	4,613	37,992	5,427	56.3
8.00	2.00	GPL	30.2	37.8	50.4	40,146	4,947	40,745	5,093	60.4
9.00	2.25	20,081	32.0	40.0	53.5	45,164	5,243	43,179	4,798	64.0
10.00	2.50		33.6	42.0	56.1	50,182	5,506	45,346	4,535	67.2
11.00	2.75	net pkg struc area / level	35.1	43.8	58.5	55,200	5,742	47,287	4,299	70.1
12.00	3.00	NPL	36.4	45.4	60.7	60,218	5,954	49,037	4,086	72.7
13.00	3.25	16,065	37.5	46.9	62.7	65,237	6,147	50,621	3,894	75.1
14.00	3.50		38.6	48.2	64.5	70,255	6,322	52,064	3,719	77.2
15.00	3.75		39.6	49.5	66.1	75,273	6,482	53,382	3,559	79.1

WARNING: These are preliminary forecasts that must not be used to make final decisions.
1) These forecasts are not a substitute for the "due diligence" research that must be conducted to support the final definition of "unbuildable areas" above and the final decision to purchase land. This research includes, but is not limited to, verification of adequate subsurface soil, zoning, environmental clearance, access, title, utilities and water pressure, clearance from deed restriction, easement and right-of-way encumbrances, clearance from existing above and below ground facility conflicts, etc.
2) The most promising forecast(s) made on the basis of data entered in the design specification from "due diligence" research must be verified at the drawing board before funds are committed and land purchase decisions are made. Actual land shape ratios, dimensions and irregularities encountered may require adjustments to the general forecasts above.
3) The software licensee shall take responsibility for the design specification values entered and any advice given or decisions made that are based on the forecast produced.

tion to the height increase, however, because the surface parking lot expands to continue to provide 0.42 surface parking spaces per dwelling unit, while the open space remains constant at 49.9% of the buildable land area. Therefore, as height FLR increases, NDU also increases. The number of underground parking levels *p* increases. The surface parking area PLA increases. The building footprint BCA decreases in relation to the surface parking increase, and the project open space *S* specified remains constant until modified in the design specification. A change to any design specification value will alter the development capacity predictions in the planning forecast panel, but not the underlying relationships just described.

Courtyard Context, High Low-Rise with No On-Site Parking in a Transition Ring Area

Townley Court and Library Park are both central business district (CBD) locations, and both illustrate the development capacity that can be achieved by high low-rise buildings when no on-site parking is included. Townley Court is illustrated by Fig. 14.4 and Library Park by Fig. 14.5; both could be claimed as historic buildings. Townley provides 45.3% open space and Library Park provides 40.9%; both are 4 stories in height. The aggregate average dwelling unit area provided by each is also similar. Townley represents 817 gross sq ft per unit and Library Park, 882 sq ft. Since Townley includes more open space but a smaller aggregate average dwelling unit area, both provide between 106 and 107 dwelling units per buildable acre. This is 9 times the Britton Woods density and is still not close to the ultimate development capacity of the site. The context record of each project is presented in Tables 14.13 and 14.14, and the design specification information distilled from this data has been entered in Exhibits 14.8 and 14.9. These exhibits show that the present density dNA of each can be more than tripled by simply building 14-story apartments to replace

(a)

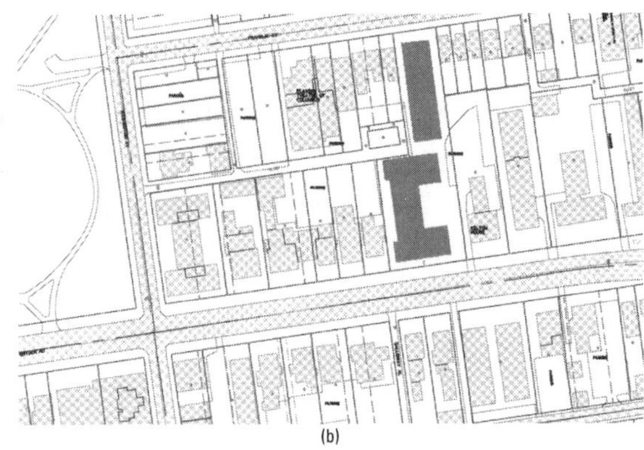

(b)

(c)

(d)

Figure 14.4 Townley Court apartment houses—courtyard context, high low-rise buildings, no on-site parking, rental; transition ring location. Townley Court produces a high housing capacity of 106.4 dwelling units per buildable acre in spite of the low height constructed because no on-site parking is provided, open space is reduced, and the aggregate average dwelling unit area is relatively small. (Aggregate average dwelling unit area is also referred to as average dwelling unit area.) It is on a mixed density residential street and is part of an emerging central business district environment. (*a*) Overview; (*b*) site plan; (*c*) and (*d*) ground-level views.

(Photos by W. M. Hosack)

these projects, but inadequate parking would result and greater potential exists.

The following example is intended to quantify the powerful incentive to demolish buildings when the shelter present significantly underutilizes the development capacity of a site. This occurs when the location is right, since modern technology makes 20, 30, or more stories easily achievable when height is not objectionable. It is also feasible to construct adjacent parking structures, parking structures above

(a)

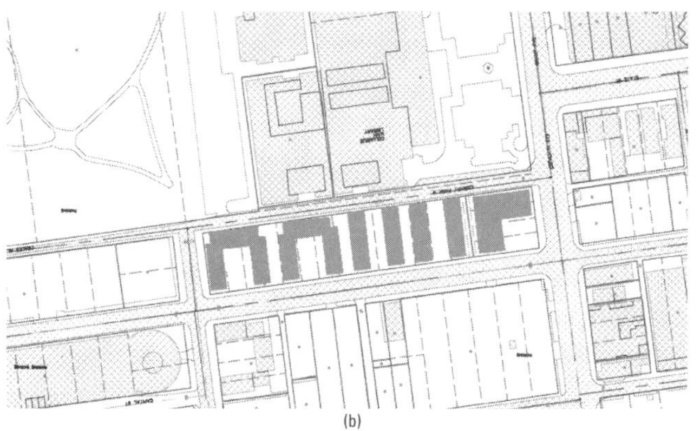

(b)

(c)

(d)

Figure 14.5 Library Park apartment houses—courtyard context, high low-rise buildings, no on-site parking, rental; transition ring location. Library Park produces a high housing capacity of 107 dwelling units per buildable acre in spite of the low height provided because no on-site parking is provided, open space is reduced, and the aggregate average dwelling unit area is relatively small. It is relatively isolated but is near employment, institutions and a large park. (*a*) Overview; (*b*) site plan; (*c*) and (*d*) street views.

(Photos by W. M. Hosack)

Table 14.13 Townley Court Context Record

Name	**Townley Court**
Address	**Town Street**
Location	**Transition ring area**
Class and Order	**Apartment house, courtyard context**
Family, Genus, Species	**High low-rise, No on-site parking, rental**
Parking System	*check box*
Surface parking around but not under building	
Surface parking around and under building	
Surface and garage parking	
No on-site parking	x

		Description	Project Measurements		Context Summary SF	%	
Site							
		Gross land area in acres	1.120	GLA	48,805		
		Public/ private ROW and paved easements in SF	7,075	W		14.5%	of GLA
		Net lot area		NLA	41,730	85.5%	of GLA
		Total unbuildable area in SF	0	U		0.0%	of GLA
		Water area within unbuildable area in SF	0	WAT			of U
		Buildable land area		BLA	41,730	85.5%	of GLA
Development Cover							
		Gross building area in SF (exclude attached garages)	83,320	GBA			
		Total building footprint in SF (exclude attached garages)	20,830	B		49.9%	of BLA
		Total bldg. support in SF (stairs, corridors, elevators, etc.)		BSU			
		Building efficiency		Be			of GBA
	Residential only	Number of dwelling units	102	DU			
		Aggregate average dwelling unit area in SF		AGG	817		
		Number of building floors	4.00	FLR			
	G2 only	Unenclosed surface parking area under building in SF		PUB			
		Total surface parking area in SF**	0	P or G		0.0%	of BLA
		Number of surface parking spaces on premise	0	NPSs			
		Building cover over surface parking area		AIR			of GPA
		Gross surface parking area per parking space in SF		ss			
		Gross building area per surface parking space in SF		as			
		Surface parking spaces per dwelling unit		us	0.00		
	Pkg. struc. only	Total parking structure footprint in SF		P		0.0%	of BLA
	Pkg. struc. only	Parking structure area below grade in SF		UNG			
	Pkg. struc. only	Parking structure area above grade under building in SF		PSC			
	Pkg. struc. only	Number of parking structure floors		p			
	Pkg. struc. only	Number of parking structure floors beneath building					
	Pkg. struc. only	Total parking structure area in SF		GPA			
	Pkg. struc. only	Total parking structure spaces		NPSg			
	Pkg. struc. only	Total pkg. struc. support in SF (stairs, elevators, etc.)		PSS			
	Pkg. struc. only	Parking structure support percentage		Pu			of GPA
	Pkg. struc. only	Gross parking structure area per space in SF		sg	0		
	Pkg. struc. only	Gross building area per structure parking space in SF		ag	0		
	Pkg. struc. only	Parking structure spaces per dwelling unit		ug	0.00		
	Residential only	Total dwelling unit garage area(s) in SF	0	GPA			
	Residential only	Dwelling unit garage above grade under builiding in SF	0	GUB			
	Residential only	Number of dwelling unit garage parking spaces	0	GPS			
	Residential only	Garage parking spaces per dwelling		Gn	0.0		
	Residential only	Total dwelling unit garage area per space in SF		Ga			
	Residential only	Total dwelling unit garage cover in addition to building cover		Gc		0.0%	of BLA
		Total parking spaces		NPS	0.0		
	Residential only	Total parking spaces per dwelling unit		u			
		Gross building area per parking space total		a			
		Number of loading spaces	0	l			
		Total loading area in SF	0	LDA			
		Gross area per loading space in SF		b			
		Loading area percentage		L		0.0%	of BLA
		Driveway areas in SF	0	R		0.0%	of BLA
		Misc. pavement and social center pvmt. in SF	2,000	M		4.8%	of BLA
		TOTAL DEVELOPMENT COVER		D	22,830	**54.7%**	of BLA
Project Open Space							
		PROJECT OPEN SPACE		S	18,900	**45.3%**	of BLA
		Water area within project open space in SF	150 0.8%	wat			of S
Summary							
		Development balance index		BNX		54.7%	
		Development intensity index		INX		4.547	
		Floor area ratio per buildable acre		BFAR		1.997	
		Capacity index (gross building sq. ft. per buildable acre)		SFAC		86,975	
		Density per buildable acre if applicable		dBA		106.47	
		Density per net acre if applicable		dNA		106.47	
		Density per gross acre if applicable		dGA		91.04	

** Include internal parking lot landscaping, circulation aisles, and private roadway parking space areas. Express private roads separately as driveway areas.

Table 14.14 Library Park Context Record

Name	**Library Park Apartments**
Address	**Library Park**
Location	**Transition ring area**
Class and Order	**Apartment house, courtyard context**
Family, Genus, Species	**High low-rise, no on-site parking, rental**
Parking System	*check box*
Surface parking around but not under building	
Surface parking around and under building	
Surface and garage parking	
No on-site parking	x

	Description	Project Measurements		Context Summary SF	Context Summary %	
Site				SF	%	
	Gross land area in acres	1.944	GLA	84,681		
	Public/ private ROW and paved easements in SF	29,720	W		35.1%	of GLA
	Net lot area		NLA	54,961	64.9%	of GLA
	Total unbuildable area in SF	0	U		0.0%	of GLA
	Water area within unbuildable area in SF	0	WAT			of U
	Buildable land area		BLA	54,961	64.9%	of GLA
Development Cover						
	Gross building area in SF (exclude attached garages)	119,040	GBA			
	Total building footprint in SF (exclude attached garages)	29,760	B		54.1%	of BLA
	Total bldg. support in SF (stairs, corridors, elevators, etc.)		BSU			
	Building efficiency		Be			of GBA
Residential only	Number of dwelling units	135	DU			
	Aggregate average dwelling unit area in SF		AGG	882		
	Number of building floors	4.00	FLR			
G2 only	Unenclosed surface parking area under building in SF		PUB			
	Total surface parking area in SF**	0	P or G		0.0%	of BLA
	Number of surface parking spaces on premise	0	NPSs			
	Building cover over surface parking area		AIR			of GPA
	Gross surface parking area per parking space in SF		ss			
	Gross building area per surface parking space in SF		as			
	Surface parking spaces per dwelling unit		us	0.00		
Pkg. struc. only	Total parking structure footprint in SF		P		0.0%	of BLA
Pkg. struc. only	Parking structure area below grade in SF		UNG			
Pkg. struc. only	Parking structure area above grade under building in SF		PSC			
Pkg. struc. only	Number of parking structure floors		p			
Pkg. struc. only	Number of parking structure floors beneath building					
Pkg. struc. only	Total parking structure area in SF		GPA			
Pkg. struc. only	Total parking structure spaces		NPSg			
Pkg. struc. only	Total pkg. struc. support in SF (stairs, elevators, etc.)		PSS			
Pkg. struc. only	Parking structure support percentage		Pu			of GPA
Pkg. struc. only	Gross parking structure area per space in SF		sg	0		
Pkg. struc. only	Gross building area per structure parking space in SF		ag	0		
Pkg. struc. only	Parking structure spaces per dwelling unit		ug	0.00		
Residential only	Total dwelling unit garage area(s) in SF	0	GPA			
Residential only	Dwelling unit garage above grade under builiding in SF	0	GUB			
Residential only	Number of dwelling unit garage parking spaces	0	GPS			
Residential only	Garage parking spaces per dwelling		Gn	0.0		
Residential only	Total dwelling unit garage area per space in SF		Ga			
Residential only	Total dwelling unit garage cover in addition to building cover		Gc		0.0%	of BLA
	Total parking spaces		NPS	0.0		
Residential only	Total parking spaces per dwelling unit		u			
	Gross building area per parking space total		a			
	Number of loading spaces	0	l			
	Total loading area in SF	0	LDA			
	Gross area per loading space in SF		b			
	Loading area percentage		L		0.0%	of BLA
	Driveway areas in SF	0	R		0.0%	of BLA
	Misc. pavement and social center pvmt. in SF	2,700	M		4.9%	of BLA
	TOTAL DEVELOPMENT COVER		D	32,460	**59.1%**	of BLA
Project Open Space						
	PROJECT OPEN SPACE		S	22,501	**40.9%**	of BLA
	Water area within project open space in SF	150 0.7%	wat			of S
Summary						
	Development balance index		BNX		59.1%	
	Development intensity index		INX		4.591	
	Floor area ratio per buildable acre		BFAR		2.166	
	Capacity index (gross building sq. ft. per buildable acre)		SFAC		94,347	
	Density per buildable acre if applicable		dBA		107.00	
	Density per net acre if applicable		dNA		107.00	
	Density per gross acre if applicable		dGA		69.44	

** Include internal parking lot landscaping, circulation aisles, and private roadway parking space areas. Express private roads separately as driveway areas.

Exhibit 14.8 Townley Court Forecast Record

Forecast Model RG1L

Development capacity forecast for ***APARTMENTS*** *based on the use of an adjacent* ***GRADE PARKING LOT*** *located on the same premises. When s and u equal zero in the design specification below, the forecast pertains to conditions when* ***NO PARKING*** *is required.*

Given: *Gross land area.* ***To Find:*** *Maximum dwelling unit capacity of the land area given based on the design specification values entered below.* ***Premise:*** *Al building floors considered equal in area.*

DESIGN SPECIFICATION

Enter values in boxed areas where text is bold and blue. Express all fractions as decimals.

Given:	**Gross land area**	GLA=	**1.120**	acres	48,787	SF
Land Variables:	Public/ private right-of-way & paved easements	W=	**0.145**	fraction of GLA	7,074	SF
	Net land area	NLA=	0.958	acres	41,713	SF
	Unbuildable and/or future expansion areas	U=	**0.000**	fraction of GLA	0	SF
	Gross land area reduction	X=	0.145	fraction of GLA	7,074	SF
	Buildable land area remaining	**BLA=**	**0.958**	acres	41,713	SF
Parking Variables:	Est. gross pkg. lot area per pkg. space in SF	s =	**0**	ENTER ZERO IF NO PARKING REQUIRED		
	Parking lot spaces planned or required per dwelling unit	u=	**0**	ENTER ZERO IF NO PARKING REQUIRED		
	Garage parking spaces planned or required per dwelling unit	Gn=	**0**	ENTER ZERO IF NO PARKING REQUIRED		
	Gross building area per garage space	Ga=	**0**	ENTER ZERO IF NO PARKING REQUIRED		
	No. of loading spaces	l =	**0**			
	Gross area per loading space	b =	**0**	SF	0	SF
Site Variables:	**Project open space as fraction of BLA**	**S=**	**0.453**		18,896	SF
	Private driveways as fraction of BLA	R=	**0.000**		0	SF
	Misc. pavement as fraction of BLA	M=	**0.048**		2,002	SF
	Loading area as fraction of BLA	L=	0.000		0	SF
	Total site support areas as a fraction of BLA	Su=	0.501		20,898	SF
Core:	**Core development area as fraction of BLA**	**C=**	0.499	C+Su must = 1	20,815	SF
Building Variables:	Building efficiency as percentage of GBA	Be=	**0.800**	must have a value entered		
	Building support as fraction of GBA	Bu=	0.200	Be + Bu must = 1		

Dwelling Unit Mix Table:

DU dwelling unit type	GDA gross du area	CDA=GDA/Be comprehensive du area	MIX du mix	PDA = (CDA)MIX Prorated du area
		data override		data override

Aggregate avg. dwelling unit area (AGG) = **817**

GBA sf per parking space a=

PLANNING FORECAST

Enter zero in the adjacent box unless you wish to override the AGG value calculated above: **817** **override**

no. of floors FLR	CORE minimum lot area for BCG & PLA	density per net acre dNA	dwelling units **NDU**	pkg. lot spaces NPS	parking lot area **PLA**	garage spaces GPS	garage area **GAR**	gross bldg area GBA no garages	footprint **BCA**	density per dBA bldable acre
1.00	20,815	26.61	25	n/a	n/a	n/a	n/a	20,815	20,815	26.61
2.00		53.21	51	n/a	n/a	n/a	n/a	41,630	20,815	53.21
3.00		79.82	76	n/a	n/a	n/a	n/a	62,444	20,815	79.82
4.00		**106.42**	**102**	**n/a**	**n/a**	**n/a**	**n/a**	**83,259**	**20,815**	**106.42**
5.00		133.03	127	n/a	n/a	n/a	n/a	104,074	20,815	133.03
6.00		159.63	153	n/a	n/a	n/a	n/a	124,889	20,815	159.63
7.00	*NOTE: Be aware when BCA becomes too small to be feasible.*	186.24	178	n/a	n/a	n/a	n/a	145,704	20,815	186.24
8.00		212.84	204	n/a	n/a	n/a	n/a	166,519	20,815	212.84
9.00		239.45	229	n/a	n/a	n/a	n/a	187,333	20,815	239.45
10.00		266.05	255	n/a	n/a	n/a	n/a	208,148	20,815	266.05
11.00		292.66	280	n/a	n/a	n/a	n/a	228,963	20,815	292.66
12.00		319.26	306	n/a	n/a	n/a	n/a	249,778	20,815	319.26
13.00		345.87	331	n/a	n/a	n/a	n/a	270,593	20,815	345.87
14.00		372.47	357	n/a	n/a	n/a	n/a	291,407	20,815	372.47

WARNING: These are preliminary forecasts that must not be used to make final decisions.

1) These forecasts are not a substitute for the "due diligence" research that must be conducted to support the final definition of "unbuildable areas" above and the final decision to purchase land. This research includes, but is not limited to, verification of adequate subsurface soil, zoning, environmental clearance, access, title, utilities and water pressure, clearance from deed restriction, easement and right-of-way encumbrances, clearance from existing above and below ground facility conflicts, etc.

2) The most promising forecast(s) made on the basis of data entered in the design specification from "due diligence" research must be verified at the drawing board before funds are committed and land purchase decisions are made. Actual land shape ratios, dimensions and irregularities encountered may require adjustments to the general forecasts above.

3) The software licensee shall take responsibility for the design specification values entered and any advice given that is based on the forecast produced.

Exhibit 14.9 Library Park Forecast Record

Forecast Model RG1L

Development capacity forecast for ***APARTMENTS*** *based on the use of an adjacent* ***GRADE PARKING LOT*** *located on the same premises. When s and u equal zero in the design specification below, the forecast pertains to conditions when* ***NO PARKING*** *is required.*

Given: *Gross land area.* ***To Find:*** *Maximum dwelling unit capacity of the land area given based on the design specification values entered below.* ***Premise:*** *All building floors considered equal in area.*

DESIGN SPECIFICATION *Enter values in boxed areas where text is bold and blue. Express all fractions as decimals.*

Given:	**Gross land area**	**GLA=**	**1.944**	acres	84,681	SF
Land Variables:	Public/ private right-of-way & paved easements	W=	**0.351**	fraction of GLA	29,723	SF
	Net land area	NLA=	1.262	acres	54,958	SF
	Unbuildable and/or future expansion areas	U=	**0.000**	fraction of GLA	0	SF
	Gross land area reduction	X=	0.351	fraction of GLA	29,723	SF
	Buildable land area remaining	**BLA=**	**1.262**	acres	54,958	SF
Parking Variables:	Est. gross pkg. lot area per pkg. space in SF	s =	**0**	ENTER ZERO IF NO PARKING REQUIRED		
	Parking lot spaces planned or required per dwelling unit	u=	**0**	ENTER ZERO IF NO PARKING REQUIRED		
	Garage parking spaces planned or required per dwelling unit	Gn=	**0**	ENTER ZERO IF NO PARKING REQUIRED		
	Gross building area per garage space	Ga=	**0**	ENTER ZERO IF NO PARKING REQUIRED		
	No. of loading spaces	l =	**0**			
	Gross area per loading space	b =	**0**	SF	0	SF
Site Variables:	**Project open space as fraction of BLA**	**S=**	**0.409**		22,478	SF
	Private driveways as fraction of BLA	R=	**0.000**		0	SF
	Misc. pavement as fraction of BLA	M=	**0.049**		2,693	SF
	Loading area as fraction of BLA	L=	0.000		0	SF
	Total site support areas as a fraction of BLA	Su=	0.458		25,171	SF
Core:	**Core development area as fraction of BLA**	**C=**	0.542	C=Su must = 1	29,787	SF
Building Variables:	Building efficiency as percentage of GBA	Be=	**0.830**	must have a value entered		
	Building support as fraction of GBA	Bu=	0.170	Be + Bu must = 1		

Dwelling Unit Mix Table:

DU dwelling unit type	GDA gross du area	CDA=GDA/Be comprehensive du area	MIX du mix	PDA = (CDA)MIX Pro-rated du area
EFF	**500**		**10%**	
1 BR	**750**		**20%**	
2 BR	**1,200**	data override	**40%**	data override
3 BR	**1,500**		**20%**	
4 BR	**1,800**		**10%**	

Aggregate avg. dwelling unit area (AGG) = **882**

GBA sf per parking space a=

PLANNING FORECAST Enter zero in the adjacent box unless you wish to override the AGG value calculated above **882** **override**

no. of floors FLR	CORE minimum lot area for BCG & PLA	density per net acre dNA	dwelling units **NDU**	pkg. lot spaces NPS	parking lot area **PLA**	garage spaces GPS	garage area **GAR**	gross bldg area GBA no garages	footprint **BCA**	density per dBA bldable acre
1.00	29,787	26.77	34	n/a	n/a	n/a	n/a	29,787	29,787	26.77
2.00		53.54	68	n/a	n/a	n/a	n/a	59,574	29,787	53.54
3.00		80.30	101	n/a	n/a	n/a	n/a	89,361	29,787	80.30
4.00		**107.07**	**135**	**n/a**	**n/a**	**n/a**	**n/a**	**119,148**	**29,787**	**107.07**
5.00		133.84	169	n/a	n/a	n/a	n/a	148,935	29,787	133.84
6.00	*NOTE:*	160.61	203	n/a	n/a	n/a	n/a	178,723	29,787	160.61
7.00	*Be aware when*	187.38	236	n/a	n/a	n/a	n/a	208,510	29,787	187.38
8.00	*BCA becomes too small to be feasible.*	214.15	270	n/a	n/a	n/a	n/a	238,297	29,787	214.15
9.00		240.91	304	n/a	n/a	n/a	n/a	268,084	29,787	240.91
10.00		267.68	338	n/a	n/a	n/a	n/a	297,871	29,787	267.68
11.00		294.45	371	n/a	n/a	n/a	n/a	327,658	29,787	294.45
12.00		321.22	405	n/a	n/a	n/a	n/a	357,445	29,787	321.22
13.00		347.99	439	n/a	n/a	n/a	n/a	387,232	29,787	347.99
14.00		374.75	473	n/a	n/a	n/a	n/a	417,019	29,787	374.75

WARNING: These are preliminary forecasts that must not be used to make final decisions.

1) These forecasts are not a substitute for the "due diligence" research that must be conducted to support the final definition of "unbuildable areas" above and the final decision to purchase land. This research includes, but is not limited to, verification of adequate subsurface soil, zoning, environmental clearance, access, title, utilities and water pressure, clearance from deed restriction, easement and right-of-way encumbrances, clearance from existing above and below ground facility conflicts, etc.

2) The most promising forecast(s) made on the basis of data entered in the design specification from "due diligence" research must be verified at the drawing board before funds are committed and land purchase decisions are made. Actual land shape ratios, dimensions and irregularities encountered may require adjustments to the general forecasts above.

3) The software licensee shall take responsibility for the design specification values entered and any advice given that is based on the forecast produced.

grade beneath these towers, and underground structures to serve the residents anticipated. In other words, the value of a site is influenced by its location and its ultimate development capacity, and this capacity can be predicted with the development forecast collection.

In both of these cases, the location is good and the ultimate development capacity of each site has hardly been considered. In the past it has been difficult to place a value on development capacity because it would be extremely time-consuming to forecast the full range of development options available at a drawing board. The development forecast collection makes it possible to predict these options in a fraction of the time without drawing a single line.

As an example, if Library Park were considered for redevelopment, it is very unlikely that the aggregate average dwelling unit area would remain at 882 sq ft. It is also unlikely that the 40.9% project open space would remain, even though this might be desirable. Exhibit 14.10 has been prepared to illustrate one of many redevelopment scenarios that could be written. It uses the RS3L forecast model to propose that a parking structure be placed above grade beneath an apartment house on 70% of the Library Park site ($S = 0.30$).[11] The exhibit shows that 1 parking space per dwelling unit is proposed and that an aggregate average dwelling unit area of 1781 sq ft is contemplated. This is very similar to the Waterford Tower design specification. Based on the entire specification included in Exhibit 14.10, the planning forecast panel predicts that 378.7 dwelling units could be constructed on 12 apartment floors above a 3-level parking structure. If the parking levels were increased to 6 and the apartment floors were increased to 24, the forecast model predicts that the 30-story structure could contain 757.3 dwelling units with an aggregate average area of 1781 sq ft per unit. It should be emphasized, however, that the development forecast collection predicts ultimate development capacity based on the values entered and is

Exhibit 14.10 Library Park Forecast with Parking Structure

Forecast Model RS3L

*Development capacity forecast for **APARTMENTS** based on the use of a **PARKING STRUCTURE** either partially or completely **ABOVE** grade **UNDER** the building or buildings.*

Given: *Gross land area.* ***To Find:*** *Maximum dwelling unit capacity of the gross land area given based on the design specification values entered below.*
Design Premise: *Parking footprint is beneath building footprint. The footprints may equal or exceed each other within the core development area. All similar floors considered equal in area.*

DESIGN SPECIFICATION *Enter values in boxed areas where text is bold and blue. Express all fractions as decimals.*

Given:	**Gross land area**	GLA=	**1.944**	acres	84,681
Land Variables:	Public/ private right-of-way & paved easements	W=	**0.000**	fraction of GLA	0
	Net land area	NLA=	1.944	acres	84,681
	Future expansion and/or unbuildalbe areas	U=	**0.000**	fraction of GLA	0
	Gross land area reduction	X=	0.000	fraction of GLA	0
	Buildable land area remaining	BLA=	**1.944**	acres	84,681
Parking Variables:	Est. net parking structure area per parking space	s =	**380**	SF	
	Parking spaces required per dwelling unit	u =	**1**		
	No. of loading spaces	l =	**2**		
	Gross area per loading space	b =	**1,000**	SF	2,000
Site Variables:	**Project open space as fraction of BLA**	S=	**0.300**		25,404
	Private driveways as fraction of BLA	R=	**0.000**		0
	Misc. pavement as fraction of BLA	M=	**0.010**		847
	Loading area as fraction of BLA	L=	0.024		2,000
	Total site support Areas as a fraction of BLA	Su=	0.334		28,251
Core:	**Core development area as fraction of BLA**	C=	0.666	C+Su must = 1	56,430
Pkg. Structure Variables:	Gross pkg. structure cover as fraction of core area (C)	P=	**1.000**	DO NOT EXCEED 1.0	56,430
	Parking support as fraction of gross parking area (GPA)	Pu=	**0.150**		
	Net area for parking & circulation as fraction of GPA	Pe=	0.850	Pu+Pe must = 1	47,965
Building Variables	Bldg. footprint over pkg structure as fraction of C	B=	**1.000**	DO NOT EXCEED 1.0	
	Building efficiency as percentage of gross bldg. area (GBA)	Be=	**0.750**	must have a value >0 entered	
	Bldg. support as fraction of GBA	Bu=	0.250	Bu+Be must equal 1	

Dwelling Unit Mix	DU dwelling unit type	GDA gross du area	CDA/Be comprehensive du area	MIX du mix	PDA = (CDA)MIX Prorated du area
	EFF	**500**	data override	**10%**	data override
	1 BR	**750**		**20%**	
	2 BR	**1,200**		**40%**	
	3 BR	**1,500**		**20%**	
	4 BR	**1,800**		**10%**	

Aggregate avg. dwelling unit area (AGG) = **1,781**

Building area planned, provided or required per parking space a = 1,781

PLANNING FORECAST Enter zero in the adjacent box unless you wish to override the AGG value calculated above **1,781** **override**

NOTE: p <= 1 = grade parking lot beneath building

no. of pkg levels **p**	minimum lot area for CORE fixed values below	net pkg area NPA for levels specified	pkg spaces NPS	pkg. struc area GPA for levels specified	gross bldg area **GBA**	no. of bldg floors **FLR**	dwelling units **NDU**	total floors **F** pkg + bldg	density per dBA bidable acre
1.00	56,430	47,965	126.2	56,430	**224,805**	**4.0**	**126.2**	**5.0**	64.93
2.00		95,930	252.4	112,859	**449,611**	**8.0**	**252.4**	**10.0**	129.86
3.00	footprint	143,896	378.7	169,289	**674,416**	**12.0**	**378.7**	**15.0**	194.79
4.00	BCA	191,861	504.9	225,719	**899,221**	**15.9**	**504.9**	**19.9**	259.72
5.00	56,430	239,826	631.1	282,148	**1,124,026**	**19.9**	**631.1**	**24.9**	324.65
6.00		287,791	757.3	338,578	**1,348,832**	**23.9**	**757.3**	**29.9**	389.58
7.00	gross pkg struc area	335,756	883.6	395,007	**1,573,637**	**27.9**	**883.6**	**34.9**	454.51
8.00	GPL	383,722	1,009.8	451,437	**1,798,442**	**31.9**	**1009.8**	**39.9**	519.44
9.00	56,430	431,687	1,136.0	507,867	**2,023,248**	**35.9**	**1136.0**	**44.9**	584.37
10.00	per level	479,652	1,262.2	564,296	**2,248,053**	**39.8**	**1262.2**	**49.8**	649.30
11.00	net pkg area per level	527,617	1,388.5	620,726	**2,472,858**	**43.8**	**1388.5**	**54.8**	714.23
12.00	NPL	575,582	1,514.7	677,156	**2,697,664**	**47.8**	**1514.7**	**59.8**	779.16
13.00	47,965	623,548	1,640.9	733,585	**2,922,469**	**51.8**	**1640.9**	**64.8**	844.09
14.00		671,513	1,767.1	790,015	**3,147,274**	**55.8**	**1767.1**	**69.8**	909.02
15.00		719,478	1,893.4	846,445	**3,372,079**	**59.8**	**1893.4**	**74.8**	973.95

WARNING: These are preliminary forecasts that must not be used to make final decisions.
1) These forecasts are not a substitute for the "due diligence" research that must be conducted to support the final definition of "unbuildable areas" above and the final decision to purchase land. This research includes, but is not limited to, verification of adequate subsurface soil, zoning, environmental clearance, access, title, utilities and water pressure, clearance from deed restriction, easement and right-of-way encumbrances, clearance from existing above and below ground facility conflicts, etc.
2) The most promising forecast(s) made on the basis of data entered in the design specification from "due diligence" research must be verified at the drawing board before funds are committed and land purchase decisions are made. Actual land shape ratios, dimensions and irregularities encountered may require adjustments to the general forecasts above.
3) The software licensee shall take responsibility for the design specification values entered and any advice given or decisions made that are based on the forecast produced.

only as good as those values. The actual capacity that can be achieved, particularly in the case of apartment forecasts, must be confirmed at the drawing board. Land shape, property restrictions, dwelling unit configuration, and parking structure dimensions can have a big influence on the values that should be used and the practical development capacity of the site. These forecasts must also be tempered with realistic appraisals of market demand, construction cost, and community acceptance. This example should illustrate, however, that the 135-dwelling-unit apartment complex with no parking and an aggregate average dwelling unit area of 882 sq ft has a lot of room to grow and is not close to its ultimate development capacity. It is also a desirable location since the site is across from a major metropolitan library, steps away from a major park, near a major hospital, and within walking distance of many corporations. This sounds like a real estate advertisement, but the objective of this discussion is not to sell land, deplete affordable housing, encourage historic demolition, or defeat historic preservation. The objective is to show how easily context records and forecast models can be used to analyze and predict housing capacity when used with care. This can easily reveal the potential that apartment houses have to improve housing capacity and preserve land if we can discover how to build them to attract a broader market.

Summary

Eight case studies are included in Table 14.3 and four are discussed in this chapter. The other four—Summit Chase, Jefferson, Lane Crest, and Waterford (presented in Figs. 14.6 to 14.9, Tables 14.15 to 14.18, and Exhibits 14.11 to 14.14, respectively)—have been introduced in previous chapters and were pulled from the library of context records in this book to define other approaches to the same issue.[12] The issue is the realistic shelter capacity of land. We cannot

(a)

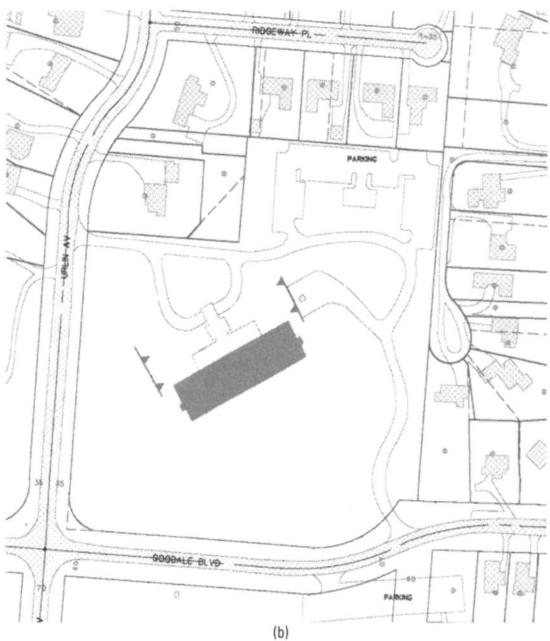

(b)

(c)

Figure 14.6 Summit Chase apartment houses—estate context, low high-rise building, S2 underground parking, condominium; first-ring suburb location. Summit Chase housing capacity is surprisingly low at 23.05 dwelling units per buildable acre because of the extensive open space provided. This low high-rise residential apartment house supplements surface parking with an underground parking structure to conserve project open space for residential enjoyment. It buffers a single-family residential neighborhood from railroad tracks and commercial land use beyond, is isolated, and stands between two worlds; but its exclusivity produces a high level of self-sufficiency. (*a*) Street view; (*b*) site plan; (*c*) overview.

(Photos by W. M. Hosack)

evaluate this issue until we can efficiently and comprehensively predict the full range and ultimate development capacity of this land. When predictions are possible—which they are with the development forecast collection—the debate over symbiotic relationships, practical development

(a)

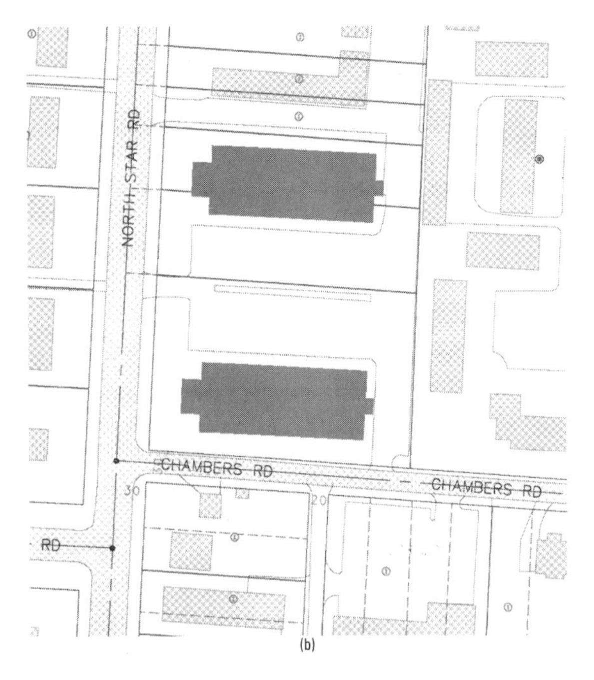

(b)

Figure 14.7 Jefferson apartment houses—transition context, high low-rise buildings, G1 surface parking, rental; inner-ring location. This high low-rise residential apartment house produces a relatively high housing capacity of 41 dwelling units per buildable acre in spite of the low building height and surface parking constructed. This results from the reduced open space and relatively small aggregate average dwelling unit area provided. It is part of a series of isolated apartment projects that are socially disconnected and have no integrated community features. (*a*) Overview; (*b*) site plan.

(Photo by W. M. Hosack)

(a)

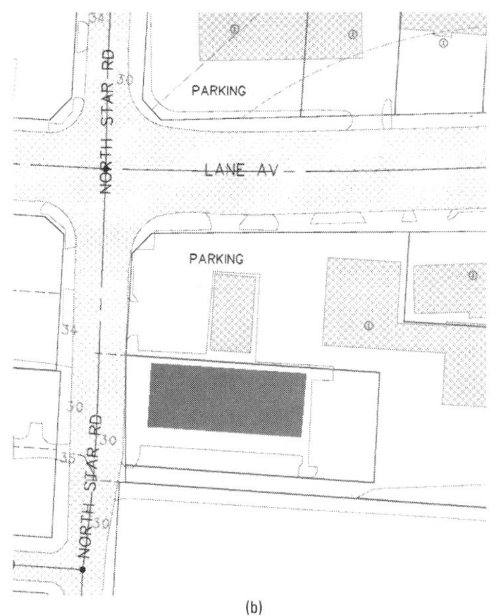

(b)

Figure 14.8 Lane apartment houses—CBD context, low-rise building, G1 surface parking, rental; inner-ring suburb location. This low-rise has a capacity of 68.5 dwelling units per buildable acre in spite of the low building height constructed because open space is held to a minimum and the aggregate average dwelling unit area provided is the smallest of those in the study. The apartment house is isolated, socially disconnected, and used to screen the strip commercial it lives with. (*a*) Street view; (*b*) site plan; (*c*) overview.

(Photos by W. M. Hosack)

(c)

(a)

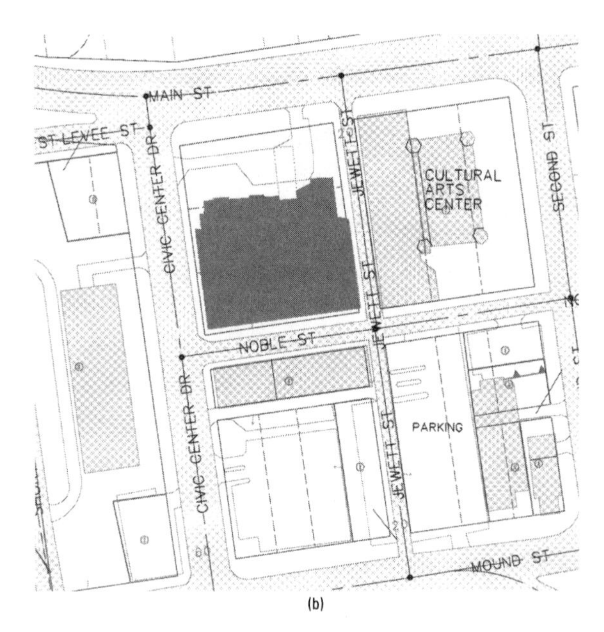

(b)

(c)

Figure 14.9 Waterford apartment houses—CBD context, high mid-rise building, S1 parking system, condominium; historic core location. Waterford has the greatest housing capacity in the group and provides 145.9 spacious dwelling units per buildable acre. It can produce this density in spite of these generous dwelling unit areas because it provides little open space and takes advantage of its high mid-rise building height. It also provides a low parking ratio per dwelling unit and uses a parking garage to multiply the number of parking spaces it can provide per acre of buildable land area. It is isolated in a business environment, but is adjacent to a park, and its exclusivity produces a high level of self-sufficiency. (*a*) Overview; (*b*) site plan; (*c*) overview.

(Photos by W. M. Hosack)

Table 14.15 Summit Chase Context Record

Name	**Summit Chase**
Address	**1000 Urlin Avenue**
Location	**First ring suburb**
Class and Order	**Apartment house, estate context**
Family, Genus, Species	**Low high-rise, S2 parking, condominium**
Parking System	*check box*
Structure parking adjacent to building	
Structure parking underground	x
Structure parking above grade under building	
Notes	**Also G1 and valet surface parking**

	Description	Project Measurements		Context Summary SF	%	
Site						
	Gross land area in acres	7.814	GLA	340,378		
	Public/ private ROW and paved easements in SF	0	W		0.0%	of GLA
	Net lot area		NLA	340,378	100.0%	of GLA
	Total unbuildable area in SF	4,000	U		1.2%	of GLA
	Water area within unbuildable area in SF	0	WAT			of U
	Buildable land area		BLA	336,378	98.8%	of GLA
Development Cover						
	Gross building area in SF	386,400	GBA			
	Total building footprint in SF	16,100	B		4.8%	of BLA
	Total bldg. support in SF (stairs, corridors, elevators, etc.)		BSU			
	Building efficiency		Be			of GBA
Residential only	Number of dwelling units	178	DU			
	Aggregate average dwelling unit area in SF		AGG	2,171		
	Number of building floors	24.00	FLR			
G2 only	Unenclosed surface parking area under building in SF	0	PUB			
	Total surface parking area in SF**	27,600	P or G		8.2%	of BLA
	Number of surface parking spaces on premise	78	NPSs			
	Building cover over surface parking area		AIR			of GPA
	Gross surface parking area per parking space in SF		ss	354		
	Gross building area per surface parking space in SF		as	4,954		
	Surface parking spaces per dwelling unit		us	0.44		
Pkg. struc. only	Total parking structure footprint in SF	33,600	P		0.0%	of BLA
Pkg. struc. only	Parking structure area below grade in SF	33,600	UNG		9.99%	of BLA
Pkg. struc. only	Parking structure area above grade under building in SF	0	PSC			
Pkg. struc. only	Number of parking structure floors	2	p			
Pkg. struc. only	Total parking structure area in SF	67,200	GPA			
Pkg. struc. only	Total parking structure spaces	188	NPSg			
Pkg. struc. only	Total pkg. struc. support % (stairs, elevators, etc.)	0.054	PSS	3,629		
Pkg. struc. only	Parking structure support SF		Pu			of GPA
Pkg. struc. only	Gross parking structure area per space in SF		sg	357		
Pkg. struc. only	Net parking structure area per space in SF		ag	338		
Pkg. struc. only	Gross building area per structure parking space in SF			2,055		
Pkg. struc. only	Parking structure spaces per dwelling unit		ug	1.06		
Residential only	Total dwelling unit garage area(s) in SF		GPA			
Residential only	Dwelling unit garage above grade under builiding in SF		GUB			
Residential only	Number of dwelling unit garage parking spaces		GPS			
Residential only	Garage parking spaces per dwelling		Gn	0.0		
Residential only	Total dwelling unit garage area per space in SF		Ga			
Residential only	Total dwelling unit garage cover in addition to building cover		Gc		0.0%	of BLA
	Total parking spaces		NPS	266.0		
Residential only	Total parking spaces per dwelling unit		u	1.5		
	Gross building area per parking space total		a	1,452.6		
	Number of loading spaces	1	l			
	Total loading area in SF	1,800	LDA			
	Gross area per loading space in SF		b	1,800		
	Loading area percentage		L		0.5%	of BLA
	Driveway areas in SF	7,900	R		2.3%	of BLA
	Misc. pavement and social center pvmt. in SF	1,185	M		0.4%	of BLA
	TOTAL DEVELOPMENT COVER		D	54,585	**16.2%**	of BLA
Project Open Space						
	PROJECT OPEN SPACE		S	281,793	**83.8%**	of BLA
	Water area within project open space in SF		wat			of S
Summary		0.0%				
	Development balance index		BNX		0.162	
	Development intensity index		INX		24.162	
	Floor area ratio per buildable acre		BFAR		1.149	
	Capacity index (gross building sq. ft. per buildable acre)		SFAC		50,038	
	Density per buildable acre if applicable		dBA		23.05	
	Density per net acre if applicable		dNA		22.78	
	Density per gross acre if applicable		dGA		22.78	

** Include internal parking lot landscaping, circulation aisles, and private roadway parking space areas. Express private roads separately as driveway areas.

Table 14.16 Jefferson Apartments Context Record

Form CR

Name	**Jefferson Apartments**
Address	**1800 N. Star**
Location	**Inner ring area**
Class and Order	**Apartment house, transition context**
Family, Genus, Species	**High low-rise, G1 parking, rental**
Parking System	*check box*
Surface parking around but not under building	x
Surface parking around and under building	
No on-site parking	

		Description	Project Measurements		Context Summary SF	Context Summary %	
Site							
		Gross land area in acres	1.756	GLA	76,491		
		Public/ private ROW and paved easements in SF	0	W		0.0%	of GLA
		Net lot area		NLA	76,491	100.0%	of GLA
		Total unbuildable area in SF	0	U		0.0%	of GLA
		Water area within unbuildable area in SF	0	WAT			of U
		Buildable land area		BLA	76,491	100.0%	of GLA
Development Cover							
		Gross building area in SF (exclude attached garages)	62,400	GBA			
		Total building footprint in SF (exclude attached garages)	20,800	B		27.2%	of BLA
		Total bldg. support in SF (stairs, corridors, elevators, etc.)		BSU			
		Building efficiency		Be			of GBA
	Residential only	Number of dwelling units	72	DU			
		Aggregate average dwelling unit area in SF		AGG	867		
		Number of building floors	3.00	FLR			
	G2 only	Unenclosed surface parking area under building in SF		PUB			
		Total surface parking area in SF**	31,725	P or G		41.5%	of BLA
		Number of surface parking spaces on premise	102	NPSs			
		Building cover over surface parking area		AIR			of GPA
		Gross surface parking area per parking space in SF		ss	311		
		Gross building area per surface parking space in SF		as	612		
		Surface parking spaces per dwelling unit		us	1.42		
	Pkg. struc. only	Total parking structure footprint in SF		P		0.0%	of BLA
	Pkg. struc. only	Parking structure area below grade in SF		UNG			
	Pkg. struc. only	Parking structure area above grade under building in SF		PSC			
	Pkg. struc. only	Number of parking structure floors		p			
	Pkg. struc. only	Number of parking structure floors beneath building					
	Pkg. struc. only	Total parking structure area in SF		GPA			
	Pkg. struc. only	Total parking structure spaces		NPSg			
	Pkg. struc. only	Total pkg. struc. support in SF (stairs, elevators, etc.)		PSS			
	Pkg. struc. only	Parking structure support percentage		Pu			of GPA
	Pkg. struc. only	Gross parking structure area per space in SF		sg	0		
	Pkg. struc. only	Gross building area per structure parking space in SF		ag	0		
	Pkg. struc. only	Parking structure spaces per dwelling unit		ug	0.00		
	Residential only	Total dwelling unit garage area(s) in SF		GPA			
	Residential only	Dwelling unit garage above grade under builiding in SF		GUB			
	Residential only	Number of dwelling unit garage parking spaces		GPS			
	Residential only	Garage parking spaces per dwelling		Gn	0.0		
	Residential only	Total dwelling unit garage area per space in SF		Ga			
	Residential only	Total dwelling unit garage cover in addition to building cover		Gc		0.0%	of BLA
		Total parking spaces		NPS	102.0		
	Residential only	Total parking spaces per dwelling unit		u	1.42		
		Gross building area per parking space total		a	611.8		
		Number of loading spaces		l			
		Total loading area in SF		LDA			
		Gross area per loading space in SF		b			
		Loading area percentage		L		0.0%	of BLA
		Driveway areas in SF	640	R		0.8%	of BLA
		Misc. pavement and social center pvmt. in SF	100	M		0.1%	of BLA
		TOTAL DEVELOPMENT COVER		D	53,265	**69.6%**	of BLA
Project Open Space							
		PROJECT OPEN SPACE		S	23,226	**30.4%**	of BLA
		Water area within project open space in SF		wat			of S
Summary			0.0%				
		Development balance index		BNX		69.6%	
		Development intensity index		INX		3.696	
		Floor area ratio per buildable acre		BFAR		0.816	
		Capacity index (gross building sq. ft. per buildable acre)		SFAC		35,535	
		Density per buildable acre if applicable		dBA		41.00	
		Density per net acre if applicable		dNA		41.00	
		Density per gross acre if applicable		dGA		41.00	

** Include internal parking lot landscaping, circulation aisles, and private roadway parking space areas. Express private roads separately as driveway areas.

Table 14.17 Lane Apartment Context Record

Form CR

Name	**LaneCrest**
Address	**Lane Avenue**
Location	**Inner ring suburb**
Class and Order	**Apartment house, CBD context**
Family, Genus, Species	**Low-rise, G1 parking, rental**
Parking System	*check box*
Surface parking around but not under building	x
Surface parking around and under building	
No on-site parking	
Notes	

	Description	Project Measurements		Context Summary SF	%	
Site						
	Gross land area in acres	0.525	GLA	22,869		
	Public/ private ROW and paved easements in SF	0	W		0.0%	of GLA
	Net lot area		NLA	22,869	100.0%	of GLA
	Total unbuildable area in SF	0	U		0.0%	of GLA
	Water area within unbuildable area in SF	0	WAT			of U
	Buildable land area		BLA	22,869	100.0%	of GLA
Development Cover						
	Gross building area in SF (exclude attached garages)	23,856	GBA			
	Total building footprint in SF (exclude attached garages)	7,952	B		34.8%	of BLA
	Total bldg. support in SF (stairs, corridors, elevators, etc.)	450	BSU			
	Building efficiency		Be		98.1%	of GBA
Residential only	Number of dwelling units	36	DU			
	Aggregate average dwelling unit area in SF		AGG	663		
	Number of building floors	3.00	FLR			
G2 only	Unenclosed surface parking area under building in SF		PUB			
	Total surface parking area in SF**	9,000	P or G		39.4%	of BLA
	Number of surface parking spaces on premise	24	NPSs			
	Building cover over surface parking area		AIR			of GPA
	Gross surface parking area per parking space in SF		ss	375		
	Gross building area per surface parking space in SF		as	994		
	Surface parking spaces per dwelling unit		us	0.67		
Pkg. struc. only	Total parking structure footprint in SF		P		0.0%	of BLA
Pkg. struc. only	Parking structure area below grade in SF		UNG			
Pkg. struc. only	Parking structure area above grade under building in SF		PSC			
Pkg. struc. only	Number of parking structure floors		p			
Pkg. struc. only	Number of parking structure floors beneath building					
Pkg. struc. only	Total parking structure area in SF		GPA			
Pkg. struc. only	Total parking structure spaces		NPSg			
Pkg. struc. only	Total pkg. struc. support in SF (stairs, elevators, etc.)		PSS			
Pkg. struc. only	Parking structure support percentage		Pu			of GPA
Pkg. struc. only	Gross parking structure area per space in SF		sg	0		
Pkg. struc. only	Gross building area per structure parking space in SF		ag	0		
Pkg. struc. only	Parking structure spaces per dwelling unit		ug	0.00		
Residential only	Total dwelling unit garage area(s) in SF		GPA			
Residential only	Dwelling unit garage above grade under builiding in SF		GUB			
Residential only	Number of dwelling unit garage parking spaces		GPS			
Residential only	Garage parking spaces per dwelling		Gn	0.0		
Residential only	Total dwelling unit garage area per space in SF		Ga			
Residential only	Total dwelling unit garage cover in addition to building cover		Gc		0.0%	of BLA
	Total parking spaces		NPS	24.0		
Residential only	Total parking spaces per dwelling unit		u	0.7		
	Gross building area per parking space total		a	994.0		
	Number of loading spaces		l			
	Total loading area in SF		LDA			
	Gross area per loading space in SF		b			
	Loading area percentage		L		0.0%	of BLA
	Driveway areas in SF		R		0.0%	of BLA
	Misc. pavement and social center pvmt. in SF	700	M		3.1%	of BLA
	TOTAL DEVELOPMENT COVER		D	17,652	**77.2%**	of BLA
Project Open Space						
	PROJECT OPEN SPACE		S	5,217	**22.8%**	of BLA
	Water area within project open space in SF	0.0%	wat			of S
Summary						
	Development balance index		BNX		77.2%	
	Development intensity index		INX		3.772	
	Floor area ratio per buildable acre		BFAR		1.043	
	Capacity index (gross building sq. ft. per buildable acre)		SFAC		45,440	
	Density per buildable acre if applicable		dBA		68.57	
	Density per net acre if applicable		dNA		68.57	
	Density per gross acre if applicable		dGA		68.57	

** Include internal parking lot landscaping, circulation aisles, and private roadway parking space areas. Express private roads separately as driveway areas.

Table 14.18 Waterford Tower Context Record

Name	**Waterford Tower**
Address	**155 W. Main Street**
Location	**Historic Core (CBD)**
Class and Order	**Apartment house, CBD context**
Family, Genus, Species	**High mid-rise, S1 parking, condominium**
Parking System	*check box*
Structure parking adjacent to building	x
Structure parking underground	
Structure parking above grade under building	
Notes	Also valet surface parking

	Description	Project Measurements		Context Summary SF	Context Summary %	
Site						
	Gross land area in acres	0.658	GLA	28,662		
	Public/ private ROW and paved easements in SF	0	W		0.0%	of GLA
	Net lot area		NLA	28,662	100.0%	of GLA
	Total unbuildable area in SF	0	U		0.0%	of GLA
	Water area within unbuildable area in SF	0	WAT			of U
	Buildable land area		BLA	28,662	100.0%	of GLA
Development Cover						
	Gross building area in SF	171,000	GBA			
	Total building footprint in SF	9,000	B		31.4%	of BLA
	Total bldg. support in SF (stairs, corridors, elevators, etc.)		BSU			
	Building efficiency		Be			of GBA
Residential only	Number of dwelling units	96	DU			
	Aggregate average dwelling unit area in SF		AGG	1,781		
	Number of building floors	19.00	FLR			
G2 only	Unenclosed surface parking area under building in SF	0	PUB			
	Total surface parking area in SF**	0	P or G		0.0%	of BLA
	Number of surface parking spaces on premise	0	NPSs			
	Building cover over surface parking area		AIR			of GPA
	Gross surface parking area per parking space in SF		ss			
	Gross building area per surface parking space in SF		as			
	Surface parking spaces per dwelling unit		us	0.00		
Pkg. struc. only	Total parking structure footprint in SF	9,000	P		31.4%	of BLA
Pkg. struc. only	Parking structure area below grade in SF	0	UNG			of BLA
Pkg. struc. only	Parking structure area above grade under building in SF	0	PSC			
Pkg. struc. only	Number of parking structure floors	3	p			
Pkg. struc. only	Total parking structure area in SF	27,000	GPA			
Pkg. struc. only	Total parking structure spaces	75	NPSg			
Pkg. struc. only	Total pkg. struc. support % (stairs, elevators, etc.)	0.050	PSS			
Pkg. struc. only	Parking structure support SF		Pu	1,350		of GPA
Pkg. struc. only	Gross parking structure area per space in SF		sg	360		
Pkg. struc. only	Net parking structure area per space in SF		ag	342		
Pkg. struc. only	Gross building area per structure parking space in SF			2,280		
Pkg. struc. only	Parking structure spaces per dwelling unit		ug	0.78		
Residential only	Total dwelling unit garage area(s) in SF		GPA			
Residential only	Dwelling unit garage above grade under building in SF		GUB			
Residential only	Number of dwelling unit garage parking spaces		GPS			
Residential only	Garage parking spaces per dwelling		Gn	0.0		
Residential only	Total dwelling unit garage area per space in SF		Ga			
Residential only	Total dwelling unit garage cover in addition to building cover		Gc		0.0%	of BLA
	Total parking spaces		NPS	75.0		
Residential only	Total parking spaces per dwelling unit		u	0.8		
	Gross building area per parking space total		a	2,280.0		
	Number of loading spaces	0	l			
	Total loading area in SF	0	LDA			
	Gross area per loading space in SF		b			
	Loading area percentage		L		0.0%	of BLA
	Driveway areas in SF	4,430	R		15.5%	of BLA
	Misc. pavement and social center pvmt. in SF	250	M		0.9%	of BLA
	TOTAL DEVELOPMENT COVER		D	22,680	**79.1%**	of BLA
Project Open Space						
	PROJECT OPEN SPACE		S	5,982	**20.9%**	of BLA
	Water area within project open space in SF	0.0%	wat			of S
Summary						
	Development balance index		BNX		0.791	
	Development intensity index		INX		19.791	
	Floor area ratio per buildable acre		BFAR		5.966	
	Capacity index (gross building sq. ft. per buildable acre)		SFAC		259,878	
	Density per buildable acre if applicable		dBA		145.90	
	Density per net acre if applicable		dNA		145.90	
	Density per gross acre if applicable		dGA		145.90	

** Include internal parking lot landscaping, circulation aisles, and private roadway parking space areas. Express private roads separately as driveway areas.

Exhibit 14.11 Summit Chase Forecast Record

Forecast Model RS2L

Development capacity forecast for ***APARTMENTS*** *based on the use of an* ***UNDERGROUND PARKING STRUCTURE.***
Given: *Gross land area.* ***To Find:*** *Dwelling unit capacity of the land area given based on the design specification values entered below.*
Design Premise: *Underground parking footprint may be larger, smaller, or equal to the building footprint above. All similar floors considered equal in area.*

DESIGN SPECIFICATION *Enter values in boxed areas where text is bold and blue. Express all fractions as decimals.*

Given:	**Gross land area**	GLA=	**7.814**	acres	340,378	SF
Land Variables:	Public/ private right-of-way & paved easements	W=	**0.000**	fraction of GLA	0	SF
	Net land area	NLA=	7.814	acres	340,378	SF
	Future expansion and/or unbuildalbe areas	U=	**0.012**	fraction of GLA	4,085	SF
	Gross land area reduction	X=	0.012	fraction of GLA	4,085	SF
	Buildable land area remaining	BLA=	**7.720**	acres	336,293	SF
Parking Variables:	Parking structure spaces planned, provided or required per dwelling unit	u =	**1.06**	**0.44**	surface parking	
	Est. net parking structure area per parking space	s =	**338**	**354**	surface parking	
	No. of loading spaces	l =	**0**			
	Gross area per loading space	b =	**0**		0	SF
Site Variables at Grade:	**Project open space as fraction of BLA**	S=	**0.838**		281,713	SF
	Private driveways as fraction of BLA	R=	**0.028**		9,416	SF
	Misc. pavement as fraction of BLA	M=	**0.004**		1,345	SF
	Loading area as fraction of BLA	L=	0.000		0	SF
	Total site support areas at grade as a fraction of BLA	Su=	0.870		292,474	SF
Core:	**Core development area at grade as fraction of BLA**	C=	0.130	C+Su must = 1	43,819	SF
Below Grade:	Gross underground parking area (UNG) as fraction of BLA	G=	**0.100**		33,596	SF
	Parking support within parking structure as fraction of UNG	Pu=	**0.054**			
	Net parking structure for pkg. & circulation as fraction of UNG	Pe=	0.946			
Building Variables:	Building efficiency as fraction of GBA	Be=	**0.850**	must have a value >0 entered		
	Building support as fraction of GBA	Bu=	0.150	Be+Bu must = 1		

Dwelling Unit Mix Table

DU dwelling unit type	GDA gross du area	CDA=GDA/Be comprehensive du area	MIX du mix	PDA = (CDA)MIX Pro-rated du area
EFF	**500**		**10%**	
1 BR	**750**		**20%**	
2 BR	**1,200**	data override	**40%**	data override
3 BR	**1,500**		**20%**	
4 BR	**1,800**		**10%**	

Aggregate avg. dwelling unit area (AGG) = **2,171**

PLANNING FORECAST Enter zero in the adjacent box unless you wish to override the AGG value calculated above **2,171** **override**

no. of bldg. floors **FLR**	no. of pkg. levels **p**	CORE	number of **NDU** dwelling units	underground NPS pkg. spaces	total TPS pkg. spaces	gross pkg. area UNG underground	surface PLA parking area	gross **GBA** building area	building BCA footprint area	density dBA per bldable acre
1.00	0.21		18.8	20.0	28.2	7,133	2,933	40,886	40,886	2.4
2.00	0.40		35.3	37.4	53.0	13,370	5,499	76,641	38,320	4.6
3.00	0.56		49.8	52.8	74.7	18,871	7,761	108,174	36,058	6.5
4.00	0.91	BCA+PLA	62.7	66.5	94.1	30,577	9,771	136,191	34,048	8.1
5.00	1.14	43,819	74.3	78.7	111.4	38,221	11,569	161,250	32,250	9.6
6.00	1.37		84.7	89.7	127.0	45,865	13,187	183,795	30,633	11.0
7.00	1.59	gross pkg struc area / level	94.1	99.7	141.1	53,510	14,650	204,187	29,170	12.2
8.00	1.82	GPL	102.6	108.7	153.9	61,154	15,979	222,719	27,840	13.3
9.00	2.05	33,596	110.4	117.0	165.6	68,798	17,193	239,636	26,626	14.3
10.00	2.28		117.5	124.6	176.3	76,442	18,305	255,139	25,514	15.2
15.00	3.41	net pkg struc area / level	145.8	154.6	218.7	114,664	22,713	316,583	21,106	18.9
20.00	4.55	NPL	165.8	175.7	248.7	152,885	25,823	359,922	17,996	21.5
24.00	**5.46**	**31,782**	**178.0**	**188.6**	**267.0**	**183,462**	**27,720**	**386,369**	**16,099**	**23.1**
25.00	5.69		180.6	191.5	270.9	191,106	28,134	392,131	15,685	23.4
30.00	6.83		192.1	203.6	288.1	229,327	29,919	417,010	13,900	24.9

WARNING: These are preliminary forecasts that must not be used to make final decisions.
1) These forecasts are not a substitute for the "due diligence" research that must be conducted to support the final definition of "unbuildable areas" above and the final decision to purchase land. This research includes, but is not limited to, verification of adequate subsurface soil, zoning, environmental clearance, access, title, utilities and water pressure, clearance from deed restriction, easement and right-of-way encumbrances, clearance from existing above and below ground facility conflicts, etc.
2) The most promising forecast(s) made on the basis of data entered in the design specification from "due diligence" research must be verified at the drawing board before funds are committed and land purchase decisions are made. Actual land shape ratios, dimensions and irregularities encountered may require adjustments to the general forecasts above.
3) The software licensee shall take responsibility for the design specification values entered and any advice given or decisions made that are based on the forecast produced.

Exhibit 14.12 Jefferson Apartment Forecast Record

Forecast Model RG1L

Development capacity forecast for ***APARTMENTS*** *based on the use of an adjacent* ***GRADE PARKING LOT*** *located on the same premises. When s and u equal zero in the design specification below, the forecast pertains to conditions when* ***NO PARKING*** *is required.*

Given: *Gross land area.* ***To Find:*** *Maximum dwelling unit capacity of the land area given based on the design specification values entered below.* ***Premise:*** *All building floors considered equal in area.*

DESIGN SPECIFICATION

Enter values in boxed areas where text is bold and blue. Express all fractions as decimals.

Given:	**Gross land area**	GLA=	**1.756**	acres	76,491	SF
Land Variables:	Public/ private right-of-way & paved easements	W=	**0.000**	fraction of GLA	0	SF
	Net land area	NLA=	1.756	acres	76,491	SF
	Unbuildable and/or future expansion areas	U=	**0.000**	fraction of GLA	0	SF
	Gross land area reduction	X=	0.000	fraction of GLA	0	SF
	Buildable land area remaining	**BLA=**	**1.756**	acres	76,491	SF
Parking Variables:	Est. gross pkg. lot area per pkg. space in SF	s =	**311**	ENTER ZERO IF NO PARKING REQUIRED		
	Parking lot spaces planned or required per dwelling unit	u=	**1.42**	ENTER ZERO IF NO PARKING REQUIRED		
	Garage parking spaces planned or required per dwelling unit	Gn=	**0**	ENTER ZERO IF NO PARKING REQUIRED		
	Gross building area per garage space	Ga=	**0**	ENTER ZERO IF NO PARKING REQUIRED		
	No. of loading spaces	l =	**0**			
	Gross area per loading space	b =	**0**	SF	0	SF
Site Variables:	**Project open space as fraction of BLA**	**S=**	**0.304**		23,215	SF
	Private driveways as fraction of BLA	R=	**0.008**		612	SF
	Misc. pavement as fraction of BLA	M=	**0.001**		76	SF
	Loading area as fraction of BLA	L=	0.000		0	SF
	Total site support Areas as a fraction of BLA	Su=	0.313		23,904	SF
Core:	**Core development area as fraction of BLA**	**C=**	0.688	C=Su must = 1	52,588	SF
Building Variables:	Building efficiency as percentage of GBA	Be=	**0.900**	must have a value entered		
	Building support as fraction of GBA	Bu=	0.100	Be + Bu must = 1		

Dwelling Unit Mix Table:

DU dwelling unit type	GDA gross du area	CDA=GDA/Be comprehensive du area	MIX du mix	PDA = (CDA)MIX Pro-rated du area
EFF	**500**		**10%**	
1 BR	**750**		**20%**	
2 BR	**1,200**	data override	**40%**	data override
3 BR	**1,500**		**20%**	
4 BR	**1,800**		**10%**	

Aggregate avg. dwelling unit area	(AGG) =	**867**
GBA sf per parking space	a=	611
Enter zero in the adjacent box unless you wish to override the AGG value calculated above		**867** **override**

PLANNING FORECAST

no. of floors FLR	CORE minimum lot area for BCG & PLA	density per net acre dNA	dwelling units **NDU**	pkg. lot spaces NPS	parking lot area **PLA**	garage spaces GPS	garage area **GAR**	gross bldg area GBA no garages	footprint **BCA**	density per dBA bldable acre
1.00	52,588	22.88	40	57.1	17,747	n/a	n/a	34,841	34,841	22.88
2.00		34.22	60	85.3	26,538	n/a	n/a	52,100	26,050	34.22
3.00		**40.99**	**72**	**102.2**	**31,786**	**n/a**	**n/a**	**62,404**	**20,801**	**40.99**
4.00		45.49	80	113.4	35,275	n/a	n/a	69,252	17,313	45.49
5.00		48.69	86	121.4	37,761	n/a	n/a	74,134	14,827	48.69
6.00		51.09	90	127.4	39,623	n/a	n/a	77,789	12,965	51.09
7.00	*NOTE: Be aware when BCA becomes too small to be feasible.*	52.96	93	132.1	41,069	n/a	n/a	80,629	11,518	52.96
8.00		54.45	96	135.8	42,226	n/a	n/a	82,898	10,362	54.45
9.00		55.67	98	138.8	43,171	n/a	n/a	84,754	9,417	55.67
10.00		56.68	100	141.3	43,958	n/a	n/a	86,299	8,630	56.68
11.00		57.54	101	143.5	44,624	n/a	n/a	87,606	7,964	57.54
12.00		58.28	102	145.3	45,194	n/a	n/a	88,726	7,394	58.28
13.00		58.92	103	146.9	45,688	n/a	n/a	89,696	6,900	58.92
14.00		59.47	104	148.3	46,120	n/a	n/a	90,545	6,467	59.47

WARNING: These are preliminary forecasts that must not be used to make final decisions.
1) These forecasts are not a substitute for the "due diligence" research that must be conducted to support the final definition of "unbuildable areas" above and the final decision to purchase land. This research includes, but is not limited to, verification of adequate subsurface soil, zoning, environmental clearance, access, title, utilities and water pressure, clearance from deed restriction, easement and right-of-way encumbrances, clearance from existing above and below ground facility conflicts, etc.
2) The most promising forecast(s) made on the basis of data entered in the design specification from "due diligence" research must be verified at the drawing board before funds are committed and land purchase decisions are made. Actual land shape ratios, dimensions and irregularities encountered may require adjustments to the general forecasts above.
3) The software licensee shall take responsibility for the design specification values entered and any advice given that is based on the forecast produced.

Exhibit 14.13 Lane Apartment Forecast Record

Forecast Model RG1L

Development capacity forecast for ***APARTMENTS*** *based on the use of an adjacent* ***GRADE PARKING LOT*** *located on the same premises. When s and u equal zero in the design specification below, the forecast pertains to conditions when* ***NO PARKING*** *is required.*

Given: *Gross land area.* ***To Find:*** *Maximum dwelling unit capacity of the land area given based on the design specification values entered below.* ***Premise:*** *All building floors considered equal in area.*

DESIGN SPECIFICATION

Enter values in boxed areas where text is bold and blue. Express all fractions as decimals.

			Value			
Given:	**Gross land area**	**GLA=**	**0.525**	acres	22,869	SF
Land Variables:	Public/ private right-of-way & paved easements	W=	**0.000**	fraction of GLA	0	SF
	Net land area	NLA=	0.525	acres	22,869	SF
	Unbuildable and/or future expansion areas	U=	**0.000**	fraction of GLA	0	SF
	Gross land area reduction	X=	0.000	fraction of GLA	0	SF
	Buildable land area remaining	**BLA=**	**0.525**	acres	22,869	SF
Parking Variables:	Est. gross pkg. lot area per pkg. space in SF	s =	**375**	ENTER ZERO IF NO PARKING REQUIRED		
	Parking lot spaces planned or required per dwelling unit	u=	**0.67**	ENTER ZERO IF NO PARKING REQUIRED		
	Garage parking spaces planned or required per dwelling unit	Gn=	**0**	ENTER ZERO IF NO PARKING REQUIRED		
	Gross building area per garage space	Ga=	**0**	ENTER ZERO IF NO PARKING REQUIRED		
	No. of loading spaces	l =	**0**			
	Gross area per loading space	b =	**0**	SF	0	SF
Site Variables:	**Project open space as fraction of BLA**	**S=**	**0.228**		5,203	SF
	Private driveways as fraction of BLA	R=	**0.000**		0	SF
	Misc. pavement as fraction of BLA	M=	**0.031**		709	SF
	Loading area as fraction of BLA	L=	0.000		0	SF
	Total site support areas as a fraction of BLA	Su=	0.259		5,912	SF
Core:	**Core development area as fraction of BLA**	**C=**	0.742	C+Su must = 1	16,957	SF
Building Variables:	Building efficiency as percentage of GBA	Be=	**0.950**	must have a value entered		
	Building support as fraction of GBA	Bu=	0.050	Be + Bu must = 1		

Dwelling Unit Mix Table:

DU dwelling unit type	GDA gross du area	CDA=GDA/Be comprehensive du area	MIX du mix	PDA = (CDA)MIX Pro-rated du area
EFF	**500**		**10%**	
1 BR	**750**		**20%**	
2 BR	**1,200**	data override	**40%**	data override
3 BR	**1,500**		**20%**	
4 BR	**1,800**		**10%**	

Aggregate avg. dwelling unit area	(AGG) =	**663**
GBA sf per parking space	a=	994
Enter zero in the adjacent box unless you wish to override the AGG value calculated above		**663** **override**

PLANNING FORECAST

no. of floors FLR	CORE minimum lot area for BCG & PLA	density per net acre dNA	dwelling units **NDU**	pkg. lot spaces NPS	parking lot area **PLA**	garage spaces GPS	garage area **GAR**	gross bldg area GBA no garages	footprint **BCA**	density per dBA bldable acre
1.00	16,957	35.37	19	12.4	4,645	n/a	n/a	12,312	12,312	35.37
2.00		55.53	29	19.4	7,292	n/a	n/a	19,330	9,665	55.53
3.00		**68.56**	**36**	**24.0**	**9,003**	**n/a**	**n/a**	**23,864**	**7,955**	**68.56**
4.00		77.67	41	27.2	10,199	n/a	n/a	27,034	6,758	77.67
5.00		84.39	44	29.6	11,082	n/a	n/a	29,375	5,875	84.39
6.00	*NOTE: Be aware when BCA becomes too small to be feasible.*	89.57	47	31.4	11,761	n/a	n/a	31,176	5,196	89.57
7.00		93.67	49	32.8	12,300	n/a	n/a	32,603	4,658	93.67
8.00		97.00	51	34.0	12,737	n/a	n/a	33,762	4,220	97.00
9.00		99.75	52	34.9	13,099	n/a	n/a	34,722	3,858	99.75
10.00		102.08	54	35.7	13,404	n/a	n/a	35,530	3,553	102.08
11.00		104.06	55	36.4	13,665	n/a	n/a	36,220	3,293	104.06
12.00		105.77	56	37.0	13,889	n/a	n/a	36,816	3,068	105.77
13.00		107.26	56	37.6	14,085	n/a	n/a	37,336	2,872	107.26
14.00		108.58	57	38.0	14,258	n/a	n/a	37,793	2,699	108.58

WARNING: These are preliminary forecasts that must not be used to make final decisions.

1) These forecasts are not a substitute for the "due diligence" research that must be conducted to support the final definition of "unbuildable areas" above and the final decision to purchase land. This research includes, but is not limited to, verification of adequate subsurface soil, zoning, environmental clearance, access, title, utilities and water pressure, clearance from deed restriction, easement and right-of-way encumbrances, clearance from existing above and below ground facility conflicts, etc.

2) The most promising forecast(s) made on the basis of data entered in the design specification from "due diligence" research must be verified at the drawing board before funds are committed and land purchase decisions are made. Actual land shape ratios, dimensions and irregularities encountered may require adjustments to the general forecasts above.

3) The software licensee shall take responsibility for the design specification values entered and any advice given that is based on the forecast produced.

Exhibit 14.14 Waterford Tower Forecast Record

Forecast Model RS1L

Development capacity forecast for ***APARTMENTS*** *using an* ***ADJACENT PARKING STRUCTURE*** *on the same premise.*
Given: *Gross land area available.* ***To Find:*** *Maximum dwelling unit capacity of the buildable land area given based on the design specification values entered below.*
Design Premise: *Building footprint adjacent to parking garage footprint within the core development area. All similar floors considered equal in area.*

DESIGN SPECIFICATION

Enter values in boxed areas where text is bold and blue. Express all fractions as decimals.

Given:	**Gross land area**	GLA=	**0.658**	acres	28,662	SF
Land Variables:	Public/ private right-of-way & paved easements	W=	**0.000**	fraction of GLA	0	SF
	Net land area	NLA=	0.658	acres	28,662	SF
	Unbuildable and/or future expansion areas	U=	**0.000**	fraction of GLA	0	SF
	Gross land area reduction	X=	0.000	fraction of GLA	0	SF
	Buildable land area remaining	BLA=	**0.658**	acres	28,662	SF
Parking Variables:	Estimated net pkg. structure area per parking space	s =	**324**	SF		
	Parking spaces required per dwelling unit	u =	**0.78**			
	No. of parking levels contemplated	p=	**3.00**			
	Parking support as fraction of gross pkg. structure area (GPA)	Pu=	**0.098**			
	Net area for parking & circulation as fraction of GPA	Pe=	0.902	Pe+Pu must = 1		
	No. of loading spaces	l=	**0**			
	Gross area per loading space	b =	**0**	SF		
Site Variables:	**Project open space as fraction of BLA**	S=	**0.209**		5,990	SF
	Private driveways as fraction of BLA	R=	**0.155**		4,443	SF
	Misc. pavement as fraction of BLA	M=	**0.009**		258	SF
	Loading area as fraction of BLA	L=	0.000		0	SF
	Total site support areas as a fraction of BLA	Su=	0.373		10,691	SF
Core:	**Core development area as fraction of BLA**	C=	0.627	C+Su must = 1	17,971	SF
Building Variables:	Building efficiency as fraction of GBA	Be=	**0.750**	must have a value >0 entered		
	Building support as fraction of GBA	Bu=	0.250	Be+Bu must = 1		

NOTE: p=1 is a grade parking lot based on design premise. Increase the number of parking levels to increase the capacity forecast below. Other variables in blue, box and bold may also be changed.

Dwelling Unit Mix Table

DU dwelling unit type	GDA gross du area	CDA=GDA/Be comprehensive du area	MIX du mix	PDA = (CDA)MIX Prorated du area
EFF	**500**		**10%**	
1 BR	**750**		**20%**	
2 BR	**1,200**	data override	**40%**	data override
3 BR	**1,500**		**20%**	
4 BR	**1,800**		**10%**	

Aggregate avg. dwelling unit area (AGG) = **1,781**

Building area planned, provided or required per parking space a = 2,283

PLANNING FORECAST

Enter zero in the adjacent box unless you wish to override the AGG value calculated above **1,781** **override**

no. of FLR bldg floors	min land area CORE for BCG & PLA	dwelling units NDU	pkg spaces NPS	net pkg area NPA	gross pkg GPA struc area	pkg. struct cover GPL based on p above	gross bldg area GBA	footprint BCA	density per dBA bldable acre	density per dNA net acre
1.00	17,971	9.6	7.5	2,423	2,686	895	17,076	17,076	14.57	14.57
2.00		18.3	14.2	4,616	5,118	1,706	32,531	16,266	27.76	27.76
3.00		26.2	20.4	6,610	7,329	2,443	46,586	15,529	39.75	39.75
4.00		33.4	26.0	8,432	9,348	3,116	59,422	14,855	50.71	50.71
5.00		40.0	31.2	10,102	11,199	3,733	71,191	14,238	60.75	60.75
6.00		46.1	35.9	11,639	12,903	4,301	82,022	13,670	69.99	69.99
7.00		51.7	40.3	13,058	14,476	4,825	92,022	13,146	78.52	78.52
8.00		56.9	44.4	14,372	15,933	5,311	101,282	12,660	86.43	86.43
9.00		61.7	48.1	15,592	17,286	5,762	109,884	12,209	93.77	93.77
10.00		66.2	51.6	16,729	18,546	6,182	117,893	11,789	100.60	100.60
11.00		70.4	54.9	17,790	19,722	6,574	125,369	11,397	106.98	106.98
12.00		74.3	58.0	18,782	20,823	6,941	132,365	11,030	112.95	112.95
15.00		84.7	66.1	21,411	23,737	7,912	150,887	10,059	128.75	128.75
19.00		**96.0**	**74.9**	**24,271**	**26,907**	**8,969**	**171,042**	**9,002**	**145.95**	**145.95**
20.00		98.5	76.8	24,894	27,599	9,200	175,436	8,772	149.70	149.70

WARNING: These are preliminary forecasts that must not be used to make final decisions.
1) These forecasts are not a substitute for the "due diligence" research that must be conducted to support the final definition of "unbuildable areas" above and the final decision to purchase land. This research includes, but is not limited to, verification of adequate subsurface soil, zoning, environmental clearance, access, title, utilities and water pressure, clearance from deed restriction, easement and right-of-way encumbrances, clearance from existing above and below ground facility conflicts, etc.
2) The most promising forecast(s) made on the basis of data entered in the design specification from "due diligence" research must be verified at the drawing board before funds are committed and land purchase decisions are made. Actual land shape ratios, dimensions and irregularities encountered may require adjustments to the general forecasts above.
3) The software licensee shall take responsibility for the design specification values entered and any advice given or decisions made that are based on the forecast produced.

capacity limits, and incentives can begin in earnest with the help of the context record system.

In residential terms, development capacity forecasts can predict the ability of land to produce an increased housing supply to support an increasing population, based on the building and parking system chosen from the development forecast collection. These predictions will not always represent desirable lifestyles, intensities, and context contributions, and must be designed, organized, related, and limited to produce sustainable and desirable urban lifestyle relationships. The development forecast collection and context record system will simply help to place this fundamental issue of shelter and survival in perspective. It is up to you to use it to provide the development capacity answers that can produce a sustainable and desirable shelter system over time. It can be a worthwhile lifetime endeavor if you choose it, and it will produce a lasting contribution.

Suburban apartment houses produce the least increase in housing capacity when densities are limited by surface parking lots, open space, and building heights that rarely exceed 3 stories. Transitional apartment houses are found closer to the urban core. They often reduce open space and may increase height as they step in from the suburbs, but generally retain surface parking lots and may use street parking and alleys to increase housing capacity and parking supply. Urban apartment houses often provide the least amount of open space, the greatest housing capacity, and the most building height; struggle to provide adequate parking; and represent the most diverse quality levels. As a group, however, apartment houses represent the only residential shelter class that has a realistic chance of consistently exceeding a housing capacity of 15 dwelling units per buildable acre when adequate on-site parking and open space is provided. This may become an issue as we attempt to shelter an increasing population.

Apartment Neighborhoods

Unfortunately, apartment houses have had a difficult time attracting market share because they have rarely been considered and built as neighborhoods.[13] They have been arranged and referred to as *buffers, barriers,* and *transitions.* Their residents and children have been considered transients without equity participation in the American Dream, and the neighborhood school system has covered too small a geographic area to accommodate this diversity without discrimination. Apartment house associations with equity participation, however, have attracted a small and exclusive market segment. This segment has freely made this choice and can afford private school systems, but the option is too select to be meaningful until school systems, costs, contexts, and investment terms can be extended to bring the formula closer to a broader market segment.

Apartment houses depend on mixed-use neighborhoods to supplement their isolated lifestyles and are at their weakest when they stand as solitary buildings and complexes in transitional settings that produce social disconnection. Zoning officials are only beginning to understand where, when, and how mixed use can produce complementary lifestyle alternatives, however. It is currently very common for zoning to prevent mixed-use environments and encourage the use of apartment houses as buffers, barriers, and transitions to suburb house neighborhoods. This has often limited desirable mixed-use neighborhoods to intense urban centers and has limited the appeal of this shelter class to the wealthy, to equity builders, and to those who have no other choice.

Apartment houses have great ability to improve housing capacity on smaller land areas, but require a different neighborhood context, movement system, social support system, financing system, and zoning format to establish broader market appeal.[14] This is possible, but will happen only when apartment house neighborhoods and communities represent legitimate lifestyle alternatives and choices rather than isolated building complexes and places of temporary shelter located to massively buffer the "real" housing supply.

Notes

1. Stairways do not constitute direct and immediate access, and a dwelling unit may not be occupied by a family, but this is its potential.
2. However, condominium ownership, or equity accumulation and association governance, may be one key to broader market acceptance.
3. In this case, the preferences are those of the author and represent an expedient substitute for market research.
4. The dark gray band indicates preference rating 2 and the lighter band indicates preference rating 3, with 1 being the highest preference.
5. Average dwelling unit area ADU is also referred to as aggregated average dwelling unit area AGG.
6. If more dwelling units are provided on the same land area using the same design specification, the average dwelling unit area could be expected to decline. Waterford has increased building height, reduced open space, and reduced parking to produce larger average dwelling unit (ADU) areas on its central business district site. Summit Chase has increased height, increased open space, and provided underground parking in its design specification to accommodate larger average dwelling areas within an open space estate setting.
7. It was based on an overall impression of the residential context and exterior quality constructed.
8. This is merely an observation and remains to be proven.
9. Exhibit 14.1 values are derived from the Britton Woods context record included as Exhibit 14.3.
10. This must be confirmed as part of forecast evaluation at the drawing board.
11. Waterford Tower provided 20.9%.
12. The phrase "library of context records" is wishful thinking at this point, and the "library" in this book consists of eight records. Increas-

ing this collection will contribute to the body of knowledge that surrounds this fundamental issue of shelter.

13. In the opinion of the author.
14. Apartment houses constitute the third of five shelter classes suggested in this book, with nonresidential (or commercial) and hybrid being the fourth and fifth. They also have the greatest potential to shelter a growing population within symbiotically sustainable limits.

Apartment House Density

In generic terms, *density* is a mathematical expression used to define the amount of building area present, planned, or permitted per acre of land available.[1] When this building area is used for housing, density is expressed in terms of dwelling units per acre. When it is used for nonresidential shelter, density is frequently expressed as the number of gross building square feet constructed per acre, or it is expressed as a floor area ratio (the gross building area present, planned, or permitted divided by the land area involved).

The land areas mentioned in density expressions are referred to as *net areas* and *gross areas*. Neither is particularly helpful in forecasting actual development capacity, however. *Gross area* does not reflect the actual buildable area present and capable of accommodating development capacity, and *net area* often includes unbuildable land such as ravines and wetlands. The term *buildable area* is used in this text to define the land that is actually available for improvement. (This is a departure from the traditional use of the term *buildable area* and is discussed in previous chapters.) This permits development forecast equations to focus on the subject. The forecast models containing these equations express the density achieved in relation to the buildable acres available (for instance, 10 dwelling units per buildable acre), but also express density in net acre terms to maintain a traditional frame of reference. All of these density indexes are useful tools, but are too imprecise to be used as recipes capable of producing desirable projects. They are more helpful in recording results than producing them.

Apartment house development objectives and zoning regulations are generally expressed in terms of the density planned, permitted, or provided, and the objective of this chapter is to explain how these density objectives can be evaluated with the development forecast collection to predict results that can actually be achieved. This evaluation shows that some density objectives are not realistic under many development conditions, even though they may be permitted by zoning ordinance legislation. When this occurs, a zoning ordinance regulation can actually encourage unrealistic expectations and overdevelopment as builders attempt to reach the levels permitted and the profit projected on this basis.

Elizabeth's Farm

The case study for this chapter is called Elizabeth's Farm. It comprises 10 gross acres adjacent to a suburban shopping center, and the owner wishes to convert this land to apartment house development under the traditional transition concept of land use arrangement. The land currently carries an agricultural zoning designation, and the owner will have to request rezoning approval for any development proposal. Three residential apartment zones exist in the community. The first permits 15 dwelling units per net acre and the second permits 35. The third permits 55, and the owner hopes to reach each density level using a surface parking lot.[2]

Surface Parking Options

Forecast model RG1D is used to evaluate the options available to Elizabeth's Farm under these density limits since it applies to surface parking lot systems. This model is used to evaluate a density limit of 15 in Exhibit 15.1. The design specification panel in this exhibit requests that the user enter the net density[3] objective and the gross land area available. It also requests entry estimates for the gross land area that will be devoted to roadways and for any unbuildable land area that may exist. These entries have produced a buildable land area forecast of 8.5 acres. Specification entries also stipulate that 1.5 surface parking spaces and 0.5 garage spaces will be provided per dwelling unit, that 400 gross sq ft of buildable land area per parking space is planned, and that 240 gross sq ft per garage parking space will be provided. No loading space values are entered. 10% has been entered for the buildable area that will be covered by private driveways, and 5% has been entered for miscellaneous pavement cover. These specification values are used by forecast model RG1D to predict the land area needed for the traditional parking and movement systems of an apartment house site plan. The land consumed by building cover and the land remaining for project open space, however, also affect the ability to achieve a given density objective.

Building cover is equal to the gross building area divided by the number of floors involved in the forecast model format. The forecast equations do not attempt to anticipate the architectural form that can occur within this gross building area envelope, or the final architectural area that will be a product of this form design. In apartment house design, the gross building area used to calculate building cover is a function of the dwelling unit areas and mix planned, as well as the building height involved. This net area is then increased by a floor plan efficiency factor to predict the gross building area involved. Efficiency represents the net

Exhibit 15.1 Elizabeth's Farm

Forecast Model RG1D: Net density objective of 15 and aggregate average dwelling unit area of 1450 sq ft.

*Development capacity forecast for **APARTMENTS** based on the use of a **GRADE PARKING LOT** located on the same premises. When s and u equal zero in the design specification below, the forecast pertains to conditions when **NO PARKING** is required.*

***Given:** Net density objective and gross land area. **To Find:** Minimum number of building floors required to achieve the net density objective given based on the design specification values entered below. **Premise:** All building floors considered equal in area.*

DESIGN SPECIFICATION

Enter values in boxed areas where text is bold and blue. Express all fractions as decimals.

Given:	**Net density objective:**	d=	**15.000**	dwelling units per net acre	653,400	SF
	Gross land area	GLA=	**10.000**	acres	435,600	SF
Land Variables:	Public/ private right-of-way & paved easements	W=	**0.150**	fraction of GLA	65,340	SF
	Net land area	NLA=	8.500	acres	370,260	SF
	Unbuildable and/or future expansion areas	U=	**0.000**	fraction of GLA	0	SF
	Gross land area reduction	X=	0.150	fraction of GLA	65,340	SF
	Buildable land area remaining	BLA=	**8.500**	acres	370,260	SF
Parking Variables:	Estimated gross pkg. lot area per pkg. space in SF	s =	**400**	ENTER ZERO IF NO PARKING REQUIRED		
	Parking spaces required per dwelling unit	u=	**1.50**	ENTER ZERO IF NO PARKING REQUIRED		
	Garage parking spaces planned or required per dwelling unit	Gn=	**0.50**	ENTER ZERO IF NO PARKING REQUIRED		
	Gross building area per garage space	Ga=	**240**	ENTER ZERO IF NO PARKING REQUIRED		
	No. of loading spaces	l=	**0**			
	Gross area per loading space	b=	**0**	SF		
Site Variables:	**Project open space as fraction of BLA**	S=	varies below			
	Private driveways as fraction of BLA	R=	**0.100**		37,026	SF
	Misc. pavement as fraction of BLA	M=	**0.050**		18,513	SF
	Loading area as fraction of BLA	L=	0.000		0	SF
	Total site support areas as a fraction of BLA	Su=	0.150		55,539	SF
Core:	**Core development area at grade as fraction of BLA**	C=	varies below			
Building Variables:	Building efficiency as fraction of GBA	Be=	**0.800**	must have a value >0 entered		
	Building support as fraction of GBA	Bu=	0.200	Be + Bu must = 1		

Dwelling Unit Mix Table

DU dwelling unit type	GDA gross du area	CDA=GDA/Be comprehensive du area	MIX du mix	PDA = (CDA)MIX Pro-rated du area
EFF	**500**	625	**10%**	63
1 BR	**750**	938	**20%**	188
2 BR	**1,200**	1,500	**40%**	600
3 BR	**1,500**	1,875	**20%**	375
4 BR	**1,800**	2,250	**10%**	225
			Aggregate avg. dwelling unit area (AGG) =	**1,450**
			GBA sf per parking space a=	725

PLANNING FORECAST

Enter zero in the adjacent box unless you wish to override the AGG value calculated above: **0**

Note: Blank values in FLR column indicate density not feasible given project open space objective in (S) column

open space **S**	min area for BCG & PLA CORE (Varies with open space provided)	no. of dwelling units **NDU**	floors needed **FLR**	gross bldg area GBA	footprint BCA	DU's per floor DUF	bldg SF per SFAC bldable acre	flr area ratio FAR function of BLA	open space S provided in acres
0.100	277,695	**127.5**	**1.0**	184,875	185,895	128.2	21,750	0.499	0.850
0.150	259,182		**1.1**		167,382	115.4			1.275
0.200	240,669	density per net acre (dNA)	**1.2**		148,869	102.7			1.700
0.250	222,156	15.0	**1.4**		130,356	89.9			2.125
0.300	203,643	density per bldable acre (dBA)	**1.7**		111,843	77.1			2.550
0.350	185,130	15.0	**2.0**		93,330	64.4			2.975
0.400	166,617	parking lot area (PLA)	**2.5**		74,817	51.6			3.400
0.450	148,104	76,500	**3.3**		56,304	38.8			3.825
0.500	129,591	pkg. lot spaces (NPS)	**4.9**		37,791	26.1			4.250
0.550	111,078	191.3	**9.6**		19,278	13.3			4.675
0.600	92,565	garage pkg. area (GPA)	**241.7**		765	0.5			5.100
0.650	74,052	15,300.0							
0.700	55,539	garage pkg. spaces (GPS)							
0.800	18,513	63.8							
0.900									

Note: Be aware when BCG becomes too small to be feasible

WARNING: These are preliminary forecasts that must not be used to make final decisions.
1) These forecasts are not a substitute for the "due diligence" research that must be conducted to support the final definition of "unbuildable areas" above and the final decision to purchase land. This research includes, but is not limited to, verification of adequate subsurface soil, zoning, environmental clearance, access, title, utilities and water pressure, clearance from deed restriction, easement and right-of-way encumbrances, clearance from existing above and below ground facility conflicts, etc.
2) The most promising forecast(s) made on the basis of data entered in the design specification from "due diligence" research must be verified at the drawing board before funds are committed and land purchase decisions are made. Actual land shape ratios, dimensions and irregularities encountered may require adjustments to the general forecasts above.
3) The software licensee shall take responsibility for the design specification values entered and any advice given that is based on the forecast produced.

usable area divided by the gross building area, which includes lobbies, hallways, elevators, stairs, mechanical rooms, etc. In the case of Exhibit 15.1, a building efficiency objective of 80% has been entered in the design specification. This is a major factor in calculating building cover, since the gross building area forecast must account not only for the dwelling unit areas and mix planned, but for the floor plan efficiency involved. Less efficiency means a larger gross building area, a larger building footprint, and less land for parking cover, miscellaneous pavement, and open space. The dwelling unit marketing strategy is included in the dwelling unit mix panel of Exhibit 15.1. Any strategy can be entered in this panel, and several examples are covered to illustrate the impact these decisions have on market attraction and the environment created. In Exhibit 15.1, 40% of the units are two-bedroom models and 30% are either above or below this midpoint. The dwelling unit areas are entered in the GDA column and are increased in the CDA column to reflect the building efficiency values entered. The aggregate average dwelling unit area AGG resulting from these mix and efficiency assumptions is calculated in the lower right-hand corner of this panel, and is shown to be 1450 sq ft in this example. The value *a* noted below the AGG value expresses the residential parking provision in nonresidential terms for cross-boundary comparison. It shows that 725 sq ft of gross residential building area has been constructed for each parking space provided. This is significantly higher than nonresidential ratios that generally range from 50 to 400 sq ft of gross building area constructed per parking space provided. It simply documents what common sense would assume—that parking requirements for residential land uses are significantly lower than those for nonresidential land uses.[4]

When the dwelling unit mix table is complete, the planning forecast panel can be used to evaluate the options available. It lists the open space context choices in the left-hand column and forecasts the building height needed to produce each context in the FLR column. For instance, by reading along the 50% open space *S* row, the panel predicts that this yard context (45 to 55% in Chap. 14) can be produced when a 4.9-story apartment house is constructed based on the net density objective of 15 and the design specification values entered.[5] The forecast panel also shows that a 2-story apartment house could be constructed to achieve the same density objective if the open space context *S* were reduced to 35%.[6]

Please note that the number of dwelling units NDU, parking lot spaces NPS, and gross building area GBA values forecast in the planning forecast panel of Exhibit 15.1 do not change with each open space increment.[7] A number of other values, including the SFAC value for building square feet predicted per buildable acre, are constant as well. This is because the net density objective is given, and many forecast values must remain constant in order to achieve this objective after the design specification is complete. However, the planning forecast panel displays a number of building height options that can be used to achieve different open space context results when the net density objective is given. A new panel of data can also be produced for evaluation by changing one or more design specification values entered in the panel above. Among the values that can be changed is the dwelling unit mix planned. Others include, but are not limited to, eliminating the garage provision, reducing surface parking from 1.5 spaces per dwelling unit, reducing the average parking area per space from 400 sq ft, and increasing the building efficiency planned. Value changes in any of these topic areas can produce a different planning forecast panel of data, but planners generally overlook dwelling unit mix changes because these options have been difficult to calculate.

Exhibit 15.2 uses the data-override feature described in Chap. 14 to assess the impact of changing the AGG of the

Exhibit 15.2 Elizabeth's Farm

Forecast Model RG1D: Net density objective of 15 and aggregate average dwelling unit area of 900 sq ft.

*Development capacity forecast for **APARTMENTS** based on the use of a **GRADE PARKING LOT** located on the same premises. When s and u equal zero in the design specification below, the forecast pertains to conditions when **NO PARKING** is required.*

***Given:** Net density objective and gross land area. **To Find:** Minimum number of building floors required to achieve the net density objective given based on the design specification values entered below. **Premise:** All building floors considered equal in area.*

DESIGN SPECIFICATION *Enter values in boxed areas where text is bold and blue. Express all fractions as decimals.*

Given:	**Net density objective:**	d=	15.000	dwelling units per net acre	653,400	SF
	Gross land area	GLA=	10.000	acres	435,600	SF
Land Variables:	Public/ private right-of-way & paved easements	W=	0.150	fraction of GLA	65,340	SF
	Net land area	NLA=	8.500	acres	370,260	SF
	Unbuildable and/or future expansion areas	U=	0.000	fraction of GLA	0	SF
	Gross land area reduction	X=	0.150	fraction of GLA	65,340	SF
	Buildable land area remaining	BLA=	8.500	acres	370,260	SF
Parking Variables:	Estimated gross pkg. lot area per pkg. space in SF	s =	400	ENTER ZERO IF NO PARKING REQUIRED		
	Parking spaces required per dwelling unit	u=	1.50	ENTER ZERO IF NO PARKING REQUIRED		
	Garage parking spaces planned or required per dwelling unit	Gn=	0.50	ENTER ZERO IF NO PARKING REQUIRED		
	Gross building area per garage space	Ga=	240	ENTER ZERO IF NO PARKING REQUIRED		
	No. of loading spaces	l=	0			
	Gross area per loading space	b=	0	SF		
Site Variables:	**Project open space as fraction of BLA**	S=	varies below			
	Private driveways as fraction of BLA	R=	0.100		37,026	SF
	Misc. pavement as fraction of BLA	M=	0.050		18,513	SF
	Loading area as fraction of BLA	L=	0.000		0	SF
	Total site support areas as a fraction of BLA	Su=	0.150		55,539	SF
Core:	**Core development area at grade as fraction of BLA**	C=	varies below			
Building Variables:	Building efficiency as fraction of GBA	Be=	0.800	must have a value >0 entered		
	Building support as fraction of GBA	Bu=	0.200	Be + Bu must = 1		

Dwelling Unit Mix Table

DU dwelling unit type	GDA gross du area	CDA=GDA/Be comprehensive du area	MIX du mix	PDA = (CDA)MIX Prorated du area
EFF	500		10%	
1 BR	750		20%	
2 BR	1,200	data override	40%	data override
3 BR	1,500		20%	
4 BR	1,800		10%	

Aggregate avg. dwelling unit area (AGG) = **900**

GBA sf per parking space a= 450

PLANNING FORECAST Enter zero in the adjacent box unless you wish to override the AGG value calculated above **900** **override**

Note: Blank values in FLR column indicate density not feasible given project open space objective in S column

open space **S**	min area for BCG & PLA CORE Varies with open space provided	no. of dwelling units **NDU**	floors needed **FLR**	gross bldg area GBA	footprint BCA	DU's per floor DUF	bldg SF per SFAC bldable acre	flr area ratio FAR function of BLA	open space S provided in acres
0.100	277,695	**127.5**	**0.6**	114,750	185,895	206.6	13,500	0.310	0.850
0.150	259,182		**0.7**		167,382	186.0			1.275
0.200	240,669	density per net acre (dNA)	**0.8**		148,869	165.4			1.700
0.250	222,156	15.0	**0.9**		130,356	144.8			2.125
0.300	203,643	density per bldable acre (dBA)	**1.0**		111,843	124.3			2.550
0.350	185,130	15.0	**1.2**		93,330	103.7			2.975
0.400	166,617	parking lot area (PLA)	**1.5**		74,817	83.1			3.400
0.450	148,104	76,500	**2.0**		56,304	62.6			3.825
0.500	129,591	pkg. lot spaces (NPS)	**3.0**		37,791	42.0			4.250
0.550	111,078	191.3	**6.0**		19,278	21.4			4.675
0.600	92,565	garage pkg. area (GPA)	**150.0**		765	0.9			5.100
0.650	74,052	15,300.0							
0.700	55,539	garage pkg. spaces (GPS)							
0.800	18,513	63.8							
0.900									

Note: Be aware when BCG becomes too small to be feasible

WARNING: These are preliminary forecasts that must not be used to make final decisions.
1) These forecasts are not a substitute for the "due diligence" research that must be conducted to support the final definition of "unbuildable areas" above and the final decision to purchase land. This research includes, but is not limited to, verification of adequate subsurface soil, zoning, environmental clearance, access, title, utilities and water pressure, clearance from deed restriction, easement and right-of-way encumbrances, clearance from existing above and below ground facility conflicts, etc.
2) The most promising forecast(s) made on the basis of data entered in the design specification from "due diligence" research must be verified at the drawing board before funds are committed and land purchase decisions are made. Actual land shape ratios, dimensions and irregularities encountered may require adjustments to the general forecasts above.
3) The software licensee shall take responsibility for the design specification values entered and any advice given that is based on the forecast produced.

project from 1450 to 900 sq ft. The override feature is a quick way to assess the implications of this change. When using this feature, keep in mind that the dwelling unit mix data will eventually have to be adjusted to ensure that the final areas and mix planned equal the override value introduced. Exhibit 15.2 shows that a 50% open space context can be produced with 3 stories instead of 5 when the aggregate average dwelling unit area is reduced from 1450 to 900 sq ft. (All other design specification values in Exhibit 15.2 equal those in Exhibit 15.1.) This achievement has a cost, however, and the implications of a 900-sq-ft average dwelling unit area should be evaluated. Exhibit 15.3 shows the adjustments that must be made to the dwelling unit mix table to produce this 900-sq-ft aggregate average dwelling unit value.[8] The percentage of efficiency apartments has been decreased to 5%, and the gross floor plan area has been reduced to 400 sq ft. The percentage of one-bedroom units has been increased to 60%, and the gross average floor plan area has been reduced to 615 sq ft. The percentage of two-bedroom units has decreased to 35%, and the gross average floor plan area has been reduced to 950 sq ft. All three- and four-bedroom dwelling units have been eliminated. Exhibit 15.3 shows that these adjustments have produced a target AGG value of 902 sq ft, which is within design tolerance limits of the 900-sq-ft objective and capable of inclusion within a 3-story building.

Exhibit 15.4 shows that the 3-story, 50% open space, 900-sq-ft dwelling unit area target could also be reached if all efficiency and four-bedroom apartments were eliminated and the remaining apartment unit areas were adjusted. The dwelling unit mix table in this exhibit contains 40% one-bedroom units that are 550 sq ft in area, 50% two-bedroom units that are 800 sq ft in area, and 10% three-bedroom units that are 1000 sq ft in area. These reductions may not be acceptable to the target market, however. If this is the case, other design specification adjustments could be reviewed

Exhibit 15.3 Elizabeth's Farm with Dwelling Unit Mix Option 1

Forecast Model RG1D: Net density objective of 15 and aggregate average dwelling unit area of 900 sq ft.

*Development capacity forecast for **APARTMENTS** based on the use of a **GRADE PARKING LOT** located on the same premises. When s and u equal zero in the design specification below, the forecast pertains to conditions when **NO PARKING** is required.*

***Given:** Net density objective and gross land area. **To Find:** Minimum number of building floors required to achieve the net density objective given based on the design specification values entered below. **Premise:** All building floors considered equal in area.*

DESIGN SPECIFICATION *Enter values in boxed areas where text is bold and blue. Express all fractions as decimals.*

Given:	**Net density objective:**	d=	**15.000**	dwelling units per net acre	653,400	SF
	Gross land area	GLA=	**10.000**	acres	435,600	SF
Land Variables:	Public/ private right-of-way & paved easements	W=	**0.150**	fraction of GLA	65,340	SF
	Net land area	NLA=	8.500	acres	370,260	SF
	Unbuildable and/or future expansion areas	U=	**0.000**	fraction of GLA	0	SF
	Gross land area reduction	X=	0.150	fraction of GLA	65,340	SF
	Buildable land area remaining	BLA=	**8.500**	acres	370,260	SF
Parking Variables:	Estimated gross pkg. lot area per pkg. space in SF	s =	**400**	ENTER ZERO IF NO PARKING REQUIRED		
	Parking spaces required per dwelling unit	u=	**1.50**	ENTER ZERO IF NO PARKING REQUIRED		
	Garage parking spaces planned or required per dwelling unit	Gn=	**0.50**	ENTER ZERO IF NO PARKING REQUIRED		
	Gross building area per garage space	Ga=	**240**	ENTER ZERO IF NO PARKING REQUIRED		
	No. of loading spaces	l=	**0**			
	Gross area per loading space	b=	**0**	SF		
Site Variables:	**Project open space as fraction of BLA**	S=	varies below			
	Private driveways as fraction of BLA	R=	**0.100**		37,026	SF
	Misc. pavement as fraction of BLA	M=	**0.050**		18,513	SF
	Loading area as fraction of BLA	L=	0.000		0	SF
	Total site support areas as a fraction of BLA	Su=	0.150		55,539	SF
Core:	**Core development area at grade as fraction of BLA**	C=	varies below			
Building Variables:	Building efficiency as fraction of GBA	Be=	**0.800**	must have a value >0 entered		
	Building support as fraction of GBA	Bu=	0.200	Be + Bu must = 1		

Dwelling Unit Mix Table

DU dwelling unit type	GDA gross du area	CDA=GDA/Be comprehensive du area	MIX du mix	PDA = (CDA)MIX Prorated du area
EFF	**400**	500	**5%**	25
1 BR	**615**	769	**60%**	461
2 BR	**950**	1,188	**35%**	416
3 BR	**1,500**	1,875	**0%**	0
4 BR	**1,800**	2,250	**0%**	0
			Aggregate avg. dwelling unit area (AGG) =	**902**
			GBA sf per parking space a=	451

PLANNING FORECAST Enter zero in the adjacent box unless you wish to override the AGG value calculated above **0**

Note: Blank values in FLR column indicate density not feasible given project open space objective in (S) column

open space **S**	min area for BCG & PLA CORE Varies with open space provided	no. of dwelling units **NDU**	floors needed **FLR**	gross bldg area GBA	footprint BCA	DU's per floor DUF	bldg SF per SFAC bldable acre	flr area ratio FAR function of BLA	open space S provided in acres
0.100	277,695	**127.5**	**0.6**	114,989	185,895	206.1	13,528	0.311	0.850
0.150	259,182		**0.7**		167,382	185.6			1.275
0.200	240,669	density per net acre (dNA)	**0.8**		148,869	165.1			1.700
0.250	222,156	15.0	**0.9**		130,356	144.5			2.125
0.300	203,643	density per bldable acre (dBA)	**1.0**		111,843	124.0			2.550
0.350	185,130	15.0	**1.2**		93,330	103.5			2.975
0.400	166,617	parking lot area (PLA)	**1.5**		74,817	83.0			3.400
0.450	148,104	76,500	**2.0**		56,304	62.4			3.825
0.500	129,591	pkg. lot spaces (NPS)	**3.0**		37,791	41.9			4.250
0.550	111,078	191.3	**6.0**		19,278	21.4			4.675
0.600	92,565	garage pkg. area (GPA)	**150.3**		765	0.8			5.100
0.650	74,052	15,300.0							
0.700	55,539	garage pkg. spaces (GPS)							
0.800	18,513	63.8							
0.900									

Note: Be aware when BCG becomes too small to be feasible

WARNING: These are preliminary forecasts that must not be used to make final decisions.

1) These forecasts are not a substitute for the "due diligence" research that must be conducted to support the final definition of "unbuildable areas" above and the final decision to purchase land. This research includes, but is not limited to, verification of adequate subsurface soil, zoning, environmental clearance, access, title, utilities and water pressure, clearance from deed restriction, easement and right-of-way encumbrances, clearance from existing above and below ground facility conflicts, etc.

2) The most promising forecast(s) made on the basis of data entered in the design specification from "due diligence" research must be verified at the drawing board before funds are committed and land purchase decisions are made. Actual land shape ratios, dimensions and irregularities encountered may require adjustments to the general forecasts above.

3) The software licensee shall take responsibility for the design specification values entered and any advice given that is based on the forecast produced.

Exhibit 15.4 Elizabeth's Farm with Dwelling Unit Mix Option 2

Forecast Model RG1D: Net density objective of 15 and aggregate average dwelling unit area of 900 sq ft.

*Development capacity forecast for **APARTMENTS** based on the use of a **GRADE PARKING LOT** located on the same premises. When s and u equal zero in the design specification below, the forecast pertains to conditions when **NO PARKING** is required.*

***Given:** Net density objective and gross land area. **To Find:** Minimum number of building floors required to achieve the net density objective given based on the design specification values entered below. **Premise:** All building floors considered equal in area.*

DESIGN SPECIFICATION *Enter values in boxed areas where text is bold and blue. Express all fractions as decimals.*

Given:	**Net density objective:**	d=	**15.000**	dwelling units per net acre	653,400	SF
	Gross land area	GLA=	**10.000**	acres	435,600	SF
Land Variables:	Public/ private right-of-way & paved easements	W=	**0.150**	fraction of GLA	65,340	SF
	Net land area	NLA=	8.500	acres	370,260	SF
	Unbuildable and/or future expansion areas	U=	**0.000**	fraction of GLA	0	SF
	Gross land area reduction	X=	0.150	fraction of GLA	65,340	SF
	Buildable land area remaining	BLA=	**8.500**	acres	370,260	SF
Parking Variables:	Estimated gross pkg. lot area per pkg. space in SF	s =	**400**	ENTER ZERO IF NO PARKING REQUIRED		
	Parking spaces required per dwelling unit	u=	**1.50**	ENTER ZERO IF NO PARKING REQUIRED		
	Garage parking spaces planned or required per dwelling unit	Gn=	**0.50**	ENTER ZERO IF NO PARKING REQUIRED		
	Gross building area per garage space	Ga=	**240**	ENTER ZERO IF NO PARKING REQUIRED		
	No. of loading spaces	l=	**0**			
	Gross area per loading space	b=	**0**	SF		
Site Variables:	**Project open space as fraction of BLA**	S=	varies below			
	Private driveways as fraction of BLA	R=	**0.100**		37,026	SF
	Misc. pavement as fraction of BLA	M=	**0.050**		18,513	SF
	Loading area as fraction of BLA	L=	0.000		0	SF
	Total site support areas as a fraction of BLA	Su=	0.150		55,539	SF
Core:	**Core development area at grade as fraction of BLA**	C=	varies below			
Building Variables:	Building efficiency as fraction of GBA	Be=	**0.800**	must have a value >0 entered		
	Building support as fraction of GBA	Bu=	0.200	Be + Bu must = 1		

Dwelling Unit Mix Table

DU (dwelling unit type)	GDA (gross du area)	CDA=GDA/Be (comprehensive du area)	MIX (du mix)	PDA = (CDA)MIX (Pro-rated du area)
EFF	**400**	500	**0%**	0
1 BR	**550**	688	**40%**	275
2 BR	**800**	1,000	**50%**	500
3 BR	**1,000**	1,250	**10%**	125
4 BR	**1,200**	1,500	**0%**	0
			Aggregate avg. dwelling unit area (AGG) =	**900**
			GBA sf per parking space a=	450
			Enter zero in the adjacent box unless you wish to override the AGG value calculated above	**0**

PLANNING FORECAST

Note: Blank values in FLR column indicate density not feasible given project open space objective in S column

open space **S**	min area for BCG & PLA CORE (Varies with open space provided)	no. of dwelling units **NDU**	floors needed **FLR**	gross bldg area GBA	footprint BCA	DU's per floor DUF	bldg SF per SFAC (bldable acre)	flr area ratio FAR (function of BLA)	open space S (provided in acres)
0.100	277,695	**127.5**	**0.6**	114,750	185,895	206.6	13,500	0.310	0.850
0.150	259,182		**0.7**		167,382	186.0			1.275
0.200	240,669	density per net acre (dNA)	**0.8**		148,869	165.4			1.700
0.250	222,156	15.0	**0.9**		130,356	144.8			2.125
0.300	203,643	density per bldable acre (dBA)	**1.0**		111,843	124.3			2.550
0.350	185,130	15.0	**1.2**		93,330	103.7			2.975
0.400	166,617	parking lot area (PLA)	**1.5**		74,817	83.1			3.400
0.450	148,104	76,500	**2.0**		56,304	62.6			3.825
0.500	129,591	pkg. lot spaces (NPS)	**3.0**		37,791	42.0			4.250
0.550	111,078	191.3	**6.0**		19,278	21.4			4.675
0.600	92,565	garage pkg. area (GPA)	**150.0**		765	0.9			5.100
0.650	74,052	15,300.0							
0.700	55,539	garage pkg. spaces (GPS)							
0.800	18,513	63.8							
0.900									

Note: Be aware when BCG becomes too small to be feasible

WARNING: These are preliminary forecasts that must not be used to make final decisions.
1) These forecasts are not a substitute for the "due diligence" research that must be conducted to support the final definition of "unbuildable areas" above and the final decision to purchase land. This research includes, but is not limited to, verification of adequate subsurface soil, zoning, environmental clearance, access, title, utilities and water pressure, clearance from deed restriction, easement and right-of-way encumbrances, clearance from existing above and below ground facility conflicts, etc.
2) The most promising forecast(s) made on the basis of data entered in the design specification from "due diligence" research must be verified at the drawing board before funds are committed and land purchase decisions are made. Actual land shape ratios, dimensions and irregularities encountered may require adjustments to the general forecasts above.
3) The software licensee shall take responsibility for the design specification values entered and any advice given that is based on the forecast produced.

based on an assessment of the variables that would not affect the market segment desired, or the density objective could be reduced. With the forecast model collection, however, the problem is no longer how much drafting time it takes to draw an alternative method of achieving a density objective, but how long it takes to enter alternate values in the design specification panel of a forecast model.

Exhibit 15.5 uses the design specification values of Exhibit 15.1 and explores the feasibility of achieving a density of 25 dwelling units per net acre. The density increase produces 212.5 dwelling units instead of the 127.5 shown in Exhibit 15.1. The same building height of 4.5 floors, however, can produce an open space context of only 25%, since the building footprint and gross building area must increase to accommodate the increased number of dwelling units. 25% open space has been referred to as a CBD context. 6 floors produce 30% and 9.6 floors produce 35% open space, which have been referred to as transitional contexts.[9] If the aggregate average dwelling unit area is reduced from 1450 to 900 sq ft, Exhibit 15.6 shows that 6 building floors (instead of 9.6) can produce 35% open space and a net density of 25 dwelling units per acre when all other design specification values are held constant. This again shows the effect of dwelling unit mix and areas on achievable residential density.

Exhibit 15.7 explores the feasibility of achieving a density of 35 dwelling units per net acre with the forecast model, parking system, and design specification of Exhibit 15.1. The density increase produces 297.5 dwelling units at an aggregate average area of 1450 sq ft, but the forecast model predicts that 6.8 building floors will be required to achieve this density, and only 10% open space will remain. The model predicts that even 54.2 floors will produce only 25% open space. This forecast implies, therefore, that a 6.8-floor apartment building would have to sit in a sea of asphalt and a school of parked cars in order to produce 35

Exhibit 15.5 Elizabeth's Farm

Forecast Model RG1D: Net density objective of 25 and aggregate average dwelling unit area of 1450 sq ft.

*Development capacity forecast for **APARTMENTS** based on the use of a **GRADE PARKING LOT** located on the same premises. When s and u equal zero in the design specification below, the forecast pertains to conditions when **NO PARKING** is required.*

***Given:** Net density objective and gross land area. **To Find:** Minimum number of building floors required to achieve the net density objective given based on the design specification values entered below. **Premise:** All building floors considered equal in area.*

DESIGN SPECIFICATION *Enter values in boxed areas where text is bold and blue. Express all fractions as decimals.*

Given:	**Net density objective:**	d=	**25.000**	dwelling units per net acre	1,089,000	SF
	Gross land area	GLA=	**10.000**	acres	435,600	SF
Land Variables:	Public/ private right-of-way & paved easements	W=	**0.150**	fraction of GLA	65,340	SF
	Net land area	NLA=	8.500	acres	370,260	SF
	Unbuildable and/or future expansion areas	U=	**0.000**	fraction of GLA	0	SF
	Gross land area reduction	X=	0.150	fraction of GLA	65,340	SF
	Buildable land area remaining	BLA=	**8.500**	acres	370,260	SF
Parking Variables:	Estimated gross pkg. lot area per pkg. space in SF	s =	**400**	ENTER ZERO IF NO PARKING REQUIRED		
	Parking spaces required per dwelling unit	u=	**1.50**	ENTER ZERO IF NO PARKING REQUIRED		
	Garage parking spaces planned or required per dwelling unit	Gn=	**0.50**	ENTER ZERO IF NO PARKING REQUIRED		
	Gross building area per garage space	Ga=	**240**	ENTER ZERO IF NO PARKING REQUIRED		
	No. of loading spaces	l=	**0**			
	Gross area per loading space	b=	**0**	SF		
Site Variables:	**Project open space as fraction of BLA**	S=	varies below			
	Private driveways as fraction of BLA	R=	**0.100**		37,026	SF
	Misc. pavement as fraction of BLA	M=	**0.050**		18,513	SF
	Loading area as fraction of BLA	L=	0.000		0	SF
	Total site support areas as a fraction of BLA	Su=	0.150		55,539	SF
Core:	**Core development area at grade as fraction of BLA**	C=	varies below			
Building Variables:	Building efficiency as fraction of GBA	Be=	**0.800**	must have a value >0 entered		
	Building support as fraction of GBA	Bu=	0.200	Be + Bu must = 1		

Dwelling Unit Mix Table

DU dwelling unit type	GDA gross du area	CDA=GDA/Be comprehensive du area	MIX du mix	PDA = (CDA)MIX Pro-rated du area
EFF	**500**	625	**10%**	63
1 BR	**750**	938	**20%**	188
2 BR	**1,200**	1,500	**40%**	600
3 BR	**1,500**	1,875	**20%**	375
4 BR	**1,800**	2,250	**10%**	225
			Aggregate avg. dwelling unit area (AGG) =	**1,450**
			GBA sf per parking space a=	725

PLANNING FORECAST Enter zero in the adjacent box unless you wish to override the AGG value calculated above **0**

Note: Blank values in FLR column indicate density not feasible given project open space objective in S column

open space **S**	min area for BCG & PLA CORE Varies with open space provided	no. of dwelling units **NDU**	floors needed **FLR**	gross bldg area GBA	footprint BCA	DU's per floor DUF	bldg SF per SFAC bldable acre	flr area ratio FAR function of BLA	open space S provided in acres
0.100	277,695	**212.5**	**2.5**	308,125	124,695	86.0	36,250	0.832	0.850
0.150	259,182		**2.9**		106,182	73.2			1.275
0.200	240,669	density per net acre (dNA)	**3.5**		87,669	60.5			1.700
0.250	222,156	25.0	**4.5**		69,156	47.7			2.125
0.300	203,643	density per bldable acre (dBA)	**6.1**		50,643	34.9			2.550
0.350	185,130	25.0	**9.6**		32,130	22.2			2.975
0.400	166,617	parking lot area (PLA)	**22.6**		13,617	9.4			3.400
0.450	148,104	127,500							
0.500	129,591	pkg. lot spaces (NPS)							
0.550	111,078	318.8							
0.600	92,565	garage pkg. area (GPA)							
0.650	74,052	25,500.0							
0.700	55,539	garage pkg. spaces (GPS)							
0.800	18,513	106.3							
0.900									

Note: Be aware when BCG becomes too small to be feasible

WARNING: These are preliminary forecasts that must not be used to make final decisions.
1) These forecasts are not a substitute for the "due diligence" research that must be conducted to support the final definition of "unbuildable areas" above and the final decision to purchase land. This research includes, but is not limited to, verification of adequate subsurface soil, zoning, environmental clearance, access, title, utilities and water pressure, clearance from deed restriction, easement and right-of-way encumbrances, clearance from existing above and below ground facility conflicts, etc.
2) The most promising forecast(s) made on the basis of data entered in the design specification from "due diligence" research must be verified at the drawing board before funds are committed and land purchase decisions are made. Actual land shape ratios, dimensions and irregularities encountered may require adjustments to the general forecasts above.
3) The software licensee shall take responsibility for the design specification values entered and any advice given that is based on the forecast produced.

Exhibit 15.6 Elizabeth's Farm

Forecast Model RG1D: Net density objective of 25 and aggregate average dwelling unit area of 900 sq ft.

*Development capacity forecast for **APARTMENTS** based on the use of a **GRADE PARKING LOT** located on the same premises. When s and u equal zero in the design specification below, the forecast pertains to conditions when **NO PARKING** is required.*

***Given:** Net density objective and gross land area. **To Find:** Minimum number of building floors required to achieve the net density objective given based on the design specification values entered below. **Premise:** All building floors considered equal in area.*

DESIGN SPECIFICATION — *Enter values in boxed areas where text is bold and blue. Express all fractions as decimals.*

Given:	**Net density objective:**	d=	**25.000**	dwelling units per net acre	1,089,000	SF
	Gross land area	GLA=	**10.000**	acres	435,600	SF
Land Variables:	Public/ private right-of-way & paved easements	W=	**0.150**	fraction of GLA	65,340	SF
	Net land area	NLA=	8.500	acres	370,260	SF
	Unbuildable and/or future expansion areas	U=	**0.000**	fraction of GLA	0	SF
	Gross land area reduction	X=	0.150	fraction of GLA	65,340	SF
	Buildable land area remaining	BLA=	**8.500**	acres	370,260	SF
Parking Variables:	Estimated gross pkg. lot area per pkg. space in SF	s =	**400**	ENTER ZERO IF NO PARKING REQUIRED		
	Parking spaces required per dwelling unit	u=	**1.50**	ENTER ZERO IF NO PARKING REQUIRED		
	Garage parking spaces planned or required per dwelling unit	Gn=	**0.50**	ENTER ZERO IF NO PARKING REQUIRED		
	Gross building area per garage space	Ga=	**240**	ENTER ZERO IF NO PARKING REQUIRED		
	No. of loading spaces	l=	**0**			
	Gross area per loading space	b=	**0**	SF		
Site Variables:	**Project open space as fraction of BLA**	S=	varies below			
	Private driveways as fraction of BLA	R=	**0.100**		37,026	SF
	Misc. pavement as fraction of BLA	M=	**0.050**		18,513	SF
	Loading area as fraction of BLA	L=	0.000		0	SF
	Total site support areas as a fraction of BLA	Su=	0.150		55,539	SF
Core:	**Core development area at grade as fraction of BLA**	C=	varies below			
Building Variables:	Building efficiency as fraction of GBA	Be=	**0.800**	must have a value >0 entered		
	Building support as fraction of GBA	Bu=	0.200	Be + Bu must = 1		

Dwelling Unit Mix Table

DU dwelling unit type	GDA gross du area	CDA=GDA/Be comprehensive du area	MIX du mix	PDA = (CDA)MIX Pro-rated du area
EFF	**500**		**10%**	
1 BR	**750**		**20%**	
2 BR	**1,200**	data override	**40%**	data override
3 BR	**1,500**		**20%**	
4 BR	**1,800**		**10%**	

Aggregate avg. dwelling unit area (AGG) = **900**

GBA sf per parking space a= 450

PLANNING FORECAST — Enter zero in the adjacent box unless you wish to override the AGG value calculated above **900** **override**

Note: Blank values in FLR column indicate density not feasible given project open space objective in (S) column

open space **S**	min area for BCG & PLA CORE Varies with open space provided	no. of dwelling units **NDU**	floors needed **FLR**	gross bldg area GBA	footprint BCA	DU's per floor DUF	bldg SF per SFAC bldable acre	flr area ratio FAR function of BLA	open space S provided in acres
0.100	277,695	**212.5**	**1.5**	191,250	124,695	138.6	22,500	0.517	0.850
0.150	259,182		**1.8**		106,182	118.0			1.275
0.200	240,669	density per net acre (dNA)	**2.2**		87,669	97.4			1.700
0.250	222,156	25.0	**2.8**		69,156	76.8			2.125
0.300	203,643	density per bldable acre (dBA)	**3.8**		50,643	56.3			2.550
0.350	185,130	25.0	**6.0**		32,130	35.7			2.975
0.400	166,617	parking lot area (PLA)	**14.0**		13,617	15.1			3.400
0.450	148,104	127,500							
0.500	129,591	pkg. lot spaces (NPS)							
0.550	111,078	318.8							
0.600	92,565	garage pkg. area (GPA)							
0.650	74,052	25,500.0							
0.700	55,539	garage pkg. spaces (GPS)							
0.800	18,513	106.3							
0.900									

Note: Be aware when BCG becomes too small to be feasible

WARNING: These are preliminary forecasts that must not be used to make final decisions.
1) These forecasts are not a substitute for the "due diligence" research that must be conducted to support the final definition of "unbuildable areas" above and the final decision to purchase land. This research includes, but is not limited to, verification of adequate subsurface soil, zoning, environmental clearance, access, title, utilities and water pressure, clearance from deed restriction, easement and right-of-way encumbrances, clearance from existing above and below ground facility conflicts, etc.
2) The most promising forecast(s) made on the basis of data entered in the design specification from "due diligence" research must be verified at the drawing board before funds are committed and land purchase decisions are made. Actual land shape ratios, dimensions and irregularities encountered may require adjustments to the general forecasts above.
3) The software licensee shall take responsibility for the design specification values entered and any advice given that is based on the forecast produced.

Exhibit 15.7 Elizabeth's Farm

Forecast Model RG1D: Net density objective of 35 and aggregate average dwelling unit area of 1450 sq ft.

*Development capacity forecast for **APARTMENTS** based on the use of a **GRADE PARKING LOT** located on the same premises. When s and u equal zero in the design specification below, the forecast pertains to conditions when **NO PARKING** is required.*

***Given:** Net density objective and gross land area. **To Find:** Minimum number of building floors required to achieve the net density objective given based on the design specification values entered below. **Premise:** All building floors considered equal in area.*

DESIGN SPECIFICATION

Enter values in boxed areas where text is bold and blue. Express all fractions as decimals.

Given:	**Net density objective:**	d=	**35.000**	dwelling units per net acre	1,524,600	SF
	Gross land area	GLA=	**10.000**	acres	435,600	SF
Land Variables:	Public/ private right-of-way & paved easements	W=	**0.150**	fraction of GLA	65,340	SF
	Net land area	NLA=	8.500	acres	370,260	SF
	Unbuildable and/or future expansion areas	U=	**0.000**	fraction of GLA	0	SF
	Gross land area reduction	X=	0.150	fraction of GLA	65,340	SF
	Buildable land area remaining	BLA=	**8.500**	acres	370,260	SF
Parking Variables:	Estimated gross pkg. lot area per pkg. space in SF	s =	**400**	ENTER ZERO IF NO PARKING REQUIRED		
	Parking spaces required per dwelling unit	u=	**1.50**	ENTER ZERO IF NO PARKING REQUIRED		
	Garage parking spaces planned or required per dwelling unit	Gn=	**0.50**	ENTER ZERO IF NO PARKING REQUIRED		
	Gross building area per garage space	Ga=	**240**	ENTER ZERO IF NO PARKING REQUIRED		
	No. of loading spaces	l=	**0**			
	Gross area per loading space	b=	**0**	SF		
Site Variables:	**Project open space as fraction of BLA**	S=	varies below			
	Private driveways as fraction of BLA	R=	**0.100**		37,026	SF
	Misc. pavement as fraction of BLA	M=	**0.050**		18,513	SF
	Loading area as fraction of BLA	L=	0.000		0	SF
	Total site support areas as a fraction of BLA	Su=	0.150		55,539	SF
Core:	**Core development area at grade as fraction of BLA**	C=	varies below			
Building Variables:	Building efficiency as fraction of GBA	Be=	**0.800**	must have a value >0 entered		
	Building support as fraction of GBA	Bu=	0.200	Be + Bu must = 1		

Dwelling Unit Mix Table

DU dwelling unit type	GDA gross du area	CDA=GDA/Be comprehensive du area	MIX du mix		PDA = (CDA)MIX Pro-rated du area
EFF	**500**	625	**10%**		63
1 BR	**750**	938	**20%**		188
2 BR	**1,200**	1,500	**40%**		600
3 BR	**1,500**	1,875	**20%**		375
4 BR	**1,800**	2,250	**10%**		225
			Aggregate avg. dwelling unit area	(AGG) =	**1,450**
			GBA sf per parking space	a=	725

PLANNING FORECAST

Enter zero in the adjacent box unless you wish to override the AGG value calculated above: **0**

Note: Blank values in FLR column indicate density not feasible given project open space objective in S column

open space **S**	min area for BCG & PLA CORE Varies with open space provided	no. of dwelling units **NDU**	floors needed **FLR**	gross bldg area GBA	footprint BCA	DU's per floor DUF	bldg SF per SFAC bldable acre	flr area ratio FAR function of BLA	open space S provided in acres
0.100	277,695	**297.5**	**6.8**	431,375	63,495	43.8	50,750	1.165	0.850
0.150	259,182		**9.6**		44,982	31.0			1.275
0.200	240,669	density per net acre (dNA)	**16.3**		26,469	18.3			1.700
0.250	222,156	35.0	**54.2**		7,956	5.5			2.125
0.300	203,643	density per bldable acre (dBA)							
0.350	185,130	35.0							
0.400	166,617	parking lot area (PLA)							
0.450	148,104	178,500							
0.500	129,591	pkg. lot spaces (NPS)							
0.550	111,078	446.3							
0.600	92,565	garage pkg. area (GPA)							
0.650	74,052	35,700.0							
0.700	55,539	garage pkg. spaces (GPS)							
0.800	18,513	148.8							
0.900									

Note: Be aware when BCG becomes too small to be feasible

WARNING: These are preliminary forecasts that must not be used to make final decisions.
1) These forecasts are not a substitute for the "due diligence" research that must be conducted to support the final definition of "unbuildable areas" above and the final decision to purchase land. This research includes, but is not limited to, verification of adequate subsurface soil, zoning, environmental clearance, access, title, utilities and water pressure, clearance from deed restriction, easement and right-of-way encumbrances, clearance from existing above and below ground facility conflicts, etc.
2) The most promising forecast(s) made on the basis of data entered in the design specification from "due diligence" research must be verified at the drawing board before funds are committed and land purchase decisions are made. Actual land shape ratios, dimensions and irregularities encountered may require adjustments to the general forecasts above.
3) The software licensee shall take responsibility for the design specification values entered and any advice given that is based on the forecast produced.

dwelling units per buildable acre based on the design specification values entered. Exhibit 15.8 is also based on a net density objective of 35, but the aggregate average dwelling unit area has been reduced to 900 sq ft. Even with this reduction however, Exhibit 15.8 predicts that a 10-story building with a surface parking lot can produce only 20% open space and will remain in essentially the same asphalt sea. It is doubtful that the permitted density of 35 was ever intended to encourage this form of surface parking lot overdevelopment, but the best of intentions must be clearly defined. When combined with the design specification of Exhibit 15.1, a density of 35 is more compatible with parking structures. It can also be feasible in contexts where parking is either not required or provided in smaller amounts per dwelling unit, and when other design specification adjustments are made.

Exhibit 15.9 does not attempt to reach a density of 55 dwelling units per net acre. It stops at 45 to show that even this density is unrealistic based on a surface parking system and the design specification of Exhibit 15.1. Exhibit 15.9 shows that a density of 45 could theoretically be reached with a 241.7-story building in the middle of a surface parking lot. 10% open space would remain. The building footprint, however, could only be 2295 sq ft. The number of units per floor permitted by this footprint area and the building height required both make this option unrealistic. Reducing the aggregate average dwelling unit area to 900 sq ft simply reduces the building height to 150 stories, as shown in Exhibit 15.10, but does not improve the feasibility of the project density.

Exhibit 15.11 is included to show that a density of 55 dwelling units per net acre produces no floor forecast in the FLR column of the planning forecast panel. This indicates that the density is too extreme to be accommodated by the surface parking system under consideration. Exhibit 15.12 simply confirms this for an aggregate average dwelling unit area of 900 sq ft.

Exhibit 15.8 Elizabeth's Farm

Forecast Model RG1D: Net density objective of 35 and aggregate average dwelling unit area of 900 sq ft.

*Development capacity forecast for **APARTMENTS** based on the use of a **GRADE PARKING LOT** located on the same premises. When s and u equal zero in the design specification below, the forecast pertains to conditions when **NO PARKING** is required.*

***Given:** Net density objective and gross land area. **To Find:** Minimum number of building floors required to achieve the net density objective given based on the design specification values entered below. **Premise:** Al building floors considered equal in area.*

DESIGN SPECIFICATION

Enter values in boxed areas where text is bold and blue. Express all fractions as decimals.

Given:	**Net density objective:**	**d=**	**35.000**	dwelling units per net acre	1,524,600	SF
	Gross land area	**GLA=**	**10.000**	acres	435,600	SF
Land Variables:	Public/ private right-of-way & paved easements	W=	**0.150**	fraction of GLA	65,340	SF
	Net land area	NLA=	8.500	acres	370,260	SF
	Unbuildable and/or future expansion areas	U=	**0.000**	fraction of GLA	0	SF
	Gross land area reduction	X=	0.150	fraction of GLA	65,340	SF
	Buildable land area remaining	**BLA=**	**8.500**	acres	370,260	SF
Parking Variables:	Estimated gross pkg. lot area per pkg. space in SF	s =	**400**	ENTER ZERO IF NO PARKING REQUIRED		
	Parking spaces required per dwelling unit	u=	**1.50**	ENTER ZERO IF NO PARKING REQUIRED		
	Garage parking spaces planned or required per dwelling unit	Gn=	**0.50**	ENTER ZERO IF NO PARKING REQUIRED		
	Gross building area per garage space	Ga=	**240**	ENTER ZERO IF NO PARKING REQUIRED		
	No. of loading spaces	l=	**0**			
	Gross area per loading space	b=	**0**	SF		
Site Variables:	**Project open space as fraction of BLA**	**S=**	varies below			
	Private driveways as fraction of BLA	R=	**0.100**		37,026	SF
	Misc. pavement as fraction of BLA	M=	**0.050**		18,513	SF
	Loading area as fraction of BLA	L=	0.000		0	SF
	Total site support areas as a fraction of BLA	Su=	0.150		55,539	SF
Core:	**Core development area at grade as fraction of BLA**	**C=**	varies below			
Building Variables:	Building efficiency as fraction of GBA	Be=	**0.800**	must have a value >0 entered		
	Building support as fraction of GBA	Bu=	0.200	Be + Bu must = 1		

Dwelling Unit Mix Table

DU dwelling unit type	GDA gross du area	CDA=GDA/Be comprehensive du area	MIX du mix	PDA = (CDA)MIX Pro-rated du area
EFF	**500**		**10%**	
1 BR	**750**		**20%**	
2 BR	**1,200**	data override	**40%**	data override
3 BR	**1,500**		**20%**	
4 BR	**1,800**		**10%**	

Aggregate avg. dwelling unit area (AGG) = **900**

GBA sf per parking space a= 450

PLANNING FORECAST

Enter zero in the adjacent box unless you wish to override the AGG value calculated above **900** **override**

Note: Blank values in FLR column indicate density not feasible given project open space objective in S column

open space **S**	min area for BCG & PLA CORE Varies with open space provided	no. of dwelling units **NDU**	floors needed **FLR**	gross bldg area GBA	footprint BCA	DU's per floor DUF	bldg SF per SFAC bldable acre	flr area ratio FAR function of BLA	open space S provided in acres
0.100	277,695	**297.5**	**4.2**	267,750	63,495	70.6	31,500	0.723	0.850
0.150	259,182		**6.0**		44,982	50.0			1.275
0.200	240,669	density per net acre (dNA)	**10.1**		26,469	29.4			1.700
0.250	222,156	35.0	**33.7**		7,956	8.8			2.125
0.300	203,643	density per bldable acre (dBA)							
0.350	185,130	35.0							
0.400	166,617	parking lot area (PLA)							
0.450	148,104	178,500							
0.500	129,591	pkg. lot spaces (NPS)							
0.550	111,078	446.3							
0.600	92,565	garage pkg. area (GPA)							
0.650	74,052	35,700.0							
0.700	55,539	garage pkg. spaces (GPS)							
0.800	18,513	148.8							
0.900									

Note: Be aware when BCG becomes too small to be feasible

WARNING: These are preliminary forecasts that must not be used to make final decisions.

1) These forecasts are not a substitute for the "due diligence" research that must be conducted to support the final definition of "unbuildable areas" above and the final decision to purchase land. This research includes, but is not limited to, verification of adequate subsurface soil, zoning, environmental clearance, access, title, utilities and water pressure, clearance from deed restriction, easement and right-of-way encumbrances, clearance from existing above and below ground facility conflicts, etc.

2) The most promising forecast(s) made on the basis of data entered in the design specification from "due diligence" research must be verified at the drawing board before funds are committed and land purchase decisions are made. Actual land shape ratios, dimensions and irregularities encountered may require adjustments to the general forecasts above.

3) The software licensee shall take responsibility for the design specification values entered and any advice given that is based on the forecast produced.

Exhibit 15.9 Elizabeth's Farm

Forecast Model RG1D: Net density objective of 45 and aggregate average dwelling unit area of 1450 sq ft.

Development capacity forecast for ***APARTMENTS*** *based on the use of a* ***GRADE PARKING LOT*** *located on the same premises. When s and u equal zero in the design specification below, the forecast pertains to conditions when* ***NO PARKING*** *is required.*

Given: *Net density objective and gross land area.* ***To Find:*** *Minimum number of building floors required to achieve the net density objective given based on the design specification values entered below.* ***Premise:*** *All building floors considered equal in area.*

DESIGN SPECIFICATION *Enter values in boxed areas where text is bold and blue. Express all fractions as decimals.*

Given:	**Net density objective:**	d=	45.000	dwelling units per net acre	1,960,200	SF
	Gross land area	GLA=	10.000	acres	435,600	SF
Land Variables:	Public/ private right-of-way & paved easements	W=	0.150	fraction of GLA	65,340	SF
	Net land area	NLA=	8.500	acres	370,260	SF
	Unbuildable and/or future expansion areas	U=	0.000	fraction of GLA	0	SF
	Gross land area reduction	X=	0.150	fraction of GLA	65,340	SF
	Buildable land area remaining	BLA=	8.500	acres	370,260	SF
Parking Variables:	Estimated gross pkg. lot area per pkg. space in SF	s =	400	ENTER ZERO IF NO PARKING REQUIRED		
	Parking spaces required per dwelling unit	u=	1.50	ENTER ZERO IF NO PARKING REQUIRED		
	Garage parking spaces planned or required per dwelling unit	Gn=	0.50	ENTER ZERO IF NO PARKING REQUIRED		
	Gross building area per garage space	Ga=	240	ENTER ZERO IF NO PARKING REQUIRED		
	No. of loading spaces	l=	0			
	Gross area per loading space	b=	0	SF		
Site Variables:	**Project open space as fraction of BLA**	S=	varies below			
	Private driveways as fraction of BLA	R=	0.100		37,026	SF
	Misc. pavement as fraction of BLA	M=	0.050		18,513	SF
	Loading area as fraction of BLA	L=	0.000		0	SF
	Total site support areas as a fraction of BLA	Su=	0.150		55,539	SF
Core:	**Core development area at grade as fraction of BLA**	C=	varies below			
Building Variables:	Building efficiency as fraction of GBA	Be=	0.800	must have a value >0 entered		
	Building support as fraction of GBA	Bu=	0.200	Be + Bu must = 1		

Dwelling Unit Mix Table

DU dwelling unit type	GDA gross du area	CDA=GDA/Be comprehensive du area	MIX du mix	PDA = (CDA)MIX Pro-rated du area
EFF	500	625	10%	63
1 BR	750	938	20%	188
2 BR	1,200	1,500	40%	600
3 BR	1,500	1,875	20%	375
4 BR	1,800	2,250	10%	225
			Aggregate avg. dwelling unit area (AGG) =	**1,450**
			GBA sf per parking space a=	725

PLANNING FORECAST Enter zero in the adjacent box unless you wish to override the AGG value calculated above: **0**

Note: Blank values in FLR column indicate density not feasible given project open space objective in S column

open space **S**	min area for BCG & PLA CORE Varies with open space provided	no. of dwelling units **NDU**	floors needed **FLR**	gross bldg area GBA	footprint BCA	DU's per floor DUF	bldg SF per SFAC bldable acre	flr area ratio FAR function of BLA	open space S provided in acres
0.100	277,695	**382.5**	**241.7**	554,625	2,295	1.6	65,250	1.498	0.850
0.150	259,182								
0.200	240,669	density per net acre (dNA)							
0.250	222,156	45.0							
0.300	203,643	density per bldable acre (dBA)							
0.350	185,130	45.0							
0.400	166,617	parking lot area (PLA)							
0.450	148,104	229,500							
0.500	129,591	pkg. lot spaces (NPS)							
0.550	111,078	573.8							
0.600	92,565	garage pkg. area (GPA)							
0.650	74,052	45,900.0							
0.700	55,539	garage pkg. spaces (GPS)							
0.800	18,513	191.3							
0.900									

Note: Be aware when BCG becomes too small to be feasible

WARNING: These are preliminary forecasts that must not be used to make final decisions.

1) These forecasts are not a substitute for the "due diligence" research that must be conducted to support the final definition of "unbuildable areas" above and the final decision to purchase land. This research includes, but is not limited to, verification of adequate subsurface soil, zoning, environmental clearance, access, title, utilities and water pressure, clearance from deed restriction, easement and right-of-way encumbrances, clearance from existing above and below ground facility conflicts, etc.

2) The most promising forecast(s) made on the basis of data entered in the design specification from "due diligence" research must be verified at the drawing board before funds are committed and land purchase decisions are made. Actual land shape ratios, dimensions and irregularities encountered may require adjustments to the general forecasts above.

3) The software licensee shall take responsibility for the design specification values entered and any advice given that is based on the forecast produced.

Exhibit 15.10 Elizabeth's Farm

Forecast Model RG1D: Net density objective of 45 and aggregate average dwelling unit area of 900 sq ft.

*Development capacity forecast for **APARTMENTS** based on the use of a **GRADE PARKING LOT** located on the same premises. When (s) and (u) equal zero in the design specification below, the forecast pertains to conditions when **NO PARKING** is required.*

***Given:** Net density objective and gross land area. **To Find:** Minimum number of building floors required to achieve the net density objective given based on the design specification values entered below. **Premise:** All building floors considered equal in area.*

DESIGN SPECIFICATION

Enter values in boxed areas where text is bold and blue. Express all fractions as decimals.

Given:	**Net density objective:**	d=	**45.000**	dwelling units per net acre	1,960,200	SF
	Gross land area	GLA=	**10.000**	acres	435,600	SF
Land Variables:	Public/ private right-of-way & paved easements	W=	**0.150**	fraction of GLA	65,340	SF
	Net land area	NLA=	8.500	acres	370,260	SF
	Unbuildable and/or future expansion areas	U=	**0.000**	fraction of GLA	0	SF
	Gross land area reduction	X=	0.150	fraction of GLA	65,340	SF
	Buildable land area remaining	BLA=	**8.500**	acres	370,260	SF
Parking Variables:	Estimated gross pkg. lot area per pkg. space in SF	s =	**400**	ENTER ZERO IF NO PARKING REQUIRED		
	Parking spaces required per dwelling unit	u=	**1.50**	ENTER ZERO IF NO PARKING REQUIRED		
	Garage parking spaces planned or required per dwelling unit	Gn=	**0.50**	ENTER ZERO IF NO PARKING REQUIRED		
	Gross building area per garage space	Ga=	**240**	ENTER ZERO IF NO PARKING REQUIRED		
	No. of loading spaces	l=	**0**			
	Gross area per loading space	b=	**0**	SF		
Site Variables:	**Project open space as fraction of BLA**	S=	varies below			
	Private driveways as fraction of BLA	R=	**0.100**		37,026	SF
	Misc. pavement as fraction of BLA	M=	**0.050**		18,513	SF
	Loading area as fraction of BLA	L=	0.000		0	SF
	Total site support areas as a fraction of BLA	Su=	0.150		55,539	SF
Core:	**Core development area at grade as fraction of BLA**	C=	varies below			
Building Variables:	Building efficiency as fraction of GBA	Be=	**0.800**	must have a value >0 entered		
	Building support as fraction of GBA	Bu=	0.200	Be + Bu must = 1		

Dwelling Unit Mix Table

DU dwelling unit type	GDA gross du area	CDA=GDA/Be comprehensive du area	MIX du mix	PDA = (CDA)MIX Pro-rated du area
EFF	**500**		**10%**	
1 BR	**750**		**20%**	
2 BR	**1,200**	data override	**40%**	data override
3 BR	**1,500**		**20%**	
4 BR	**1,800**		**10%**	

Aggregate avg. dwelling unit area (AGG) = **900**

GBA sf per parking space a= 450

Enter zero in the adjacent box unless you wish to override the AGG value calculated above **900** **override**

PLANNING FORECAST

Note: Blank values in FLR column indicate density not feasible given project open space objective in S column

open space **S**	min area for BCG & PLA CORE Varies with open space provided	no. of dwelling units **NDU**	floors needed **FLR**	gross bldg area GBA	footprint BCA	DU's per floor DUF	bldg SF per SFAC bldable acre	flr area ratio FAR function of BLA	open space S provided in acres
0.100	277,695	**382.5**	**150.0**	344,250	2,295	2.6	40,500	0.930	0.850
0.150	259,182								
0.200	240,669	density per net acre (dNA)							
0.250	222,156	45.0							
0.300	203,643	density per bldable acre (dBA)							
0.350	185,130	45.0							
0.400	166,617	parking lot area (PLA)							
0.450	148,104	229,500							
0.500	129,591	pkg. lot spaces (NPS)							
0.550	111,078	573.8							
0.600	92,565	garage pkg. area (GPA)							
0.650	74,052	45,900.0							
0.700	55,539	garage pkg. spaces (GPS)							
0.800	18,513	191.3							
0.900									

Note: Be aware when BCG becomes too small to be feasible

WARNING: These are preliminary forecasts that must not be used to make final decisions.
1) These forecasts are not a substitute for the "due diligence" research that must be conducted to support the final definition of "unbuildable areas" above and the final decision to purchase land. This research includes, but is not limited to, verification of adequate subsurface soil, zoning, environmental clearance, access, title, utilities and water pressure, clearance from deed restriction, easement and right-of-way encumbrances, clearance from existing above and below ground facility conflicts, etc.
2) The most promising forecast(s) made on the basis of data entered in the design specification from "due diligence" research must be verified at the drawing board before funds are committed and land purchase decisions are made. Actual land shape ratios, dimensions and irregularities encountered may require adjustments to the general forecasts above.
3) The software licensee shall take responsibility for the design specification values entered and any advice given that is based on the forecast produced.

Exhibit 15.11 Elizabeth's Farm

Forecast Model RG1D: Net density objective of 55 and aggregate average dwelling unit area of 1450 sq ft.

*Development capacity forecast for **APARTMENTS** based on the use of a **GRADE PARKING LOT** located on the same premises. When s and u equal zero in the design specification below, the forecast pertains to conditions when **NO PARKING** is required.*

***Given:** Net density objective and gross land area. **To Find:** Minimum number of building floors required to achieve the net density objective given based on the design specification values entered below. **Premise:** All building floors considered equal in area.*

DESIGN SPECIFICATION *Enter values in boxed areas where text is bold and blue. Express all fractions as decimals.*

Given:	**Net density objective:**	d=	**55.000**	dwelling units per net acre	2,395,800	SF
	Gross land area	GLA=	**10.000**	acres	435,600	SF
Land Variables:	Public/ private right-of-way & paved easements	W=	**0.150**	fraction of GLA	65,340	SF
	Net land area	NLA=	8.500	acres	370,260	SF
	Unbuildable and/or future expansion areas	U=	**0.000**	fraction of GLA	0	SF
	Gross land area reduction	X=	0.150	fraction of GLA	65,340	SF
	Buildable land area remaining	BLA=	**8.500**	acres	370,260	SF
Parking Variables:	Estimated gross pkg. lot area per pkg. space in SF	s =	**400**	ENTER ZERO IF NO PARKING REQUIRED		
	Parking spaces required per dwelling unit	u=	**1.50**	ENTER ZERO IF NO PARKING REQUIRED		
	Garage parking spaces planned or required per dwelling unit	Gn=	**0.50**	ENTER ZERO IF NO PARKING REQUIRED		
	Gross building area per garage space	Ga=	**240**	ENTER ZERO IF NO PARKING REQUIRED		
	No. of loading spaces	l=	**0**			
	Gross area per loading space	b=	**0**	SF		
Site Variables:	**Project open space as fraction of BLA**	S=	varies below			
	Private driveways as fraction of BLA	R=	**0.100**		37,026	SF
	Misc. pavement as fraction of BLA	M=	**0.050**		18,513	SF
	Loading area as fraction of BLA	L=	0.000		0	SF
	Total site support areas as a fraction of BLA	Su=	0.150		55,539	SF
Core:	**Core development area at grade as fraction of BLA**	C=	varies below			
Building Variables:	Building efficiency as fraction of GBA	Be=	**0.800**	must have a value >0 entered		
	Building support as fraction of GBA	Bu=	0.200	Be + Bu must = 1		

Dwelling Unit Mix Table

DU dwelling unit type	GDA gross du area	CDA=GDA/Be comprehensive du area	MIX du mix		PDA = (CDA)MIX Pro-rated du area
EFF	**500**	625	**10%**		63
1 BR	**750**	938	**20%**		188
2 BR	**1,200**	1,500	**40%**		600
3 BR	**1,500**	1,875	**20%**		375
4 BR	**1,800**	2,250	**10%**		225
			Aggregate avg. dwelling unit area	(AGG) =	**1,450**
			GBA sf per parking space	a=	725

PLANNING FORECAST Enter zero in the adjacent box unless you wish to override the AGG value calculated above **0**

Note: Blank values in FLR column indicate density not feasible given project open space objective in S column

open space **S**	min area for BCG & PLA CORE Varies with open space provided	no. of dwelling units **NDU**	floors needed **FLR**	gross bldg area GBA	footprint BCA	DU's per floor DUF	bldg SF per SFAC bldable acre	flr area ratio FAR function of BLA	open space S provided in acres
0.100	277,695	**467.5**		677,875			79,750	1.831	
0.150	259,182								
0.200	240,669	density per net acre (dNA)							
0.250	222,156	55.0							
0.300	203,643	density per bldable acre (dBA)							
0.350	185,130	55.0							
0.400	166,617	parking lot area (PLA)							
0.450	148,104	280,500							
0.500	129,591	pkg. lot spaces (NPS)							
0.550	111,078	701.3							
0.600	92,565	garage pkg. area (GPA)							
0.650	74,052	56,100.0							
0.700	55,539	garage pkg. spaces (GPS)							
0.800	18,513	233.8							
0.900									

Note: Be aware when BCG becomes too small to be feasible

WARNING: These are preliminary forecasts that must not be used to make final decisions.

1) These forecasts are not a substitute for the "due diligence" research that must be conducted to support the final definition of "unbuildable areas" above and the final decision to purchase land. This research includes, but is not limited to, verification of adequate subsurface soil, zoning, environmental clearance, access, title, utilities and water pressure, clearance from deed restriction, easement and right-of-way encumbrances, clearance from existing above and below ground facility conflicts, etc.

2) The most promising forecast(s) made on the basis of data entered in the design specification from "due diligence" research must be verified at the drawing board before funds are committed and land purchase decisions are made. Actual land shape ratios, dimensions and irregularities encountered may require adjustments to the general forecasts above.

3) The software licensee shall take responsibility for the design specification values entered and any advice given that is based on the forecast produced.

Exhibit 15.12 Elizabeth's Farm

Forecast Model RG1D: Net density objective of 55 and aggregate average dwelling unit area of 900 sq ft.

Development capacity forecast for ***APARTMENTS*** *based on the use of a* ***GRADE PARKING LOT*** *located on the same premises. When s and u equal zero in the design specification below, the forecast pertains to conditions when* ***NO PARKING*** *is required.*

Given: *Net density objective and gross land area.* ***To Find:*** *Minimum number of building floors required to achieve the net density objective given based on the design specification values entered below.* ***Premise:*** *All building floors considered equal in area.*

DESIGN SPECIFICATION *Enter values in boxed areas where text is bold and blue. Express all fractions as decimals.*

Given:	**Net density objective:**	d=	**55.000**	dwelling units per net acre	2,395,800	SF
	Gross land area	GLA=	**10.000**	acres	435,600	SF
Land Variables:	Public/ private right-of-way & paved easements	W=	**0.150**	fraction of GLA	65,340	SF
	Net land area	NLA=	8.500	acres	370,260	SF
	Unbuildable and/or future expansion areas	U=	**0.000**	fraction of GLA	0	SF
	Gross land area reduction	X=	0.150	fraction of GLA	65,340	SF
	Buildable land area remaining	BLA=	**8.500**	acres	370,260	SF
Parking Variables:	Estimated gross pkg. lot area per pkg. space in SF	s =	**400**	ENTER ZERO IF NO PARKING REQUIRED		
	Parking spaces required per dwelling unit	u=	**1.50**	ENTER ZERO IF NO PARKING REQUIRED		
	Garage parking spaces planned or required per dwelling unit	Gn=	**0.50**	ENTER ZERO IF NO PARKING REQUIRED		
	Gross building area per garage space	Ga=	**240**	ENTER ZERO IF NO PARKING REQUIRED		
	No. of loading spaces	l=	**0**			
	Gross area per loading space	b=	**0**	SF		
Site Variables:	**Project open space as fraction of BLA**	**S=**	varies below			
	Private driveways as fraction of BLA	R=	**0.100**		37,026	SF
	Misc. pavement as fraction of BLA	M=	**0.050**		18,513	SF
	Loading area as fraction of BLA	L=	0.000		0	SF
	Total site support areas as a fraction of BLA	Su=	0.150		55,539	SF
Core:	**Core development area at grade as fraction of BLA**	**C=**	varies below			
Building Variables:	Building efficiency as fraction of GBA	Be=	**0.800**	must have a value >0 entered		
	Building support as fraction of GBA	Bu=	0.200	Be + Bu must = 1		

Dwelling Unit Mix Table

DU dwelling unit type	GDA gross du area	CDA=GDA/Be comprehensive du area	MIX du mix	PDA = (CDA)MIX Pro-rated du area
EFF	**500**		**10%**	
1 BR	**750**		**20%**	
2 BR	**1,200**	data override	**40%**	data override
3 BR	**1,500**		**20%**	
4 BR	**1,800**		**10%**	

Aggregate avg. dwelling unit area (AGG) = **900**

GBA sf per parking space a= 450

PLANNING FORECAST Enter zero in the adjacent box unless you wish to override the AGG value calculated above **900** **override**

Note: Blank values in FLR column indicate density not feasible given project open space objective in S column

open space **S**	min area for BCG & PLA CORE Varies with open space provided	no. of dwelling units **NDU**	floors needed **FLR**	gross bldg area GBA	footprint BCA	DU's per floor DUF	bldg SF per SFAC bldable acre	flr area ratio FAR function of BLA	open space S provided in acres
0.100	277,695	**467.5**		420,750			49,500	1.136	
0.150	259,182								
0.200	240,669	density per net acre (dNA)							
0.250	222,156	55.0							
0.300	203,643	density per bldable acre (dBA)							
0.350	185,130	55.0							
0.400	166,617	parking lot area (PLA)							
0.450	148,104	280,500							
0.500	129,591	pkg. lot spaces (NPS)							
0.550	111,078	701.3							
0.600	92,565	garage pkg. area (GPA)							
0.650	74,052	56,100.0							
0.700	55,539	garage pkg. spaces (GPS)							
0.800	18,513	233.8							
0.900									

Note: Be aware when BCG becomes too small to be feasible

WARNING: These are preliminary forecasts that must not be used to make final decisions.

1) These forecasts are not a substitute for the "due diligence" research that must be conducted to support the final definition of "unbuildable areas" above and the final decision to purchase land. This research includes, but is not limited to, verification of adequate subsurface soil, zoning, environmental clearance, access, title, utilities and water pressure, clearance from deed restriction, easement and right-of-way encumbrances, clearance from existing above and below ground facility conflicts, etc.

2) The most promising forecast(s) made on the basis of data entered in the design specification from "due diligence" research must be verified at the drawing board before funds are committed and land purchase decisions are made. Actual land shape ratios, dimensions and irregularities encountered may require adjustments to the general forecasts above.

3) The software licensee shall take responsibility for the design specification values entered and any advice given that is based on the forecast produced.

Underground Parking Structure Options

A density of 55 dwelling units per net acre is not feasible with a surface parking lot system, but it can be reached when the focus is shifted to a parking structure. In fact, a density of 150 dwelling units per net acre can easily be reached when parking is not limited to surface pavement. The example chosen for this alternative is an underground parking structure that preserves the greatest amount of grade-level open space with the least amount of building height and mass. The forecast is based on the design specification values entered in the top panel of Exhibit 15.13. The specification notes that 3.2 acres of Elizabeth's Farm will be split off for residential development using an underground parking structure. No roadways are anticipated for this area, and no unbuildable areas exist, leaving a 3.2-acre buildable area for construction. The parking allocation remains at 1.5 spaces per dwelling unit, but the net area per parking structure space has been reduced to 375 sq ft. Private driveways have been reduced to 2% of the buildable land area, since it is assumed that they will be used only for garage access, formal entry, and service areas. Miscellaneous pavement has been retained at 5% to anticipate outdoor activity areas. The underground parking area will cover 80% of the buildable area below grade to hold the number of predicted levels to 1. The garage efficiency objective is 85%, and the building efficiency objective is 75%. The dwelling unit mix table has been completed, and an AGG of 1547 sq ft is the result. This is somewhat less than the Waterford Tower aggregate average condominium dwelling unit area of 1781 sq ft, and substantially less than the Summit Chase average of 2171 sq ft.

The planning forecast panel in Exhibit 15.13 predicts that 176.0 dwelling units can be constructed on 3.2 buildable acres when a 1-level underground parking structure is constructed, based on the design specification values

Exhibit 15.13 Elizabeth's Farm with Underground Parking Garage

Forecast Model RS2D: Net density objective of 55 and aggregate average dwelling unit area of 1547 sq ft.

Development capacity forecast for ***APARTMENTS*** *based on the use of an* ***UNDERGROUND PARKING STRUCTURE.***
Given: *Net density objective and gross land area.* ***To Find:*** *Minimum number of building floors and underground parking structure levels required to achieve the net density objective given based on the design specification values entered below.*
Design Premise: *Underground parking footprint may be larger, smaller, or equal to the building footprint above. All similar floors considered equal in area.*

DESIGN SPECIFICATION

Enter values in boxed areas where text is bold and blue. Express all fractions as decimals.

Given:	**Net density objective**	**d=**	**55.000**	dwelling units per net acre		
	Gross land area	**GLA=**	**3.200**	acres	139,392	SF
Land Variables:	Public/ private right-of-way & paved easements	W=	**0.000**	fraction of GLA	0	SF
	Net land area	NLA=	3.200	acres	139,392	SF
	Future expansion and/or unbuildalbe areas	U=	**0.000**	fraction of GLA	0	SF
	Gross land area reduction	X=	0.000	fraction of GLA	0	SF
	Buildable land area remaining	BLA=	**3.200**	acres	139,392	SF
Parking Variables:	Est. net pkg. structure area per pkg. space	s =	**375**	SF		
	Parking spaces required per dwelling unit	u =	**1.5**			
	No. of loading spaces	l =	**0**			
	Gross area per loading space	b =	**0**		0	SF
Site Variables at Grade:	Private driveways as fraction of BLA	R=	**0.020**		2,788	SF
	Misc. pavement as fraction of BLA	M=	**0.050**		6,970	SF
	Loading area as fraction of BLA	L=	0.000		0	SF
	Total site support areas above grade as a fraction of BLA	Su=	0.070		9,757	SF
Core:	**Core area + open space at grade as fraction of BLA**	**CS=**	0.930	CS+Su must = 1	129,635	SF
Below Grade:	Gross underground pkg. area (UNG) as fraction of BLA	G=	**0.800**			
	Pkg. support within parking structure as fraction of UNG	Pu=	**0.150**			
	Net pkg. area for parking & circulation as fraction of BLA	Pe=	0.850	Pe+Pu must = 1		
Building Variables:	Building efficiency as fraction of GBA	Be=	**0.750**	must have a value >0 entered		
	Building support as fraction of GBA	Bu=	0.250	Be+Bu must = 1		

Dwelling Unit Mix Table

DU dwelling unit type	GDA gross du area	CDA=GDA/Be comprehensive du area	MIX du mix	PDA = (CDA)MIX Prorated du area
EFF	**500**	667	**10%**	67
1 BR	**750**	1,000	**20%**	200
2 BR	**1,200**	1,600	**40%**	640
3 BR	**1,500**	2,000	**20%**	400
4 BR	**1,800**	2,400	**10%**	240
			Aggregate avg. dwelling unit area (AGG) =	**1,547** SF
			Enter zero in the adjacent box unless you wish to override the AGG value calculated above	**0**

PLANNING FORECAST

net pkg area NPA	gross bldg area GBA	underground **p** parking levels needed
99,000	272,213	1.04

pkg. spaces NPS	dwelling units **NDU**	pkg spaces / 1000 BSF UPR
264.0	**176.0**	0.970

gross pkg struc area GPA	bldg SF / BLAC SFAC	bldg SF / pkg space a
116,471 needed	85,067	1,031

gross pkg area UNG	flr area ratio FAR	density per dBA
111,514 available / level	1.953	55.0 bldable acre

open space at grade **S**	footprint BCA equals CORE	bldg. floors **FLR**	total floors **F**	open space OSAC buildable acres
0.000	129,635	**2.10**	**3.14**	0.000
0.050	122,665	**2.22**	**3.26**	0.160
0.100	115,695	**2.35**	**3.40**	0.320
0.150	108,726	**2.50**	**3.55**	0.480
0.200	101,756	**2.68**	**3.72**	0.640
0.250	94,787	**2.87**	**3.92**	0.800
0.300	87,817	**3.10**	**4.14**	0.960
0.350	80,847	**3.37**	**4.41**	1.120
0.400	73,878	**3.68**	**4.73**	1.280
0.450	66,908	**4.07**	**5.11**	1.440
0.500	59,939	**4.54**	**5.59**	1.600
0.550	52,969	**5.14**	**6.18**	1.760
0.600	45,999	**5.92**	**6.96**	1.920
0.650	39,030	**6.97**	**8.02**	2.080
0.700	32,060	**8.49**	**9.54**	2.240

WARNING: These are preliminary forecasts that must not be used to make final decisions.
1) These forecasts are not a substitute for the "due diligence" research that must be conducted to support the final definition of "unbuildable areas" above and the final decision to purchase land. This research includes, but is not limited to, verification of adequate subsurface soil, zoning, environmental clearance, access, title, utilities and water pressure, clearance from deed restriction, easement and right-of-way encumbrances, clearance from existing above and below ground facility conflicts, etc.
2) The most promising forecast(s) made on the basis of data entered in the design specification from "due diligence" research must be verified at the drawing board before funds are committed and land purchase decisions are made. Actual land shape ratios, dimensions and irregularities encountered may require adjustments to the general forecasts above.
3) The software licensee shall take responsibility for the design specification values entered and any advice given or decisions made that are based on the forecast produced.

entered. A yard context (45 to 55% open space) can be created for these units when 4.07 to 5.14 building floors are constructed (high low-rise building). A lawn context (55 to 65% open space) can be created if 5.14 to 6.97 building floors are constructed (low mid-rise building), and a garden context (65 to 75% open space) can be created if 8.49 building floors are constructed (low mid-rise building). All options would contain the same 176 dwelling units, but increasing building height decreases the footprint needed and collects the land area saved as open space. The forecast panel is set up, therefore, to explain the open space quantities that will result as building height increases while the density objective remains constant.[10]

If the AGG is reduced from 1547 to 1200 sq ft, Exhibit 15.14 predicts that an open space yard context can be created with building heights ranging from 3.16 to 3.99 floors. A lawn context can be created with a 3.99- to 5.41-floor building height range, and a garden context can be created using a 6.59-floor building height. Therefore, when the aggregate average dwelling unit area is reduced, the yard and lawn contexts can each be created with 1 less building floor, and the garden context can be produced with 2 less building floors.

Exhibit 15.15 evaluates what it would take to construct a density of 100 dwelling units per net acre and an AGG value of 1547 sq ft on the same 3.2-acre area. The specification and dwelling unit mix values are the same as those used in Exhibit 15.13, except for an underground parking area that is now 75% of the buildable area instead of 80%. This reduction produces an underground structure forecast of 2.03 levels and was chosen for that reason. The result is 320 dwelling units on 3.2 buildable acres with a 2-level parking structure under 80% of the buildable area. The development capacity SFAC would be 154,667 sq ft per acre, and the universal parking ratio *a* would remain at 0.97 parking spaces per 1000 sq ft of buildable area. A 50%

Exhibit 15.14 Elizabeth's Farm with Underground Parking Garage

Forecast Model RS2D: Net density objective of 55 and aggregate average dwelling unit area of 1200 sq ft.

Development capacity forecast for ***APARTMENTS*** *based on the use of an* ***UNDERGROUND PARKING STRUCTURE.***
Given: *Net density objective and gross land area.* ***To Find:*** *Minimum number of building floors and underground parking structure levels required to achieve the net density objective given based on the design specification values entered below.*
Design Premise: *Underground parking footprint may be larger, smaller, or equal to the building footprint above. All similar floors considered equal in area.*

DESIGN SPECIFICATION

Enter values in boxed areas where text is bold and blue. Express all fractions as decimals.

Given:	**Net density objective**	d=	**55.000**	dwelling units per net acre		
	Gross land area	GLA=	**3.200**	acres	139,392	SF
Land Variables:	Public/ private right-of-way & paved easements	W=	**0.000**	fraction of GLA	0	SF
	Net land area	NLA=	3.200	acres	139,392	SF
	Future expansion and/or unbuildalbe areas	U=	**0.000**	fraction of GLA	0	SF
	Gross land area reduction	X=	0.000	fraction of GLA	0	SF
	Buildable land area remaining	BLA=	**3.200**	acres	139,392	SF
Parking Variables:	Est. net pkg. structure area per pkg. space	s =	**375**	SF		
	Parking spaces required per dwelling unit	u =	**1.5**			
	No. of loading spaces	l =	**0**			
	Gross area per loading space	b =	**0**		0	SF
Site Variables at Grade:	Private driveways as fraction of BLA	R=	**0.020**		2,788	SF
	Misc. pavement as fraction of BLA	M=	**0.050**		6,970	SF
	Loading area as fraction of BLA	L=	0.000		0	SF
	Total site support areas above grade as a fraction of BLA	Su=	0.070		9,757	SF
Core:	**Core area + open space at grade as fraction of BLA**	CS=	0.930	CS+Su must = 1	129,635	SF
Below Grade:	Gross underground pkg. area (UNG) as fraction of BLA	G=	**0.800**			
	Pkg. support within parking structure as fraction of UNG	Pu=	**0.150**			
	Net pkg. area for parking & circulation as fraction of BLA	Pe=	0.850	Pe+Pu must = 1		
Building Variables:	Building efficiency as fraction of GBA	Be=	**0.750**	must have a value >0 entered		
	Building support as fraction of GBA	Bu=	0.250	Be+Bu must = 1		

Dwelling Unit Mix Table

DU dwelling unit type	GDA gross du area	CDA=GDA/Be comprehensive du area	MIX du mix	PDA = (CDA)MIX Prorated du area
EFF	**500**		**10%**	
1 BR	**750**		**20%**	
2 BR	**1,200**	data override	**40%**	data override
3 BR	**1,500**		**20%**	
4 BR	**1,800**		**10%**	

Aggregate avg. dwelling unit area (AGG) = **1,200** SF

PLANNING FORECAST

Enter zero in the adjacent box unless you wish to override the AGG value calculated above **1,200** **override**

net pkg area NPA	gross bldg area GBA	underground p parking levels needed
99,000	211,200	1.04
pkg. spaces NPS	**dwelling units NDU**	**pkg spaces / 1000 BSF UPR**
264.0	**176.0**	1.250
gross pkg struc area GPA	**bldg SF / BLAC SFAC**	**bldg SF / pkg space a**
116,471 needed	66,000	800
gross pkg area UNG	**flr area ratio FAR**	**density per dBA**
111,514 available / level	1.515	55.0 bldable acre

open space at grade S	footprint BCA equals CORE	bldg. floors FLR	total floors F	open space OSAC buildable acres
0.000	129,635	**1.63**	**2.67**	0.000
0.050	122,665	**1.72**	**2.77**	0.160
0.100	115,695	**1.83**	**2.87**	0.320
0.150	108,726	**1.94**	**2.99**	0.480
0.200	101,756	**2.08**	**3.12**	0.640
0.250	94,787	**2.23**	**3.27**	0.800
0.300	87,817	**2.41**	**3.45**	0.960
0.350	80,847	**2.61**	**3.66**	1.120
0.400	73,878	**2.86**	**3.90**	1.280
0.450	66,908	**3.16**	**4.20**	1.440
0.500	59,939	**3.52**	**4.57**	1.600
0.550	52,969	**3.99**	**5.03**	1.760
0.600	45,999	**4.59**	**5.64**	1.920
0.650	39,030	**5.41**	**6.46**	2.080
0.700	32,060	**6.59**	**7.63**	2.240

WARNING: These are preliminary forecasts that must not be used to make final decisions.
1) These forecasts are not a substitute for the "due diligence" research that must be conducted to support the final definition of "unbuildable areas" above and the final decision to purchase land. This research includes, but is not limited to, verification of adequate subsurface soil, zoning, environmental clearance, access, title, utilities and water pressure, clearance from deed restriction, easement and right-of-way encumbrances, clearance from existing above and below ground facility conflicts, etc.
2) The most promising forecast(s) made on the basis of data entered in the design specification from "due diligence" research must be verified at the drawing board before funds are committed and land purchase decisions are made. Actual land shape ratios, dimensions and irregularities encountered may require adjustments to the general forecasts above.
3) The software licensee shall take responsibility for the design specification values entered and any advice given or decisions made that are based on the forecast produced.

Exhibit 15.15 Elizabeth's Farm with Underground Parking Garage

Forecast Model RS2D: Net density objective of 100 and aggregate average dwelling unit area of 1547 sq ft.

Development capacity forecast for ***APARTMENTS*** *based on the use of an* ***UNDERGROUND PARKING STRUCTURE.***
Given: *Net density objective and gross land area.* ***To Find:*** *Minimum number of building floors and underground parking structure levels required to achieve the net density objective given based on the design specification values entered below.*
Design Premise: *Underground parking footprint may be larger, smaller, or equal to the building footprint above. All similar floors considered equal in area.*

DESIGN SPECIFICATION *Enter values in boxed areas where text is bold and blue. Express all fractions as decimals.*

Given:	**Net density objective**	d=	**100.000**	dwelling units per net acre		
	Gross land area	GLA=	**3.200**	acres	139,392	SF
Land Variables:	Public/ private right-of-way & paved easements	W=	**0.000**	fraction of GLA	0	SF
	Net land area	NLA=	3.200	acres	139,392	SF
	Future expansion and/or unbuildalbe areas	U=	**0.000**	fraction of GLA	0	SF
	Gross land area reduction	X=	0.000	fraction of GLA	0	SF
	Buildable land area remaining	BLA=	**3.200**	acres	139,392	SF
Parking Variables:	Est. net pkg. structure area per pkg. space	s =	**375**	SF		
	Parking spaces required per dwelling unit	u =	**1.5**			
	No. of loading spaces	l =	**0**			
	Gross area per loading space	b =	**0**		0	SF
Site Variables at Grade:	Private driveways as fraction of BLA	R=	**0.020**		2,788	SF
	Misc. pavement as fraction of BLA	M=	**0.050**		6,970	SF
	Loading area as fraction of BLA	L=	0.000		0	SF
	Total site support areas above grade as a fraction of BLA	Su=	0.070		9,757	SF
Core:	**Core area + open space at grade as fraction of BLA**	**CS=**	0.930	CS+Su must = 1	129,635	SF
Below Grade:	Gross underground pkg. area (UNG) as fraction of BLA	G=	**0.750**			
	Pkg. support within parking structure as fraction of UNG	Pu=	**0.150**			
	Net pkg. area for parking & circulation as fraction of BLA	Pe=	0.850	Pe+Pu must = 1		
Building Variables:	Building efficiency as fraction of GBA	Be=	**0.750**	must have a value >0 entered		
	Building support as fraction of GBA	Bu=	0.250	Be+Bu must = 1		

Dwelling Unit Mix Table

DU dwelling unit type	GDA gross du area	CDA=GDA/Be comprehensive du area	MIX du mix	PDA = (CDA)MIX Prorated du area
EFF	**500**	667	**10%**	67
1 BR	**750**	1,000	**20%**	200
2 BR	**1,200**	1,600	**40%**	640
3 BR	**1,500**	2,000	**20%**	400
4 BR	**1,800**	2,400	**10%**	240

Aggregate avg. dwelling unit area (AGG) = **1,547** SF

PLANNING FORECAST Enter zero in the adjacent box unless you wish to override the AGG value calculated above **0**

open space at grade **S**	footprint BCA equals CORE	bldg. floors **FLR**	total floors **F**	open space OSAC buildable acres
0.000	129,635	**3.82**	**5.84**	0.000
0.050	122,665	**4.03**	**6.06**	0.160
0.100	115,695	**4.28**	**6.30**	0.320
0.150	108,726	**4.55**	**6.58**	0.480
0.200	101,756	**4.86**	**6.89**	0.640
0.250	94,787	**5.22**	**7.25**	0.800
0.300	87,817	**5.64**	**7.66**	0.960
0.350	80,847	**6.12**	**8.15**	1.120
0.400	73,878	**6.70**	**8.72**	1.280
0.450	66,908	**7.40**	**9.42**	1.440
0.500	59,939	**8.26**	**10.28**	1.600
0.550	52,969	**9.34**	**11.37**	1.760
0.600	45,999	**10.76**	**12.79**	1.920
0.650	39,030	**12.68**	**14.71**	2.080
0.700	32,060	**15.44**	**17.46**	2.240

net pkg area NPA	gross bldg area GBA	underground **p** parking levels needed
180,000	494,933	2.03

pkg. spaces NPS	dwelling units **NDU**	pkg spaces / 1000 BSF UPR
480.0	**320.0**	0.970

gross pkg struc area GPA	bldg SF / BLAC SFAC	bldg SF / pkg space a
211,765 needed	154,667	1,031

gross pkg area UNG	flr area ratio FAR	density per dBA
104,544 available / level	3.551	100.0 bldable acre

WARNING: These are preliminary forecasts that must not be used to make final decisions.
1) These forecasts are not a substitute for the "due diligence" research that must be conducted to support the final definition of "unbuildable areas" above and the final decision to purchase land. This research includes, but is not limited to, verification of adequate subsurface soil, zoning, environmental clearance, access, title, utilities and water pressure, clearance from deed restriction, easement and right-of-way encumbrances, clearance from existing above and below ground facility conflicts, etc.
2) The most promising forecast(s) made on the basis of data entered in the design specification from "due diligence" research must be verified at the drawing board before funds are committed and land purchase decisions are made. Actual land shape ratios, dimensions and irregularities encountered may require adjustments to the general forecasts above.
3) The software licensee shall take responsibility for the design specification values entered and any advice given or decisions made that are based on the forecast produced.

open space yard context would require an 8.26-story building. A 60% lawn context would require a 10.76-story building, and a 70% garden context would require a 15.44-story building. Exhibit 15.16 shows that if the aggregate average dwelling unit area is reduced to 1200 sq ft, the yard and lawn contexts can be achieved with 2 less building floors, and the garden context can be achieved with 3.5 less floors.

Exhibit 15.17 repeats the process of Exhibit 15.15 for a density objective of 150 dwelling units per net acre and an AGG value of 1547 sq ft on the same 3.2-acre area. The specification and dwelling unit mix values are repeated from Exhibit 15.13, and the underground parking area remains at 75% of the buildable area above. Based on these data, the number of parking levels increases to 3 to meet the need of the increased density objective, and 480 dwelling units are produced. The development capacity SFAC is 232,000 sq ft per acre, and the parking ratio remains constant. A 50% yard context would require a 12.39-story building. A 60% lawn context would require a 16.14-story building, and a 70% garden context would require a 23.16-story building. Exhibit 15.18 shows that when the AGG value is reduced to 1200 sq ft, the number of building floors required to produce yard, lawn, and garden open space contexts declines to 9.6, 12.5, and 17.97 floors, respectively.

The pattern of Exhibits 15.13 through 15.18 documents the intuition of the design and construction professions. It also confirms what members of the general public would call common sense if they stopped to think about the issue. All have known that parking structures increase development capacity and that increased parking levels multiply this capacity. They have also known that each parking space constructed permits a proportional increase in building area, and that increased building height permits accommodating this building area on a smaller land area. They also know that increased open space changes the context and appearance of a development project and that it also

Exhibit 15.16 Elizabeth's Farm with Underground Parking Garage

Forecast Model RS2D: Net density objective of 100 and aggregate average dwelling unit area of 1200 sq ft.

*Development capacity forecast for **APARTMENTS** based on the use of an **UNDERGROUND PARKING STRUCTURE.***
***Given:** Net density objective and gross land area. **To Find:** Minimum number of building floors and underground parking structure levels required to achieve the net density objective given based on the design specification values entered below.*
***Design Premise:** Underground parking footprint may be larger, smaller, or equal to the building footprint above. All similar floors considered equal in area.*

DESIGN SPECIFICATION *Enter values in boxed areas where text is bold and blue. Express all fractions as decimals.*

			Value	Unit	SF	
Given:	**Net density objective**	d=	100.000	dwelling units per net acre		
	Gross land area	GLA=	3.200	acres	139,392	SF
Land Variables:	Public/ private right-of-way & paved easements	W=	0.000	fraction of GLA	0	SF
	Net land area	NLA=	3.200	acres	139,392	SF
	Future expansion and/or unbuildalbe areas	U=	0.000	fraction of GLA	0	SF
	Gross land area reduction	X=	0.000	fraction of GLA	0	SF
	Buildable land area remaining	BLA=	3.200	acres	139,392	SF
Parking Variables:	Est. net pkg. structure area per pkg. space	s =	375	SF		
	Parking spaces required per dwelling unit	u =	1.5			
	No. of loading spaces	l =	0			
	Gross area per loading space	b =	0		0	SF
Site Variables at Grade:	Private driveways as fraction of BLA	R=	0.020		2,788	SF
	Misc. pavement as fraction of BLA	M=	0.050		6,970	SF
	Loading area as fraction of BLA	L=	0.000		0	SF
	Total site support areas above grade as a fraction of BLA	Su=	0.070		9,757	SF
Core:	**Core area + open space at grade as fraction of BLA**	CS=	0.930	CS+Su must = 1	129,635	SF
Below Grade:	Gross underground pkg. area (UNG) as fraction of BLA	G=	0.750			
	Pkg. support within parking structure as fraction of UNG	Pu=	0.150			
	Net pkg. area for parking & circulation as fraction of BLA	Pe=	0.850	Pe+Pu must = 1		
Building Variables:	Building efficiency as fraction of GBA	Be=	0.750	must have a value >0 entered		
	Building support as fraction of GBA	Bu=	0.250	Be+Bu must = 1		

Dwelling Unit Mix Table

DU dwelling unit type	GDA gross du area	CDA=GDA/Be comprehensive du area	MIX du mix	PDA = (CDA)MIX Prorated du area
EFF	500		10%	
1 BR	750		20%	
2 BR	1,200	data override	40%	data override
3 BR	1,500		20%	
4 BR	1,800		10%	

Aggregate avg. dwelling unit area (AGG) = 1,200 SF

PLANNING FORECAST Enter zero in the adjacent box unless you wish to override the AGG value calculated above 1,200 **override**

open space at grade S	net pkg area NPA	gross bldg area GBA	underground p parking levels needed	footprint BCA equals CORE	bldg. floors FLR	total floors F	open space OSAC buildable acres
0.000	180,000	384,000	2.03	129,635	2.96	4.99	0.000
0.050				122,665	3.13	5.16	0.160
0.100	pkg. spaces	dwelling units	pkg spaces / 1000 BSF	115,695	3.32	5.34	0.320
0.150	NPS	**NDU**	UPR	108,726	3.53	5.56	0.480
0.200	480.0	**320.0**	1.250	101,756	3.77	5.80	0.640
0.250				94,787	4.05	6.08	0.800
0.300	gross pkg struc area	bldg SF / BLAC	bldg SF / pkg space	87,817	4.37	6.40	0.960
0.350	GPA	SFAC	a	80,847	4.75	6.78	1.120
0.400	211,765 needed	120,000	800	73,878	5.20	7.22	1.280
0.450				66,908	5.74	7.76	1.440
0.500	gross pkg area	flr area ratio	density per	59,939	6.41	8.43	1.600
0.550	UNG	FAR	dBA	52,969	7.25	9.28	1.760
0.600	104,544 available / level	2.755	100.0 bldable acre	45,999	8.35	10.37	1.920
0.650				39,030	9.84	11.86	2.080
0.700				32,060	11.98	14.00	2.240

WARNING: These are preliminary forecasts that must not be used to make final decisions.
1) These forecasts are not a substitute for the "due diligence" research that must be conducted to support the final definition of "unbuildable areas" above and the final decision to purchase land. This research includes, but is not limited to, verification of adequate subsurface soil, zoning, environmental clearance, access, title, utilities and water pressure, clearance from deed restriction, easement and right-of-way encumbrances, clearance from existing above and below ground facility conflicts, etc.
2) The most promising forecast(s) made on the basis of data entered in the design specification from "due diligence" research must be verified at the drawing board before funds are committed and land purchase decisions are made. Actual land shape ratios, dimensions and irregularities encountered may require adjustments to the general forecasts above.
3) The software licensee shall take responsibility for the design specification values entered and any advice given or decisions made that are based on the forecast produced.

Exhibit 15.17 Elizabeth's Farm with Underground Parking Garage

Forecast Model RS2D: Net density objective of 150 and aggregate average dwelling unit area of 1547 sq ft.

Development capacity forecast for ***APARTMENTS*** *based on the use of an* ***UNDERGROUND PARKING STRUCTURE.***
Given: *Net density objective and gross land area.* ***To Find:*** *Minimum number of building floors and underground parking structure levels required to achieve the net density objective given based on the design specification values entered below.*
Design Premise: *Underground parking footprint may be larger, smaller or equal to the building footprint above. All similar floors considered equal in area.*

DESIGN SPECIFICATION

Enter values in boxed areas where text is bold and blue. Express all fractions as decimals.

			Value	Unit	SF	
Given:	**Net density objective**	d=	**150.000**	dwelling units per net acre		
	Gross land area	GLA=	**3.200**	acres	139,392	SF
Land Variables:	Public/ private right-of-way & paved easements	W=	**0.000**	fraction of GLA	0	SF
	Net land area	NLA=	3.200	acres	139,392	SF
	Future expansion and/or unbuildalbe areas	U=	**0.000**	fraction of GLA	0	SF
	Gross land area reduction	X=	0.000	fraction of GLA	0	SF
	Buildable land area remaining	BLA=	**3.200**	acres	139,392	SF
Parking Variables:	Est. net pkg. structure area per pkg. space	s =	**375**	SF		
	Parking spaces required per dwelling unit	u =	**1.5**			
	No. of loading spaces	l =	**0**			
	Gross area per loading space	b =	**0**		0	SF
Site Variables at Grade:	Private driveways as fraction of BLA	R=	**0.020**		2,788	SF
	Misc. pavement as fraction of BLA	M=	**0.050**		6,970	SF
	Loading area as fraction of BLA	L=	0.000		0	SF
	Total site support areas above grade as a fraction of BLA	Su=	0.070		9,757	SF
Core:	**Core area + open space at grade as fraction of BLA**	CS=	0.930	CS+Su must = 1	129,635	SF
Below Grade:	Gross underground pkg. area (UNG) as fraction of BLA	G=	**0.750**			
	Pkg. support within parking structure as fraction of UNG	Pu=	**0.150**			
	Net pkg. area for parking & circulation as fraction of BLA	Pe=	0.850	Pe+Pu must = 1		
Building Variables:	Building efficiency as fraction of GBA	Be=	**0.750**	must have a value >0 entered		
	Building support as fraction of GBA	Bu=	0.250	Be+Bu must = 1		

Dwelling Unit Mix Table

DU dwelling unit type	GDA gross du area	CDA=GDA/Be comprehensive du area	MIX du mix	PDA = (CDA)MIX Prorated du area
EFF	**500**	667	**10%**	67
1 BR	**750**	1,000	**20%**	200
2 BR	**1,200**	1,600	**40%**	640
3 BR	**1,500**	2,000	**20%**	400
4 BR	**1,800**	2,400	**10%**	240

Aggregate avg. dwelling unit area (AGG) = **1,547** SF

PLANNING FORECAST

Enter zero in the adjacent box unless you wish to override the AGG value calculated above: **0**

open space at grade **S**	footprint BCA equals CORE	bldg. floors **FLR**	total floors **F**	open space OSAC buildable acres
0.000	129,635	**5.73**	**8.77**	0.000
0.050	122,665	**6.05**	**9.09**	0.160
0.100	115,695	**6.42**	**9.46**	0.320
0.150	108,726	**6.83**	**9.87**	0.480
0.200	101,756	**7.30**	**10.33**	0.640
0.250	94,787	**7.83**	**10.87**	0.800
0.300	87,817	**8.45**	**11.49**	0.960
0.350	80,847	**9.18**	**12.22**	1.120
0.400	73,878	**10.05**	**13.09**	1.280
0.450	66,908	**11.10**	**14.13**	1.440
0.500	59,939	**12.39**	**15.42**	1.600
0.550	52,969	**14.02**	**17.05**	1.760
0.600	45,999	**16.14**	**19.18**	1.920
0.650	39,030	**19.02**	**22.06**	2.080
0.700	32,060	**23.16**	**26.19**	2.240

net pkg area NPA	gross bldg area GBA	underground **p** parking levels needed
270,000	742,400	3.04
pkg. spaces NPS	dwelling units **NDU**	pkg spaces / 1000 BSF UPR
720.0	**480.0**	0.970
gross pkg struc area GPA	bldg SF / BLAC SFAC	bldg SF / pkg space a
317,647 needed	232,000	1,031
gross pkg area UNG	flr area ratio FAR	density per dBA
104,544 available / level	5.326	150.0 bldable acre

WARNING: These are preliminary forecasts that must not be used to make final decisions.
1) These forecasts are not a substitute for the "due diligence" research that must be conducted to support the final definition of "unbuildable areas" above and the final decision to purchase land. This research includes, but is not limited to, verification of adequate subsurface soil, zoning, environmental clearance, access, title, utilities and water pressure, clearance from deed restriction, easement and right-of-way encumbrances, clearance from existing above and below ground facility conflicts, etc.
2) The most promising forecast(s) made on the basis of data entered in the design specification from "due diligence" research must be verified at the drawing board before funds are committed and land purchase decisions are made. Actual land shape ratios, dimensions and irregularities encountered may require adjustments to the general forecasts above.
3) The software licensee shall take responsibility for the design specification values entered and any advice given or decisions made that are based on the forecast produced.

Exhibit 15.18 Elizabeth's Farm with Underground Parking Garage

Forecast Model RS2D: Net density objective of 150 and aggregate average dwelling unit area of 1200 sq ft.

*Development capacity forecast for **APARTMENTS** based on the use of an **UNDERGROUND PARKING STRUCTURE.***
***Given:** Net density objective and gross land area. **To Find:** Minimum number of building floors and underground parking structure levels required to achieve the net density objective given based on the design specification values entered below.*
***Design Premise:** Underground parking footprint may be larger, smaller or equal to the building footprint above. All similar floors considered equal in area.*

DESIGN SPECIFICATION

Enter values in boxed areas where text is bold and blue. Express all fractions as decimals.

Given:	**Net density objective**	d=	150.000	dwelling units per net acre		
	Gross land area	GLA=	3.200	acres	139,392	SF
Land Variables:	Public/ private right-of-way & paved easements	W=	0.000	fraction of GLA	0	SF
	Net land area	NLA=	3.200	acres	139,392	SF
	Future expansion and/or unbuildalbe areas	U=	0.000	fraction of GLA	0	SF
	Gross land area reduction	X=	0.000	fraction of GLA	0	SF
	Buildable land area remaining	BLA=	3.200	acres	139,392	SF
Parking Variables:	Est. net pkg. structure area per pkg. space	s =	375	SF		
	Parking spaces required per dwelling unit	u =	1.5			
	No. of loading spaces	l =	0			
	Gross area per loading space	b =	0		0	SF
Site Variables at Grade:	Private driveways as fraction of BLA	R=	0.020		2,788	SF
	Misc. pavement as fraction of BLA	M=	0.050		6,970	SF
	Loading area as fraction of BLA	L=	0.000		0	SF
	Total site support areas above grade as a fraction of BLA	Su=	0.070		9,757	SF
Core:	**Core area + open space at grade as fraction of BLA**	CS=	0.930	CS+Su must = 1	129,635	SF
Below Grade:	Gross underground pkg. area (UNG) as fraction of BLA	G=	0.750			
	Pkg. support within parking structure as fraction of UNG	Pu=	0.150			
	Net pkg. area for parking & circulation as fraction of BLA	Pe=	0.850	Pe+Pu must = 1		
Building Variables:	Building efficiency as fraction of GBA	Be=	0.750	must have a value >0 entered		
	Building support as fraction of GBA	Bu=	0.250	Be+Bu must = 1		

Dwelling Unit Mix Table

DU dwelling unit type	GDA gross du area	CDA=GDA/Be comprehensive du area	MIX du mix	PDA = (CDA)MIX Prorated du area
EFF	**500**		**10%**	
1 BR	**750**		**20%**	
2 BR	**1,200**	data override	**40%**	data override
3 BR	**1,500**		**20%**	
4 BR	**1,800**		**10%**	

Aggregate avg. dwelling unit area (AGG) = **1,200** SF

PLANNING FORECAST

Enter zero in the adjacent box unless you wish to override the AGG value calculated above **1,200** **override**

open space at grade **S**	footprint BCA equals CORE	bldg. floors **FLR**	total floors **F**	open space OSAC buildable acres
0.000	129,635	4.44	7.48	0.000
0.050	122,665	4.70	7.73	0.160
0.100	115,695	4.98	8.02	0.320
0.150	108,726	5.30	8.34	0.480
0.200	101,756	5.66	8.70	0.640
0.250	94,787	6.08	9.12	0.800
0.300	87,817	6.56	9.60	0.960
0.350	80,847	7.12	10.16	1.120
0.400	73,878	7.80	10.84	1.280
0.450	66,908	8.61	11.65	1.440
0.500	59,939	9.61	12.65	1.600
0.550	52,969	10.87	13.91	1.760
0.600	45,999	12.52	15.56	1.920
0.650	39,030	14.76	17.80	2.080
0.700	32,060	17.97	21.00	2.240

net pkg area NPA	gross bldg area GBA	underground **p** parking levels needed
270,000	576,000	3.04
pkg. spaces NPS	dwelling units **NDU**	pkg spaces / 1000 BSF UPR
720.0	**480.0**	1.250
gross pkg struc area GPA	bldg SF / BLAC SFAC	bldg SF / pkg space a
317,647 needed	180,000	800
gross pkg area UNG	flr area ratio FAR	density per dBA
104,544 available / level	4.132	150.0 bldable acre

WARNING: These are preliminary forecasts that must not be used to make final decisions.
1) These forecasts are not a substitute for the "due diligence" research that must be conducted to support the final definition of "unbuildable areas" above and the final decision to purchase land. This research includes, but is not limited to, verification of adequate subsurface soil, zoning, environmental clearance, access, title, utilities and water pressure, clearance from deed restriction, easement and right-of-way encumbrances, clearance from existing above and below ground facility conflicts, etc.
2) The most promising forecast(s) made on the basis of data entered in the design specification from "due diligence" research must be verified at the drawing board before funds are committed and land purchase decisions are made. Actual land shape ratios, dimensions and irregularities encountered may require adjustments to the general forecasts above.
3) The software licensee shall take responsibility for the design specification values entered and any advice given or decisions made that are based on the forecast produced.

reduces the development capacity of a given land area if building height is not increased to compensate for the core area reduction. These are design principles that have simply lacked a mathematical foundation, and that can be proven with the development forecast collection. What we know very little about, and where intuition has been proven inadequate, is the relationship of development cover, building mass, and open space to the intensity, context, quality of life, personal association, environmental protection, and sustainable lifestyle produced. On the topic of personal association alone, several public housing projects have dramatically proven that intuition is inadequate and can actually exacerbate the social disassociation it seeks to overcome.

Exhibits 15.13 through 15.18 not only document intuition for comparison and evaluation, they also reveal relationships. The parking structure portion of this case tracks the performance of two aggregate average dwelling unit areas through a density range beginning at 55 and ending at 150 dwelling units per acre. Table 15.1 shows that as the density increases, the smaller aggregate average dwelling unit area requires increasingly less building height to accommodate the density objective. This is again simply common sense that is quantitatively predicted by the development forecast collection. The steady accumulation of smaller dwelling unit areas within a given footprint area requires less building height as density increases. This is not news to design professionals. What is new is the ability to quantitatively predict these relationships to produce a broad spectrum of options for evaluation in a brief period of time. This has the potential to turn intuition into documented awareness and comparative evaluation that can produce accumulated knowledge over time.

Table 15.1 displays the pattern just mentioned and is based on Exhibits 15.13 through 15.18. The table compares two different average dwelling unit areas and the building height that each requires to achieve a given density

Table 15.1 Building Height Implications of Open Space, Parking System, Dwelling Unit Area, and Density Objectives

	Open Space Percentage and Context											
	Surface Parking						Underground Structure Parking					
	30%—Transition		40%—Courtyard		50%—Yard		50%—Yard		60%—Lawn		70%—Garden	
	AGG Value in sq ft											
	1450	900	1450	900	1450	900	1547	1200	1547	1200	1547	1200
Density in Dwelling Units Per Net Acre	Building Height in Floors											
15	1.7	1.0	2.5	1.5	4.9	3.0						
25	6.1	3.8	22.6	14.0	—	—						
35	—	—	—	—	—	—						
55							4.54	3.52	5.92	4.59	8.49	6.59
100							8.26	6.41	10.76	8.35	15.44	11.98
150							12.39	9.61	16.14	12.52	23.16	17.97

when an open space context objective is established. The density alternatives are listed in the left-hand column. The parking system, aggregate average dwelling unit area, and open space context objective are listed across the top of the table. The table itself summarizes the data from the forecast models and illustrates the lower building heights required by smaller AGG values when the open space objective and parking system are given.

As a side note, the design specifications that have driven the forecasts in Exhibits 15.13 to 15.18 and the summary in Table 15.1 are based on a parking requirement of 1.5 spaces per dwelling unit. This value could be reduced to produce an entirely new series of forecasts. If 1.5 were changed to 1.0, for instance, the increase in development capacity would be significant, and the motivation for parking variance requests would become very apparent.

Massing Index

This book creates an intensity index to function as a simplified recipe capable of producing planned context results; it is an expression that combines building height in floors with the total development cover percentage planned or present. Building mass is a three-dimensional component of development cover that produces urban form. The amount of building mass present within a total development cover area can be expressed with a *massing index* that is a simple variation of the intensity index concept. The massing index does not predict or describe the building form itself, which can take an infinite variety of shapes and appearances. It describes the simple geometric volume containing this form, and can be used to plan the relationship of this volume to the land area available and the neighborhood present or planned.[11] This is particularly important where residential development is concerned, since the open space context framed by massing and pavement determines our relationship to the land.

A massing index might read 10.20, which represents a 10-story building with a footprint covering 20% of the buildable area available. An intensity index might read 10.70, which represents a 10-story building within a development cover area equal to 70% of the total buildable land area. The 50% difference between building cover and development cover is occupied by parking areas and miscellaneous pavement.[12] The remaining 30% would constitute the project open space present.

The intensity index and the massing index are not substitutes for a design specification, but they can be used as short-form prescriptions. These two indexes can also be assembled to produce a *combined index* simply by adding the massing index to the intensity index. A combined index would express the two previous examples as 10.70.20. This value represents a 10-story building with 70% development cover and 20% building cover. This index defines the context produced by a set of design specification values. It can also be used to prescribe design specification limits. It does not obviate the need for design specification values, however, since these values define the final project in greater detail that is more easily converted to anticipated results. The combined index simply establishes the parameters that may not be exceeded by the combined effect of the specification values assigned and the project that results. Since these values represent pressure points of development, however, a development capacity and project context policy can be specified with these three simple values. They can also be used to specify larger neighborhood objectives and land use policies within a master plan and zoning ordinance.

Notes

1. This is a generic definition that includes all forms of shelter. Buildings provide shelter, but they have not been the issue. What they shelter has been the focus, since some activities are incompatible and their arrange-

ment must be carefully considered to protect the public health, safety, and welfare. The sustainable issue, however, is the capacity of land to accommodate shelter, the environmental intensity that is produced, and the ecological balance that results. If urban land use is internally arranged to produce compatible land use activity and its Cambrian ring is spreading without restraint, we will have solved one issue and overlooked the fact that this growth does not take place in a vacuum. The capacity of land to accommodate shelter has been referred to as *development capacity.* Development capacity analysis will not produce ecological answers. It is simply designed to assess the ability of land to accommodate shelter. It is up to others to establish ecological limits, to use development capacity analysis to evaluate the shelter that can be produced within these limits, and to evaluate the lifestyles and quality of life that will result. Density expressions such as "dwelling units per acre" correlate an activity with development capacity. The capacity index SFAC indicates the gross building area constructed per acre and is a subtle reminder that there is more to city planning than the two-dimensional location and arrangement of compatible land use activities.

2. Since this book is concerned with development capacity, it ignores the compatibility issues that could be represented by the densities permitted.

3. *Net density* is retained as a commonly recognized expression, but is adjusted for use with the buildable land area available by the equations within a forecast model.

4. This inverse relationship may be confusing. A higher ratio means a lower parking requirement. For instance, given a 10,000-sq-ft building, a parking requirement of 1 space per 100 sq ft of building area would result in 1000 parking spaces. A parking requirement of 1 space per 1000 sq ft of building area would result in 100 parking spaces. The trend is to reverse this ratio to read 10 parking spaces per 1000 sq ft of building area and 1 parking space per 1000 sq ft of building area in order to eliminate this inverse relationship. Using this direct relationship index, nonresidential parking requirements generally range from 2.5 to 20 spaces per 1000 sq ft of building area and residential requirements will rarely exceed 2.0 parking spaces per 1000 sq ft of building area. The historic inverse parking expression has been used in this text, however, for its convenience and compatibility with the forecast model equations introduced.

5. The mathematical precision of the forecast must be adjusted to represent practical architectural solutions. In this case, there is enough design tolerance to easily accommodate an adjustment from 4.9 to 5.0 stories. This height is referred to as a *high low-rise* building in Chap. 14.
6. The implications of such context reductions require the attention of many diverse disciplines. The 25 to 35% open space range is referred to as a *transitional context* in Chap. 14.
7. Keep in mind that Elizabeth's Farm is composed of 10 gross acres and that the gross building forecast is for the entire area. This building area can be divided into any number of separate buildings. The land area can also be divided into separate parcels, and the development capacity of each can be evaluated with a separate forecast model.
8. This does not mean that this is the only set of adjustments that can be made. It does mean that a set of adjustments must be introduced to produce an AGG of 900 sq ft.
9. This again illustrates the decreasing benefit of increased building height when surface parking is used. See Chap. 1.
10. Exhibit 15.13 shows that the SFAC value for building area per buildable acre produced by this density and dwelling unit mix is 85,067 sq ft. The floor area ratio is 1.953. The universal parking ratio *a* can be found by dividing the residential AGG value of 1547 sq ft by the residential parking ratio of 1.5. This division reveals that 1031 sq ft of building area is provided per parking space. This inverse ratio means that 0.97 parking spaces are provided per 1000 sq ft of building area.
11. This is often referred to as *urban form*. If a massing index is used to regulate urban form, it will need to be adjusted in relation to height when mid-rise, high-rise, and sky-rise buildings are involved, but these adjustments are beyond the scope of this book.
12. Remember that roadways and unbuildable areas have previously been subtracted from the gross land area to define the buildable land area available.

Economic Development

An economic development objective is to attract, retain, and/or expand business activity that has the potential to increase employment and improve community income. The average tax yield from these business acres, and the ratio of these acres to a jurisdiction's entire geographic area, directly affect the prosperity of that jurisdiction. These business acres can be thought of as the high-yield portion of a community's land use portfolio. Low yield acres must be offset by the high-yield portion of this portfolio in order to maintain a balanced budget and stable cash flow[1] that is equal to the cost of maintaining a community's lifestyle[2] over an extended period of time.

Yield per acre is an agricultural term that has rarely been used in the planning, design, and construction disciplines, but is a term that is meant to associate the economic performance of development with the land it consumes. Economic performance is a function of the building area that occupies a site and the success of the activity that is sheltered within. The real estate tax, income tax, and miscellaneous service fees associated with this activity produce economic yield, or community income. The building area on the site determines the number of people and scope of activity that can take place, as well as the real estate value of the property. Total economic yield is limited by this building area, which may or may not reflect the actual development capacity of the site, but which is often difficult to change once established. A farmer can change crops, improve crop performance, and reallocate crop acreage within a farm to improve its economic yield. Cities cannot easily change zoning, reallocate land use patterns, and improve development capacity to increase their yield, so they sprawl when they can and struggle when they can't. Unfortunately, sprawl can produce a decline in a community's average yield per acre when it does not include a high-yield component. This shortfall becomes apparent only when the yield from the annexed construction proves inadequate to meet its maintenance cost years later, however. Until that time, all physical improvements accepted by a city involve no cost, and the yield appears to be a windfall.

Land use allocation and economic stability are introduced in Chap. 6. This chapter adds several site-specific case studies to amplify the capacity of the development forecast collection to predict and evaluate the potential yield from individual projects. This is the cornerstone of economic development, since the yield from larger land use areas will not improve until project yield combines to improve the average. Before these case studies are introduced, however, it might be helpful to paint a general picture of the economic landscape in which they can be useful.

Economic Landscape

A community's economic context is often described by a government budget that is too complex for landowners and taxpayers to comprehend. The details in this budget confuse the fact that it is very similar to a personal budget. It contains annual operating expenses (bills), capital maintenance costs (repairs and remodeling), debt service (mortgage or installment debt payments), and capital investment expense (additions and new construction). These total expenses purchase a community lifestyle. Dividing total annual expense by a city's total geographic area produces the annual cost per acre for this lifestyle. This cost is offset by the annual income received from each acre within the city, which varies by land use activity. Table 16.1 represents a typical upper-middle-class suburb and is repeated from Chap. 6 to illustrate the yield from various land use areas. Data circled and labeled #1 in this table are associated with a 3-acre residential zone (where at least 3 acres per lot are required). 34 homes are located in this zone, at an actual density of 6.19 gross acres per home. They produced gross estimated municipal revenue of $109,140. (Additional yield went to the county, the schools, and the library.) This produced an average municipal yield of $518 per gross acre.

Data labeled #2 represent a residential zone within this city with a maximum permitted net density of 4.84 dwelling units per net acre. The zone contains 2336.1 gross acres and 6162 homes, for an actual density of 2.64 homes per gross acre. These gross acres yielded an estimated total of $6,593,340, for an average yield of $2822 per gross acre. These data reveal that the 3-acre zone contains far more expensive houses, but when the yield is divided by the land consumed, the yield per acre is a fraction of that of the higher-density, less prestigious zone. Neither zone produced enough, however, to equal the city's total annual expense per acre.

Data labeled #3 show that the city received $2,772,071 of income that was not related to land use, such as interest received on its fund balance. This was approximately 11.8% of its budget. The remaining 88.2% illustrates that the economies of most cities depend heavily on the yield produced by the allocation of activities within their land use portfolios.

Data labeled #4 shows that the average residential yield per gross acre was $2412 and the average nonresidential,

Table 16.1 Economic Implications of Geographic Land Use Allocation

	Effective Rates	Commercial
MILLAGES		
1 City: 6.29	6.138015	6.290000
2 County: 14.57	11.830000	13.329492
3 School: 74.66	39.549064	43.042970
4 Library: 1.00	0.850000	1.000000
Total millage	58.367079	63.662462
Personal prop. millage		96.52
VARIABLES		
Income tax %	0.02	
Real prop. valuation %	0.35	
Total income tax paid	$13,347,366	
Personal prop. valuation	0.25	
		avg. res. density
No. of dwelling units	13,956	2.61
RESIDENTIAL REVENUE		
Avg. real estate tax	$4,169,874	
Avg. income tax	$2,808,321	
OTHER RESIDENTIAL REVENUE		
Estate tax	$3,028,710	
Permits & fees	$399,468	
Other	$3,081,022	
Total other	$6,509,200	
COMMERCIAL VARIABLES		
Income Tax	$5,367,211	
Real Estate Tax	$605,893	
Other	$258,130	
AREAS		
Residential AC	5,353	
Institutional AC	589	
Res+Inst AC	5,942	
Commercial AC	296	
Total	6,238	
LAND VALUATION		personal prop.
Residential	$721,903,410	$93,051,224
Commercial	$88,434,130	
EXPENSE		
Operations	$20,171,314	$3,234
Depreciated assets	$838,813	$134
Old debt service	$2,336,870	$375
1994 Improvements		
Proj.1994 debt service		
1993 COST/ AC		$3,743
REVENUE		Rev/ AC
Residential revenue	$13,487,395	
Commercial revenue	$6,231,234	
Institutional revenue	$843,239	
Other revenue	$2,772,071	$3,741

6

3

Zone	RA	RB	RC	RD	R1	R2	R3	R4, R5 & R6				R7	R7	Commercial	Totals
Land use %	3.37%	0.52%	4.58%	0.13%	1.62%	38.64%	37.45%	1.34%	3.56%	3.56%	0.45%	0.03%	0.00%		
Acres/DU required	3	2	1	0.5	0.34	0.27	0.21	0.21	0.10	0.09	0.07	0.05	0.03		
Est. road allocation %	0.2	0.2	0.2	0.15	0.15	0.15	0.1	0.1	0.1	0.1	0.1	0.1	0.1		
Net density	0.33	0.50	1.00	2.00	2.90	3.63	4.84	4.84	9.68	10.89	14.52	21.78	34.85		
Gross density	0.27	0.40	0.80	1.70	2.47	3.09	4.36	4.36	8.71	9.80	13.07	19.60	31.37		
Land use AC	210.46	32.36	285.81	8.24	101.14	2410.59	2336.10	83.59	222.07	222.07	28.07	1.62	0.00		5,942
Est. no. of DU	34	8	138	14	151	4498	6162	220	1171	1318	222	19	0		13,956
Avg REtax paid/ DU	**$9,035**	**$8,192**	**$5,421**	**$4,217**	**$3,614**	**$3,313**	**$3,012**	**$2,409**	**$1,807**	**$1,205**	**$964**	**$843**			$39,651,804
Income tax to city/ DU	**$585**	**$487**	**$390**	**$341**	**$292**	**$234**	**$195**	**$175**	**$156**	**$136**	**$117**	**$97**			$2,808,321
School property tax/ DU	$5,478	$4,967	$3,287	$2,557	$2,191	$2,009	$1,826	$1,461	$1,096	$730	$584	$511			$24,042,271
County property tax/ DU	$1,639	$1,486	$983	$765	$655	$601	$546	$437	$328	$218	$175	$153			$7,191,575
Library property tax/ DU	$118	$107	$71	$55	$47	$43	$39	$31	$24	$16	$13	$11			$516,723
City property tax/ DU	$950	$861	$570	$443	$380	$348	$317	$253	$190	$127	$101	$89		$2,047	$4,169,874
City other tax/ DU	$1,483	$1,345	$890	$692	$593	$544	$494	$396	$297	$198	$158	$138		$872	$6,509,200
City income tax/ DU	$585	$487	$390	$341	$292	$234	$195	$175	$156	$136	$117	$97		$18,132	$2,808,321
Institutional revenue/ DU	$192	$174	$115	$90	$77	$70	$64	$51	$38	$26	$20	$18			$843,239
TOTAL CITY REV/ DU	**$3,210**	**$2,868**	**$1,965**	**$1,566**	**$1,343**	**$1,197**	**$1,070**	**$876**	**$681**	**$486**	**$397**	**$342**			
TOTAL CITY REV/ AC	$518	$695	$952	$2,663	$2,004	$2,233	$2,822	$2,309	$3,592	$2,887	$3,141	$4,065	$2,412	$21,051	
SCHOOL, COUNTY, & LIBRARY REV per DU	$8,085	$7,331	$4,851	$3,773	$3,234	$2,965	$2,695	$2,156	$1,617	$1,078	$862	$755			
per AC	$1,305	$1,775	$2,350	$6,414	$4,827	$5,532	$7,108	$5,687	$8,530	$6,397	$6,824	$8,956	$5,971	$12,835	

1

2

Avg. City Revenue per Residential AC

4

Avg. City Revenue per Commercial AC

NOTE: 4.7% commercial land use
22.7% of total city revenue provided by commercial land use

Avg. School, County & Library Revenue per Commercial AC

5

NOTE: Institutional = income tax & stormwater revenue from city, school, and other institutions

Avg. School, County, & Library Revenue per Residential AC

or commercial, yield was $21,051 per gross acre, or 9 times the residential yield.

Data labeled #5 shows that the average municipal expense was $3743 per gross acre and the average income was $3741. This break-even position was possible because the city's high-yielding nonresidential land use allocation is offsetting its low-yield acreage to produce an average income equal to the expense of the city's current lifestyle. (Of course, *lifestyle* and *quality of life* are not necessarily synonymous.)

Data labeled #6 show that only 296 of the city's 6238 acres are devoted to nonresidential land use. This 4.74% land use allocation, when combined with other sources of reliable and unreliable annual income, is meeting current operating expenses and a limited maintenance program. The city's extensive residential land use allocation, however, makes it less able to undertake deferred maintenance, construct capital improvements, and benefit from positive business cycles that enhance annual revenue.

The underlying message from this brief review is that cities are not divorced from the land, and the way they use it has great agricultural similarities. Most, if not all, government budgets overlook this correlation and do not report or evaluate the yield per acre produced by their various land use components. This is only natural, since income is received from taxpayers, and the land they consume for these activities is not correlated to evaluate yield. This is surprising, since a city can be thought of as a farm with various land use "crops." Each crop produces a yield, and the sum of these crop yields establishes the prosperity of the farm. Unfortunately, the farm is not contained, and the crops are not consciously balanced to produce the yield needed. Since government is not correlating land use and yield, urban development conservation is overlooked. Many local economies have unstable land use allocation ratios that are unrecognized, and sprawl results from a congeries of land use laws, policies, and practices that ignore yield and assume that new land will always be available to compensate for past practice. At least part of this systemic problem may result from a relatively vague recognition that community income, or revenue, is "grown" from the land, and that development capacity affects the economic yield produced. Economic development, therefore, is about getting the most from planned high-yield land use areas so that the average produced is adequate to serve the physical, social, and economic needs of a growing population and an aging infrastructure.

The development forecast collection can assist in assessing the development capacity of potential business locations by predicting the answers to two questions:

- How much land is needed to meet a given building area objective?
- How much gross building area can be located on a given site?

Remember that building area determines the number of people and scope of activity that can take place on a given land area, and that economic yield is a function of the activity sheltered within and the real estate value present. When an accurate prediction of development capacity can be produced, it is a short step to forecasting construction cost, population capacity, traffic generation, tax yield, minimum lease rates, returns on investment, etc.[3]

MetroCenter

MetroCenter represents a speculative subdivision for business that many economic developers dream about, and illustrates that development capacity and environmental context can be coordinated to produce high-yield acres within a city's land use portfolio. The nearly complete project is shown in Fig. 16.1, and is an image of success that does not convey the

Figure 16.1 MetroCenter overview from the west. The internal loop road, retention pond concept, freeway exposure, interchange access, and residential neighborhood relationship of MetroCenter is clearly visible in this photograph. (Photo by W. M. Hosack)

risk taken when only the loop road, pond retention system, and "seed" office buildings were present. MetroCenter has office, retail, and hotel space. It also has an environmentally friendly image. Figure 16.2 illustrates the loop road at the heart of this image, which contains approximately 42.42 acres. Building cover consumes 2.31 acres, or 5.4% of this land. Parking cover consumes 17.12 acres, or 40.4% of the land. Open space and miscellaneous pavement preserve 22.99 acres, or 54.2%, for people, and ponds represent 22.5% of the open space provided. This recipe produces the image in Fig. 16.2 and is promoted by all real estate interests in the region, thus multiplying the economic development effort of the community.

Figure 16.2 MetroCenter loop road. The 54.2% open space allocation, including stormwater retention ponds, and the influence of this open space on curb appeal and pedestrian amenity benefits the entire center.

(Photo by W. M. Hosack)

MetroFive

MetroFive is located within MetroCenter and represents the one-to-one relationship that an economic developer can face with a potential business interest as the developer works to attract occupancy, preserve context, and maximize yield. MetroFive is a 287,175-sq-ft gross building area with an average building height of 8.23 floors; it is used here to illustrate the comparative process of development capacity evaluation. The case is stated in terms of a building area objective rather than the lot area available, because the land in an office subdivision is not necessarily composed of fixed lot areas. These areas can be customized to meet an owner's objectives and are generally sold on a per-acre basis. Fixed lot areas may be left after early development choices are made, however, which means that an economic developer should also be prepared to forecast the development capacity of these remaining areas as discussed in previous chapters.

The MetroFive project is shown abutting the freeway in Fig. 16.3. The evaluation scenario leading to its construction in this case study is written to illustrate the potential use of the development forecast collection, but is entirely fictional. The interior loop of MetroCenter provides 54% open space, with 22.5% as common retention pond area, and much of the perimeter borrows from this central amenity. An attempt was made to mirror the signature pond features of the development along the southern leg of the loop, as shown in Fig. 16.1, but the pattern was not completed and the effort was compromised with adjacent parking. The development concept and unique identity of MetroCenter, however, is based on storm retention ponds in common areas along the northern and southern legs of the interior loop road. For this case study, we also assume that this is combined with a 35% open space requirement for private development areas and a maximum permitted height of 10 stories.

Existing Conditions

The actual project constructed has a footprint of 34,875 sq ft. Seven building floors and atrium levels contain 93,450 gross sq ft. (Open atrium levels without floors are still included in the total area per floor.) Nine adjacent building floors contain 193,725 sq ft, for a gross building area of 287,175 sq ft. This area produces a yield of 19,356 building sq ft per acre of buildable land available. The average building height of 8.234 floors is found by dividing the gross building area by the building footprint.

(a)

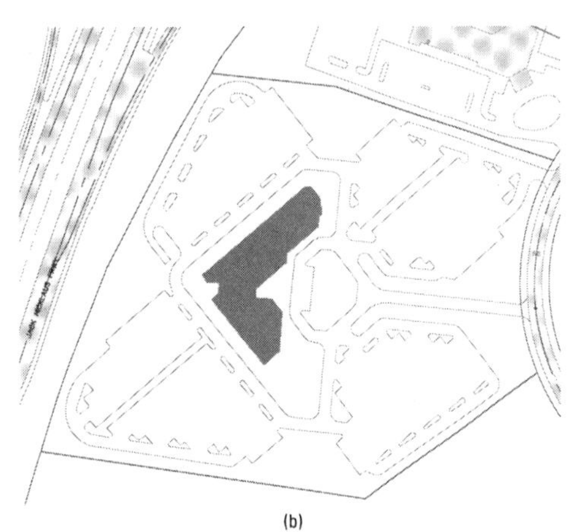

(b)

(c)

(d)

Figure 16.3 MetroFive from the east. MetroFive emphasizes freeway exposure and driveway separation from parking lot circulation aisles to establish a high-profile entrance and office image. The property perimeter has received less attention to the relationship between development cover, open space, and landscape design. (*a*) Overview; (*b*) site plan; (*c*) and (*d*) ground-level views.

Design Specification

The design specification panel of Exhibit 16.1 defines this project and notes that one parking space is provided for every 352 sq ft of building area. The planning forecast panel notes that this ratio produces 816 surface parking spaces. The total parking area divided by the number of spaces provided produces an average of 350 sq ft of surface area per space. The independent driveways serving these lots consume 14.9% of the buildable land area. Miscellaneous pavement consumes approximately 0.5% of the land area, and 35% is left as project open space.

It is unlikely that an economic developer would have anticipated the preceding construction data as values for entry in the design specification panel of Exhibit 16.2 (forecast model CG1B). A more likely set of values would include the MetroCenter open space standard of 35% and a parking requirement equal to the zoning ordinance in place. This is often 1 parking space for every 250 sq ft of gross building area (i.e., 4 spaces per 1000 gross sq ft of office development). A conservative entry for the amount of surface parking lot area per parking space would equal 400 sq ft. This does not imply a lot of interior landscaping, but most basic parking lots can be designed using this allotment. Estimates for driveways and miscellaneous pavement could vary greatly for a typical development, but it is unlikely that 14.9% would be entered unless the owner made it clear that interior vehicular access driveways were to be separated from parking circulation aisles.

Forecast

The specification values entered in Exhibit 16.2 are therefore typical values that might be used to produce a preliminary forecast for MetroFive. A 10% expansion value has also been included in the design specification panel as a contingency to account for the unknown. (If future expansion

Exhibit 16.1 MetroFive Existing Conditions

Forecast Model CG1B

Development capacity forecast for ***NONRESIDENTIAL BUILDINGS*** *based on the use of a* ***GRADE PARKING LOT*** *located on the same premises. When s and a equal zero in the design specification below, the forecast pertains to conditions when* ***NO PARKING*** *is required.*

Given: *Gross building area objective.* ***To Find:*** *Minimum buildable land area needed to achieve the gross building area objective given based on the design specification values entered below.* ***Premise:*** *All building floors considered equal in area.*

DESIGN SPECIFICATION *Enter values in boxed areas where text is bold and blue. Express all fractions as decimals.*

Given:	**Gross building area objective**	GBA=	**287,175**	SF
Land Variables:	Public/ private right-of-way & paved easements	W=	**0.000**	fraction of GLA
	Future expansion and/or unbuildalbe areas	U=	**0.000**	
	Gross land area reduction	X=	0.000	
Parking Variables:	Estimated gross pkg. lot area per parking space in SF	s =	**350**	ENTER ZERO IF NO PARKING REQUIRED
	Building SF permitted per parking space	a =	**352**	ENTER ZERO IF NO PARKING REQUIRED
	No. of loading spaces	l=	**0**	
	Gross area per loading space	b=	**0**	SF
Site Variables: BLA= buildable land area	**Project open dpace as fraction of BLA**	S=	**0.350**	
	Private driveways as fraction of BLA	R=	**0.149**	
	Misc. pavement as fraction of BLA	M=	**0.0052**	
	Total site support areas as a fraction of BLA	Su=	0.504	
Core:	**Core development area at grade as fraction of BLA:**	C=	0.496	C+Su must = 1

PLANNING FORECAST

NOTE: Be aware when BCG becomes too small to be feasible.

net pkg. area PLA	pkg. spaces NPS
285,543	816

NOTE:
Be aware when BCG becomes too small to be feasible.

no. of floors **FLR**	footprint BCA	minimum lot area CORE for BCG & PLA	buildable land BLA area forecast	buildable acre **BAC** forecast	BLA with expansion BLX forecast	BAC with expansion **BAX** forecast	bldg SF per SFAC BAX acre	flr area ratio FAR function of BLX
1.00	287,175	572,718	1,155,140	26.518	1,155,140	26.518	10,829	0.249
2.00	143,588	429,131	865,532	19.870	865,532	19.870	14,453	0.332
3.00	95,725	381,268	768,996	17.654	768,996	17.654	16,267	0.373
4.00	71,794	357,337	720,728	16.546	720,728	16.546	17,357	0.398
5.00	57,435	342,978	691,767	15.881	691,767	15.881	18,083	0.415
6.00	47,863	333,406	672,460	15.438	672,460	15.438	18,602	0.427
7.00	41,025	326,568	658,669	15.121	658,669	15.121	18,992	0.436
8.234	**34,877**	**320,420**	**646,269**	**14.836**	**646,269**	**14.836**	**19,356**	**0.444**
9.00	31,908	317,452	640,282	14.699	640,282	14.699	19,537	0.449
10.00	28,718	314,261	633,846	14.551	633,846	14.551	19,736	0.453
11.00	26,107	311,650	628,580	14.430	628,580	14.430	19,901	0.457
12.00	23,931	309,475	624,192	14.329	624,192	14.329	20,041	0.460
13.00	22,090	307,634	620,479	14.244	620,479	14.244	20,161	0.463
14.00	20,513	306,056	617,297	14.171	617,297	14.171	20,265	0.465
15.00	19,145	304,688	614,539	14.108	614,539	14.108	20,356	0.467

WARNING: These are preliminary forecasts that must not be used to make final decisions.
1) These forecasts are not a substitute for the "due diligence" research that must be conducted to support the final definition of "unbuildable areas" above and the final decision to purchase land. This research includes, but is not limited to, verification of adequate subsurface soil, zoning, environmental clearance, access, title, utilities and water pressure, clearance from deed restriction, easement and right-of-way encumbrances, clearance from existing above and below ground facility conflicts, etc.
2) The most promising forecast(s) made on the basis of data entered in the design specification from "due diligence" research must be verified at the drawing board before funds are committed and land purchase decisions are made. Actual land shape ratios, dimensions and irregularities encountered may require adjustments to the general forecasts above.
3) The software licensee shall take responsibility for the design specification values entered and any advice given that is based on the forecast produced.

Exhibit 16.2 MetroFive Forecast Specifications

Forecast Model CG1B

Development capacity forecast for ***NONRESIDENTIAL BUILDINGS*** *based on the use of a* ***GRADE PARKING LOT*** *located on the same premises. When s and a equal zero in the design specification below, the forecast pertains to conditions when* ***NO PARKING*** *is required.*

Given: *Gross building area objective.* ***To Find:*** *Minimum buildable land area needed to achieve the gross building area objective given based on the design specification values entered below.* ***Premise:*** *All building floors considered equal in area.*

DESIGN SPECIFICATION *Enter values in boxed areas where text is bold and blue. Express all fractions as decimals.*

Given:	**Gross building area objective**	GBA=	**287,175**	SF
Land Variables:	Public/ private right-of-way & paved easements	W=	**0.000**	fraction of GLA
	Future expansion and/or unbuildalbe areas	U=	**0.100**	
	Gross land area reduction	X=	0.100	
Parking Variables:	Estimated gross pkg. lot area per parking space in SF	s =	**400**	ENTER ZERO IF NO PARKING REQUIRED
	Building SF permitted per parking space	a =	**250**	ENTER ZERO IF NO PARKING REQUIRED
	No. of loading spaces	l=	**0**	
	Gross area per loading space	b=	**0**	SF
Site Variables: BLA= buildable land area	**Project open dpace as fraction of BLA**	S=	**0.350**	
	Private driveways as fraction of BLA	R=	**0.030**	
	Misc. pavement as fraction of BLA	M=	**0.020**	
	Total site support areas as a fraction of BLA	Su=	0.400	
Core:	**Core development area at grade as fraction of BLA:**	C=	0.600	C+Su must = 1

PLANNING FORECAST

NOTE: Be aware when BCG becomes too small to be feasible.

net pkg. area PLA	pkg. spaces NPS
459,480	1,149

NOTE: Be aware when BCG becomes too small to be feasible.

no. of floors **FLR**	footprint BCA	minimum lot area CORE for BCG & PLA	buildable land BLA area forecast	buildable acre **BAC** forecast	BLA with expansion BLX forecast	BAC with expansion **BAX** forecast	bldg SF per SFAC BAX acre	flr area ratio FAR function of BLX
1.00	287,175	746,655	1,244,425	28.568	1,382,694	31.742	9,047	0.208
2.00	143,588	603,068	1,005,113	23.074	1,116,792	25.638	11,201	0.257
3.00	95,725	555,205	925,342	21.243	1,028,157	23.603	12,167	0.279
4.00	71,794	531,274	885,456	20.327	983,840	22.586	12,715	0.292
5.00	57,435	516,915	861,525	19.778	957,250	21.975	13,068	0.300
6.00	47,863	507,343	845,571	19.412	939,523	21.568	13,315	0.306
7.00	41,025	500,505	834,175	19.150	926,861	21.278	13,496	0.310
8.23	**34,877**	**494,357**	**823,928**	**18.915**	**915,475**	**21.016**	**13,664**	**0.314**
9.00	31,908	491,388	818,981	18.801	909,978	20.890	13,747	0.316
10.00	28,718	488,198	813,663	18.679	904,069	20.755	13,837	0.318
11.00	26,107	485,587	809,311	18.579	899,235	20.644	13,911	0.319
12.00	23,931	483,411	805,685	18.496	895,206	20.551	13,974	0.321
13.00	22,090	481,570	802,617	18.426	891,797	20.473	14,027	0.322
14.00	20,513	479,993	799,988	18.365	888,875	20.406	14,073	0.323
15.00	19,145	478,625	797,708	18.313	886,343	20.348	14,113	0.324

WARNING: These are preliminary forecasts that must not be used to make final decisions.

1) These forecasts are not a substitute for the "due diligence" research that must be conducted to support the final definition of "unbuildable areas" above and the final decision to purchase land. This research includes, but is not limited to, verification of adequate subsurface soil, zoning, environmental clearance, access, title, utilities and water pressure, clearance from deed restriction, easement and right-of-way encumbrances, clearance from existing above and below ground facility conflicts, etc.

2) The most promising forecast(s) made on the basis of data entered in the design specification from "due diligence" research must be verified at the drawing board before funds are committed and land purchase decisions are made. Actual land shape ratios, dimensions and irregularities encountered may require adjustments to the general forecasts above.

3) The software licensee shall take responsibility for the design specification values entered and any advice given that is based on the forecast produced.

possibilities were actually involved, this value could easily equal or exceed 50%.) Exhibit 16.2 shows that this conservative set of design specification values produces a buildable land area forecast BAX of 21.016 acres for MetroFive. The yield, however, is only 13,664 building sq ft per acre because of the conservative values used.

Exhibit 16.3 predicts that only 13.69 acres would be required to support MetroFive, including a 10% contingency, if the prospective business were able to obtain the following concessions from the economic developer and the city. First, each parking space would be permitted to serve 350 sq ft of gross building area instead of 250, a reduction from 4 to 2.857 spaces per 1000 sq ft of gross building area. Second, the average parking lot area per space would be reduced from 400 to 350 sq ft to eliminate the internal landscape islands intended. Third, driveway separation from parking lot circulation aisles would be deleted and the percentage reduced from 14.9% to 3%. This would increase the yield to 20,915 building sq ft per acre and produce a total land area forecast less than that actually used.

Evaluation

The buildable land area currently forecast in this example to support MetroFive stretches from 13.69 to 21.01 acres, and these forecasts are a function of the design specification values assigned in each forecast model. When land can easily cost $100,000 to $200,000 per acre and more, a 7.32-acre spread becomes significant, and the relationship of design specification values to project cost and economic yield can easily be seen. However, design specification values also represent context recipes. If economic development has created a subdivision to sell to high-yield business activity with a design concept as its signature and a design specification as its recipe, the degree of departure from this specification becomes a serious issue. The development forecast collection can help with these negotiations and with a forecast of the result that will be produced with compromise.

Exhibit 16.3 MetroFive Forecast with Reduced Specifications

Forecast Model CG1B

*Development capacity forecast for **NONRESIDENTIAL BUILDINGS** based on the use of a **GRADE PARKING LOT** located on the same premises. When s and a equal zero in the design specification below, the forecast pertains to conditions when **NO PARKING** is required.*

***Given:** Gross building area objective. **To Find:** Minimum buildable land area needed to achieve the gross building area objective given based on the design specification values entered below. **Premise:** All building floors considered equal in area.*

DESIGN SPECIFICATION *Enter values in boxed areas where text is bold and blue. Express all fractions as decimals.*

Given:	**Gross building area objective**	GBA=	**287,175**	SF
Land Variables:	Public/ private right-of-way & paved easements	W=	**0.000**	fraction of GLA
	Future expansion and/or unbuildalbe areas	U=	**0.100**	
	Gross land area reduction	X=	0.100	
Parking Variables:	Estimated gross pkg. lot area per parking space in SF	s =	**350**	ENTER ZERO IF NO PARKING REQUIRED
	Building SF permitted per parking space	a =	**350**	ENTER ZERO IF NO PARKING REQUIRED
	No. of loading spaces	l=	**0**	
	Gross area per loading space	b=	**0**	SF
Site Variables: BLA= buildable land area	**Project open dpace as fraction of BLA**	S=	**0.350**	
	Private driveways as fraction of BLA	R=	**0.030**	
	Misc. pavement as fraction of BLA	M=	**0.020**	
	Total site support areas as a fraction of BLA	Su=	0.400	
Core:	**Core development area at grade as fraction of BLA:**	C=	0.600	C+Su must = 1

PLANNING FORECAST

NOTE: Be aware when BCG becomes too small to be feasible.

net pkg. area PLA	pkg. spaces NPS
287,175	821

NOTE: Be aware when BCG becomes too small to be feasible.

no. of floors **FLR**	footprint BCA	minimum lot area CORE for BCG & PLA	buildable land BLA area forecast	buildable acre **BAC** forecast	BLA with expansion BLX forecast	BAC with expansion **BAX** forecast	bldg SF per SFAC BAX acre	flr area ratio FAR function of BLX
1.00	287,175	574,350	957,250	21.975	1,063,611	24.417	11,761	0.270
2.00	143,588	430,763	717,938	16.482	797,708	18.313	15,682	0.360
3.00	95,725	382,900	638,167	14.650	709,074	16.278	17,642	0.405
4.00	71,794	358,969	598,281	13.735	664,757	15.261	18,818	0.432
5.00	57,435	344,610	574,350	13.185	638,167	14.650	19,602	0.450
6.00	47,863	335,038	558,396	12.819	620,440	14.243	20,162	0.463
7.00	41,025	328,200	547,000	12.557	607,778	13.953	20,582	0.473
8.23	**34,877**	**322,052**	**536,753**	**12.322**	**596,392**	**13.691**	**20,975**	**0.482**
9.00	31,908	319,083	531,806	12.209	590,895	13.565	21,170	0.486
10.00	28,718	315,893	526,488	12.086	584,986	13.429	21,384	0.491
11.00	26,107	313,282	522,136	11.987	580,152	13.318	21,562	0.495
12.00	23,931	311,106	518,510	11.903	576,123	13.226	21,713	0.498
13.00	22,090	309,265	515,442	11.833	572,714	13.148	21,842	0.501
14.00	20,513	307,688	512,813	11.773	569,792	13.081	21,954	0.504
15.00	19,145	306,320	510,533	11.720	567,259	13.022	22,052	0.506

WARNING: These are preliminary forecasts that must not be used to make final decisions.

1) These forecasts are not a substitute for the "due diligence" research that must be conducted to support the final definition of "unbuildable areas" above and the final decision to purchase land. This research includes, but is not limited to, verification of adequate subsurface soil, zoning, environmental clearance, access, title, utilities and water pressure, clearance from deed restriction, easement and right-of-way encumbrances, clearance from existing above and below ground facility conflicts, etc.

2) The most promising forecast(s) made on the basis of data entered in the design specification from "due diligence" research must be verified at the drawing board before funds are committed and land purchase decisions are made. Actual land shape ratios, dimensions and irregularities encountered may require adjustments to the general forecasts above.

3) The software licensee shall take responsibility for the design specification values entered and any advice given that is based on the forecast produced.

Options

Exhibit 16.4 shows that the 287,175-sq-ft MetroFive project could actually have been built on 11.92 acres instead of 14.83 acres with 35% open space and an 8.23-story average building height. This would have produced a yield of 24,082 sq ft per buildable acre. To achieve this yield, however, the formal entry road, separated driveways, and parking lot islands would all have to be sacrificed. Parking would move closer to the building, all circulation would be through parking aisles, and the spacious lawns would shrink as the 35% open space value became a function of a smaller total land area. These are all compatibility factors that must be kept in mind when forecasts are made and yield is considered, since the results implied by a single change to the design specification recipe can not only affect cost, return on investment, and yield but context, quality, and future attraction of additional business. This is where negotiation based on the common language of development capacity analysis can help all to understand the objectives, compromises, and contributions of others. This may encourage everyone to work toward a mutually beneficial agreement that defines the land required to preserve the environmental context desired.

Yield

It is easier to think of yield in terms of the SFAC development capacity index produced and the associated activity that will occupy the building square feet per buildable acre because fewer variables are involved.[4] Financial forecasts are a function of this building area and activity, but are more unpredictable because they depend on the intricacies of private financing arrangements. They can be helpful, however, when attempting to forecast general revenue expectations and incentives, and they improve with the level of information given.

Exhibit 16.4 MetroFive Minimum Forecast

Forecast Model CG1B

Development capacity forecast for ***NONRESIDENTIAL BUILDINGS*** *based on the use of a* ***GRADE PARKING LOT*** *located on the same premises. When s and a equal zero in the design specification below, the forecast pertains to conditions when* ***NO PARKING*** *is required.*

Given: *Gross building area objective.* ***To Find:*** *Minimum buildable land area needed to achieve the gross building area objective given based on the design specification values entered below.* ***Premise:*** *All building floors considered equal in area.*

DESIGN SPECIFICATION *Enter values in boxed areas where text is bold and blue. Express all fractions as decimals.*

Given:	**Gross building area objective**	GBA=	**287,175**	SF
Land Variables:	Public/ private right-of-way & paved easements	W=	**0.000**	fraction of GLA
	Future expansion and/or unbuildalbe areas	U=	**0.000**	
	Gross land area reduction	X=	0.000	
Parking Variables:	Estimated gross pkg. lot area per parking space in SF	s =	**350**	ENTER ZERO IF NO PARKING REQUIRED
	Building SF permitted per parking space	a =	**350**	ENTER ZERO IF NO PARKING REQUIRED
	No. of loading spaces	l=	**0**	
	Gross area per loading space	b=	**0**	SF
Site Variables: BLA= buildable land area	**Project open dpace as fraction of BLA**	S=	**0.350**	
	Private driveways as fraction of BLA	R=	**0.020**	
	Misc. pavement as fraction of BLA	M=	**0.010**	
	Total site support areas as a fraction of BLA	Su=	0.380	
Core:	**Core development area at grade as fraction of BLA:**	C=	0.620	C+Su must = 1

PLANNING FORECAST

NOTE: Be aware when BCG becomes too small to be feasible.

net pkg. area PLA	pkg. spaces NPS
287,175	821

NOTE:
Be aware when BCG becomes too small to be feasible.

no. of floors **FLR**	footprint BCA	minimum lot area CORE for BCG & PLA	buildable land BLA area forecast	buildable acre **BAC** forecast	BLA with expansion BLX forecast	BAC with expansion **BAX** forecast	bldg SF per SFAC BAX acre	flr area ratio FAR function of BLX
1.00	287,175	574,350	926,371	21.267	926,371	21.267	13,504	0.310
2.00	143,588	430,763	694,778	15.950	694,778	15.950	18,005	0.413
3.00	95,725	382,900	617,581	14.178	617,581	14.178	20,255	0.465
4.00	71,794	358,969	578,982	13.292	578,982	13.292	21,606	0.496
5.00	57,435	344,610	555,823	12.760	555,823	12.760	22,506	0.517
6.00	47,863	335,038	540,383	12.405	540,383	12.405	23,149	0.531
7.00	41,025	328,200	529,355	12.152	529,355	12.152	23,631	0.543
8.23	**34,877**	**322,052**	**519,438**	**11.925**	**519,438**	**11.925**	**24,082**	**0.553**
9.00	31,908	319,083	514,651	11.815	514,651	11.815	24,306	0.558
10.00	28,718	315,893	509,504	11.697	509,504	11.697	24,552	0.564
11.00	26,107	313,282	505,293	11.600	505,293	11.600	24,757	0.568
12.00	23,931	311,106	501,784	11.519	501,784	11.519	24,930	0.572
13.00	22,090	309,265	498,815	11.451	498,815	11.451	25,078	0.576
14.00	20,513	307,688	496,270	11.393	496,270	11.393	25,207	0.579
15.00	19,145	306,320	494,065	11.342	494,065	11.342	25,319	0.581

WARNING: These are preliminary forecasts that must not be used to make final decisions.
1) These forecasts are not a substitute for the "due diligence" research that must be conducted to support the final definition of "unbuildable areas" above and the final decision to purchase land. This research includes, but is not limited to, verification of adequate subsurface soil, zoning, environmental clearance, access, title, utilities and water pressure, clearance from deed restriction, easement and right-of-way encumbrances, clearance from existing above and below ground facility conflicts, etc.
2) The most promising forecast(s) made on the basis of data entered in the design specification from "due diligence" research must be verified at the drawing board before funds are committed and land purchase decisions are made. Actual land shape ratios, dimensions and irregularities encountered may require adjustments to the general forecasts above.
3) The software licensee shall take responsibility for the design specification values entered and any advice given that is based on the forecast produced.

Property Value

I doubt that many would disagree with the basic premise that government collects taxes to reinvest in the community, although the term *spending* is far more familiar than the term *reinvestment*. Fewer make the connection that this reinvestment affects property value, and instead tend to attribute all such value and appreciation to the school system. However, a city with sewer backups, crumbling streets and curbs, low water pressure, inadequate trash collection, overgrown parks, and inadequate fire and police protection does not contribute to property value and does not establish the proper foundation for property appreciation. This foundation of value must be in place before a school system can improve the rate of appreciation realized. Building, maintaining, and improving this foundation requires annual reinvestment from each shareholder in the form of taxes. This is the crux of the problem, since the public rarely agrees with the amount of taxes collected and has a poor understanding of the real estate value maintained by this investment. It often disagrees with the budget priorities established for this reinvestment and is taught to suspect the motives of all elected and appointed representatives in charge of it. These are political issues that are beyond the scope of this book, but all of these issues are affected by the amount of annual income received by the community for reinvestment. This income is a function of the revenue-producing activities within the community and the real estate value of the shelter they occupy. The scope of these activities is a function of the building area present, which in turn is a function of the development capacity of land. The more we sprawl to produce building area, the less development capacity we use, and the more land we consume.

Sprawl

Sprawl ignores development capacity and conservation. It consumes indiscriminately and can actually erode a city's financial position when the income received from its land use allocation does not equal its cost of maintenance and operation over an extended period of time. Sprawl is simply one method of producing the building area needed to shelter community activities. The value of these buildings and the revenue produced by these activities represent every community's investment portfolio. More can be received from fewer acres when development capacity is taken into account, and the development forecast collection can help everyone evaluate both the quantity and quality of the investment yield produced in relation to the land consumed and the environmental context created (see Chaps. 3 and 6). Ignoring development capacity will simply continue a pattern of sprawl that ignores the fact that we must eventually learn to live within limits.

Doing More with Less

If the 14.836-acre MetroFive tract had been left after the development of adjacent properties, an economic developer could have chosen forecast model CG1L to predict its development capacity. If the developer had guessed the exact design specification of MetroFive, the model would have produced the forecast shown as Exhibit 16.5. As mentioned, however, its very unlikely that anyone would guess this precise specification. An economic developer would be more likely to produce the forecast shown in Exhibit 16.6. This forecast is based on the open space standard of 35%, the more generic parking and miscellaneous pavement values shown in the design specification, and the maximum permitted building height of 10 stories. Based on these values, the gross building area (GBA) projection could have been 228,090 sq ft, with 912 parking spaces and a yield of 15,374 sq ft per buildable acre. The fact that 287,175 sq ft was actually produced with an average building height of 8.234 stories is a function of the revised design specification values shown in Exhibit 16.1. This again demonstrates

Exhibit 16.5 MetroFive Exact Forecast

Forecast Model CG1L

*Development capacity forecast for **NONRESIDENTIAL BUILDINGS** based on the use of an adjacent **GRADE PARKING LOT** located on the same premises. When s and a equal zero in the design specification below, the forecast pertains to conditions when **NO PARKING** is required.*

Given: *Gross land area.* ***To Find:*** *Maximum development capacity of the land area (gross building area potential) based on the design specification values entered below.* ***Premise:*** *All building floors considered equal in area.*

DESIGN SPECIFICATION

Enter values in boxed areas where text is bold and blue. Express all fractions as decimals.

Given:	**Gross land area**	GLA=	**14.836**	acres	646,256	SF
Land Variables:	Public/ private right-of-way & paved easements	W=	**0.000**	fraction of GLA	0	SF
	Net land area	NLA=	14.836	acres	646,256	SF
	Unbuildable and/or future expansion areas	U=	**0.000**	fraction of GLA	0	SF
	Gross land area reduction	X=	0.000	fraction of GLA	0	SF
	Buildable land area remaining	BLA=	14.836	acres	646,256	SF
Parking Variables:	Est. gross pkg. lot area per space in SF	s =	**350**	ENTER ZERO IF NO PARKING REQUIRED		
	Building SF permitted per parking space	a =	**352**	ENTER ZERO IF NO PARKING REQUIRED		
	No. of loading spaces	l=	**0**			
	Gross area per loading space	b =	**0**	SF	0	SF
Site Variables:	**Project open space as fraction of BLA**	S=	**0.350**	⬅	226,190	SF
	Private driveways as fraction of BLA	R=	**0.149**		96,292	SF
	Misc. pavement as fraction of BLA	M=	**0.005**		3,361	SF
	Loading area as fraction of BLA	L=	0.000		0	SF
	Total site support areas as a fraction of BLA	Su=	0.504		325,842	SF
Core:	**Core development area as fraction of BLA**	C=	0.496	C+Su must = 1	320,414	SF

PLANNING FORECAST

no. of floors **FLR**	CORE minimum land area for BCG & PLA	gross building area **GBA**	parking lot area PLA	pkg spaces NPS	footprint **BCA**	bldg SF / acre SFAC function of BLA	flr area ratio FAR function of BLA
1.00	320,414	160,663	159,750	456.4	160,663	10,829	0.249
2.00		214,421	213,203	609.2	107,211	14,453	0.332
3.00		241,339	239,968	685.6	80,446	16,267	0.373
4.00		257,502	256,038	731.5	64,375	17,357	0.398
5.00		268,282	266,757	762.2	53,656	18,083	0.415
6.00		275,984	274,416	784.0	45,997	18,602	0.427
7.00		281,763	280,162	800.5	40,252	18,992	0.436
8.234		**287,169**	**285,538**	**815.8**	**34,876**	**19,356**	**0.444**
9.00		289,855	288,208	823.5	32,206	19,537	0.449
10.00	*NOTE:*	292,798	291,134	831.8	29,280	19,736	0.453
11.00	*Be aware when BCA*	295,250	293,573	838.8	26,841	19,901	0.457
12.00	*becomes too small to be feasible.*	297,326	295,637	844.7	24,777	20,041	0.460
13.00		299,105	297,406	849.7	23,008	20,161	0.463
14.00		300,647	298,939	854.1	21,475	20,265	0.465
15.00		301,997	300,281	857.9	20,133	20,356	0.467

WARNING: These are preliminary forecasts that must not be used to make final decisions.
1) These forecasts are not a substitute for the "due diligence" research that must be conducted to support the final definition of "unbuildable areas" above and the final decision to purchase land. This research includes, but is not limited to, verification of adequate subsurface soil, zoning, environmental clearance, access, title, utilities and water pressure, clearance from deed restriction, easement and right-of-way encumbrances, clearance from existing above and below ground facility conflicts, etc.
2) The most promising forecast(s) made on the basis of data entered in the design specification from "due diligence" research must be verified at the drawing board before funds are committed and land purchase decisions are made. Actual land shape ratios, dimensions and irregularities encountered may require adjustments to the general forecasts above.
3) The software licensee shall take responsibility for the design specification values entered and any advice given that is based on the forecast produced.

Exhibit 16.6 MetroFive General Forecast

Forecast Model CG1L

Development capacity forecast for ***NONRESIDENTIAL BUILDINGS*** *based on the use of an adjacent* ***GRADE PARKING LOT*** *located on the same premises. When s and a equal zero in the design specification below, the forecast pertains to conditions when* ***NO PARKING*** *is required.*

Given: *Gross land area.* ***To Find:*** *Maximum development capacity of the land area (gross building area potential) based on the design specification values entered below.* ***Premise:*** *All building floors considered equal in area.*

DESIGN SPECIFICATION

Enter values in boxed areas where text is bold and blue. Express all fractions as decimals.

Given:	**Gross land area**	GLA=	14.836	acres	646,256	SF
Land Variables:	Public/ private right-of-way & paved easements	W=	0.000	fraction of GLA	0	SF
	Net land area	NLA=	14.836	acres	646,256	SF
	Unbuildable and/or future expansion areas	U=	0.000	fraction of GLA	0	SF
	Gross land area reduction	X=	0.000	fraction of GLA	0	SF
	Buildable land area remaining	BLA=	14.836	acres	646,256	SF
Parking Variables:	Est. gross pkg. lot area per space in SF	s =	400	ENTER ZERO IF NO PARKING REQUIRED		
	Building SF permitted per parking space	a =	250	ENTER ZERO IF NO PARKING REQUIRED		
	No. of loading spaces	l=	0			
	Gross area per loading space	b =	0	SF	0	SF
Site Variables:	**Project open space as fraction of BLA**	S=	0.350		226,190	SF
	Private driveways as fraction of BLA	R=	0.030		19,388	SF
	Misc. pavement as fraction of BLA	M=	0.020		12,925	SF
	Loading area as fraction of BLA	L=	0.000		0	SF
	Total site support areas as a fraction of BLA	Su=	0.400		258,502	SF
Core:	**Core development area as fraction of BLA**	C=	0.600	C+Su must = 1	387,754	SF

PLANNING FORECAST

no. of floors **FLR**	CORE minimum land area for BCG & PLA	gross building area **GBA**	parking lot area PLA	pkg spaces NPS	footprint **BCA**	bldg SF / acre SFAC function of BLA	flr area ratio FAR function of BLA
1.00	387,754	149,136	238,618	596.5	149,136	10,052	0.231
2.00		184,645	295,431	738.6	92,322	12,446	0.286
3.00	*NOTE:*	200,562	320,900	802.2	66,854	13,519	0.310
4.00	*Be aware when BCA*	209,597	335,355	838.4	52,399	14,128	0.324
5.00	*becomes too small to be feasible.*	215,419	344,670	861.7	43,084	14,520	0.333
6.00		219,483	351,173	877.9	36,581	14,794	0.340
7.00		222,482	355,971	889.9	31,783	14,996	0.344
8.00		224,785	359,656	899.1	28,098	15,151	0.348
9.00		226,609	362,575	906.4	25,179	15,274	0.351
10.00		**228,090**	**364,945**	**912.4**	**22,809**	**15,374**	**0.353**
11.00		229,317	366,907	917.3	20,847	15,457	0.355
12.00		230,349	368,558	921.4	19,196	15,526	0.356
13.00		231,229	369,967	924.9	17,787	15,586	0.358
14.00		231,989	371,183	928.0	16,571	15,637	0.359
15.00		232,652	372,244	930.6	15,510	15,682	0.360

WARNING: These are preliminary forecasts that must not be used to make final decisions.

1) These forecasts are not a substitute for the "due diligence" research that must be conducted to support the final definition of "unbuildable areas" above and the final decision to purchase land. This research includes, but is not limited to, verification of adequate subsurface soil, zoning, environmental clearance, access, title, utilities and water pressure, clearance from deed restriction, easement and right-of-way encumbrances, clearance from existing above and below ground facility conflicts, etc.

2) The most promising forecast(s) made on the basis of data entered in the design specification from "due diligence" research must be verified at the drawing board before funds are committed and land purchase decisions are made. Actual land shape ratios, dimensions and irregularities encountered may require adjustments to the general forecasts above.

3) The software licensee shall take responsibility for the design specification values entered and any advice given that is based on the forecast produced.

the impact of these values and the need to either negotiate or mandate these specifications as part of a development process written to improve the yield from a city's land use allocation portfolio.

Redevelopment

Redevelopment is another way to improve the yield from a city's land use portfolio. The term generally means either a significant modification of existing real estate or the replacement of these assets. When demolition and reconstruction is involved, the process involves additional cost but is no different than building on a vacant piece of land. Development capacity analysis in this case would simply involve the use of the development forecast collection as it is described in this book.

The modification and improvement of existing assets involves circumstances that may not be anticipated by the development forecast collection. The mixed-use series has been previously introduced to address some of these issues, and an expansion model is introduced in this chapter to address others.

Kingsdale Shopping Center

Kingsdale Shopping Center is a fictional case study involving a significant expansion of existing assets with a mixed-use development that optimizes the use of a shopping center parking lot. Most shopping centers are built with surface parking lots, and a desirable parking standard is 5 spaces per 1000 gross sq ft of building area (5/1000 GBA), or 1 space for every 200 sq ft of building area (1/200 sq ft). This is often reduced to 4 spaces per 1000 GBA when land is in short supply. When it is reduced below 3.5 spaces per 1000 GBA, serious parking capacity problems can emerge when the center is healthy. When the center is not healthy, any parking ratio represents a hidden asset.

A retail parking ratio of 1/200 sq ft or 5/1000 GBA is 4 times an apartment parking ratio of 1/800 sq ft.[5] This means that 1 shopping center parking space can support either 800 sq ft of gross dwelling unit area or 200 sq ft of retail space. From this standpoint, a shopping center parking space represents a hidden asset when the center itself is not performing well for any number of reasons. This asset can be used if the retail area is reduced to meet the shrinking market demand it faces. Every parking space released from retail service can support 800 sq ft or more of dwelling unit area, depending on the dwelling unit mix planned. This can produce a combined yield from the land available and additional customers for the retail space remaining when apartments are introduced as a mixed use supported by the excess retail parking.

Kingsdale is shown in Fig. 16.4. The site plan is somewhat unusual, since the northern 5.7 acres of the 26.38-acre tract is under separate ownership. This 5.7-acre tract is visible in Fig. 16.4 as the lighter shade of parking lot. The remaining 20.68-acre tract has been used for this vertically integrated mixed-use case study. Forecast model MG1L has been selected for this example since it addresses surface parking lots (category G1) when the land area (category L) is known and the concept is vertically mixed use (category M). This forecast model is displayed in Exhibit 16.7, and its design specification table has been completed with values that pertain to both the retail area and the anticipated mixed-use apartment area. The design specification shows that the parking lot area forecast will be based on 375 sq ft per parking space s. It also shows that 1 parking space will be provided per apartment house dwelling unit u, and that retail parking will be provided at a ratio of 1 space for every 285 gross sq ft of building area a (3.5/1000 GBA). Three loading dock areas l will be provided, and no separated driveways will be included, thus leaving the miscellaneous driveway R and pavement M estimate at 2%. The shopping

(a)

(b)

(c)

Figure 16.4 Kingsdale Shopping Center. Kingsdale represents a strip shopping mall converted to an outdoor atrium shopping plaza. It substituted as a regional shopping center for years, but has encountered recent competition from two new and much larger covered shopping centers designed to serve the region. This may indicate further change in this center's future. (*a*) Overview; (*b*) site plan; (*c*) overview.

(Photos by W. M. Hosack)

Exhibit 16.7 Kingsdale with Residential Allocation Percentage of 55%

Forecast Model MG1L

*Development capacity forecast for **MIXED USE** based on an adjacent **GRADE PARKING LOT** located on the same premises.*
***Given:** Gross land area. **To Find:** Maximum commercial building area and apartment dwelling unit capacity of the land area given when the residential land use allocation varies. **Premise:** All building floors considered equal in area.*

DESIGN SPECIFICATION — *Enter values in boxed areas where text is bold and blue. Express all fractions as decimals.*

Given:	**Gross land area**	GLA=	**20.680**	acres	900,821	SF
Land Variables:	Public/ private right-of-way & paved easements	W=	**0.000**	fraction of GLA	0	SF
	Net land area	NLA=	20.680	acres	900,821	SF
	Unbuildable and/or future expansion areas	U=	**0.000**	fraction of GLA	0	SF
	Gross land area reduction	X=	0.000	fraction of GLA	0	SF
	Buildable land area remaining	BLA=	**20.680**	acres	900,821	SF
Parking Variables:	Est. gross pkg. lot area per pkg. space in SF	s =	**375**			
	Parking lot spaces planned or required per dwelling unit	u=	**1**			
	Garage parking spaces planned or required per dwelling unit	Gn=	**0.00**			
	Gross building area per garage space	Ga=	**0**			
	Nonresidential building SF permitted per parking space	a=	**285**			
	No. of loading spaces	l =	**3**			
	Gross area per loading space	b =	**1,000**	SF	3,000	SF
Site Variables:	**Project open space as fraction of BLA**	S=	**0.400**		360,328	SF
	Private driveways as fraction of BLA	R=	**0.010**		9,008	SF
	Misc. pavement as fraction of BLA	M=	**0.010**		9,008	SF
	Loading area as fraction of BLA	L=	0.003		3,000	SF
	Total site support areas as a fraction of BLA	Su=	0.423		381,345	SF
Core:	**Core development area as fraction of BLA**	C=	0.577	C+Su must = 1	519,476	SF
Building Variables:	Res. bldg. efficiency as percentage of GBA	Be=	**0.800**			
	Building support as fraction of GBA	Bu=	0.200	Be + Bu must = 1		

Dwelling Unit Mix Table:

DU dwelling unit type	GDA gross du area	CDA=GDA/Be comprehensive du area	MIX du mix	PDA = (CDA)MIX Pro-rated du area
EFF	**400**	500	**5%**	25
1 BR	**500**	625	**25%**	156
2 BR	**800**	1,000	**50%**	500
3 BR	**1,000**	1,250	**20%**	250
4 BR	**1,200**	1,500	**0%**	0
			Aggregate avg. dwelling unit area (AGG) =	**931**
			GBA sf per parking space a=	931

MIXED-USE PLANNING FORECAST

55.00% =(RAP): residential land use allocation percentage
45.00% =(CAP): nonresidential land use allocation percentage

total floors FLR	CGBA nonres GBA	RGBA res GBA	total bldg MBCA cover area	nonres CFLR floors	total parking MPLA lot area	total parking MNPS spaces	total dwelling MNDU units	density per dBA bidable acre	EFF	1 BR	2 BR	3 BR	4 BR
1.00	100,944	204,219	305,162	0.331	215,056	573	219.3	10.6	11.0	54.8	109.6	43.9	0.0
2.00	128,740	317,336	223,038	0.577	297,181	792	340.8	16.5	17.0	85.2	170.4	68.2	0.0
3.00	141,751	389,195	176,982	0.801	343,237	915	417.9	20.2	20.9	104.5	209.0	83.6	0.0
4.00	**149,295**	**438,886**	**147,045**	**1.015**	**373,173**	**995**	**471.3**	**22.8**	**23.6**	**117.8**	**235.6**	**94.3**	**0.0**
5.00	154,219	475,297	125,903	1.225	394,315	1,052	510.4	24.7	25.5	127.6	255.2	102.1	0.0
6.00	157,687	503,124	110,135	1.432	410,083	1,094	540.3	26.1	27.0	135.1	270.1	108.1	0.0
7.00	160,261	525,082	97,906	1.637	422,312	1,126	563.8	27.3	28.2	141.0	281.9	112.8	0.0
8.00	162,247	542,851	88,137	1.841	432,081	1,152	582.9	28.2	29.1	145.7	291.5	116.6	0.0
9.00	163,827	557,526	80,150	2.044	440,068	1,174	598.7	28.9	29.9	149.7	299.3	119.7	0.0
10.00	165,112	569,849	73,496	2.247	446,722	1,191	611.9	29.6	30.6	153.0	306.0	122.4	0.0
11.00	166,179	580,344	67,866	2.449	452,353	1,206	623.2	30.1	31.2	155.8	311.6	124.6	0.0
12.00	167,079	589,391	63,039	2.650	457,179	1,219	632.9	30.6	31.6	158.2	316.5	126.6	0.0
13.00	167,848	597,268	58,855	2.852	461,363	1,230	641.4	31.0	32.1	160.3	320.7	128.3	0.0
14.00	168,513	604,190	55,193	3.053	465,026	1,240	648.8	31.4	32.4	162.2	324.4	129.8	0.0
15.00	169,093	610,320	51,961	3.254	468,258	1,249	655.4	31.7	32.8	163.8	327.7	131.1	0.0

(dwelling unit breakdown by type: EFF, 1 BR, 2 BR, 3 BR, 4 BR)

WARNING: These are preliminary forecasts that must not be used to make final decisions.
1) These forecasts are not a substitute for the "due diligence" research that must be conducted to support the final definition of "unbuildable areas" above and the final decision to purchase land. This research includes, but is not limited to, verification of adequate subsurface soil, zoning, environmental clearance, access, title, utilities and water pressure, clearance from deed restriction, easement and right-of-way encumbrances, clearance from existing above and below ground facility conflicts, etc.
2) The most promising forecast(s) made on the basis of data entered in the design specification from "due diligence" research must be verified at the drawing board before funds are committed and land purchase decisions are made. Actual land shape ratios, dimensions and irregularities encountered may require adjustments to the general forecasts above.
3) The software licensee shall take responsibility for the design specification values entered and any advice given that is based on the forecast produced.

center's current project open space allocation is virtually nil, but since it is introducing a mixed-use residential component *S*, the design specification value for this element has been increased to 40%.

Apartment house efficiency Be has been estimated at 80%, and the dwelling unit mix table displays the apartment areas anticipated within the building and the percentage of each that will be included in the mix. Based on this dwelling unit mix table, the aggregate average dwelling unit area AGG is 931 sq ft, and the average parking ratio is 1 space for every 931 sq ft of gross building area.

The planning forecast panel of Exhibit 16.7 contains a value in the upper left-hand corner identified as RAP. This value specifies that the data predicted in the forecast panel is based on the assumption that 55% of the buildable land area will be used to support the residential component of this mixed-use project. This assumption is a variable that may be changed to produce alternate forecasts. Based on the current assumption and the design specification values entered, the planning forecast panel shows that a 4-story building with a 1-story retail area CFLR will produce a footprint MBCA of 147,045 sq ft that can shelter an MNDU of 471 apartments on 3 floors above. The number of dwelling types in the mix is forecast in the right-hand section of the planning forecast panel. In order to produce this forecast, however, the current retail footprint of 218,800 sq ft would have to be reduced by at least 71,655 sq ft in order to increase building height and provide the areas forecast from the design specification values entered.[6] The combined 4-story result would produce 588,181 sq ft of total building area, with an apartment density of 22.8 dwelling units per buildable acre and a nonresidential yield of 7,110.5 sq ft per acre. The total building area increase of 369,481 sq ft represents a combined yield of 28,442 sq ft per acre. It can only result, however, when the retail footprint area is reduced by

71,655 sq ft and building height is increased to 4 floors, or when design specification values are altered in other ways to achieve the same result.

If the reduction in retail area is not acceptable, the design specification variables in forecast model MG1L can be changed at will to find a combination that can meet the shopping center's market objectives. Once this is found, the model will also reveal the redevelopment adjustments required. For instance, assume the center wishes to retain 186,000 sq ft of first-floor retail area instead of 147,045 sq ft. Exhibit 16.8 shows that an MBCA of 186,802 sq ft of retail space can be produced on 1 floor (CFLR column) when an RAP of only 30% of the buildable area is devoted to apartment house activity and the rest of the design specification remains intact. This produces a building with an FLR of 2 stories, a footprint MBCA of 186,802 sq ft, a parking lot MNPS of 889 cars, 40% open space *S,* and an MNDU of 186 dwelling units on the second floor. It still requires a footprint reduction of 31,898 sq ft, however. If the center wished to retain only 100,000 sq ft of first-floor retail area, Exhibit 16.9 shows that this area could be produced on 1 floor when an RAP of 71.5% of the land area is devoted to the support of apartment house activity. This could produce an 8-story building with a 100,974-sq-ft footprint, 40% open space, and 757 dwelling units on 7 floors above the retail level. Exhibit 16.9 shows that the resulting apartment density would be 36.6 dwelling units per buildable acre. The nonresidential yield would be 4882.7 sq ft per acre. The combined gross building area would be 807,793 sq ft, and the combined yield would be 39,061.5 building sq ft per acre with 40% open space.

All of these projects require demolition and reconstruction of significant parts of the project area in order to accomplish their vertically integrated mixed-use objective, however. In some cases, it is possible to adapt rather than reconstruct when land is available.

Exhibit 16.8 Kingsdale with Residential Allocation Percentage of 30%

Forecast Model MG1L

*Development capacity forecast for **MIXED USE** based on an adjacent **GRADE PARKING LOT** located on the same premises.*
***Given:** Gross land area. **To Find:** Maximum commercial building area and apartment dwelling unit capacity of the land area given when the residential land use allocation varies. **Premise:** All building floors considered equal in area.*

DESIGN SPECIFICATION

Enter values in boxed areas where text is bold and blue. Express all fractions as decimals.

Given:	**Gross land area**	GLA=	**20.680**	acres	900,821	SF
Land Variables:	Public/ private right-of-way & paved easements	W=	**0.000**	fraction of GLA	0	SF
	Net land area	NLA=	20.680	acres	900,821	SF
	Unbuildable and/or future expansion areas	U=	**0.000**	fraction of GLA	0	SF
	Gross land area reduction	X=	0.000	fraction of GLA	0	SF
	Buildable land area remaining	BLA=	**20.680**	acres	900,821	SF
Parking Variables:	Est. gross pkg. lot area per pkg. space in SF	s=	**375**			
	Parking lot spaces planned or required per dwelling unit	u=	**1**			
	Garage parking spaces planned or required per dwelling unit	Gn=	**0.00**			
	Gross building area per garage space	Ga=	**0**			
	Nonresidential building SF permitted per parking space	a=	**285**			
	No. of loading spaces	l=	**3**			
	Gross area per loading space	b=	**1,000**	SF	3,000	SF
Site Variables:	**Project open space as fraction of BLA**	S=	**0.400**		360,328	SF
	Private driveways as fraction of BLA	R=	**0.010**		9,008	SF
	Misc. pavement as fraction of BLA	M=	**0.010**		9,008	SF
	Loading area as fraction of BLA	L=	0.003		3,000	SF
	Total site support areas as a fraction of BLA	Su=	0.423		381,345	SF
Core:	**Core development area as fraction of BLA**	C=	0.577	C+Su must = 1	519,476	SF
Building Variables:	Res. bldg. efficiency as percentage of GBA	Be=	**0.800**			
	Building support as fraction of GBA	Bu=	0.200	Be + Bu must = 1		

Dwelling Unit Mix Table:

DU dwelling unit type	GDA gross du area	CDA=GDA/Be comprehensive du area	MIX du mix	PDA = (CDA)MIX Pro-rated du area
EFF	**400**	500	**5%**	25
1 BR	**500**	625	**25%**	156
2 BR	**800**	1,000	**50%**	500
3 BR	**1,000**	1,250	**20%**	250
4 BR	**1,200**	1,500	**0%**	0
			Aggregate avg. dwelling unit area (AGG) =	**931**
			GBA sf per parking space a=	931

MIXED USE PLANNING FORECAST

30.00% =(RAP): residential land use allocation percentage
70.00% =(CAP): nonresidential land use allocation percentage

total floors FLR	CGBA nonres GBA	RGBA res GBA	total bldg MBCA cover area	nonres CFLR floors	total parking MPLA lot area	total parking MNPS spaces	total dwelling MNDU units	density per dBA bldable acre	EFF	1 BR	2 BR	3 BR	4 BR
									dwelling unit breakdown by type				
1.00	157,023	111,552	268,576	0.585	251,530	671	119.8	5.8	6.0	29.9	59.9	24.0	0.0
2.00	**200,262**	**173,342**	**186,802**	**1.072**	**333,304**	**889**	**186.1**	**9.0**	**9.3**	**46.5**	**93.1**	**37.2**	**0.0**
3.00	220,501	212,594	144,365	1.527	375,741	1,002	228.3	11.0	11.4	57.1	114.1	45.7	0.0
4.00	232,236	239,737	117,993	1.968	402,113	1,072	257.4	12.4	12.9	64.4	128.7	51.5	0.0
5.00	239,897	259,626	99,905	2.401	420,201	1,121	278.8	13.5	13.9	69.7	139.4	55.8	0.0
6.00	245,291	274,827	86,686	2.830	433,420	1,156	295.1	14.3	14.8	73.8	147.6	59.0	0.0
7.00	249,295	286,821	76,588	3.255	443,518	1,183	308.0	14.9	15.4	77.0	154.0	61.6	0.0
8.00	252,385	296,527	68,614	3.678	451,492	1,204	318.4	15.4	15.9	79.6	159.2	63.7	0.0
9.00	254,841	304,543	62,154	4.100	457,952	1,221	327.0	15.8	16.4	81.8	163.5	65.4	0.0
10.00	256,841	311,274	56,812	4.521	463,294	1,235	334.3	16.2	16.7	83.6	167.1	66.9	0.0
11.00	258,501	317,007	52,319	4.941	467,787	1,247	340.4	16.5	17.0	85.1	170.2	68.1	0.0
12.00	259,901	321,949	48,487	5.360	471,619	1,258	345.7	16.7	17.3	86.4	172.9	69.1	0.0
13.00	261,097	326,252	45,181	5.779	474,925	1,266	350.3	16.9	17.5	87.6	175.2	70.1	0.0
14.00	262,131	330,033	42,297	6.197	477,809	1,274	354.4	17.1	17.7	88.6	177.2	70.9	0.0
15.00	263,034	333,381	39,761	6.615	480,345	1,281	358.0	17.3	17.9	89.5	179.0	71.6	0.0

WARNING: These are preliminary forecasts that must not be used to make final decisions.
1) These forecasts are not a substitute for the "due diligence" research that must be conducted to support the final definition of "unbuildable areas" above and the final decision to purchase land. This research includes, but is not limited to, verification of adequate subsurface soil, zoning, environmental clearance, access, title, utilities and water pressure, clearance from deed restriction, easement and right-of-way encumbrances, clearance from existing above and below ground facility conflicts, etc.
2) The most promising forecast(s) made on the basis of data entered in the design specification from "due diligence" research must be verified at the drawing board before funds are committed and land purchase decisions are made. Actual land shape ratios, dimensions and irregularities encountered may require adjustments to the general forecasts above.
3) The software licensee shall take responsibility for the design specification values entered and any advice given that is based on the forecast produced.

Exhibit 16.9 Kingsdale with Residential Allocation Percentage of 71.5%

Forecast Model MG1L

Development capacity forecast for ***MIXED USE*** *based on an adjacent* ***GRADE PARKING LOT*** *located on the same premises.*
Given: *Gross land area.* ***To Find:*** *Maximum commercial building area and apartment dwelling unit capacity of the land area given when the residential land use allocation varies.* ***Premise:*** *All building floors considered equal in area.*

DESIGN SPECIFICATION — *Enter values in boxed areas where text is bold and blue. Express all fractions as decimals.*

Category	Item	Symbol	Value	Unit	SF value	
Given:	**Gross land area**	GLA=	20.680	acres	900,821	SF
Land Variables:	Public/ private right-of-way & paved easements	W=	0.000	fraction of GLA	0	SF
	Net land area	NLA=	20.680	acres	900,821	SF
	Unbuildable and/or future expansion areas	U=	0.000	fraction of GLA	0	SF
	Gross land area reduction	X=	0.000	fraction of GLA	0	SF
	Buildable land area remaining	BLA=	20.680	acres	900,821	SF
Parking Variables:	Est. gross pkg. lot area per pkg. space in SF	s=	375			
	Parking lot spaces planned or required per dwelling unit	u=	1			
	Garage parking spaces planned or required per dwelling unit	Gn=	0.00			
	Gross building area per garage space	Ga=	0			
	Non-residential building SF permitted per parking space	a=	285			
	No. of loading spaces	l=	3			
	Gross area per loading space	b=	1,000	SF	3,000	SF
Site Variables:	**Project open space as fraction of BLA**	S=	0.400		360,328	SF
	Private driveways as fraction of BLA	R=	0.010		9,008	SF
	Misc. pavement as fraction of BLA	M=	0.010		9,008	SF
	Loading area as fraction of BLA	L=	0.003		3,000	SF
	Total site support areas as a fraction of BLA	Su=	0.423		381,345	SF
Core:	**Core development area as fraction of BLA**	C=	0.577	C+Su must = 1	519,476	SF
Building Variables:	Res. bldg. efficiency as percentage of GBA	Be=	0.800			
	Building support as fraction of GBA	Bu=	0.200	Be + Bu must = 1		

Dwelling Unit Mix Table:

DU dwelling unit type	GDA gross du area	CDA=GDA/Be comprehensive du area	MIX du mix	PDA = (CDA)MIX Pro-rated du area
EFF	**400**	500	**5%**	25
1 BR	**500**	625	**25%**	156
2 BR	**800**	1,000	**50%**	500
3 BR	**1,000**	1,250	**20%**	250
4 BR	**1,200**	1,500	**0%**	0
			Aggregate avg. dwelling unit area (AGG) =	**931**
			GBA sf per parking space a=	931

MIXED USE PLANNING FORECAST

71.50% =(RAP): residential land use allocation percentage
28.50% =(CAP): nonresidential land use allocation percentage

total floors FLR	CGBA nonres GBA	RGBA res GBA	total bldg MBCA cover area	nonres CFLR floors	total parking MPLA lot area	total parking MNPS spaces	total dwelling MNDU units	density per dBA bldable acre	EFF	1 BR	2 BR	3 BR	4 BR
1.00	63,931	265,232	329,163	0.194	190,925	509	284.8	13.8	14.2	71.2	142.4	57.0	0.0
2.00	81,535	412,145	246,840	0.330	273,247	729	442.6	21.4	22.1	110.6	221.3	88.5	0.0
3.00	89,775	505,472	198,416	0.452	321,671	858	542.8	26.2	27.1	135.7	271.4	108.6	0.0
4.00	94,553	570,010	166,141	0.569	353,947	944	612.1	29.6	30.6	153.0	306.0	122.4	0.0
5.00	97,672	617,299	142,994	0.683	377,093	1,006	662.9	32.1	33.1	165.7	331.4	132.6	0.0
6.00	99,869	653,440	125,551	0.795	394,536	1,052	701.7	33.9	35.1	175.4	350.8	140.3	0.0
7.00	101,499	681,958	111,922	0.907	408,165	1,088	732.3	35.4	36.6	183.1	366.2	146.5	0.0
8.00	**102,757**	**705,036**	**100,974**	**1.018**	**419,113**	**1,118**	**757.1**	**36.6**	**37.9**	**189.3**	**378.5**	**151.4**	**0.0**
9.00	103,757	724,095	91,983	1.128	428,104	1,142	777.6	37.6	38.9	194.4	388.8	155.5	0.0
10.00	104,571	740,100	84,467	1.238	435,620	1,162	794.7	38.4	39.7	198.7	397.4	158.9	0.0
11.00	105,247	753,731	78,089	1.348	441,999	1,179	809.4	39.1	40.5	202.3	404.7	161.9	0.0
12.00	105,817	765,479	72,608	1.457	447,479	1,193	822.0	39.7	41.1	205.5	411.0	164.4	0.0
13.00	106,304	775,711	67,847	1.567	452,240	1,206	833.0	40.3	41.6	208.2	416.5	166.6	0.0
14.00	106,725	784,700	63,673	1.676	456,414	1,217	842.6	40.7	42.1	210.7	421.3	168.5	0.0
15.00	107,092	792,662	59,984	1.785	460,104	1,227	851.2	41.2	42.6	212.8	425.6	170.2	0.0

(dwelling unit breakdown by type: EFF, 1 BR, 2 BR, 3 BR, 4 BR)

WARNING: These are preliminary forecasts that must not be used to make final decisions.
1) These forecasts are not a substitute for the "due diligence" research that must be conducted to support the final definition of "unbuildable areas" above and the final decision to purchase land. This research includes, but is not limited to, verification of adequate subsurface soil, zoning, environmental clearance, access, title, utilities and water pressure, clearance from deed restriction, easement and right-of-way encumbrances, clearance from existing above and below ground facility conflicts, etc.
2) The most promising forecast(s) made on the basis of data entered in the design specification from "due diligence" research must be verified at the drawing board before funds are committed and land purchase decisions are made. Actual land shape ratios, dimensions and irregularities encountered may require adjustments to the general forecasts above.
3) The software licensee shall take responsibility for the design specification values entered and any advice given that is based on the forecast produced.

The Expansion Model

Another form of economic development involves the evaluation of existing lands and buildings for expansion potential without intrusive demolition requirements. The expansion model is written to address this issue. It applies to nonresidential construction when the land area is known and surface pavement is used for parking. It contains the following five panels:

EG1A: Existing conditions

EG1B: Development capacity forecast

EG1C: Construction cost forecast

EG1D: Revenue forecast

EG1E: Return-on-investment forecast

Each panel requires data entry, but only the first two are required to produce a development capacity forecast. The remaining panels involve additional variables that extend the implications of the development capacity forecast.

Existing Conditions

The EG1A panel is illustrated by Exhibit 16.10 and requests values that define existing conditions. The subpanel titled Reduction in Land Area seeks to identify the expansion potential of the land by first identifying all facilities and features that will remain. In the example shown, 50.9% of the site is unavailable for new construction, and 8.6% of this total is open space. The second subpanel, titled Accommodation Costs, needs to be completed only if the user intends to proceed to the panels involving construction cost, revenue, and return on investment. Theoretical values have been entered in this panel for the sake of this case study, but they should not be considered development standards or relevant for use beyond this example. Accommodation costs should not be overlooked, however, since redevelopment often involves damage, repair, and replacement expense.

Exhibit 16.10 Existing Conditions
Forecast Model EG1A

EXISTING FACILITIES and CONDITIONS

REDUCTION IN LAND AREA

Item	Value	Ratio	Note
Gross land area (GLA) in acres	**8.000**		sent to Exhibit 16.11
Total existing building footprint to remain	**32,162**	0.092	
Total existing building area to remain	**43,480**		
Existing building area subject to parking ratio	**40,369**		
Parking area to remain in SF	**58,000**	0.166	
Parking to remain in spaces	**145**		
Existing parking ratio to remain (SF / pkg. space)	278		check zoning adequacy
Existing parking ratio to remain (pkg. spaces / 1000 SF)	3.6		check zoning adequacy
Existing project open space to remain	**30,000**	0.086	
Existing road area to remain	**4,000**	0.011	
Miscellaneous pavement to remain	**2,000**	0.006	
Unbuildable area such as ravines, ponds, wetlands, etc.	**51,243**	0.147	
Other area to remain	**0**	0.000	
Unbuildable area including facilities and features to remain	177,405	50.9%	sent to Exhibit 16.11

ACCOMMODATION COSTS

Item	area to replace in sq. ft.	unit price $ / SF	lump sum	Note
Demolition cost (lump sum)			**$50,000**	sent to Exhibit 16.12
Other replacement costs				
Parking lot replacement cost	**58,000**	**$2.50**	$145,000	
Site/ landscape replacement	**30,000**	**$3.00**	$90,000	
Road replacement	**4,000**	**$5.00**	$20,000	
Other replacement costs (lump sum)	**0**		**$0**	
If required, subtotal other replacement costs below and carry up to lump sum cell above.	92,000		**$255,000**	sent to Exhibit 16.12

WARNING: These are preliminary forecasts that must not be used to make final decisions.
1) These forecasts are not a substitute for the "due diligence" research that must be conducted to support the final definition of "unbuildable areas" above and the final decision to purchase land. This research includes, but is not limited to, verification of adequate subsurface soil, zoning, environmental clearance, access, title, utilities and water pressure, clearance from deed restriction, easement and right-of-way encumbrances, clearance from existing above and below ground facility conflicts, etc.
2) The most promising forecast(s) made on the basis of data entered in the design specification from "due diligence" research must be verified at the drawing board before funds are committed and land purchase decisions are made. Actual land shape ratios, dimensions and irregularities encountered may require adjustments to the general forecasts above.
3) The software licensee shall take responsibility for the design specification values entered and any advice given that is based on the forecast produced.

Development Capacity Forecast

The EG1B panel illustrated by Exhibit 16.11 should be familiar, since it is the standard CG1L format included in the development forecast collection. The difference is that it is linked to the existing conditions panel of the adaptation model. The data that has been sent from the existing conditions panel is noted as linked in the design specification subpanel of Exhibit 16.11. Additional values have been entered in this panel to complete the data needed for a development capacity forecast. The parking requirement a is noted as 1 space for every 250 sq ft of new building area. Each parking space s is programmed to consume 375 sq ft of surface area, and 1 new loading space l is assumed to require

Exhibit 16.11 Development Capacity Forecast

Forecast Model EG1B

Development capacity forecast for ***NONRESIDENTIAL BUILDINGS*** *based on the use of an adjacent* ***GRADE PARKING LOT*** *located on the same premises. When s and a equal zero in the design specification below, the forecast pertains to conditions when* ***NO PARKING*** *is required.*

Given: *Gross land area.* ***To Find:*** *Maximum development capacity of the land area (gross building area potential) based on the design specification values entered below.* ***Premise:*** *All building floors considered equal in area.*

DESIGN SPECIFICATION

Enter values in boxed areas where text is bold and blue. Express all fractions as decimals.

Given:	**Gross land area**	**GLA=**	**8.000**	acres	348,480	SF	*received from Exhibit 16.10*
Land Variables:	Public/ private right-of-way & paved easements	W=	**0.150**	fraction of GLA	52,272	SF	
	Net land area	NLA=	6.800	acres	296,208	SF	
	Unbuildable areas such as ravines, easements, etc.	U=	**0.509**	fraction of GLA	177,405	SF	*received from Exhibit 16.10*
	Gross land area reduction	X=	0.659	fraction of GLA	229,677	SF	
	Buildable land area remaining	**BLA=**	2.727	acres	118,803	SF	
Parking Variables:	Est. gross pkg. lot area per space in SF	s =	**375**	ENTER ZERO IF NO PARKING REQUIRED			
	Building SF permitted per parking space	a =	**250**	ENTER ZERO IF NO PARKING REQUIRED			
	No. of loading spaces	l=	**1**				
	Gross area per loading space	b =	**1,000**	SF	1,000	SF	
Site Variables:	**Project open space as fraction of BLA**	**S=**	**0.600**		71,282	SF	
	Private driveways as fraction of BLA	R=	**0.030**		3,564	SF	
	Misc. pavement as fraction of BLA	M=	**0.020**		2,376	SF	
	Loading area as fraction of BLA	L=	0.008		1,000	SF	
	Total site support areas as a fraction of BLA	Su=	0.658		78,222	SF	
Core:	**Core development area as fraction of BLA**	**C=**	0.342	C+Su must = 1	40,581	SF	

DEVELOPMENT CAPACITY FORECAST

no. of floors **FLR**	CORE (minimum land area for BCG & PLA)	gross building area **GBA**	parking lot area PLA	pkg. spaces NPS	footprint **BCA**	bldg SF / acre SFAC (function of BLA)	flr area ratio FAR (function of BLA)
1.00	40,581	**16,232**	24,349	64.9	**16,232**	5,952	0.137
2.00		**20,291**	30,436	81.2	**10,145**	7,440	0.171
3.00		**22,135**	33,203	88.5	**7,378**	8,116	0.186
4.00		**23,189**	34,784	92.8	**5,797**	8,502	0.195
5.00		**23,871**	35,807	95.5	**4,774**	8,753	0.201
6.00	*NOTE: Be aware when BCG becomes too small to be feasible.*	**24,349**	36,523	97.4	**4,058**	8,928	0.205
7.00		**24,702**	37,052	98.8	**3,529**	9,057	0.208
8.00		**24,973**	37,459	99.9	**3,122**	9,157	0.210
9.00		**25,188**	37,782	100.8	**2,799**	9,235	0.212
10.00		**25,363**	38,045	101.5	**2,536**	9,300	0.213
11.00		**25,508**	38,262	102.0	**2,319**	9,353	0.215
12.00		**25,630**	38,445	102.5	**2,136**	9,397	0.216
13.00		**25,734**	38,601	102.9	**1,980**	9,436	0.217
14.00		**25,824**	38,736	103.3	**1,845**	9,469	0.217
15.00		**25,903**	38,854	103.6	**1,727**	9,497	0.218

WARNING: These are preliminary forecasts that must not be used to make final decisions.
1) These forecasts are not a substitute for the "due diligence" research that must be conducted to support the final definition of "unbuildable areas" above and the final decision to purchase land. This research includes, but is not limited to, verification of adequate subsurface soil, zoning, environmental clearance, access, title, utilities and water pressure, clearance from deed restriction, easement and right-of-way encumbrances, clearance from existing above and below ground facility conflicts, etc.
2) The most promising forecast(s) made on the basis of data entered in the design specification from "due diligence" research must be verified at the drawing board before funds are committed and land purchase decisions are made. Actual land shape ratios, dimensions and irregularities encountered may require adjustments to the general forecasts above.
3) The software licensee shall take responsibility for the design specification values entered and any advice given that is based on the forecast produced.

1000 sq ft of land area *b*. Driveways *R* and miscellaneous pavement *M* are predicted to require 5% of the land area available. Project open space, including yard areas required by the local zoning ordinance, is planned for 60% of the land area available. Based on this data, the development capacity forecast subpanel of Exhibit 16.11 predicts potential building area capacity when building height ranges from 1 to 15 stories. For instance, a 2-story building would produce a 20,291-sq-ft building with a 10,145-sq-ft footprint and a yield of 7440 sq ft per acre. A 5-story building would produce 23,871 sq ft with a 4774-sq-ft footprint and a yield of 8753 sq ft per acre. A 10-story building would produce a 25,363-sq-ft building with a 2536-sq-ft footprint and a yield of 9300 sq ft per acre.

Chapter 1 points out that two things happen as building height increases when a design specification is constant and a surface parking lot is used. First, the rate of increase in gross building area (GBA) per floor rapidly declines above five floors; second, the footprint area declines as building height increases. This decline can reach a point where it is no longer feasible to accommodate a desired floor plan. Both of these characteristics are visible in the GBA and BCA columns of Exhibit 16.11. The point at which a floor plan is no longer feasible, however, is not firmly established. It is a function of each owner's activity and the values entered in the design specification. The options presented in Exhibit 16.11 are therefore some of many that can be explored by changing the values entered in the design specification.

Construction Cost Forecast

The construction cost of the option defined in Exhibits 16.10 and 16.11 can be evaluated with forecast model EG1C. The model is presented as Exhibit 16.12 and is linked to both Exhibits 16.10 and 16.11. The building design subpanel in this model requests information that will enable the model to differentiate between building shell

Exhibit 16.12 Construction Cost Forecast

Forecast Model EG1C

BUILDING DESIGN OBJECTIVES

Building skin	Bs=	**0.05**
Net building area	Bn=	0.95
Building core	Bc=	**0.05**
Building mechanical area	Bm=	**0.02**
Building circulation area	Bc=	**0.08**
BUILDING SUPPORT	Bs=	0.20
BUILDING EFFICIENCY	Be=	0.80

NOTE: Tenant finish costs generally apply to speculative office buildings. Interior finish and equipment costs generally apply to custom buildings. Both interior and tenant finish costs may not be needed on the same project.

COST SPECIFICATION

Land cost per acre in $/AC	**$100,000**	$2.30
Structure / surface parking in $/SF	Pc=	**$2.50**
Loading area in $/SF	Lc=	**$5.00**
Open space grading & landscaping in $/SF	Sc=	**$3.00**
Public / private rights-of-way in $/SF	Rc=	**$7.50**
Driveways in $/SF	Dc=	**$2.50**
Miscellaneous pavement in $/SF	Mc=	**$2.50**
Interior finishes if applicable in $/SF	Ic=	**$0.00**
Furnishing, fixtures & equipment if applicable in $/SF	Ffe=	**$0.00**
Demolition costs (lump sum)	Dc=	$50,000
Replacement costs (lump sum)	Rc=	$255,000

received from Exhibit 16.10 (Demolition and Replacement costs)

0.380	=Tf	Tenant finishes if applicable as % of shell cost
0.120	=Df	Preconstruction fees as % of shell cost
0.085	=Mf	Marketing fees as % of shell cost
0.100	=Lf	Legal fees as % of shell cost
0.005	=Zf	Zoning and permit fees as % of shell cost
0.000	=Ff	Financing costs as % of shell cost
0.000	=Cf	Contingency as % of shell cost
0.690	=Tc	TOTAL ADMINISTRATION & OVERHEAD

sum from data in Exhibit 16.11

$1,650,896 — included in TCF below

CONSTRUCTION COST FORECAST

Site acquisition, cemolition, preparation, replacement, roadway, and landscape improvement

received from Exhibit 16.11 (FLR, GBA: CG1L) — *sum from Exhibit 16.11* (PLD)

no. of floors FLR	gross building area GBA	shell cost forecast SHC per sq. ft.	total shell TSC cost forecast	parking & loading PLD cost forecast	interior finish INF cost forecast	tenant finishes TNF cost forecast	fees, demolition ADM replacement & admin.	total cost TCF forecast	cost forecast $SF per sq. ft. of GBA
1.00	16,232	**$45.00**	$730,459	$65,872	see TNF	$277,574	$504,017	**$3,228,817**	**$198.91**
2.00	20,291	**$45.00**	913,074	81,089	see TNF	346,968	630,021	**$3,622,048**	**178.51**
3.00	22,135	**$50.00**	1,106,756	88,007	see TNF	420,567	763,662	**$4,029,887**	**182.06**
4.00	23,189	**$50.00**	1,159,459	91,959	see TNF	440,594	800,026	**$4,142,934**	**178.66**
5.00	23,871	**$55.00**	1,312,916	94,517	see TNF	498,908	905,912	**$4,463,150**	**186.97**
6.00	24,349	**$55.00**	1,339,175	96,307	see TNF	508,886	924,031	**$4,519,295**	**185.61**
7.00	24,702	**$60.00**	1,482,091	97,631	see TNF	563,194	1,022,642	**$4,816,454**	**194.99**
8.00	24,973	**$60.00**	1,498,377	98,649	see TNF	569,383	1,033,880	**$4,851,185**	**194.26**
9.00	25,188	**$65.00**	1,637,235	99,456	see TNF	622,149	1,129,692	**$5,139,429**	**204.04**
10.00	25,363	**$65.00**	1,648,605	100,112	see TNF	626,470	1,137,538	**$5,163,620**	**203.59**
11.00	25,508	**$70.00**	1,785,566	100,655	see TNF	678,515	1,232,041	**$5,447,673**	**213.57**
12.00	25,630	**$70.00**	1,794,110	101,113	see TNF	681,762	1,237,936	**$5,465,816**	**213.26**
13.00	25,734	**$75.00**	1,930,074	101,504	see TNF	733,428	1,331,751	**$5,747,653**	**223.35**
14.00	25,824	**$75.00**	1,936,823	101,841	see TNF	735,993	1,336,408	**$5,761,960**	**223.12**
15.00	25,903	**$80.00**	$2,072,224	$102,135	see TNF	$787,445	$1,429,834	**$6,042,535**	**$233.28**

WARNING: These are preliminary forecasts that must not be used to make final decisions.

1) These forecasts are not a substitute for the "due diligence" research that must be conducted to support the final definition of "unbuildable areas" above and the final decision to purchase land. This research includes, but is not limited to, verification of adequate subsurface soil, zoning, environmental clearance, access, title, utilities and water pressure, clearance from deed restriction, easement and right-of-way encumbrances, clearance from existing above and below ground facility conflicts, etc.

2) The most promising forecast(s) made on the basis of data entered in the design specification from "due diligence" research must be verified at the drawing board before funds are committed and land purchase decisions are made. Actual land shape ratios, dimensions and irregularities encountered may require adjustments to the general forecasts above.

3) The software licensee shall take responsibility for the design specification values entered and any advice given that is based on the forecast produced.

and building finish costs. The cost specification subpanel requests information regarding construction costs and other administrative costs that combine to produce total project cost. Building cost forecasts are entered in terms of dollars per square foot of construction. Project costs are entered as percentages of the shell cost. Land cost may not apply if the evaluation involves existing property, and accommodation costs involving demolition and replacement are linked from the existing conditions panel.

Shell cost values are entered in the construction cost forecast panel and increase with increases in building height. They may be altered to evaluate construction options, and the values shown should not be considered development standards. The forecasts produced in this panel are somewhat unusual, since most construction cost forecasts do not attempt to predict total project cost. This model includes these costs in its forecast when administrative values are entered in the right-hand column of the cost specification panel. This may make the total cost forecast TCF and the cost per square foot forecast $SF look high. If 0 is entered for each of these values, more typical construction cost values will be forecast.

Revenue Forecast

Exhibit 16.13 is based on forecast model EG1D and forecasts potential revenue from the project defined in Exhibits 16.10, 16.11, and 16.12. The building design subpanel within is linked from Exhibit 16.12, and also contains an additional field requesting the user to enter an estimate for the average number of people that will occupy each 1000 sq ft of habitable building area. (Habitable area = Be × GBA.) In addition to this, the revenue specification subpanel requests basic financial data and millage rates used to forecast tax yield. A tax abatement panel is also included to offer the opportunity to evaluate these incentives in relation to the revenue potential of the project. The revenue forecast

Exhibit 16.13 Revenue Forecast
Forecast Model EG1D

BUILDING DESIGN OBJECTIVES

Building skin	Bs=	0.05	*received from Exhibit 16.12*
Net building area	Bn=	0.95	
Building core	Bc=	0.05	
Building mechanical area	Bm=	0.02	
Building circulation area	Bc=	0.08	
BUILDING SUPPORT	Bs=	0.20	
BUILDING EFFICIENCY	Be=	0.80	
Avg. pop. per 1000 sf	Pop=	5.5	

REVENUE SPECIFICATION

Avg. income per person	Inc=	$30,000
Avg. daily trips per person	Adt=	5.0
Income tax rate		0.02
Real estate tax millage		67.082324
Fraction of real estate value assessed		0.35
Personal property millage		100.92
Personal property as % of building value		0.10
Fraction of personal property value assessed		0.25

effective commercial real estate tax millage rates

6.838642	10.2%	Fraction of REtax to local government
45.490353	67.8%	Fraction of REtax to local schools
13.754687	20.5%	Fraction of REtax to county
0.998642	1.5%	Fraction of REtax to library
0.000000	0.0%	Fraction of REtax to other
67.082324	100.0%	Total effective millage rate

ABATEMENT SPECIFICATION

First year real estate tax abatement request	1.00	58,728	Total local government revenue
Fraction of local school tax abated	0.50	7,728	Local government real estate tax abated
Effective first year property tax abatement	66.1%	13.2%	Total reduction in local government income

REVENUE FORECAST

no. of floors FLR (16.11)	gross building area GBA (16.11)	total project GBA$ cost forecast (16.12)	total building POP population	income tax to ITX local government	total real estate REX tax	real estate tax to RGX local government	real estate tax to RSX local schools	personal property PPX tax to local gov.	yield to YTG local government	max. first year FYA tax abatement
1.00	16,232	$3,228,817	71	$42,854	$75,809	$7,728	$51,408	$8,146	$58,728	$50,105
2.00	20,291	3,622,048	89	53,567	85,041	8,669	57,669	9,138	71,375	56,207
3.00	22,135	4,029,887	97	58,437	94,617	9,646	64,162	10,167	78,250	62,536
4.00	23,189	4,142,934	102	61,219	97,271	9,916	65,962	10,453	81,588	64,290
5.00	23,871	4,463,150	105	63,020	104,789	10,683	71,061	11,261	84,963	69,259
6.00	24,349	4,519,295	107	64,280	106,108	10,817	71,955	11,402	86,500	70,130
7.00	24,702	4,816,454	109	65,212	113,085	11,528	76,686	12,152	88,892	74,742
8.00	24,973	4,851,185	110	65,929	113,900	11,611	77,239	12,240	89,780	75,281
9.00	25,188	5,139,429	111	66,497	120,668	12,301	81,828	12,967	91,765	79,754
10.00	25,363	5,163,620	112	66,959	121,236	12,359	82,213	13,028	92,346	80,129
11.00	25,508	5,447,673	112	67,341	127,905	13,039	86,736	13,744	94,125	84,537
12.00	25,630	5,465,816	113	67,664	128,331	13,083	87,025	13,790	94,536	84,819
13.00	25,734	5,747,653	113	67,939	134,948	13,757	91,512	14,501	96,197	89,192
14.00	25,824	5,761,960	114	68,176	135,284	13,791	91,740	14,537	96,505	89,414
15.00	25,903	$6,042,535	114	$68,383	$141,872	$14,463	$96,207	$15,245	$98,092	$93,768

panel at the bottom of Exhibit 16.13 uses these values to produce yield forecasts in relation to the height, area, value, and population of the building area forecast. It also predicts tax abatement and the relationship of this abatement to the yield that will be received by all institutional and government entities.

Return-on-Investment Forecast

Exhibit 16.14 is the final panel in this model, but its debt service variables do not anticipate the full range of financing options available. It is included to complete a picture of the implications that spin off from a series of design specification decisions from Exhibits 16.10 and 16.11, but is by no means conclusive and is based on the assumption that the building involved is a speculative office building. Operating costs are requested in the operations subpanel. Debt service values are requested in the financing subpanel, and investment predictions are produced in the return-on-investment forecast subpanel based on the values entered. Total owner cost per net square foot CST is equal to the sum of debt service DBT and real estate tax per net square foot RXA. (Real estate tax is discounted by the tax abatement projected in Exhibit 16.13.) The average lease rate ALR results from increasing owner cost CST by the cost markup noted. The risk-adjusted lease rate RAL increases this rate to compensate for the building occupancy rate predicted in the financing subpanel. The gross annual profit forecast GPR simply subtracts total project cost from the total project lease income forecast. Return on investment is found by dividing this gross annual profit by the capital invested in the financing subpanel. The resulting percentage produces a very rough prediction of the potential health of the project, but should not be relied upon to make financial decisions.

The Barley Block

The retention and expansion of existing business enterprise gets less attention than the attraction of new business operations to a community, but is equally important since it has the potential to occur more frequently. If a community incubates a new business venture, it must have a place for it to grow. If it doesn't, the improved revenue stream is lost when the company relocates, and the community must begin again with another new and more risky venture. Landlocked incu-

Exhibit 16.14 Return on Investment
Forecast Model EG1E

BUILDING DESIGN OBJECTIVES

Building skin	Bs=	0.05	received from Exhibit 16.12
Net building area	Bn=	0.95	
Building core	Bc=	0.05	
Building mechanical area	Bm=	0.02	
Building circulation area	Bc=	0.08	
Building support	Bs=	0.20	
Building efficiency	Be=	0.80	

OPERATION SPECIFICATION

Utilities	$1.50	per net sq. ft.
Maintenance	$2.50	per net sq. ft.
Management	$2.50	per net sq. ft.
Land lease/ year	$0.00	per net sq. ft.
Other/ year	$0.00	per net sq. ft.
Total not included in owner cost per net sq. ft.	$6.50	per net sq. ft.

FINANCING SPECIFICATION

Percent down	25.0%
Annual interest	7.5%
Term in years	$30.00
Cost markup for lease rate	60.0%
Building occupancy rate	80.0%

Exhibit 16.12 data used in calculations

RETURN on INVESTMENT FORECAST 16.3%

no. of floors FLR (16.11)	gross GBA building area (16.11)	total project GBA$ cost forecast (16.12)	net leaseable NBA building area	debt service DBT per nsf	real estate tax RXA per nsf with abate	owner cost CST per nsf	average lease ALR rate no operating	risk adjusted RAL lease rate	gross GPR annual profit
1.00	16,232	$3,228,817	15,421	$12.63	$1.58	$14.22	$22.74	$28.43	$131,525
2.00	20,291	3,622,048	19,276	11.34	1.42	12.76	20.41	25.51	147,543
3.00	22,135	4,029,887	21,028	11.56	1.45	13.01	20.82	26.02	164,156
4.00	23,189	4,142,934	22,030	11.35	1.42	12.77	20.43	25.54	168,761
5.00	23,871	4,463,150	22,678	11.87	1.49	13.36	21.38	26.72	181,805
6.00	24,349	4,519,295	23,131	11.79	1.48	13.26	21.22	26.53	184,092
7.00	24,702	4,816,454	23,466	12.38	1.55	13.93	22.30	27.87	196,196
8.00	24,973	4,851,185	23,724	12.34	1.55	13.88	22.21	27.76	197,611
9.00	25,188	5,139,429	23,929	12.96	1.62	14.58	23.33	29.16	209,353
10.00	25,363	5,163,620	24,095	12.93	1.62	14.55	23.28	29.10	210,338
11.00	25,508	5,447,673	24,233	13.56	1.70	15.26	24.42	30.52	221,909
12.00	25,630	5,465,816	24,349	13.54	1.70	15.24	24.38	30.48	222,648
13.00	25,734	5,747,653	24,448	14.18	1.78	15.96	25.54	31.92	234,128
14.00	25,824	5,761,960	24,533	14.17	1.78	15.95	25.51	31.89	234,711
15.00	25,903	$6,042,535	24,608	$14.81	$1.86	$16.67	$26.67	$33.34	$246,140

WARNING: These are preliminary forecasts that must not be used to make final decisions.
1) These forecasts are not a substitute for the "due diligence" research that must be conducted to support the final definition of "unbuildable areas" above and the final decision to purchase land. This research includes, but is not limited to, verification of adequate subsurface soil, zoning, environmental clearance, access, title, utilities and water pressure, clearance from deed restriction, easement and right-of-way encumbrances, clearance from existing above and below ground facility conflicts, etc.
2) The most promising forecast(s) made on the basis of data entered in the design specification from "due diligence" research must be verified at the drawing board before funds are committed and land purchase decisions are made. Actual land shape ratios, dimensions and irregularities encountered may require adjustments to the general forecasts above.
3) The software licensee shall take responsibility for the design specification values entered and any advice given that is based on the forecast produced.

bator communities are destined to ride this roller coaster until they adjust their land use allocations and development capacities to accommodate the land and building areas needed by mature businesses. The simplest approach is to annex farmland to provide these areas. In the case of land-locked communities, this is not an option, as they are forced to adjust by evaluating their expenses, lifestyle, and shared values in relation to the yield produced by their land use allocation ratios.

Adjusting land use allocation with rezoning and redevelopment, however, involves major disruption, conflict, and political risk. Fortunately, both development capacity and yield can also be improved by reexamining a community's existing assets and land use relationships. Barley's, a fictional example that involves an historic preservation component, is used to illustrate this option. Barley's represents a point in time rather than a stylistic masterpiece and is more significant for its age than for its appearance. Unfortunately, its relevance has diminished with age because it does not adequately accommodate the automobile. The collection of buildings is shown in Fig. 16.5. The buildings range from 2 to 5 stories in height and occupy a footprint area of 20,900 sq ft. Their combined gross building area is 78,980 sq ft, and their average building height is 3.8 stories. Their first floors are used for small retail activities, and their upper floors, when occupied, are used for apartments. The Barley Block sits on a major artery close to the central business district (CBD) between a redeveloping university community and a redeveloping CBD fringe business district. There is a surface parking lot behind that combines with the building footprint area to produce a 2.051-acre tract. The Barley Block asset is 78,980 sq ft of existing building area in a convenient location. Its liability is a surface parking lot that cannot begin to supply the 316 parking spaces that an office/retail building of this size should have. Its upper floors are used for apartments in a nonresidential environ-

(a)

(b)

Figure 16.5 The Barley Block. This is a series of hitching-post buildings that struggle to accommodate the automobile along a major urban artery. The block is, however, adjacent to a Big 10 campus and across the street from a nationally competitive convention center. (*a*) Street view; (*b*) site plan.

(Photo by B. Higgins)

ment because this requires less parking. In other words, the development capacity of the property is underutilized, the yield to both owner and government is a fraction of its potential, and its building area capacity does not represent a potential home for a growing business.

The development capacity potential of the Barley Block can be evaluated with the development forecast collection. Many different concepts can be considered, but the one chosen for this case study involves preserving the Barley Block buildings, expanding the total building area available, and increasing the block's usefulness with an adjacent parking structure. The site plan concept for this approach is shown in Fig. 16.6, but it should not be taken literally since the

(a)

(b)

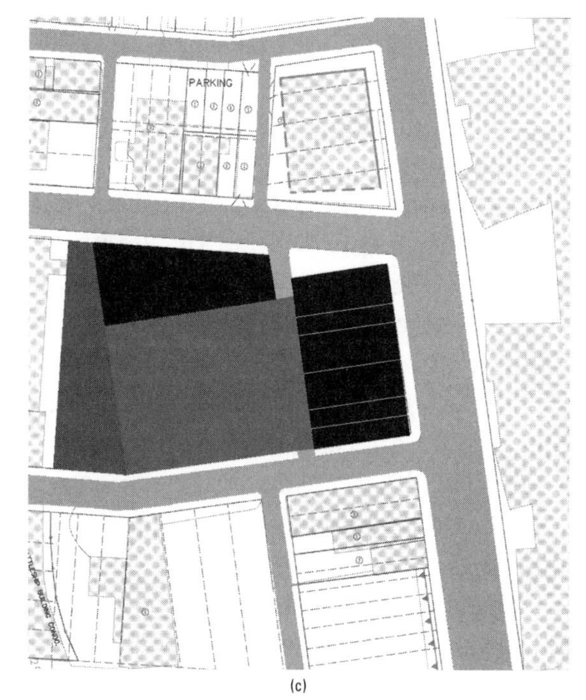

(c)

Figure 16.6 The Barley Block concept diagram. The Barley Block, when combined with an adjacent parking garage, has great potential not only to double its size and produce well-served office, apartment, and retail activities, but also to use its office parking spaces as a catalyst for surrounding evening activities. (*a*) Street view; (*b*) site plan; (*c*) concept diagram.

(Photo by B. Higgins)

form and arrangement of the areas shown may take any shape.

Forecast model CS1L pertains to nonresidential land use when the land area is known and an adjacent parking structure is contemplated. Since the Barley Block buildings are to be preserved, 1.57 acres of the 2.051-acre tract remains for evaluation, but the final analysis must take the parking needs and building area of the block into consideration. Exhibit 16.15 is based on a 4-level parking garage *p* within a 1.57-acre tract. Parking garage area is calculated based on 1 space for every 250 sq ft of building area, an estimate of 350 sq ft of parking garage area per space *a* and a garage efficiency of 85%. No loading areas are provided, and 18% open space is planned. Based on this design specification, Exhibit 16.15 predicts that a 4-level parking garage with a land cover GPL of 45,121 sq ft will support an adjacent 10-story building with a footprint BCA of 10,958 sq ft and a gross building area of 109,580 sq ft. Remember, however, that the Barley Block already represents 78,980 sq ft of this gross building area and also needs a parking ratio of 1 space for every 250 sq ft of building area if it is to adequately function as office space. This means that new office area construction can be found by subtracting 78,980 sq ft from the 109,580-sq-ft total. This produces 30,600 sq ft of new office potential. When this potential is divided by the building footprint area BCA of 10,958 sq ft, 2.79 floors result. This is a rather small expansion of the existing Barley Block area and can be improved with additional parking levels.[7]

The number of parking levels *p* in Exhibit 16.15 has been increased to 5 in Exhibit 16.16 and to 6 in Exhibit 16.17. Table 16.2 summarizes the results and shows the expanded building area that can be constructed in the NEW column. The table identifies the number of parking levels involved in the PSL column and summarizes the total building area that can be supported, including the Barley Block, in the GBA column. Each of these gross building areas is divided by the

Exhibit 16.15 The Barley Block with Four Parking Structure Levels
Forecast Model CS1L

Development capacity forecast for ***NONRESIDENTIAL BUILDINGS*** *using an* ***ADJACENT PARKING STRUCTURE*** *located on the same premise.*
Given: *Gross land area available.* ***To Find:*** *Maximum development capacity of the land area given (gross building area potential) based on the design specification values entered below.*
Design Premise: *Building footprint adjacent to parking garage footprint within the core development area. All similar floors considered equal in area.*

Enter values in boxed areas where text is bold and blue. Express all fractions as decimals.

DESIGN SPECIFICATION

Given:	**Gross land area**	GLA=	1.570	acres	68,389	SF
Land Variables:	Public/ private right-of-way & paved easements	W=	0.000	fraction of GLA	0	SF
	Net land area	NLA=	1.570	acres	68,389	SF
	Unbuildable and/or future expansion areas	U=	0.000	fraction of GLA	0	SF
	Gross land area reduction	X=	0.000	fraction of GLA	0	SF
	Buildable land area remaining	BLA=	1.570	acres	68,389	SF
Parking Variables:	Est. net pkg. structure area per pkg. space	s =	350	SF		
	Building SF permitted per parking space	a =	250	SF		
	Number of parking levels to be evaluated	p=	4.00			
	Pkg. support as fraction of gross pkg. structure area (GPA)	Pu=	0.150			
	Net area for parking & circulation as fraction of GPA	Pe=	0.850	Pu+Pe must = 1		
	No. of loading spaces	l=	0			
	Gross area per loading space	b =	1,000	SF	0	SF
Site Variables:	**Project open space as fraction of BLA**	S=	0.180		12,310	SF
	Private driveways as fraction of BLA	R=	0.000		0	SF
	Misc. pavement as fraction of BLA	M=	0.000		0	SF
	Loading area as fraction of BLA	L=	0.000		0	SF
	Total site support areas as a fraction of BLA	Su=	0.180		12,310	SF
Core:	**Core development area as fraction of BLA**	C=	0.820	C+Su must = 1	56,079	SF

NOTE: p=1 is a grade parking lot based on design premise. Increase the number of parking levels to increase the capacity forecast forecast below. Other variables in blue, box and bold may also be changed.

PLANNING FORECAST

no. of FLR bldg floors	min land area CORE for BCG & PSA	gross building area GBA	gross pkg GPA struc area	pkg struct cover GPL per level	net parking NPA area	pkg spaces NPS	bldg cover BCA footprint	bldg SF per SFAC bldable acre	flr area ratio FAR function of BLA
1.00	56,079	39,723	65,426	16,356	55,612	158.89	39,723	25,301	0.581
2.00		61,506	101,304	25,326	86,109	246.02	30,753	39,176	0.899
3.00		75,264	123,964	30,991	105,370	301.06	25,088	47,939	1.101
4.00		84,742	139,575	34,894	118,639	338.97	21,185	53,976	1.239
5.00		91,668	150,982	37,746	128,335	366.67	18,334	58,387	1.340
6.00		96,950	159,683	39,921	135,731	387.80	16,158	61,752	1.418
7.00		101,112	166,538	41,635	141,557	404.45	14,445	64,403	1.478
8.00		104,476	172,078	43,020	146,267	417.90	13,060	66,545	1.528
9.00		107,251	176,649	44,162	150,152	429.01	11,917	68,313	1.568
10.00		109,580	180,485	45,121	153,412	438.32	10,958	69,796	1.602
11.00		111,562	183,749	45,937	156,186	446.25	10,142	71,058	1.631
12.00		113,269	186,560	46,640	158,576	453.08	9,439	72,146	1.656
13.00		114,755	189,007	47,252	160,656	459.02	8,827	73,092	1.678
14.00		116,059	191,157	47,789	162,483	464.24	8,290	73,923	1.697
15.00		117,215	193,059	48,265	164,100	468.86	7,814	74,659	1.714

WARNING: These are preliminary forecasts that must not be used to make final decisions.
1) These forecasts are not a substitute for the "due diligence" research that must be conducted to support the final definition of "unbuildable areas" above and the final decision to purchase land. This research includes, but is not limited to, verification of adequate subsurface soil, zoning, environmental clearance, access, title, utilities and water pressure, clearance from deed restriction, easement and right-of-way encumbrances, clearance from existing above and below ground facility conflicts, etc.
2) The most promising forecast(s) made on the basis of data entered in the design specification from "due diligence" research must be verified at the drawing board before funds are committed and land purchase decisions are made. Actual land shape ratios, dimensions and irregularities encountered may require adjustments to the general forecasts above.
3) The software licensee shall take responsibility for the design specification values entered and any advice given that is based on the forecast produced.

Exhibit 16.16 The Barley Block with Five Parking Structure Levels

Forecast Model CS1L

Development capacity forecast for ***NONRESIDENTIAL BUILDINGS*** *using an* ***ADJACENT PARKING STRUCTURE*** *located on the same premise.*
Given: *Gross land area available.* ***To Find:*** *Maximum development capacity of the land area given (gross building area potential) based on the design specification values entered below.*
Design Premise: *Building footprint adjacent to parking garage footprint within the core development area. All similar floors considered equal in area.*

Enter values in boxed areas where text is bold and blue. Express all fractions as decimals.

DESIGN SPECIFICATION

Given:	**Gross land area**	GLA=	1.570	acres	68,389	SF
Land Variables:	Public/ private right-of-way & paved easements	W=	0.000	fraction of GLA	0	SF
	Net land area	NLA=	1.570	acres	68,389	SF
	Unbuildable and/or future expansion areas	U=	0.000	fraction of GLA	0	SF
	Gross land area reduction	X=	0.000	fraction of GLA	0	SF
	Buildable land area remaining	BLA=	1.570	acres	68,389	SF
Parking Variables:	Est. net pkg. structure area per pkg. space	s =	350	SF		
	Building SF permitted per parking space	a =	250	SF		
	Number of parking levels to be evaluated	p=	5.00			
	Pkg. support as fraction of gross pkg. structure area (GPA)	Pu=	0.150			
	Net area for parking & circulation as fraction of GPA	Pe=	0.850	Pu+Pe must = 1		
	No. of loading spaces	l=	0			
	Gross area per loading space	b =	1,000	SF	0	SF
Site Variables:	**Project open space as fraction of BLA**	S=	0.180		12,310	SF
	Private driveways as fraction of BLA	R=	0.000		0	SF
	Misc. pavement as fraction of BLA	M=	0.000		0	SF
	Loading area as fraction of BLA	L=	0.000		0	SF
	Total site support areas as a fraction of BLA	Su=	0.180		12,310	SF
Core:	**Core development area as fraction of BLA**	C=	0.820	C+Su must = 1	56,079	SF

NOTE: p=1 is a grade parking lot based on design premise. Increase the number of parking levels to increase the capacity forecast forecast below. Other variables in blue, box and bold may also be changed.

PLANNING FORECAST

no. of **FLR** bldg floors	min land area CORE for BCG & PSA	gross building area **GBA**	gross pkg GPA struc area	pkg struct cover **GPL** per level	net parking NPA area	pkg spaces NPS	bldg cover **BCA** footprint	bldg SF per SFAC bldable acre	flr area ratio FAR function of BLA
1.00	56,079	**42,183**	69,479	**13,896**	59,057	168.73	**42,183**	26,868	0.617
2.00		**67,613**	111,363	**22,273**	94,658	270.45	**33,807**	43,066	0.989
3.00		**84,616**	139,368	**27,874**	118,463	338.47	**28,205**	53,896	1.237
4.00		**96,786**	159,413	**31,883**	135,501	387.15	**24,197**	61,647	1.415
5.00		**105,927**	174,468	**34,894**	148,298	423.71	**21,185**	67,470	1.549
6.00		**113,045**	186,192	**37,238**	158,263	452.18	**18,841**	72,003	1.653
7.00		**118,744**	195,579	**39,116**	166,242	474.98	**16,963**	75,633	1.736
8.00		**123,410**	203,264	**40,653**	172,775	493.64	**15,426**	78,605	1.805
9.00		**127,301**	209,673	**41,935**	178,222	509.21	**14,145**	81,084	1.861
10.00		**130,595**	215,098	**43,020**	182,833	522.38	**13,060**	83,182	1.910
11.00		**133,420**	219,750	**43,950**	186,788	533.68	**12,129**	84,981	1.951
12.00		**135,869**	223,784	**44,757**	190,216	543.47	**11,322**	86,541	1.987
13.00		**138,012**	227,314	**45,463**	193,217	552.05	**10,616**	87,906	2.018
14.00		**139,904**	230,430	**46,086**	195,866	559.62	**9,993**	89,111	2.046
15.00		**141,586**	233,200	**46,640**	198,220	566.34	**9,439**	90,182	2.070

WARNING: These are preliminary forecasts that must not be used to make final decisions.
1) These forecasts are not a substitute for the "due diligence" research that must be conducted to support the final definition of "unbuildable areas" above and the final decision to purchase land. This research includes, but is not limited to, verification of adequate subsurface soil, zoning, environmental clearance, access, title, utilities and water pressure, clearance from deed restriction, easement and right-of-way encumbrances, clearance from existing above and below ground facility conflicts, etc.
2) The most promising forecast(s) made on the basis of data entered in the design specification from "due diligence" research must be verified at the drawing board before funds are committed and land purchase decisions are made. Actual land shape ratios, dimensions and irregularities encountered may require adjustments to the general forecasts above.
3) The software licensee shall take responsibility for the design specification values entered and any advice given that is based on the forecast produced.

Exhibit 16.17 The Barley Block with Six Parking Structure Levels

Forecast Model CS1L

Development capacity forecast for ***NONRESIDENTIAL BUILDINGS*** *using an* ***ADJACENT PARKING STRUCTURE*** *located on the same premise.*
Given: *Gross land area available.* ***To Find:*** *Maximum development capacity of the land area given (gross building area potential) based on the design specification values entered below.*
Design Premise: *Building footprint adjacent to parking garage footprint within the core development area. All similar floors considered equal in area.*

Enter values in boxed areas where text is bold and blue. Express all fractions as decimals.

DESIGN SPECIFICATION

Given:	**Gross land area**	GLA=	**1.570**	acres	68,389	SF
Land Variables:	Public/ private right-of-way & paved easements	W=	**0.000**	fraction of GLA	0	SF
	Net land area	NLA=	1.570	acres	68,389	SF
	Unbuildable and/or future expansion areas	U=	**0.000**	fraction of GLA	0	SF
	Gross land area reduction	X=	0.000	fraction of GLA	0	SF
	Buildable land area remaining	BLA=	1.570	acres	68,389	SF
Parking Variables:	Est. net pkg. structure area per pkg. space	s =	**350**	SF		
	Building SF permitted per parking space	a =	**250**	SF		
	Number of parking levels to be evaluated	p=	**6.00**			
	Pkg. support as fraction of gross pkg. structure area (GPA)	Pu=	**0.150**			
	Net area for parking & circulation as fraction of GPA	Pe=	0.850	Pu+Pe must = 1		
	No. of loading spaces	l=	**0**			
	Gross area per loading space	b =	**1,000**	SF	0	SF
Site Variables:	**Project open space as fraction of BLA**	S=	**0.180**		12,310	SF
	Private driveways as fraction of BLA	R=	**0.000**		0	SF
	Misc. pavement as fraction of BLA	M=	**0.000**		0	SF
	Loading area as fraction of BLA	L=	0.000		0	SF
	Total site support areas as a fraction of BLA	Su=	0.180		12,310	SF
Core:	**Core development area as fraction of BLA**	C=	0.820	C+Su must = 1	56,079	SF

NOTE: p=1 is a grade parking lot based on design premise. Increase the number of parking levels to increase the capacity forecast forecast below. Other variables in blue, box and bold may also be changed.

PLANNING FORECAST

no. of **FLR** bldg floors	min land area CORE for BCG & PSA	gross building area **GBA**	gross pkg GPA struc area	pkg struct cover **GPL** per level	net parking NPA area	pkg spaces NPS	bldg cover **BCA** footprint	bldg SF per SFAC bldable acre	flr area ratio FAR function of BLA
1.00	56,079	**44,001**	72,472	**12,079**	61,601	176.00	**44,001**	28,026	0.643
2.00		**72,406**	119,257	**19,876**	101,368	289.62	**36,203**	46,118	1.059
3.00		**92,259**	151,956	**25,326**	129,163	369.04	**30,753**	58,764	1.349
4.00		**106,917**	176,099	**29,350**	149,684	427.67	**26,729**	68,100	1.563
5.00		**118,183**	194,655	**32,442**	165,457	472.73	**23,637**	75,276	1.728
6.00		**127,113**	209,362	**34,894**	177,958	508.45	**21,185**	80,964	1.859
7.00		**134,364**	221,306	**36,884**	188,110	537.46	**19,195**	85,582	1.965
8.00		**140,370**	231,197	**38,533**	196,518	561.48	**17,546**	89,408	2.053
9.00		**145,426**	239,524	**39,921**	203,596	581.70	**16,158**	92,628	2.126
10.00		**149,740**	246,631	**41,105**	209,636	598.96	**14,974**	95,376	2.190
11.00		**153,465**	252,766	**42,128**	214,852	613.86	**13,951**	97,749	2.244
12.00		**156,714**	258,118	**43,020**	219,400	626.86	**13,060**	99,818	2.292
13.00		**159,573**	262,826	**43,804**	223,402	638.29	**12,275**	101,639	2.333
14.00		**162,107**	267,000	**44,500**	226,950	648.43	**11,579**	103,253	2.370
15.00		**164,370**	270,727	**45,121**	230,118	657.48	**10,958**	104,694	2.403

WARNING: These are preliminary forecasts that must not be used to make final decisions.
1) These forecasts are not a substitute for the "due diligence" research that must be conducted to support the final definition of "unbuildable areas" above and the final decision to purchase land. This research includes, but is not limited to, verification of adequate subsurface soil, zoning, environmental clearance, access, title, utilities and water pressure, clearance from deed restriction, easement and right-of-way encumbrances, clearance from existing above and below ground facility conflicts, etc.
2) The most promising forecast(s) made on the basis of data entered in the design specification from "due diligence" research must be verified at the drawing board before funds are committed and land purchase decisions are made. Actual land shape ratios, dimensions and irregularities encountered may require adjustments to the general forecasts above.
3) The software licensee shall take responsibility for the design specification values entered and any advice given that is based on the forecast produced.

Table 16.2 New Office Floors in Addition to the Barley Block

Exhibit	Parking Structure Levels PSL	Gross Building Area for 10-Story Building, sq ft GBA	Footprint for 10-Story Building, sq ft BCA	Barley's Gross Building Area, sq ft BGBA	New Building Area, sq ft NEW	Number of New Office Floors, NEW/BCA CFLR	Total Building sq ft per Acre SFAC	Number of New Building Floors, Rounded RFL	New Building Footprint, sq ft NEW/RFL BCA
16.15	4	109,580	10,958	78,980	**30,600**	**2.79**	53,428	3	10,200
16.16	5	130,595	13,060	78,980	**51,615**	**3.95**	63,674	4	12,904
16.17	6	149,740	14,974	78,980	**70,760**	**4.73**	73,008	5	14,152

NOTE: Data summarized from Exhibits 16.15, 16.16, and 16.17.

total project area of 2.051 acres to forecast the relevant capacity index in the SFAC column. This index ranges from 53,427.6 to 73,008.3 building sq ft per acre and includes an historic Barley Block that has been rejuvenated to Class A office standards. The block is now an eligible home for any business and includes new office area that can be either attached to or detached from the Barley Block. It also includes open space designed to humanize the gridiron street pattern and unrelieved block faces of the past.

Barley Block Office/Residential Mixed Use

It might be helpful to explain that the data in Table 16.2 can also be used to evaluate a mixed-use residential and office option for the Barley Block. To pursue this option, Exhibit 16.18 isolates the dwelling unit mix table from a typical residential forecast model as a separate forecast model. The demonstration mix shown in Exhibit 16.18 produces an aggregate average dwelling unit area AGG of 938 sq ft. This area is used to evaluate the mixed-use potential of the project.

Table 16.3 draws its information from Exhibits 16.15, 16.16, and 16.17, based on a 10-story building. Any building height could be chosen, but the 10-story height produces a reasonable footprint area in each of these examples. This table shows that a 4-level parking garage has 122 parking spaces left after supporting the Barley Block with 1 parking space for every 250 sq ft of gross building area.[8] It also shows that a 5-level garage has 206 extra spaces, and a 6-level garage has 283 extra spaces. Since 1.5 parking spaces are required for every dwelling unit in this example, the table shows that the extra parking spaces can support 81.3, 137.3, and 188.6 dwelling units, respectively. Since the AGG from Exhibit 16.18 is 938 sq ft, these dwelling units represent gross building areas RGBA of 76,259, 128,787, and 176,906 sq ft. Table 16.3 also repeats information from Exhibit 16.15 to note that a 4-level garage and

Exhibit 16.18 Dwelling Unit Mix Forecast Model

DWELLING UNIT MIX TABLE

Parking lot spaces planned or required per dwelling unit	u=	1.5	
Garage parking spaces planned or required per dwelling unit	Gn=	0	
Building efficiency as percentage of GBA	Be=	0.800	must have a value >0 entered
Building support as fraction of GBA	Bu=	0.200	Be + Bu must = 1

DU dwelling unit type	GDA gross du area	CDA=GDA/Be comprehensive du area	MIX du mix		PDA = (CDA)MIX Prorated du area
EFF	400	500	0%		0
1 BR	500	625	25%		156
2 BR	750	938	50%		469
3 BR	1,000	1,250	25%		313
4 BR	1,200	1,500	0%		0
		Aggregate avg. dwelling unit area		(AGG) =	938
		GBA sf per parking space		a=	625

a 10-story building leave 10,958 sq ft of footprint area BCA available. It repeats from Exhibit 16.16 that a 5-level garage leaves 13,060 sq ft and repeats from Exhibit 16.17 that a 6-level garage leaves 14,974 sq ft. When each gross building area forecast is divided by the footprint area available, the RFLR column in Table 16.3 shows that the garage levels can respectively support 7.0, 9.9, and 11.8 apartment floors in addition to the Barley Block office area.

The dwelling unit capacity from the NDU column in Table 16.3 can be multiplied by the dwelling unit mix percentages entered in Exhibit 16.18 to find the dwelling unit breakdown implied. Table 16.4 shows that the number of parking levels identified in the NPL column can support 20 to 47 one-bedroom units, 41 to 94 two-bedroom units, and 20 to 47 three-bedroom units. This is in addition to the Barley Block building area, but is only one of a number of mixed-use relationships that can be explored with the development forecast collection.

Table 16.3 New Apartment Floors in Addition to the Barley Block

Exhibit	Parking Structure Levels PSL	Total Parking Spaces for 10-Story Building TPS	Parking Spaces for Barley Block BPS	Remaining Parking Spaces OPS	Number of Dwelling Units, OPS/1.5 NDU	Aggregate Average Dwelling Unit Area AGG	Apartment Gross Building Area, NDU × AGG RGBA	Footprint for 10-Story Building BCA	Number of New Apartment Floors, RGBA/BCA RFLR
16.15	4	438	316	122	**81.3**	938	76,291	10,958	**7.0**
16.16	5	522	316	206	**137.3**	938	128,819	13,060	**9.9**
16.17	6	599	316	283	**188.7**	938	176,969	14,974	**11.8**

NOTE: Data summarized from Exhibits 16.15, 16.16, and 16.17.

Table 16.4 Dwelling Unit Breakdown Based on Dwelling Unit Mix

Parking Structure Levels	Total Number of Dwelling Units*	Dwelling Unit Mix†					
		1 BR—25%		2 BR—50%		3 BR—25%	
		Calculated	Rounded	Calculated	Rounded	Calculated	Rounded
4	81.3	20.3	**20**	40.7	**41**	20.3	**20**
5	137.3	34.3	**34**	68.7	**69**	34.3	**34**
6	188.7	47.2	**47**	94.4	**94**	47.2	**47**

*Data from Table 16.3.
†Percentages from Exhibit 16.18.

Another mixed-use option is made possible by the parking ratio used, since 4 parking spaces per 1000 gross sq ft of building area (1 space per 250 sq ft) is also adequate for many retail services. Since this is the case, the first floor of the Barley Block could easily be used for retail activity served by the parking garage, since it is across the street from a nationally competitive convention center.

Summary

Economic development is helping to make it clear that land use planning not only affects the compatibility, relationship, and organization of activities within a city; it also establishes the land use allocation ratios that make high-yield land available. The high-yield portion of a city's land use portfolio directly affects its average yield per acre and its ability to support a desired lifestyle and basic quality of life. When a city sprawls, it changes its land use allocation ratios, but it does not necessarily improve its average yield per gross acre. In fact, the chances are that this yield is reduced even though the initial gross revenue increases. This new revenue is often used to support older areas of the city. As the land use combination ages and maintenance costs increase, even larger new areas are annexed to provide the new revenue required. The process is not really this sophisticated, however; annexation generally takes place at the request of a landowner and is often used to house an expanding population through suburban sprawl. There is generally little objection to this request, because the city rarely understands the dilution in average yield per acre that can take place. This dilution can reduce the city's ability to afford the operations, improvement, and maintenance costs that increase with age and population. The new revenue produced by the new development is difficult to resist because the city rarely understands or predicts the current and future average yield per acre that it will need to support the obligations assumed as the combination ages. The income from new development looks and acts like increased revenue because it is, but the development can easily represent a deficit as it ages.

Future deficits are easiest to visualize by imagining that other incorporated areas surround a city and that its growth corridors are cut off. Imagine that this city is forced to live within its current limits with inadequate land use allocation ratios. As the cost of aging increases, the city's average yield per acre not only becomes less adequate to meet expenses, it can decline with the average age of the city's population. This age shift can also produce a change in voting patterns and a reduction in population that further exacerbates the problem. The problem, however, can be less a function of inadequate voter support than of inadequate land use allocation ratios that do not reduce the burden on individual taxpayers.

Therefore, when we see economic development as simply the attraction and retention of business activity, we miss the larger picture. Economic development is about the economic stability of cities, and this is driven by the land use allocation ratios that make economic development possible. We currently know very little about these ratios and the yield they produce, but it is clear that they influence the income a community receives to support its lifestyle. Yield, however, is also a function of development capacity within the ratios provided. We can organize current accounting information to correlate land use activities or categories with the current revenue and expense related to each, and this can be invaluable planning information. Unfortunately, we will still have difficulty predicting the future development capacity of land areas and their ability to improve the yield from a city's land use portfolio without the development forecast collection. For instance, Table 16.1 shows that this city received $6,231,234 in revenue from 296 commercial acres, for a yield of $21,051 per acre. This table does not explain the condition, quality, and location of the buildings that produced this revenue. The yield appears significant when compared to residential acres; however, the fact is that many of the office buildings involved are not competitive with the market, but their location is attractive. If the yield from these acres were doubled, annual community income would increase by 21%. Tax abatement based on development capacity forecasts and analysis could be used to stimulate this redevelopment with clearly defined objectives written to benefit all interests. The development forecast collection has been assembled to help with these forecasts.

Notes

1. In this context stable cash flow means reliable annual income that a community can expect to consistently receive without expiration to meet the annual expense of its lifestyle. Estate tax revenue, interest from a fluctuating fund balance and proceeds from a limited term bond issue requiring voter renewal are examples of more unreliable cash flow. The degree of budget dependency on these more unreliable revenue streams is one indication of economic stability.
2. Lifestyle and quality of life are not necessarily synonymous terms.
3. This assumes that there is a site to evaluate, or that there is a location that can accommodate the land area requirement forecast. If neither possibility exists in a community, and if existing building area is not

available, then it is unlikely that most new business contacts will consider the community a path of least resistance and greatest opportunity. In this case, a community must work to define its shared values, the cost of these values, and the sources of income that may represent options to meet the expense of these values, including redevelopment plans that can improve the yield of its land use pattern with less burden on individual taxpayers.

4. Than in terms of construction cost, lease rates, return on investment, tax yield, abatement incentives, etc.
5. This is a general value used for the sake of simplicity. It is a function of the dwelling unit areas and mix planned and can vary widely.
6. Building cover BCA = 3.375 acres. Parking cover PCA = 8.566 acres. Miscellaneous pavement $M + R = 0.482$ acres. Project open space $S =$ 8.257 acres.
7. It also shows the degree to which the existing building is underserved and relatively undesirable as a result.
8. Since the Barley Block is 78,980 sq ft in area, this ratio produces a parking requirement of 78,980/250 = 315.92 spaces, which has been rounded to 316.

CHAPTER 17

Conclusion

Building area represents *shelter.* Calling it *architecture* has somehow confused the public perception of this essential presence. It is one of the five foundations of life, and is needed for most forms of human activity. The capacity of land to accommodate shelter has been called *development capacity.* The wise use of development capacity will not only help us earn a living, make a profit, shelter a growing population, conserve land, and support the lifestyle of our communities; it will also affect our sustainable future. The development forecast collection has been written to forecast and help evaluate the wisdom of the choices we face. If this collection makes sense to the building-oriented professions, the study of development capacity and the evaluation of these choices could be called *tegmenistics* after the Latin word for shelter. (City planning, urban design, architecture, landscape architecture, real estate development, real estate appraisal, building construction, civil engineering, land use law, economic development, urban geography, city management, and investment banking represent fragments of the effort to plan, design, and construct shelter.) We have assumed that land is an inexhaustible resource that will always be available for the expansion of our shelter requirements, and have randomly consumed this resource with our fragmented approach. We may be entering a new age of awareness, however. The development forecast collection does not provide this awareness, but it may help us build it.

It does not take a scientific genius to recognize that populations seek food and shelter. They grow and move in response to this need and expand until they exceed the supply. This produces more movement, which the Earth has sustained with its superabundance. As we have continued to grow and consume, however, we have coined the word *ecology* to mirror a fundamental principle of physics—that all actions have a reaction and are part of what American Indians have long called the *great circle*. We should offer profound respect and admiration for the wisdom and vision of the great North American Indian nations. We will never completely learn what they have always understood. It is up to us to use the gifts we have been given to recognize our role and protect the great circle for our own self-interest, if for no other reason.

CHAPTER 18

Exercises

The following exercises are suggested as a way to become more familiar with each model within the development forecast collection. There are no wrong answers until you define them with research and experience, however, so the objective is not to explain how to solve a problem but how to recognize an issue, select a model, and forecast options for evaluation. Each set of exercises relates to the land use family noted. All exercises are preceded by a process statement that should be followed when completing any exercise. The process involves context record research and evaluation that supports the design specification values chosen. The exercises demonstrate how these design specification values can be entered and altered within a forecast model to increase or reduce the development capacity predictions produced. The models that apply to each exercise are listed following the description. You need only select this model from the development forecast collection on the included CD-ROM to conduct the exercise suggested.

Nonresidential Development Capacity Exercises

Process

Select one of the parking systems and exercises in this section and print the context record spreadsheet and forecast model associated with your choice for reference. Ignore any data that may already be entered.

Visit two existing projects that use the parking system represented by the exercise. Choose these projects based on the environmental objective you wish to achieve. Photograph each in context. *Context* means the property under study and its surrounding environment. Aerial photographs are best for this purpose and can be supplemented with ground-level shots for emphasis. Do not focus on building appearance, since this is only one aspect of the environmental design associated with the word *context*. Single-engine planes are relatively inexpensive to charter and can be shared by two photographers for this purpose. With only four pictures to obtain, the flying time should be brief.

Conduct field measurements as required to complete a context record spreadsheet for each project. The applicable context record is noted with each exercise and can be found on the included CD-ROM. Most context record data requested, if not all, will be included on plans available at the jurisdiction having authority for building regulation in the area chosen. Other sources, such as the county auditor and the architectural design firm of record, are also available to reduce the need for field measurement. (It would be wonderful to establish a national clearinghouse to collect and distribute the information collected and evaluations produced, since this would increase the foundation of knowledge available regarding the wise use of land for shelter.)

From this research, prepare several alternate design specification "recipes" and be prepared to explain their purpose, since each recipe will have a different implication. Enter these design specifications in the forecast model identified to produce at least three development capacity forecasts for a new land area size and location. Summarize the purpose behind each alternate specification and the implications of the forecast produced. You must be able to explain the design specification values used, since each set represents some form of limitation on the ultimate development capacity of the site. It is not realistically possible to reach this ultimate capacity, since potential building height is limited by current technology, but the reasons for limitation should be clear and defensible.

Present your results verbally by explaining the existing projects chosen for evaluation and comparison, the context records that define these existing conditions, the adjustments made to produce alternate design specifications, the reasons for these adjustments, the forecasts produced by these specifications, and the implications represented by these adjustments.

Surface Parking

EXERCISE 1

Given a specific land area, how much can be built using different design specification options?

Forecast model CG1L
Context record CR

EXERCISE 2

Given a specific gross building area, how much land is needed to serve this building area using different design specification options?

Forecast model CG1B
Context record CR

No Parking Required

EXERCISE 3

Given a specific land area, how much can be built using different design specification options?

Forecast model CG2L
Context record CR

EXERCISE 4

Given a specific gross building area, how much land is needed to serve this building area using different design specification options?

Forecast model CG2B
Context record CR

Adjacent Parking Structure

EXERCISE 5

Given a specific land area, how much can be built using different design specification options?

Forecast model CS1L
Context record CR

EXERCISE 6

Given a specific gross building area, how much land is needed to serve this building area using different design specification options?

Forecast model CS1B
Context record CR

Underground Parking Structure

EXERCISE 7

Given a specific land area, how much can be built using different design specification options?

Forecast model CS2L
Context record CR

EXERCISE 8

Given a specific gross building area, how much land is needed to serve this building area using different design specification options?

Forecast model CS2B
Context record CR

Parking Structure Above Grade Under Building

EXERCISE 9

Given a specific land area, how much can be built using different design specification options?

Forecast model CS3L
Context record CR

EXERCISE 10

Given a specific gross building area, how much land is needed to serve this building area using different design specification options?

Forecast model CS3B
Context record CR

Apartment House Development Capacity Exercises

Process

Use the same research and evaluation process for each exercise as that described for nonresidential development capacity topics.

Surface Parking

EXERCISE 11

Given a specific gross land area, dwelling unit mix, and dwelling unit area schedule, what is the maximum number of dwelling units that can be built using different design specification values?

Forecast model RG1L
Context record CR

EXERCISE 12

Given a specific gross building area objective, dwelling unit mix, and dwelling unit area schedule, how much land is required and how many dwelling units can be created when using different design specification values?

Forecast model RG1B
Context record CR

EXERCISE 13

Given a specific net density objective, land area, dwelling unit mix, and dwelling unit area schedule, how many floors are required to reach the density objective when using different design specification values and project open space requirements?

Forecast model RG1D
Context record CR

No Parking Required

EXERCISE 14

Given a specific gross land area, dwelling unit mix, and dwelling unit area schedule, what is the maximum number of dwelling units that can be built using different design specification values?

Forecast model RG2L
Context record CR

EXERCISE 15

Given a specific gross building area objective, dwelling unit mix, and dwelling unit area schedule, how much land is required and how many dwelling units can be created when using different design specification values?

Forecast model RG2B
Context record CR

EXERCISE 16

Given a specific net density objective, land area, dwelling unit mix, and dwelling unit area schedule, how many floors are required to reach the density objective when using different design specification values and project open space requirements?

Forecast model RG2D
Context record CR

Adjacent Parking Structure

EXERCISE 17

Given a specific gross land area, dwelling unit mix, and dwelling unit area schedule, what is the maximum number of dwelling units that can be built using different design specification values?

Forecast model RS1L
Context record CR

EXERCISE 18

Given a specific gross building area objective, dwelling unit mix, and dwelling unit area schedule, how much land is required and how many dwelling units can be created when using different design specification values?

Forecast model RS1B
Context record CR

EXERCISE 19

Given a specific net density objective, land area, dwelling unit mix, and dwelling unit area schedule, how many floors are required to reach the density objective when using different design specification values and project open space requirements?

Forecast model RS1D
Context record CR

Underground Parking Structure

EXERCISE 20

Given a specific gross land area, dwelling unit mix, and dwelling unit area schedule, what is the maximum number of dwelling units that can be built using different design specification values?

Forecast model RS2L
Context record CR

EXERCISE 21

Given a specific gross building area objective, dwelling unit mix, and dwelling unit area schedule, how much land is required and how many dwelling units can be created when using different design specification values?

Forecast model RS2B
Context record CR

EXERCISE 22

Given a specific net density objective, land area, dwelling unit mix, and dwelling unit area schedule, how many floors are required to reach the density objective when using different design specification values and project open space requirements?

Forecast model RS2D
Context record CR

Parking Structure Above Grade Under Building

EXERCISE 23

Given a specific gross land area, dwelling unit mix, and dwelling unit area schedule, what is the maximum number of dwelling units that can be built using different design specification values?

Forecast model RS3L
Context record CR

EXERCISE 24

Given a specific gross building area objective, dwelling unit mix, and dwelling unit area schedule, how much land is required and how many dwelling units can be created when using different design specification values?

Forecast model RS3B
Context record CR

EXERCISE 25

Given a specific net density objective, land area, dwelling unit mix, and dwelling unit area schedule, how many floors are required to reach the density objective when using different design specification values and project open space requirements?

Forecast model RS3D
Context record CR

Suburb House Development Capacity Exercises

Process

Use the same research and evaluation process for each exercise as that described for nonresidential development capacity topics.

Surface Parking

Standard lot.

Cluster lot.

EXERCISE 26

Given a specific gross land area and net density objective, how many single-family lots can be created?

Forecast model RSFD
Context record SF

EXERCISE 27

Given a specific gross land area, how many single-family lots can be created, and what maximum dwelling unit sizes are feasible?

Forecast model RSFL
Context record SF

EXERCISE 28

Given a specific number of lots and a minimum area of each, how many buildable acres are required, and what maximum dwelling unit sizes are feasible?

Forecast model RSFN
Context record SF

Urban House Development Capacity Exercises

Process

Use the same research and evaluation process for each exercise as that described for nonresidential development capacity topics.

Surface Parking

EXERCISE 29

Given a specific land area and net den-sity objective, what dwelling unit specifications are required to meet the density objective?

Forecast model RGTD
Context record GT

EXERCISE 30

Given a specific land area, what is the dwelling unit capacity?

Forecast model RGTL
Context record GT

Mixed-Use Development Capacity Exercises

Process

Use the same research and evaluation process for each exercise as that described for nonresidential development capacity topics.

Surface Parking

EXERCISE 31

Given a specific land area, how many dwelling units and how much nonresidential building area can be constructed?

Forecast model MG1L
Context record CR

EXERCISE 32

Given a specific footprint area, parking lot cover percentage, and gross land area, how many dwelling units and how much nonresidential building area can be constructed when the residential land use allocation varies within the total?

Forecast model MG1B
Context record CR

Adjacent Parking Structure

EXERCISE 33

Given a specific gross land area, how many dwelling units and how much nonresidential building area can be constructed when the residential land use allocation varies within the total?

Forecast model MS1L
Context record CR

Underground Parking Structure

EXERCISE 34

Given a specific gross land area, how many dwelling units and how much nonresidential building area can be constructed when the residential land use allocation varies within the total?

Forecast model MS2L
Context record CR

Parking Structure Above Grade Under Building

EXERCISE 35

Given a specific gross land area, how many dwelling units and how much nonresidential building area can be constructed when the residential land use allocation varies within the total?

Forecast model MS3L
Context record CR

Summary

The preceding exercises can also be used as an index to the development forecast collection. Each exercise is written to describe the issue pertaining to the forecast model. This may help amplify the decision guides contained in Chap. 1. This chapter also includes mixed-use forecast models that are not included in the Chap. 1 decision guides in order to avoid confusion at that early stage of explanation.

Suggested Divisions of the Built Environment

Shelter

Category	Shelter type Residential, nonresidential, and hybrid
Class	Building type Urban house, suburb house,[1] apartment house, nonresidential,[2] and hybrid
Order	Open space context category or percentage (lot organization for suburb house—standard plan versus cluster plan)
Family	Building height (lot area range for suburb house) Height range in feet or stories: low-rise, 1 to 3 stories; high low-rise, 4 to 5 stories; low mid-rise, 6 to 10 stories; mid-rise, 11 to 15 stories; high mid-rise, 16 to 20 stories; low high-rise, 21 to 30 stories; high-rise, 30 to 50 stories; sky-rise, >50 stories
Genus	Parking system (lot frontage for suburb house) Surface parking, structure parking, no parking, garage parking, covered parking, or hybrid parking

Species	Activity and/or ownership status Office, retail, institutional, business, condominium, rental, lease, own, etc.
Subspecies	Balance index (building cover and development cover percentages) Intensity index (number of floors + development cover percentage)[3] Development capacity index SFAC Appearance or style Building code use group, construction class, and occupancy limits

Movement

Category	Airborne, terrestrial, aquatic, and subterranean
Class	Roads, railways, air traffic, water traffic, subways, etc.
Order	Intercontinental railways, short-haul railways, international airlines, local airlines, intercontinental shipping, local ferries, intercontinental tunnels, local subway systems, etc.
Family	Method of movement within order (e.g., local, collector, arterial and freeway road systems)
Genus	Type of construction
Species	Level of demand or service
Subspecies	Ownership status

Open Space

Category	Preservation, participation, defense, and buffer
Class	Active, passive, restricted, or prohibited
Order	Federal, state, regional, local, and project

Family	Full use, limited use, and restricted use
Genus	Type of accommodation (campgrounds, day visits, lodges, proving grounds, etc.)
Species	Level of demand or service
Subspecies	Ownership status

Life Support

Category	Energy, agriculture, communication, and health
Class	Airborne, terrestrial, aquatic, or subterranean
Order	Energy sources, sewer systems, water systems, power systems, communication systems, and food systems
Family	Supply, return, recycle and discovery
Genus	Type of system
Species	Level of demand or service
Subspecies	Ownership status

Notes

1. This building type has separate order, family, genus, and species criteria. See Chap. 13.
2. Often referred to as *commercial* even though the class includes shelter for nonresidential land uses that are not commercial in nature.
3. Increment definition is needed based on research into the context, lifestyle, and quality of life represented by these indexes.

CD Software Information and File Organization

The development capacity forecast models and context record forms on this disk require Microsoft Windows 95 and Microsoft Excel 97 software for operation.

This is an autostart CD. It is best viewed with your monitor resolution set to 800 × 600. In the event it does not start, locate and open the CD folder using Windows Explorer. Browse this folder for the subfolder entitled Executables. Open this folder and double-click on the file entitled Development Capacity Evaluation for a manual start.

If you wish to store and run this CD from your hard drive successfully, you must take the following steps and have at least 22 MB of free disk space:

1. Exit this CD and use Windows Explorer to locate the drive letter on your PC that contains the Development Capacity Evaluation CD. Right-click on the icon associated with this drive.
2. Choose Explore from the context menu that pops up.
3. Locate the Executables folder in the CD file list.
4. Drag a copy of the Executables folder to the Desktop icon at the top of your Windows Explorer file list and

drop it there. *Note:* It is very important that the folder be dropped on the Desktop icon. Do not drop it on your C:\ drive icon or place it in any other file or folder location on your PC. Drag the Executables folder *only* to the Desktop icon and drop it there.

5. Open the Executables folder and find the file entitled Development Capacity Evaluation. At this point you have two options. To manually start the program, double-click on the file name. To place a shortcut on your desktop screen for easy access to the program in the future, right-click on the file name to reach a second menu. Click on the "send to" option in this menu and then click on the "Desktop (create shortcut)" option that next appears. This will place the Development Capacity Evaluation icon on your desktop screen. Exit Windows Explorer and return to the Windows desktop screen. Double-click on the Development Capacity Evaluation icon that now appears to start the program. This icon will allow you to avoid manual starts from Windows Explorer in the future.

Forecast Limitations

Forecasts made by this software are subject to the values entered by a user in each design specification and to site shapes, slopes, conditions, and characteristics of land that cannot be anticipated by each forecast model. These shapes, slopes, conditions, and other characteristics may significantly reduce the forecast results that can be achieved. The user must anticipate the nature and character of a site when choosing values for entry in each design specification. Failure to choose realistic values may result in unrealistic forecasts.

Each building floor is assumed equal in area by the development capacity forecast collection, and parking requirements are based on gross building area with no exceptions.

Final site plans, floor plans, massing, and appearance are more complex than this simple geometry, but should remain within the areas and envelopes established to achieve the environmental relationships forecast. The forecast may have to be revised after detailed surveys, geotechnical evaluation, and drawing-board analysis are complete. These final revisions can be determined only after this level of technical evaluation and consultation is concluded.

Because of this, there is no express or implied guarantee or promise that the land areas forecast will be adequate, that the building areas forecast will be achieved, or that the development capacities predicted will be reached by the final construction plans produced. The software is intended to help the user focus on the most promising site plan options worthy of further investigation, but should not be considered a substitute for this detailed analysis and should not be used as the basis for making final development decisions.

Locating Files with Windows Explorer

There are three primary folders on the CD. The first is a collection of development forecast models entitled the Forecast Collection by Workbook Series, which contains all development capacity forecast models arranged by design premise. The second, entitled Context Record Forms, is a collection of context record forms used for existing project measurement and design specification definition. The third, Dwelling Unit Forecast Model, pertains to the evaluation of residential dwelling unit area and mix options separate from the development forecast model format. These folders and their subfolders are more extensively defined under CD File Organization.

Decision Guides

Figures 1.1 and 1.2 and Tables 3.2, 5.1, and 5.2 in the book *Land Development Calculations* are decision trees that can help select of one of the forecast models contained on the CD. These models are contained in the folders and subfolders identified under CD File Organization. They are also included in bulk without folder and subfolder identification for simplicity if the user knows the forecast model name desired.

Opening Files from Windows Explorer

- Make sure there is a copy of Microsoft Excel 97 on your hard drive.
- Place the CD included with this book in your CD-ROM drive.
- Open Windows Explorer.
- Open this CD by right-clicking on the drive letter representing the CD location in your computer and choose "Explore" from the pop-up menu.
- Choose a forecast model from the bulk list or from the folder and subfolder organization described under CD File Organization.
- Once you have a forecast model on the screen, enter values in all boxes with blue and bold numerals. These are the variables within each forecast model. The model will not produce intended results until all cells are complete.
- All fixed text and forecast data produced from the variables entered is in black.
- Modify the variable values entered to produce alternate forecasts for comparison.
- Repeat this process with other forecast models to compare the results produced by alternate design premises represented by these models.

CD File Organization

Development Capacity Forecast Collection

Figures 1.1 and 1.2 and Tables 5.1 and 5.2 are decision trees that can help to lead to the selection of one of the following forecast models.

Subfolder	Model	Description
G1 Design Premise		**Grade parking around but not under building(s) *or* no parking required**
	CG1B	Nonresidential building type. *Given:* Gross building area objective. *To find:* Land area forecast based on building height options.
	CG1L	Nonresidential building type. *Given:* Gross land area. *To find:* Gross building area forecast based on building height options.
	RG1B	Apartment house building type. *Given:* Gross building area objective. *To find:* Land area purchase options.
	RG1D	Apartment house building type. *Given:* gross land area and net density objective. *To find:* Development capacity options.
	RG1L	Apartment house building type. *Given:* Gross land area. *To find:* Gross building area forecast based on building height options.
G2 Design Premise		**Grade parking around and under building(s)**
	CG2B	Nonresidential building type. *Given:* Building area objective. *To find:* Land area forecast based on building height options.
	CG2L	Nonresidential building type. *Given:* Gross land area. *To find:* Gross building area forecast based on building height options.
	RG2B	Apartment house building type. *Given:* Gross building area objective. *To find:* Land area forecast based on building height options.
	RG2D	Apartment house building type. *Given:* Gross land area and net density objective. *To find:* Development capacity options.
	RG2L	Apartment house building type. *Given:* Gross land area. *To find:* Gross building area forecast based on building height options.
S1 Design Premise		**Parking structure adjacent to building on same premise**
	CS1B	Nonresidential building type. *Given:* Gross building area objective. *To find:* Land area forecast based on building height options.
	CS1L	Nonresidential building type. *Given:* Gross land area. *To find:* Gross building area forecast based on building height options.
	RS1B	Apartment house building type. *Given:* Gross building area objective. *To find:* Land area forecast based on building height options.
	RS1D	Apartment house building type. *Given:* Gross land area and net density objective. *To find:* Development capacity options.
	RS1L	Apartment house building type. *Given:* Net density objective. *To find:* Gross building area forecast based on building height options.

Subfolder	Model	Description
S2 Design Premise		**Parking structure underground**
	CS2B	Nonresidential building type. *Given:* Gross building area objective. *To find:* Land area forecast based on building height options.
	CS2L	Nonresidential building type. *Given:* Gross land area. *To find:* Gross building area forecast based on building height options.
	RS2B	Apartment house building type. *Given:* Gross building area objective. *To find:* Land area forecast based on building height options.
	RS2D	Apartment house building type. *Given:* Gross land area and net density objective. *To find:* Development capacity options.
	RS2L	Apartment house building type. *Given:* Net density objective. *To find:* Gross building area forecast based on building height options.
S3 Design Premise		**Parking structure below building and at least partially above grade**
	CS3B	Nonresidential building type. *Given:* Gross building area objective. *To find:* Land area forecast based on building height options.
	CS3L	Nonresidential building type. *Given:* Gross land area. *To find:* Gross building area forecast based on building height options.
	RS3B	Apartment house building type. *Given:* Gross building area objective. *To find:* Land area forecast based on building height options.
	RS3D	Apartment house building type. *Given:* Gross land area and net density objective. *To find:* Development capacity options.
	RS3L	Apartment house building type. *Given:* Net density objective. *To find:* Gross building area forecast based on building height options.
SF and GT Design Premises		**SF: Single-family *detached* residential dwelling units with grade parking (suburb houses)** **GT: Single family *attached* residential dwelling units with grade parking (urban houses)**
	RSFD	Suburb house building type. *Given:* Gross land area and net density objective. *To find:* Maximum number of lots that can be created.
	RSFN	Suburb house building type. *Given:* Number of lots desired. *To find:* (1) Minimum buildable acres required, (2) maximum gross dwelling unit area that can be built on each lot.
	RSFL	Suburb house building type. *Given:* Gross land area. *To find:* (1) Maximum number of lots that can be created, (2) maximum gross dwelling unit area that can be built on each lot.
	RGTL	Urban house building type. *Given:* Gross land area. *To find:* Maximum dwelling unit capacity of the land.
	RGTD	Urban house building type. *Given:* Gross land area and net density objective. *To find:* A design specification that can meet the net density objective given.

Subfolder	Model	Description
MX Design Strategy		**Vertically integrated mixed land use**
G1 Design Premise		Adjacent parking lot located on same premise
	MG1B	Vertically integrated mixed-use building type. *Given:* Gross building area objective. *To find:* The maximum commercial building area and apartment dwelling unit capacity of an existing building footprint within an existing land area when the land allocation between residential and nonresidential activities varies and when building height varies.
	MG1L	Vertically integrated mixed-use building type. *Given:* Gross land area available. *To find:* The maximum commercial building area and apartment dwelling unit capacity of a given land area when the land allocation between residential and nonresidential activities varies, and when building height varies.
S1 Design Premise		Adjacent parking structure located on same premise
	MS1B	Vertically integrated mixed-use building type. *Given:* Gross building objective. *To find:* The maximum commercial building area and apartment dwelling unit capacity of a given land area when the land allocation between residential and nonresidential activities varies, and when building height varies.
S2 Design Premise		Underground parking structure
	MS2L	Vertically integrated mixed-use building type. *Given:* Gross land area available. *To find:* The maximum commercial building area and apartment dwelling unit capacity of a given land area when the land allocation between residential and nonresidential activities varies, and when building height varies.
S3 Design Premise		Parking structure partially or completely above grade under building
	MS3L	Vertically integrated mixed-use building type. *Given:* Gross land area available. *To find:* The maximum commercial building area and apartment dwelling unit capacity of a given land area when the land allocation between residential and nonresidential activities varies, and when building height varies.
EX Design Strategy		**Expansion of an existing facility**
G1 Design Premise		Adjacent parking lot located on same premise
	EG1L	Nonresidential building type. *Given:* Gross land area and existing conditions. *To find:* Expansion options and implications.

Context Record Forms

These forms record and convert site plan measurements of existing projects to a project design specification that defines the essential ingredients and quantities representing the development intensity under study.

Forecast Model	Description
CR	Use for nonresidential and apartment house development projects.
GT	Use for urban house development projects.
SF	Use for suburb house development projects.

Special Model

This model can be used to evaluate the development capacity implications of various dwelling unit area and mix combinations. It is included separately and is also contained in each apartment house forecast model. It will generally be used within a forecast model, but is also included separately for use in some special circumstances.

Forecast Model	Description
DU	Use for separate evaluation that is not attached to a chosen forecast model.

Development Capacity Evaluation Logic

1. Buildings represent shelter.
2. Buildings and their associated site development represent cells[1] that combine to form neighborhoods, districts, communities, and regions within the built environment.
3. The built environment contains shelter, movement, open space, and life-support cells.
4. These cells provide safety, security, mobility, life support, enjoyment, value, profit, and yield.[2]
5. The built environment expands one cell at a time, and each of us lives within one of the cells created.
6. The plan of a cell is referred to as a *site plan*.
7. Each site plan has an ultimate development capacity[3] that is rarely, if ever, reached.
8. The building cover, parking cover, loading cover, driveway cover, and miscellaneous pavement introduced in a site plan equals the development cover.

9. Development cover has both two-dimensional and three-dimensional components.
10. Two-dimensional development cover is equal to building cover plus the pavement constructed for surface parking, loading, circulation, and miscellaneous sidewalks, patios, plazas, entrances, etc.
11. The relationship between open space and development cover in a site plan determines the two-dimensional balance present.
12. Three-dimensional development cover is produced when building cover and building height are combined to produce building mass and form.
13. The relationship between building mass, pavement, and open space in a site plan determines the development capacity utilized and the relative intensity introduced.
14. Relative intensity can be defined by values assigned to the components of a design specification.[4]
15. Each design specification is related to a design premise.[5]
16. Assigning different values to the components of this specification produces development intensity options.
17. The development intensity of a site plan is a function of the design premise chosen and the values assigned to the components of its design specification.
18. The development intensity produced by the design premise and specification values chosen affects the capacity, cost, compatibility, profitability, and yield of the project.
19. Attempting to reach the ultimate development capacity of a site produces excessive intensity and overdevelopment.

20. Ignoring the development capacity of a site produces sprawl and excessive consumption of our natural environment.
21. Underdevelopment, overdevelopment, and excessive intensity can be avoided by studying the design specifications associated with existing building projects.
22. The context record system has been written to receive and convert existing project measurements to design specifications and intensity definitions.
23. Future development intensity options can be forecast, compared, and evaluated in relation to existing context record histories with the development forecast collection.
24. Intensity comparisons are relevant when considering construction cost, investment options, profit, economic stability, safety, security, sustainability, compatibility, life support, quality of life, and land conservation.
25. The development forecast collection and the context record system are tools that can be used to forecast, compare, and evaluate the capacity of land to shelter the activities of an expanding population within a context that is enjoyable, profitable, self-supporting, and sustainable.

Notes

1. The word *cell* is a single term of convenience for methods of land subdivision variously referred to as lots, land areas, parcels, tracts, easements, rights-of-way, etc.
2. *Yield* means the revenue produced to support the community services provided by authorized taxing authorities.
3. *Development capacity* means the amount of gross building area that can be placed on a given lot or land area. It is restricted only by current technology that limits the ultimate achievable building height.

4. *Design specification* means a collection of design components that control development results without dictating appearance. Each component can be assigned a broad range of values that have direct development capacity implications.

5. *Design premise* means one of six generic approaches to the development of land for shelter.

Glossary

aggregate average dwelling unit area—AGG The aggregate average dwelling unit area is equal to the gross building area divided by the total number of dwelling units provided. It is also referred to as *average dwelling unit area—ADU*. It is a function of the floor plan area allocated per dwelling unit type and the dwelling unit mix of types introduced. The AGG value is a major factor affecting development capacity. It varies by development project and is a planning policy decision that directly affects capacity and intensity.

aggregate average land allocation per dwelling unit type—AFP The average amount of development cover and private yard area allocated per dwelling unit based on the dwelling unit mix proposed.

aggregate average lot area—AAL A single lot area calculated to represent the average lot areas and mix proposed.

average lot area—ALT The average lot area that can be provided within a given subdivision area.

basement area—BSA The gross basement area expressed in square feet. A floor level with a floor elevation that is

3 ft or more below grade at any point along its exterior perimeter.

basement percentage—BSM The gross basement area expressed as a fraction of the habitable footprint above.

buildable area—BLA The total land area available for improvement on a given land area, excluding only unbuildable areas of the site such as public or private rights-of-way, ravines, lakes, etc. This term should not be confused with the more common zoning expression that often means the land area located within the building setback lines on a given lot.

building area, gross—GBA The total building area as measured along the exterior perimeter, including exterior wall thicknesses.

building area, gross square feet per buildable acre—SFAC The total number of building square feet provided per buildable acre of land available.

building area per parking space—a The gross building area in square feet per parking space provided.

building cover area—BCA The area enclosed by the building foundation perimeter. This area is often referred to as the building *footprint*.

building cover area including garage—BCG The area covered by the building floor plan, including attached or detached garages.

building efficiency—Be The fraction of the gross building area that is available for occupancy.

building floors—FLR The number of building floors planned, provided, or required.

building plane area—BPA The imaginary horizontal area above grade within a given lot or land area that is planned, permitted, or provided for an elevated building floor plan.

building plane percentage—B The fraction of the core development area that is covered by a building above.

building support—Bu A fraction of the gross building area devoted to support functions such as circulation and mechanical and plumbing services.

cluster plan A subdivision plan that reduces lot sizes in order to assemble the remainder as shared neighborhood open space.

common open space—COS Open space shared and maintained by a number of dwelling units, generally involving association fees or condominium ownership. Common open space is added to private yard areas and unbuildable areas such as ravines and lakes to produce the gross project open space available.

context The relationship among paved surfaces, parking areas, open space, and building mass that affects the meaning and quality of the lifestyle within.

context record A form that pertains to a single development project or section of a project. The form requests a series of project measurements that are translated by equations within to produce context values. Some of these values can be entered in the design specification of a forecast model to reproduce the development capacity measured and to forecast alternate possibilities.

core development area—CORE The land area exclusively available, or used, for building cover and parking cover.

core development area percentage—C A fraction of the buildable land area that is exclusively available for building cover and parking cover.

core + open space area—CS An area including the core development area and the project open space provided.

core + open space + loading area—CSL An area including the core development area, the total loading area, and the total project open space provided.

crawl space—CRW Building area at least partially below grade that has a headroom of less than 4 ft.

design specification A collection of design component values that control development results without dictating appearance. Each component can be assigned a broad range of values that have direct development capacity implications.

development balance The relative relationship between the total development cover introduced and the project open space that remains.

development capacity The capacity of land to accommodate building area or square footage. Development capacity calculations are based on the maximum core area present, planned, or permitted.

development cover area, total—DCA The sum of all impervious cover introduced, including, but not limited to, building cover, parking cover, driveway cover, and miscellaneous pavement within the buildable area of a given lot or land area. Within the gross land area available, development cover also includes all public and private roadway cover.

development intensity A function of the building height introduced and the total development cover present in relation to the project open space that remains.

driveway cover—DRV Miscellaneous internal project roadway surfaces that are separate from parking lot circulation aisles but intended to provide access to them.

driveways, private—R A fraction of the buildable land area that will be used for driveway cover that is not a public maintenance responsibility.

dwelling unit allocation—MIX The percentage of each dwelling unit type planned or provided within the total mix.

dwelling unit area, comprehensive—CDA The total building area constructed per dwelling unit type.

dwelling unit area, gross—GDA The gross dwelling area provided per dwelling unit type, including wall thicknesses.

dwelling unit area, prorated—PDA A comprehensive dwelling unit area prorated according to the dwelling unit mix anticipated.

dwelling unit footprint with yard and garage—DYG The total private land area allocation per dwelling unit.

dwelling units, number of—NDU The number of dwelling units that can be placed on a given land area.

dwelling units per buildable acre—dBA The total number of dwelling units provided per buildable acre available.

dwelling units per floor—DUF The total number of dwelling units that can be placed on a single building floor.

dwelling units per net acre—d or dNA The total number of dwelling units planned, permitted, or required per net acre of available land area.

floor area ratio—FAR As used and calculated in this book, the gross building area planned, permitted, produced or forecast in square feet divided by the gross buildable land area available in square feet.

floor total—F The total number of floors provided above grade, including parking structure levels.

footprint, garage—GAR The garage foundation area measured to the outside of all perimeter walls.

footprint, habitable area—FTP The total land area covered by the habitable portion of a residential building floor plan.

garage spaces, number of—NGS The number of parking spaces that will be included within one or more totally enclosed garage buildings.

habitable area, total—HAB The gross dwelling unit area to be occupied. This does not include basement, crawl space, and garage.

index, balance—BLX A mathematical notation that expresses the two-dimensional relationship between development cover and project open space on a given buildable land area. For instance, the expression 70.30 means that 70% of the site is consumed by development cover of all types and 30% remains as open space. This notation can also be used to express the relationship between development cover and the gross open space present on a given land area.

index, capacity—SFAC The gross building area constructed, in square feet, divided by the number of buildable acres available.

index, combined—CIX A mathematical notation that expresses the intensity index and massing index in one combined expression. For instance, the expression 10.70.20 represents a 10-story building with 70% development cover and 20% building cover.

index, intensity—INX A mathematical notation that expresses building height and development cover percentages in a combined whole-number expression. For instance, the index 10.70 represents a 10-story building that has covered 70% of the buildable land area with development cover.

index, massing—MSX A mathematical notation that expresses building height and building cover in a combined whole-number expression. For instance, the expression 10.20 represents a 10-story building with a footprint that covers 20% of the buildable area.

land area, buildable—BLA The total land area available for improvement on a given land area, excluding only unbuildable areas of the site such as public or private rights-of-way, ravines, lakes, etc. This term should not be confused with the more common zoning expression that often means the land area located within the building setback lines on a given lot.

land area, buildable acres—BAC The total land area available for improvement, in acres. It does not include public and private rights-of-way, paved easements, and unbuildable areas.

land area, gross—GLA The gross land area in acres, as defined by recorded property lines.

land area, gross acres—GAC The total land area given or forecast, in acres.

land area, gross reduction—X The fraction of the gross land area that is either unbuildable or devoted to public or private rights-of-way and/or paved easements.

land area, net—NLA The fraction of the gross land area that is present after rights-of-way and paved easements are subtracted.

land donation—Do The amount of gross land area subtracted from the total for one or more public or semipublic purposes.

land donation for parks—Dp The amount of gross land area subtracted from the total for one or more future park sites.

land donation for schools—Ds The amount of gross land area subtracted from the total for one or more future school sites.

loading area, gross—LDA The land area devoted to the delivery and pickup of materials, supplies, and equipment associated with building services and operations. It is often referred to as a *loading dock area*.

loading area percentage—L A fraction of the buildable land area devoted to truck maneuvering, loading, and parking.

loading space, gross pavement area per—b The gross truck maneuvering, loading, and parking surface area per loading space provided.

loading spaces, number of—l The number of loading spaces planned or required.

lot area, tabular—TLA A lot area calculated from data entered in a given spreadsheet.

lot area, total—LOT The land area devoted to a single dwelling unit and defined by recorded property lines.

lot number, total—NLT The number of lots planned, permitted, or provided within a given land area.

lot type mix—LTX The percentage of each lot area planned or provided.

miscellaneous building and pavement areas—Msf The area covered by miscellaneous structures plus the area covered by sidewalks, terraces, swimming pools, swimming pool decks, and other miscellaneous single-family residential structures and pavement.

miscellaneous pavement—M Miscellaneous pavement cover includes walks, patios, terraces, fountains, and other impervious landscape amenities on a lot.

mixed use Building or development plans that combine land use activities that have historically been separated by zoning district.

nonresidential allocation percentage—CAP A value calculated in a vertically integrated mixed-use forecast model (M series) to indicate the amount of development capacity to be allocated to the primary first-floor use.

nonresidential gross building area—CGBA A forecast associated with vertically integrated mixed-use project forecast models (M series) to anticipate the total building area predicted for the primary first-floor use.

open space, acres—OSAC Project open space expressed in acres.

open space, gross—Sg Gross project open space includes project open space, yard areas, and unbuildable areas such as, but not limited to, ecologically fragile areas, unstable soil areas, ravines, ponds, marshes, and existing improved areas that are to remain within the gross land area available.

open space, project—S Unless specifically referred to as *gross project open space* or *gross open space,* project open space means *net project open space.* Net project open space is the area among buildings within the buildable lot area that is unpaved. It can be used for expansion, but is initially used for landscape and fire separation. It does not include unbuildable areas on a lot such as, but not limited to, rights-of-way, ecologically fragile areas, unstable soil areas, ravines, ponds, marshes, and existing improved areas to remain. It does include restricted setback areas, mounds, gardens, and other unpaved landscape features.

overdevelopment An unsatisfactory level of development intensity, or an unsatisfactory relationship between development cover, building height, and project open space. This relationship may involve a given building type within a given land area, a given neighborhood within a community, a given city within a region, etc. The level at which these relationships become unsatisfactory remains to be defined.

parking, roadway space number—RDS The number of parking spaces served from a public and/or private right-of-way.

parking cover—PCA The total amount of land covered by parking structures. It can be inadvertently used to indicate the amount of land covered by parking lots PLA, including internal parking lot landscaping and circulation aisles.

parking lot area, gross—GPA When a building or structure is present above the parking lot, the gross parking lot area including areas for building structure, service cores, and remote exit stairs.

parking lot area, net parking—NPA That portion of a parking lot area devoted to spaces and circulation aisles.

parking lot area, total—PLA The total surface parking lot area, including all circulation aisles and internal landscape islands, when no building is located above the parking lot.

parking lot area per space, gross—s The gross parking lot area, including circulation aisles and internal landscaped areas, divided by the number of parking spaces present.

parking spaces, number of—NPS The number of parking spaces planned, provided, or required.

parking spaces per dwelling unit—u The number of parking spaces planned, provided, or required per dwelling unit.

parking structure area, gross per level—GPL Total parking structure area per level, including, but not limited to, building structure, wall thickness, service cores, remote exit stairs, parking spaces, and circulation aisles.

parking structure area, net per level—NPL That portion of a parking structure area devoted to parking spaces and circulation aisles.

parking structure cover—P A percentage of the core development area that is covered by the gross parking structure floor plan.

parking structure efficiency—Pe The fraction of the gross parking structure area devoted exclusively to parking spaces and circulation aisles.

parking structure support areas—Pu A fraction of the gross parking structure area that is not devoted to parking spaces or circulation aisles.

residential allocation percentage—RAP A value specified in a vertically integrated mixed-use forecast model (M series) to indicate the amount of development capacity to be allocated to the secondary use.

residential garage area per parking space—Ga The total garage area per parking space.

residential garage spaces per dwelling unit—Gn The number of garage spaces to be provided per dwelling unit.

residential gross building area—RGBA A forecast associated with vertically integrated mixed-use project forecast models (M series) to indicate the total building area anticipated for the secondary use.

right-of-way, private residential—Wp The fraction of the gross land area allocated for private residential roadway easements and circulation rights-of-way that are not dedicated to public use.

right-of-way, public and private—W The amount of land dedicated for the construction of public roads and improvements with the land area under consideration. Also include any dedicated easement areas that are or will be paved. Do not include unpaved easement areas dedicated for public purposes that contribute to the project open space provided.

roadway cover—RDA The amount of land covered by public and private rights-of-way and paved easements that are part of a primary circulation system. It does not include parking lot circulation aisles serving parking spaces nor separate driveway cover that is isolated to reduce internal movement conflict, but is not defined by right-of-way or easement.

roadway parking spaces, net area per space—Rs The net parking space pavement excluding roadway pavement.

roadway parking spaces per dwelling unit—Rn The number of roadway parking spaces planned or provided per dwelling unit. These spaces are served by a public or private right-of-way or easement, as opposed to parking lot spaces served by a parking lot circulation aisle.

rowhouse An urban house that implies rental status and less defined private open space. It can be similar in appearance to a townhouse.

social center, gross building area—Sb The total building area devoted to gathering and shared entertainment in a multifamily housing development.

social center, gross outdoor area—Sp The common or shared outdoor social center area devoted to social activities and entertainment in a multifamily housing development.

social center, number of parking spaces—Sn The number of parking spaces allocated to serve the functions of a multifamily housing social center.

social center footprint—Sf The building footprint area devoted to gathering and shared entertainment in a multifamily housing development.

standard plan A subdivision plan that allocates all or nearly all available land area to private suburb house lots and street rights-of-way.

suburb house A single-family *detached* dwelling unit (home) intended for the use of only one family unit. A suburb house is an imitation farmhouse and shares many of the same characteristics in different quantities.

support areas of site—Su The buildable land area minus all areas devoted to building cover and parking cover. Support areas can include, but are not limited to, project open space, miscellaneous pavement, driveways, and loading areas.

total building cover area—MBCA A forecast associated with vertically integrated mixed-use project forecast models (M series) to predict the gross building footprint anticipated.

total building cover area—MGBA A forecast associated with vertically integrated mixed-use project forecast models (M series) that anticipates the combined gross building area that will be produced for both primary and secondary use group areas.

total dwelling units—MNDU A forecast associated with vertically integrated mixed-use project forecast models (M series) that predicts the total number of dwelling units anticipated for the residential use within the vertically integrated mixed-use assembly.

total nonresidential floors—CFLR A forecast associated with vertically integrated mixed-use project forecast models (M series) that represents the number of primary use floors predicted.

total parking cover—MPLA A forecast associated with vertically integrated mixed-use project forecast models (M series) to predict the combined parking cover area anticipated for both primary and secondary use groups within the vertical mixture.

total parking spaces—MNPS A forecast associated with vertically integrated mixed-use project forecast models (M series) that represents the combined parking space count anticipated for both primary and secondary use groups within the vertical mixture.

townhouse An urban house that implies separate ownership and defined private open space. It can be similar in appearance to a rowhouse.

unbuildable area—U A fraction of the gross land area that cannot be used for development cover or project open space improvements.

underground parking structure area—G A fraction of the buildable area that will be used for a parking structure floor plan below grade.

urban house A single-family *attached* home that includes twin-singles, three-families, four-families, rowhouses, townhouses, garden apartments, etc. *or* a single-family detached home on a lot that is less than or equal to 40 ft wide.

yard area, private—YRD Lawn, courtyard, or other outdoor space that is owned or privately allocated to an individual dwelling unit. It is referred to by some as *defensible space*.

Index

About the Author

WALTER MARTIN HOSACK is a certified city planner and registered architect. He has management and design experience in architecture, engineering, urban design, city planning, and zoning in both public agency and private consulting practice. Over the past 32 years he has planned, designed, and managed projects ranging in size from residential homes to $100 million commercial projects and has participated and led development efforts at both the state and local levels. He has also had extensive public involvement requiring client, public, and professional coordination in the search for common points of agreement, and has published several articles relating to this work. He brings his experience to bear in this book, which is designed to present a common tool that will be useful to all development students, disciplines, and professionals.

SOFTWARE AND INFORMATION LICENSE

The attached CD-ROM contains proprietary software, data, and information (together, the "Product") owned by Walter Martin Hosack ("Hosack") and his licensors. This product is distributed by the McGraw-Hill Companies, Inc. ("McGraw-Hill") by arrangement with Hosack. Your right to use and use of the product is governed by the terms and contitions of this agreement.

LICENSE

Throughout this License Agreement, "you" shall mean either the individual or the entity whose agent opens this package. You are granted a limited, nonexclusive and nontransferable license to use the Product subject to the following terms:

(i) The Product may only be used on a single computer (i.e., a single CPU).

(ii) You may make one copy of the Product for backup purposes only and you must maintain an accurate record as to the location of the backup at all times.

PROPRIETARY RIGHTS; RESTRICTIONS ON USE AND TRANSFER

All rights (including patent and copyright) in and to the Product are owned by Hosack and his licensors. You are the owner of the enclosed disc on which the Product is recorded. You may not use, copy, decompile, disassemble, reverse engineer, modify, reproduce, create derivative works, transmit, distribute, sublicense, store in a database or retrieval system of any kind, rent, or transfer the Product, or any portion thereof, in any form or by any means (including electronically or otherwise) except as expressly provided for in this License Agreement. You must reproduce the copyright notices, trademark notices, legends, and logos of Hosack, McGraw-Hill, and their respective licensors that appear on the Product on the backup copy of the Product which you are permitted to make hereunder. All rights in the Product not expressly granted herein are reserved by Hosack, McGraw-Hill, and their respective licensors.

TERM

This License Agreement is effective until terminated. It will terminate if you fail to comply with any term or condition of this License Agreement. Upon termination, you are obligated to return to McGraw-Hill the Product together with all copies thereof and to purge and destroy all copies of the Product included in any and all systems, servers, and facilities.

DISCLAIMER OF WARRANTY

The Product and the backup copy of the Product are licensed "as is." Hosack, McGraw-Hill, and their respective licensors make no warranties, express or implied, as to results to be obtained by any person or entity from use of the Product and/or any information or data included therein. Hosack, McGraw-Hill, and their respective licensors make no express or implied warranties of merchantability or fitness for a particular purpose or use with respect to the Product and/or any information or data included therein. In additon, Hosack, McGraw-Hill, and their respective licensors make no warranty regarding the accuracy, adequacy, or completeness of the Product and/or any information or data included therein. Neither Hosack, McGraw-Hill, nor any of their licensors warrant that the functions contained in the Product will meet your requirements or that the operation of the Product will be uninterrupted or error free. You assume the entire risk with respect to the quality and performance of the Product.

LIMITED WARRANTY FOR DISC

To the original licensee only, McGraw-Hill warrants that the enclosed disc on which the Product is recorded is free from defects in materials and workmanship under normal use and service for a period of ninety (90) days from the date of purchase. In the event of a defect in the disc covered by the foregoing warranty, McGraw-Hill will replace the disc.

LIMITATION OF LIABILITY

Neither Hosack, McGraw-Hill, nor any of their licensors shall be liable for any indirect, incidental, special, punitive, consequential, or similar damages, such as but not limited to, loss of anticipated profits or benefits, resulting from the use or inability to use the Product even if any of them has been advised of the possibility of such damages. This limitation or liability shall apply to any claim or cause whatsoever whether such claim or cause arises in contract, tort, or otherwise.

Some states do not allow the exclusion or limitation of indirect, special or consequential damages, so the above limitation may not apply to you.

GENERAL

This License Agreement constitutes the entire agreement between the parties relating to the Product. The terms of any Purchase Order shall have no effect on the terms of this License Agreement. Failure of Licensor and/or McGraw-Hill to insist at any time on strict compliance with this License Agreement shall not constitute a waiver of any rights under this License Agreement. This License Agreement shall be construed and governed in accordance with the laws of the State of New York. If any provision of this License Agreement is held to be contrary to law, that provision will be enforced to the maximum extent permissible and the remaining provisions will remain in full force and effect. **If you do not agree with this License Agreement, please return this product to the place of purchase for a refund.**